Finite Volume Methods for Hyperbolic Problems

This book contains an introduction to hyperbolic partial differential equations and a powerful class of numerical methods for approximating their solution, including both linear problems and nonlinear conservation laws. These equations describe a wide range of wave-propagation and transport phenomena arising in nearly every scientific and engineering discipline. Several applications are described in a self-contained manner, along with much of the mathematical theory of hyperbolic problems. High-resolution versions of Godunov's method are developed, in which Riemann problems are solved to determine the local wave structure and limiters are then applied to eliminate numerical oscillations. These methods were originally designed to capture shock waves accurately, but are also useful tools for studying linear wave-propagation problems, particularly in heterogenous material. The methods studied are implemented in the CLAWPACK software package. Source code for all the examples presented can be found on the web, along with animations of many time-dependent solutions. This provides an excellent learning environment for understanding wave-propagation phenomena and finite volume methods.

Randall J. LeVeque is a Professor of Applied Mathematics at the University of Washington.

Cambridge Texts in Applied Mathematics

Finite Volume Methods for Hyperbolic Problems

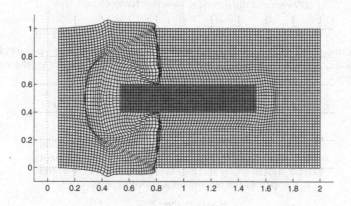

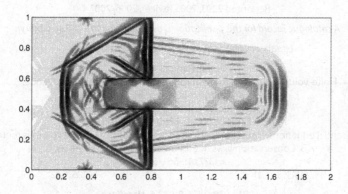

RANDALL J. LEVEQUE
University of Washington

CAMBRIDGE
UNIVERSITY PRESS

CAMBRIDGE
UNIVERSITY PRESS

32 Avenue of the Americas, New York NY 10013-2473, USA

Cambridge University Press is part of the University of Cambridge.

It furthers the University's mission by disseminating knowledge in the pursuit of
education, learning and research at the highest international levels of excellence.

www.cambridge.org
Information on this title: www.cambridge.org/9780521009249

First published 2002
Reprinted 2003, 2005 (twice), 2006, 2007

A catalogue record for this publication is available from the British Library

Library of Congress Cataloguing in Publication data

LaVeque, Randall J., 1995–
Finite-volume methods for hyperbolic problems / Randall J. LeVeque.
p.   cm. – (Cambridge texts in applied mathematics)
Includes bibliographical references and index.
ISBN 0-521-81087-6 – ISBN 0-521-00924-3 (pbk.)
1. Differential equations, Hyperbolic – Numerical solutions.   2. Finite volumes method
3. Conservation laws (Mathematics)   I. Title.   II. Series.
QA377.L41566   2002
515'.353-dc21   2001052642

ISBN 978-0-521-81087-6 Hardback
ISBN 978-0-521-00924-9 Paperback

To Loyce and Benjamin

Contents

Preface

Hyperbolic partial differential equations arise in a broad spectrum of disciplines where wave motion or advective transport is important: gas dynamics, acoustics, elastodynamics, optics, geophysics, and biomechanics, to name but a few. This book is intended to serve as an introduction to both the theory and the practical use of high-resolution finite volume methods for hyperbolic problems. These methods have proved to be extremely useful in modeling a broad set of phenomena, and I believe that there is need for a book introducing them in a general framework that is accessible to students and researchers in many different disciplines.

Historically, many of the fundamental ideas were first developed for the special case of compressible gas dynamics (the Euler equations), for applications in aerodynamics, astrophysics, detonation waves, and related fields where shock waves arise. The study of simpler equations such as the advection equation, Burgers' equation, and the shallow water equations has played an important role in the development of these methods, but often only as model problems, the ultimate goal being application to the Euler equations. This orientation is still reflected in many of the texts on these methods. Of course the Euler equations remain an extremely important application, and are presented and studied in this book, but there are also many other applications where challenging problems can be successfully tackled by understanding the basic ideas of high-resolution finite volume methods. Often it is *not* necessary to understand the Euler equations in order to do so, and the complexity and peculiarities of this particular system may obscure the more basic ideas.

In particular, the Euler equations are *nonlinear*. This nonlinearity, and the consequent shock formation seen in solutions, leads to many of the computational challenges that motivated the development of these methods. The mathematical theory of nonlinear hyperbolic problems is also quite beautiful, and the development and analysis of finite volume methods requires a rich interplay between this mathematical theory, physical modeling, and numerical analysis. As a result it is a challenging and satisfying field of study, and much of this book focuses on nonlinear problems.

However, all of Part I and much of Part III (on multidimensional problems) deals entirely with linear hyperbolic systems. This is partly because many of the concepts can be introduced and understood most easily in the linear case. A thorough understanding of linear hyperbolic theory, and the development of high-resolution methods in the linear case, is extremely useful in fully understanding the nonlinear case. In addition, I believe there are many linear wave-propagation problems (e.g., in acoustics, elastodynamics, or electromagnetics) where these methods have a great deal of potential that has not been fully exploited, particularly for problems in heterogeneous media. I hope to encourage students to explore some of these areas, and researchers in these areas to learn about finite volume methods. I have tried to make it possible to do so without delving into the additional complications of the nonlinear theory.

Studying these methods in the context of a broader set of applications has other pedagogical advantages as well. Identifying the common features of various problems (as unified by the hyperbolic theory) often leads to a better understanding of this theory and greater ability to apply these techniques later to new problems. The finite volume approach can itself lead to greater insight into the physical phenomena and mathematical techniques. The derivation of most conservation laws gives first an integral formulation that is then converted to a differential equation. A finite volume method is based on the integral formulation, and hence is often closer to the physics than is the partial differential equation.

Mastering a set of numerical methods in conjunction with learning the related mathematics and physics has a further advantage: it is possible to apply the methods immediately in order to observe the behavior of solutions to the equations, and thereby gain intuition for how these solutions behave. To facilitate this hands-on approach to learning, virtually every example in the book (and many examples not in the book) can be solved by the reader using programs and data that are easy to download from the web. The basis for most of these programs is the CLAWPACK software package, which stands for "conservation-law-package." This package was originally developed for my own use in teaching and so is intimately linked with the methods studied in this book. By having access to the source code used to generate each figure, it is possible for the interested reader to delve more deeply into implementation details that aren't presented in the text. Animations of many of the figures are also available on the webpages, making it easier to visualize the time-dependent nature of these solutions. By downloading and modifying the code, it is also possible to experiment with different initial or boundary conditions, with different mesh sizes or other parameters, or with different methods on the same problem.

CLAWPACK has been freely available for several years and is now extensively used for research as well as teaching purposes. Another function of this book is to serve as a reference to users of the software who desire a better understanding of the methods employed and the ways in which these methods can be adapted to new applications. The book is not, however, designed to be a user's manual for the package, and it is not necessary to do any computing in order to follow the presentation.

There are many different approaches to developing and implementing high-resolution finite volume methods for hyperbolic equations. In this book I concentrate primarily on one particular approach, the *wave-propagation algorithm* that is implemented in CLAWPACK, but numerous other methods and the relation between them are discussed at least briefly. It would be impossible to survey all such methods in any detail, and instead my aim is to provide enough understanding of the underlying ideas that the reader will have a good basis for learning about other methods from the literature. With minor modifications of the CLAWPACK code it is possible to implement many different methods and easily compare them on the same set of problems.

This book is the result of an evolving set of lecture notes that I have used in teaching this material over the past 15 years. An early version was published in 1989 after giving the course at ETH in Zürich [281]. That version has proved popular among instructors and students, perhaps primarily because it is short and concise. Unfortunately, the same claim cannot be made for the present book. I have tried, however, to write the book in such a way that self-contained subsets can be extracted for teaching (and learning) this material. The latter part of many chapters gets into more esoteric material that may be useful to have available for reference but is not required reading. In addition, many whole chapters can be omitted without loss of continuity in a course that stresses certain aspects of the material. In particular, to focus on linear hyperbolic problems and heterogeneous media, a suggested set of chapters might be 1–9 and 18–21, omitting the sections in the multidimensional chapters that deal with nonlinearity. Other chapters may also be of interest, but can be omitted without loss of continuity. To focus on nonlinear conservation laws, the basic theory can be found in Chapters 1–8, 11–15, and 18–21. Again, other topics can also be covered if time permits, or the course can be shortened further by concentrating on scalar equations or one-dimensional problems, for example.

This book may also be useful in a course on hyperbolic problems where the focus is not on numerical methods at all. The mathematical theory in the context of physical applications is developed primarily in Chapters 1–3, 9, 11, 13, 14, 16, 18, and 22, chapters that contain little discussion of numerical issues. It may still be advantageous to use CLAWPACK to further explore these problems and develop physical intuition, but this can be done without a detailed study of the numerical methods employed.

Many topics in this book are closely connected to my own research. Repeatedly teaching this material, writing course notes, and providing students with sample programs has motivated me to search for more general formulations that are easier to explain and more broadly applicable. This work has been funded for many years by the National Science Foundation, the Department of Energy, and the University of Washington. Without their support the present form of this book would not have been possible.

I am indebted to the many students and colleagues who have taught me so much about hyperbolic problems and numerical methods over the years. I cannot begin to thank everyone by name, and so will just mention a few people who had a particular impact on what is presented in this book. Luigi Quartapelle deserves high honors for carefully reading every word of several drafts, finding countless errors, and making numerous suggestions for substantial improvement. Special thanks are also due to Mike Epton, Christiane Helzel, Jan Olav Langseth, Sorin Mitran, and George Turkiyyah. Along with many others, they helped me to avoid a number of blunders and present a more polished manuscript. The remaining errors are, of course, my own responsibility.

I would also like to thank Cambridge University Press for publishing this book at a reasonable price, especially since it is intended to be used as a textbook. Many books are priced exorbitantly these days, and I believe it is the responsibility of authors to seek out and support publishers that serve the community well.

Most importantly, I would like to thank my family for their constant encouragement and support, particularly my wife and son. They have sacrificed many evenings and weekends of family time for a project that, from my nine-year old's perspective at least, has lasted a lifetime.

Seattle, Washington, August, 2001

Introduction

Hyperbolic systems of partial differential equations can be used to model a wide variety of phenomena that involve wave motion or the advective transport of substances. This chapter contains a brief introduction to some of the fundamental concepts and an overview of the primary issues discussed in this book.

The problems we consider are generally time-dependent, so that the solution depends on time as well as one or more spatial variables. In one space dimension, a homogeneous first-order system of partial differential equations in x and t has the form

$$q_t(x, t) + Aq_x(x, t) = 0 \tag{1.1}$$

in the simplest constant-coefficient linear case. Here $q : \mathbb{R} \times \mathbb{R} \to \mathbb{R}^m$ is a vector with m components representing the unknown functions (pressure, velocity, etc.) we wish to determine, and A is a constant $m \times m$ real matrix. In order for this problem to be *hyperbolic*, the matrix must satisfy certain properties discussed below. Note that subscripts are used to denote partial derivatives with respect to t and x.

The simplest case is the constant-coefficient *scalar* problem, in which $m = 1$ and the matrix A reduces to a scalar value. This problem is hyperbolic provided the scalar A is real. Already this simple equation can model either advective transport or wave motion, depending on the context.

Advective transport refers to a substance being carried along with fluid motion. For example, consider a contaminant being advected downstream with some fluid flowing through a one-dimensional pipe at constant velocity $\bar{u}$. Then the concentration or density $q(x, t)$ of the contaminant satisfies a scalar advection equation of the form

$$q_t(x, t) + \bar{u}q_x(x, t) = 0, \tag{1.2}$$

as derived in Chapter 2. It is easy to verify that this equation admits solutions of the form

$$q(x, t) = \tilde{q}(x - \bar{u}t) \tag{1.3}$$

for any function $\tilde{q}(\xi)$. The concentration profile (or waveform) specified by $\tilde{q}$ simply propagates with constant speed $\bar{u}$ and unchanged shape. In this context the equation (1.2) is generally called the *advection equation*.

The phenomenon of *wave motion* is observed in its most basic form if we model a sound wave traveling down a tube of gas or through an elastic solid. In this case the molecules of

the gas or solid barely move, and yet a distinct wave can propagate through the material
with its shape essentially unchanged over long distances, and at a speed c (the speed of
sound in the material) that is much larger than the velocity of material particles. We will see
in Chapter 2 that a sound wave propagating in one direction (to the right with speed $c > 0$)
can be modeled by the equation

$$w_t(x, t) + cw_x(x, t) = 0, \tag{1.4}$$

where $w(x, t)$ is an appropriate combination of the pressure and particle velocity. This again
has the form of a scalar first-order hyperbolic equation. In this context the equation (1.4) is
sometimes called the *one-way wave equation* because it models waves propagating in one
particular direction.

Mathematically the advection equation (1.2) and the one-way wave equation (1.4) are
identical, which suggests that advective transport and wave phenomena can be handled by
similar mathematical and numerical techniques.

To model acoustic waves propagating in both directions along a one-dimensional medium,
we must consider the full acoustic equations derived in Chapter 2,

$$p_t(x, t) + Ku_x(x, t) = 0,$$
$$u_t(x, t) + (1/\rho)p_x(x, t) = 0, \tag{1.5}$$

where $p(x, t)$ is the pressure (or more properly the perturbation from some background
constant pressure), and $u(x, t)$ is the particle velocity. These are the unknown functions to
be determined. The material is described by the constants K (the bulk modulus of compressibility) and ρ (the density). The system (1.5) can be written as the first-order system
$q_t + Aq_x = 0$, where

$$q = \begin{bmatrix} p \\ u \end{bmatrix}, \qquad A = \begin{bmatrix} 0 & K \\ 1/\rho & 0 \end{bmatrix}. \tag{1.6}$$

To connect this with the one-way wave equation (1.4), let

$$w^1(x, t) = p(x, t) + \rho cu(x, t),$$

where $c = \sqrt{K/\rho}$. Then it is easy to check that $w^1(x, t)$ satisfies the equation

$$w_t^1 + cw_x^1 = 0$$

and so we see that c can be identified as the speed of sound. On the other hand, the function

$$w^2(x, t) = p(x, t) - \rho cu(x, t)$$

satisfies the equation

$$w_t^2 - cw_x^2 = 0.$$

This is also a one-way wave equation, but with propagation speed $-c$. This equation has
solutions of the form $q^2(x, t) = \tilde{q}(x + ct)$ and models acoustic waves propagating to the
left at the speed of sound, rather than to the right.

The system (1.5) of two equations can thus be decomposed into two scalar equations modeling the two distinct acoustic waves moving in different directions. This is a fundamental theme of hyperbolic equations and crucial to the methods developed in this book. We will see that this type of decomposition is possible more generally for hyperbolic systems, and in fact the definition of "hyperbolic" is directly connected to this. We say that the constant-coefficient system (1.1) is *hyperbolic* if the matrix A has real eigenvalues and a corresponding set of m linearly independent eigenvectors. This means that any vector in $\mathbb{R}^m$ can be uniquely decomposed as a linear combination of these eigenvectors. As we will see in Chapter 3, this provides the decomposition into distinct waves. The corresponding eigenvalues of A give the wave speeds at which each wave propagates. For example, the acoustics matrix A of (1.6) has eigenvalues $-c$ and $+c$, the speeds at which acoustic waves can travel in this one-dimensional medium.

For simple acoustic waves, some readers may be more familiar with the *second-order wave equation*

$$p_{tt} = c^2 p_{xx}. \tag{1.7}$$

This equation for the pressure can be obtained from the system (1.5) by differentiating the first equation with respect to t and the second with respect to x, and then eliminating the u_{xt} terms. The equation (1.7) is also called a *hyperbolic* equation according to the standard classification of second-order linear equations into hyperbolic, parabolic, and elliptic equations (see [234], for example). In this book we only consider first-order hyperbolic systems as described above. This form is more fundamental physically than the derived second-order equation, and is more amenable to the development of high-resolution finite volume methods.

In practical problems there is often a coupling of advective transport and wave motion. For example, we will see that the speed of sound in a gas generally depends on the density and pressure of the gas. If these properties of the gas vary in space and the gas is flowing, then these variations will be advected with the flow. This will have an effect on any sound waves propagating through the gas. Moreover, these variations will typically cause acceleration of the gas and have a direct effect on the fluid motion itself, which can also be modeled as wave-propagation phenomena. This coupling leads to *nonlinearity* in the equations.

1.1 Conservation Laws

Much of this book is concerned with an important class of homogeneous hyperbolic equations called *conservation laws*. The simplest example of a one-dimensional conservation law is the partial differential equation (PDE)

$$q_t(x, t) + f(q(x, t))_x = 0, \tag{1.8}$$

where $f(q)$ is the *flux function*. Rewriting this in the *quasilinear form*

$$q_t + f'(q)q_x = 0 \tag{1.9}$$

suggests that the equation is hyperbolic if the flux Jacobian matrix $f'(q)$ satisfies the conditions previously given for the matrix A. In fact the linear problem (1.1) is a conservation law with the linear flux function $f(q) = Aq$. Many physical problems give rise to *nonlinear conservation laws* in which $f(q)$ is a nonlinear function of q, a vector of *conserved quantities*.

1.1.1 Integral Form

Conservation laws typically arise most naturally from physical laws in an integral form as developed in Chapter 2, stating that for any two points x_1 and x_2,

$$\frac{d}{dt} \int_{x_1}^{x_2} q(x, t)\, dx = f(q(x_1, t)) - f(q(x_2, t)). \tag{1.10}$$

Each component of q measures the density of some conserved quantity, and the equation (1.10) simply states that the "total mass" of this quantity between any two points can change only due to the flux past the endpoints. Such conservation laws naturally hold for many fundamental physical quantities. For example, the advection equation (1.2) for the density of a contaminant is derived from the fact that the total mass of the contaminant is conserved as it flows down the pipe and the flux function is $f(q) = \bar{u}q$. If the total mass of contaminant is not conserved, because of chemical reactions taking place, for example, then the conservation law must also contain *source terms* as described in Section 2.5, Chapter 17, and elsewhere.

The constant-coefficient linear acoustics equations (1.5) can be viewed as conservation laws for pressure and velocity. Physically, however, these are not conserved quantities except approximately in the case of very small amplitude disturbances in uniform media. In Section 2.7 the acoustics equations are derived from the *Euler equations* of gas dynamics, the nonlinear conservation laws that model more general disturbances in a compressible gas. These equations model the conservation of mass, momentum, and energy, and the laws of physics determine the flux functions. See Section 2.6 and Chapter 14 for these derivations. These equations have been intensively studied and used in countless computations because of their importance in aerodynamics and elsewhere.

There are many other systems of conservation laws that are important in various applications, and several are used in this book as examples. However, the Euler equations play a special role in the historical development of the techniques discussed in this book. Much of the mathematical theory of nonlinear conservation laws was developed with these equations in mind, and many numerical methods were developed specifically for this system. So, although the theory and methods are applicable much more widely, a good knowledge of the Euler equations is required in order to read much of the available literature and benefit from these developments. A brief introduction is given in Chapter 14. It is a good idea to become familiar with these equations even if your primary interest is far from gas dynamics.

1.1.2 Discontinuous Solutions

The differential equation (1.8) can be derived from the integral equation (1.10) by simple manipulations (see Chapter 2) *provided that q and f(q) are sufficiently smooth*. This proviso is important because in practice many interesting solutions are not smooth, but contain discontinuities such as shock waves. A fundamental feature of nonlinear conservation laws is that these discontinuities can easily develop spontaneously even from smooth initial data, and so they must be dealt with both mathematically and computationally.

At a discontinuity in q, the partial differential equation (1.8) does not hold in the classical sense and it is important to remember that the integral conservation law (1.10) is the more fundamental equation which does continue to hold. A rich mathematical theory of shock-wave solutions to conservation laws has been developed. This theory is introduced starting in Chapter 11.

1.2 Finite Volume Methods

Discontinuities lead to computational difficulties and the main subject of this book is the accurate approximation of such solutions. Classical finite difference methods, in which derivatives are approximated by finite differences, can be expected to break down near discontinuities in the solution where the differential equation does not hold. This book concerns finite volume methods, which are based on the integral form (1.10) instead of the differential equation. Rather than pointwise approximations at grid points, we break the domain into *grid cells* and approximate the total integral of q over each grid cell, or actually the *cell average* of q, which is this integral divided by the volume of the cell. These values are modified in each time step by the flux through the edges of the grid cells, and the primary problem is to determine good *numerical flux functions* that approximate the correct fluxes reasonably well, based on the approximate cell averages, the only data available. We will concentrate primarily on one class of *high-resolution* finite volume methods that have proved to be very effective for computing discontinuous solutions. See Section 6.3 for an introduction to the properties of these methods.

Other classes of methods have also been applied to hyperbolic equations, such as finite element methods and spectral methods. These are not discussed directly in this book, although much of the material presented here is good background for understanding high-resolution versions.

1.2.1 Riemann Problems

A fundamental tool in the development of finite volume methods is the *Riemann problem*, which is simply the hyperbolic equation together with special initial data. The data is piecewise constant with a single jump discontinuity at some point, say $x = 0$,

$$q(x, 0) = \begin{cases} q_l & \text{if } x < 0, \\ q_r & \text{if } x > 0. \end{cases} \tag{1.11}$$

If Q_{i-1} and Q_i are the cell averages in two neighboring grid cells on a finite volume grid, then by solving the Riemann problem with $q_l = Q_{i-1}$ and $q_r = Q_i$, we can obtain information that can be used to compute a numerical flux and update the cell averages over a time step. For hyperbolic problems the solution to the Riemann problem is typically a similarity solution, a function of x/t alone, and consists of a finite set of waves that propagate away from the origin with constant wave speeds. For linear hyperbolic systems the Riemann problem is easily solved in terms of the eigenvalues and eigenvectors of the matrix A, as developed in Chapter 3. This simple structure also holds for nonlinear systems of equations and the exact solution (or arbitrarily good approximations) to the Riemann problem can be constructed even for nonlinear systems such as the Euler equations. The theory of nonlinear Riemann solutions for scalar problems is developed in Chapter 11 and extended to systems in Chapter 13.

Computationally, the exact Riemann solution is often too expensive to compute for nonlinear problems and *approximate Riemann solvers* are used in implementing numerical methods. These techniques are developed in Section 15.3.

1.2.2 Shock Capturing vs. Tracking

Since the PDEs continue to hold away from discontinuities, one possible approach is to combine a standard finite difference or finite volume method in smooth regions with some explicit procedure for tracking the location of discontinuities. This is the numerical analogue of the mathematical approach in which the PDEs are supplemented by jump conditions across discontinuities. This approach is often called *shock tracking* or front tracking. In more than one space dimension, discontinuities typically lie along curves (in two dimensions) or surfaces (in three dimensions), and such algorithms typically become quite complicated. Moreover, in realistic problems there may be many such surfaces that interact in complicated ways as time evolves. This approach will not be discussed further in this book. For some examples and discussion, see [41], [66], [103], [153], [154], [171], [207], [289], [290], [321], [322], [371], [372].

Instead we concentrate here on *shock-capturing* methods, where the goal is to capture discontinuities in the solution automatically, without explicitly tracking them. Discontinuities must then be smeared over one or more grid cells. Success requires that the method implicitly incorporate the correct jump conditions, reduce smearing to a minimum, and not introduce nonphysical oscillations near the discontinuities. High-resolution finite volume methods based on Riemann solutions often perform well and are much simpler to implement than shock-tracking methods.

1.3 Multidimensional Problems

The Riemann problem is inherently one-dimensional, but is extensively used also in the solution of multidimensional hyperbolic problems. A two-dimensional finite volume grid typically consists of polygonal grid cells; quadrilaterals or triangles are most commonly used. A Riemann problem normal to each edge of the cell can be solved in order to determine the flux across that edge. In three dimensions each face of a finite volume cell can be approximated by a plane, and a Riemann problem normal to this plane solved in order to compute the flux. Multidimensional problems are discussed in the Part III of the book, starting with an introduction to the mathematical theory in Chapter 18.

If the finite volume grid is rectangular, or at least logically rectangular, then the simplest way to extend one-dimensional high-resolution methods to more dimensions is to use *dimensional splitting*, a fractional-step approach in which one-dimensional problems along each coordinate direction are solved in turn. This approach, which is often surprisingly effective in practice, is discussed in Section 19.5. In some cases a more fully multidimensional method is required, and one approach is developed starting in Chapter 20, which again relies heavily on our ability to solve one-dimensional Riemann problems.

1.4 Linear Waves and Discontinuous Media

High-resolution methods were originally developed for nonlinear problems in order to accurately capture discontinuous solutions such as shock waves. Linear hyperbolic equations often arise from studying small-amplitude waves, where the physical nonlinearities of the true equations can be safely ignored. Such waves are often smooth, since shock waves can only appear from nonlinear phenomena. The acoustic waves we are most familiar with arise from oscillations of materials at the molecular level and are typically well approximated by linear combinations of sinusoidal waves at various frequencies. Similarly, most familiar electromagnetic waves, such as visible light, are governed by the linear Maxwell equations (another hyperbolic system) and again consist of smooth sinusoidal oscillations.

For many problems in acoustics or optics the primary computational difficulty arises from the fact that the domain of interest is many orders of magnitude larger than the wavelengths of interest, and so it is important to use a method that can resolve smooth solutions with a very high order of accuracy in order to keep the number of grid points required manageable. For problems of this type, the methods developed in this book may not be appropriate. These finite volume high-resolution methods are typically at best second-order accurate, resulting in the need for many points per wavelength for good accuracy. Moreover they have a high cost per grid cell relative to simpler finite difference methods, because of the need to solve Riemann problems for each pair of grid cells every time step. The combination can be disastrous if we need to compute over a domain that spans thousands of wavelengths. Instead methods with a higher order of accuracy are typically used, e.g., fourth-order finite difference methods or spectral methods. For some problems it is hopeless to try to resolve individual wavelengths, and instead *ray-tracing* methods such as geometrical optics are used to determine how rays travel without discretizing the hyperbolic equations directly.

However, there are some situations in which high-resolution methods based on Riemann solutions may have distinct advantages even for linear problems. In many applications wave-propagation problems must be solved in materials that are not homogeneous and isotropic. The heterogeneity may be smoothly varying (e.g., acoustics in the ocean, where the sound speed varies with density, which may vary smoothly with changes in salinity, for example). In this case high-order methods may still be applicable. In many cases, however, there are sharp interfaces between different materials. If we wish to solve for acoustic or seismic waves in the earth, for example, the material parameters typically have jump discontinuities where soil meets rock or at the boundaries between different types of rock. Ultrasound waves in the human body also pass through many interfaces, between different organs or tissue and bone. Even in ocean acoustics there may be distinct layers of water with different salinity, and hence jump discontinuities in the sound speed, as well as the interface at the ocean floor where waves pass between water and earth. With wave-tracing methods it may be possible to use reflection and transmission coefficients and Snell's law to trace rays and reflected rays at interfaces, but for problems with many interfaces this can be unwieldy. If we wish to model the wave motion directly by solving the hyperbolic equations, many high-order methods can have difficulties near interfaces, where the solution is typically not smooth.

For these problems, high-resolution finite volume methods based on solving Riemann problems can be an attractive alternative. Finite volume methods are a natural choice for heterogeneous media, since each grid cell can be assigned different material properties via an appropriate averaging of the material parameters over the volume enclosed by the cell. The idea of a Riemann problem is easily extended to the case where there is a discontinuity in the medium at $x = 0$ as well as a discontinuity in the initial data. Solving the Riemann problem at the interface between two cells then gives a decomposition of the data into waves moving into each cell, including the effects of reflection and transmission as waves move between different materials. Indeed, the classical reflection and transmission coefficients for various problems are easily derived and understood in terms of particular Riemann solutions. Variable-coefficient linear problems are discussed in Chapter 9 and Section 21.5.

Hyperbolic equations with variable coefficients may not be in conservation form, and so the methods are developed here in a form that applies more generally. These *wave-propagation methods* are based directly on the waves arising from the solution of the Riemann problem rather than on numerical fluxes at cell interfaces. When applied to conservation laws, there is a natural connection between these methods and more standard flux-differencing methods, which will be elucidated as we go along. But many of the shock-capturing ideas that have been developed in the context of conservation laws are valuable more broadly, and one of my goals in writing this book is to present these methods in a more general framework than is available elsewhere, and with more attention to applications where they have not traditionally been applied in the past.

This book is organized in such a way that all of the ideas required to apply the methods on linear problems are introduced first, before discussing the more complicated nonlinear theory. Readers whose primary interest is in linear waves should be able to skip the nonlinear parts entirely by first studying Chapters 2 through 9 (on linear problems in one dimension) and then the preliminary parts of Chapters 18 through 23 (on multidimensional problems).

For readers whose primary interest is in nonlinear problems, I believe that this organization is still sensible, since many of the fundamental ideas (both mathematical and algorithmic) arise already with linear problems and are most easily understood in this context. Additional issues arise in the nonlinear case, but these are most easily understood if one already has a firm foundation in the linear theory.

1.5 CLAWPACK Software

The CLAWPACK software ("conservation-laws package") implements the various wave-propagation methods discussed in this book (in Fortran). This software was originally developed as a teaching tool and is intended to be used in conjunction with this book. The use of this software is briefly described in Chapter 5, and additional documentation is available online, from the webpage

```
http://www.amath.washington.edu/~claw
```

Virtually all of the computational examples presented in the book were created using CLAWPACK, and the source code used is generally available via the website

```
http://www.amath.washington.edu/~claw/book.html
```

A parenthetical remark in the text or figure captions of the form

`[claw/book/chapN/examplename]`

is an indication that accompanying material is available at

`http://www.amath.washington.edu/~claw/book/chapN/examplename/www`

often including an animation of time-dependent solutions. From this webpage it is generally possible to download a CLAWPACK directory of the source code for the example. Downloading the tarfile and unpacking it in your `claw` directory results in a subdirectory called `claw/book/chapN/examplename`. (You must first obtain the basic CLAWPACK routines as described in Chapter 5.)

You are encouraged to use this software actively, both to develop an intuition for the behavior of solutions to hyperbolic equations and also to develop direct experience with these numerical methods. It should be easy to modify the examples to experiment with different parameters or initial conditions, or with the use of different methods on the same problem.

These examples can also serve as templates for developing codes for other problems. In addition, many problems not discussed in this book have already been solved using CLAWPACK and are often available online. Some pointers can be found on the webpages for the book, and others are collected within the CLAWPACK software in the `applications` subdirectory; see

`http://www.amath.washington.edu/~claw/apps.html`

1.6 References

Some references for particular applications and methods are given in the text. There are thousands of papers on these topics, and I have not attempted to give an exhaustive survey of the literature by any means. The references cited have been chosen because they are particularly relevant to the discussion here or provide a good entrance point to the broader literature. Listed below are a few books that may be of general interest in understanding this material, again only a small subset of those available.

An earlier version of this book appeared as a set of lecture notes [281]. This contains a different presentation of some of the same material and may still be of interest. My contribution to [287] also has some overlap with this book, but is directed specifically towards astrophysical flows and also contains some description of hyperbolic problems arising in magnetohydrodynamics and relativistic flow, which are not discussed here.

The basic theory of hyperbolic equations can be found in many texts, for example John [229], Kevorkian [234]. The basic theory of nonlinear conservation laws is neatly presented in the monograph of Lax [263]. Introductions to this material can also be found in many other books, such as Liu [311], Whitham [486], or Chorin & Marsden [68]. The book of Courant & Friedrichs [92] deals almost entirely with gas dynamics and the Euler equations, but includes much of the general theory of conservation laws in this context and is very useful. The books by Bressan [46], Dafermos [98], Majda [319], Serre [402], Smoller [420], and Zhang & Hsiao [499] present many more details on the mathematical theory of nonlinear conservation laws.

For general background on numerical methods for PDEs, the books of Iserles [211], Morton & Mayers [333], Strikwerda [427], or Tveito & Winther [461] are recommended. The book of Gustafsson, Kreiss & Oliger [174] is aimed particularly at hyperbolic problems and contains more advanced material on well-posedness and stability of both initial- and initial–boundary-value problems. The classic book of Richtmyer & Morton [369] contains a good description of many of the mathematical techniques used to study numerical methods, particularly for linear equations. It also includes a large section on methods for nonlinear applications including fluid dynamics, but is out of date by now and does not discuss many of the methods we will study.

A number of books have appeared recently on numerical methods for conservation laws that cover some of the same techniques discussed here, e.g., Godlewski & Raviart [156], Kröner [245], and Toro [450]. Several other books on computational fluid dynamics are also useful supplements, including Durran [117], Fletcher [137], Hirsch [198], Laney [256], Oran & Boris [348], Peyret & Taylor [359], and Tannehill, Anderson & Pletcher [445]. These books discuss the fluid dynamics in more detail, generally with emphasis on specific applications.

For an excellent collection of photographs illustrating a wide variety of interesting fluid dynamics, including shock waves, Van Dyke's *Album of Fluid Motion* [463] is highly recommended.

Many more references on these topics can easily be found these days by searching on the web. In addition to using standard web search engines, there are preprint servers that contain collections of preprints on various topics. In the field of conservation laws, the *Norwegian preprint server* at

```
http://www.math.ntnu.no/conservation/
```

is of particular note. Online citation indices and bibliographic databases are extremely useful in searching the literature, and students should be encouraged to learn to use them. Some useful links can be found on the webpage [claw/book/chap1/].

1.7 Notation

Some nonstandard notation is used in this book that may require explanation. In general I use q to denote the solution to the partial differential equation under study. In the literature the symbol u is commonly used, so that a general one-dimensional conservation law has the form $u_t + f(u)_x = 0$, for example. However, most of the specific problems we will study involve a velocity (as in the acoustics equations (1.5)), and it is very convenient to use u for this quantity (or as the x-component of the velocity vector $\vec{u} = (u, v)$ in two dimensions).

The symbol Q_i^n (in one dimension) or Q_{ij}^n (in two dimensions) is used to denote the numerical approximation to the solution q. Subscripts on Q denote spatial locations (e.g., the ith grid cell), and superscript n denotes time level t_n. Often the temporal index is suppressed, since we primarily consider one-step methods where the solution at time t_{n+1} is determined entirely by data at time t_n. When Q or other numerical quantities lack a temporal superscript it is generally clear that the current time level t_n is intended.

For a system of m equations, q and Q are m-vectors, and superscripts are also used to denote the components of these vectors, e.g., q^p for $p = 1, 2, \ldots, m$. It is more convenient to use superscripts than subscripts for this purpose to avoid conflicts with spatial indices. Superscripts are also used to enumerate the eigenvalues λ^p and eigenvectors r^p of an $m \times m$ matrix. Luckily we generally do not need to refer to specific components of the eigenvectors. Of course superscripts must also be used for exponents at times, and this will usually be clear from context. Initial data is denoted by a circle *above* the variable, e.g., $\mathring{q}(x)$, rather than by a subscript or superscript, in order to avoid further confusion.

Several symbols play multiple roles in different contexts, since there are not enough letters and familiar symbols to go around. For example, ψ is used in different places for the entropy flux, for source terms, and for stream functions. For the most part these different uses are well separated and should be clear from context, but some care is needed to avoid confusion. In particular, the index p is generally used for indexing eigenvalues and eigenvectors, as mentioned above, but is also used for the pressure in acoustics and gas dynamics applications, often in close proximity. Since the pressure is never a superscript, I hope this will be clear.

One new symbol I have introduced is $q^{\vee}(q_l, q_r)$ (pronounced perhaps "q Riemann") to denote the value that arises in the similarity solution to a Riemann problem along the ray $x/t = 0$, when the data q_l and q_r is specified (see Section 1.2.1). This value is often used in defining numerical fluxes in finite volume methods, and it is convenient to have a general symbol for the function that yields it. This symbol is meant to suggest the spreading of waves from the Riemann problem, as will be explored starting in Chapter 3. Some notation specific to multidimensional problems is introduced in Section 18.1.

Part one
Linear Equations

Conservation Laws and Differential Equations

To see how conservation laws arise from physical principles, we will begin by considering the simplest possible fluid dynamics problem, in which a gas or liquid is flowing through a one-dimensional pipe with some known velocity $u(x, t)$, which is assumed to vary only with x, the distance along the pipe, and time t. Typically in fluid dynamics problems we must determine the motion of the fluid, i.e., the velocity function $u(x, t)$, as part of the solution, but let's assume this is already known and we wish to simply model the concentration or density of some chemical present in this fluid (in very small quantities that do not affect the fluid dynamics). Let $q(x, t)$ be the density of this chemical *tracer*, the function that we wish to determine.

In general the density should be measured in units of mass per unit volume, e.g., grams per cubic meter, but in studying the one-dimensional pipe with variations only in x, it is more natural to assume that q is measured in units of mass per unit length, e.g., grams per meter. This density (which is what is denoted by q here) can be obtained by multiplying the three-dimensional density function by the cross-sectional area of the pipe (which has units of square meters). Then

$$\int_{x_1}^{x_2} q(x, t)\, dx \tag{2.1}$$

represents the total mass of the tracer in the section of pipe between x_1 and x_2 at the particular time t, and has the units of mass. In problems where chemical kinetics is involved, it is often necessary to measure the "mass" in terms of moles rather than grams, and the density in moles per meter or moles per cubic meter, since the important consideration is not the mass of the chemical but the number of molecules present. For simplicity we will speak in terms of mass, but the conservation laws still hold in these other units.

Now consider a section of the pipe $x_1 < x < x_2$ and the manner in which the integral (2.1) changes with time. If we are studying a substance that is neither created nor destroyed within this section, then the total mass within this section can change only due to the *flux* or flow of particles through the endpoints of the section at x_1 and x_2. Let $F_i(t)$ be the rate at which the tracer flows past the fixed point x_i for $i = 1, 2$ (measured in grams per second, say). We use the convention that $F_i(t) > 0$ corresponds to flow to the right, while $F_i(t) < 0$ means a leftward flux, of $|F_i(t)|$ grams per second. Since the total mass in the section $[x_1, x_2]$

changes only due to fluxes at the endpoints, we have

$$\frac{d}{dt} \int_{x_1}^{x_2} q(x, t)\, dx = F_1(t) - F_2(t). \tag{2.2}$$

Note that $+F_1(t)$ and $-F_2(t)$ both represent fluxes *into* this section.

The equation (2.2) is the basic *integral form* of a conservation law, and equations of this type form the basis for much of what we will study. The rate of change of the total mass is due only to fluxes through the endpoints – this is the basis of *conservation*. To proceed further, we need to determine how the flux functions $F_j(t)$ are related to $q(x, t)$, so that we can obtain an equation that might be solvable for q. In the case of fluid flow as described above, the flux at any point x at time t is simply given by the product of the density $q(x, t)$ and the velocity $u(x, t)$:

$$\text{flux at } (x, t) = u(x, t)q(x, t). \tag{2.3}$$

The velocity tells how rapidly particles are moving past the point x (in meters per second, say), and the density q tells how many grams of chemical a meter of fluid contains, so the product, measured in grams per second, is indeed the rate at which chemical is passing this point.

Since $u(x, t)$ is a known function, we can write this flux function as

$$\text{flux} = f(q, x, t) = u(x, t)q. \tag{2.4}$$

In particular, if the velocity is independent of x and t, so $u(x, t) = \bar{u}$ is some constant, then we can write

$$\text{flux} = f(q) = \bar{u}q. \tag{2.5}$$

In this case the flux at any point and time can be determined directly from the value of the conserved quantity at that point, and does not depend at all on the location of the point in space–time. In this case the equation is called *autonomous*. Autonomous equations will occupy much of our attention because they arise in many applications and are simpler to deal with than nonautonomous or *variable-coefficient* equations, though the latter will also be studied.

For a general autonomous flux $f(q)$ that depends only on the value of q, we can rewrite the conservation law (2.2) as

$$\frac{d}{dt} \int_{x_1}^{x_2} q(x, t)\, dx = f(q(x_1, t)) - f(q(x_2, t)). \tag{2.6}$$

The right-hand side of this equation can be rewritten using standard notation from calculus:

$$\frac{d}{dt} \int_{x_1}^{x_2} q(x, t)\, dx = -f(q(x, t)) \Big|_{x_1}^{x_2}. \tag{2.7}$$

This shorthand will be useful in cases where the flux has a complicated form, and also suggests the manipulations performed below, leading to the differential equation for q.

Once the flux function $f(q)$ is specified, e.g., by (2.5) for the simplest case considered above, we have an equation for q that we might hope to solve. This equation should hold over every interval $[x_1, x_2]$ for arbitrary values of x_1 and x_2. It is not clear how to go about finding a function $q(x, t)$ that satisfies such a condition. Instead of attacking this problem directly, we generally transform it into a partial differential equation that can be handled by standard techniques. To do so, we must assume that the functions $q(x, t)$ and $f(q)$ are sufficiently smooth that the manipulations below are valid. This is very important to keep in mind when we begin to discuss nonsmooth solutions to these equations.

If we assume that q and f are smooth functions, then this equation can be rewritten as

$$\frac{d}{dt} \int_{x_1}^{x_2} q(x, t)\, dx = - \int_{x_1}^{x_2} \frac{\partial}{\partial x} f(q(x, t))\, dx, \tag{2.8}$$

or, with some further modification, as

$$\int_{x_1}^{x_2} \left[\frac{\partial}{\partial t} q(x, t)' + \frac{\partial}{\partial x} f(q(x, t)) \right] dx = 0. \tag{2.9}$$

Since this integral must be zero for all values of x_1 and x_2, it follows that the integrand must be identically zero. This gives, finally, the differential equation

$$\frac{\partial}{\partial t} q(x, t) + \frac{\partial}{\partial x} f(q(x, t)) = 0. \tag{2.10}$$

This is called the *differential form* of the conservation laws. Partial differential equations (PDEs) of this type will be our main focus. Partial derivatives will usually be denoted by subscripts, so this will be written as

$$q_t(x, t) + f(q(x, t))_x = 0. \tag{2.11}$$

2.1 The Advection Equation

For the flux function (2.5), the conservation law (2.10) becomes

$$q_t + \bar{u} q_x = 0. \tag{2.12}$$

This is called the *advection equation*, since it models the advection of a tracer along with the fluid. By a *tracer* we mean a substance that is present in very small concentrations within the fluid, so that the magnitude of the concentration has essentially no effect on the fluid dynamics. For this one-dimensional problem the concentration (or density) q can be measured in units such as grams per meter along the length of the pipe, so that $\int_{x_1}^{x_2} q(x, t)\, dx$ measures the total mass (in grams) within this section of pipe. In Section 9.1 we will consider more carefully the manner in which this is measured and the form of the resulting advection equation in more complicated cases where the diameter of the pipe and the fluid velocity need not be constant.

Equation (2.12) is a scalar, linear, constant-coefficient PDE of hyperbolic type. The general solution of this equation is very easy to determine. Any smooth function of the form

$$q(x, t) = \tilde{q}(x - \bar{u}t) \tag{2.13}$$

satisfies the differential equation (2.12), as is easily verified, and in fact any solution to (2.12) is of this form for some $\tilde{q}$. Note that $q(x, t)$ is constant along any ray in space–time for which $x - \bar{u}t = $ constant. For example, all along the ray $X(t) = x_0 + \bar{u}t$ the value of $q(X(t), t)$ is equal to $\tilde{q}(x_0)$. Values of q simply advect (i.e., translate) with constant velocity $\bar{u}$, as we would expect physically, since the fluid in the pipe (and hence the density of tracer moving with the fluid) is simply advecting with constant speed. These rays $X(t)$ are called the *characteristics* of the equation. More generally, characteristic curves for a PDE are curves along which the equation simplifies in some particular manner. For the equation (2.12), we see that along $X(t)$ the time derivative of $q(X(t), t)$ is

$$\frac{d}{dt} q(X(t), t) = q_t(X(t), t) + X'(t)q_x(X(t), t)$$
$$= q_t + \bar{u}q_x$$
$$= 0. \tag{2.14}$$

and the equation (2.12) reduces to a trivial ordinary differential equation $\frac{d}{dt} Q = 0$, where $Q(t) = q(X(t), t)$. This again leads to the conclusion that q is constant along the characteristic.

To find the particular solution to (2.12) of interest in a practical problem, we need more information in order to determine the particular function $\tilde{q}$ in (2.13): *initial conditions* and perhaps *boundary conditions* for the equation. First consider the case of an infinitely long pipe with no boundaries, so that (2.12) holds for $-\infty < x < \infty$. Then to determine $q(x, t)$ uniquely for all times $t > t_0$ we need to know the initial condition at time t_0, i.e., the initial density distribution at this particular time. Suppose we know

$$q(x, t_0) = \overset{\circ}{q}(x), \tag{2.15}$$

where $\overset{\circ}{q}(x)$ is a given function. Then since the value of q must be constant on each characteristic, we can conclude that

$$q(x, t) = \overset{\circ}{q}(x - \bar{u}(t - t_0))$$

for $t \geq t_0$. The initial profile $\overset{\circ}{q}$ simply translates with speed $\bar{u}$.

If the pipe has finite length, $a < x < b$, then we must also specify the density of tracer entering the pipe as a function of time, at the inflow end. For example, if $\bar{u} > 0$ then we must specify a boundary condition at $x = a$, say

$$q(a, t) = g_0(t) \quad \text{for } t \geq t_0$$

in addition to the initial condition

$$q(x, t) = \overset{\circ}{q}(x) \quad \text{for } a < x < b.$$

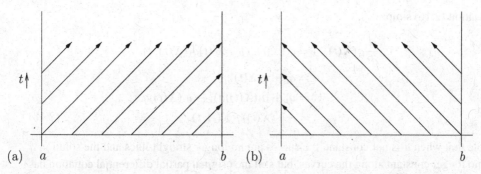

(a) a b (b) a b

Fig. 2.1. The solution to the advection equation is constant along characteristics. When solving this equation on the interval $[a, b]$, we need boundary conditions at $x = a$ if $\bar{u} > 0$ as shown in (a), or at $x = b$ if $\bar{u} < 0$ as shown in (b).

The solution is then

$$q(x, t) = \begin{cases} g_0(t - (x - a)/\bar{u}) & \text{if } a < x < a + \bar{u}(t - t_0), \\ \overset{\circ}{q}(x - \bar{u}(t - t_0)) & \text{if } a + \bar{u}(t - t_0) < x < b. \end{cases}$$

Note that we do not need to specify a boundary condition at the outflow boundary $x = b$ (and in fact cannot, since the density there is entirely determined by the data given already).

If on the other hand $\bar{u} < 0$, then flow is to the left and we would need a boundary condition at $x = b$ rather than at $x = a$. Figure 2.1 indicates the flow of information along characteristics for the two different cases. The proper specification of boundary conditions is always an important part of the setup of a problem.

From now on, we will generally take the initial time to be $t = 0$ to simplify notation, but everything extends easily to general t_0.

2.1.1 Variable Coefficients

If the fluid velocity u varies with x, then the flux (2.4) leads to the conservation law

$$q_t + (u(x)q)_x = 0. \tag{2.16}$$

In this case the characteristic curves $X(t)$ are solutions to the ordinary differential equations

$$X'(t) = u(X(t)). \tag{2.17}$$

Starting from an arbitrary initial point x_0, we can solve the equation (2.17) with initial condition $X(0) = x_0$ to obtain a particular characteristic curve $X(t)$. Note that these curves track the motion of particular material particles carried along by the fluid, since their velocity at any time matches the fluid velocity. Along a characteristic curve we find that the advection equation (2.16) simplifies:

$$\frac{d}{dt}q(X(t), t) = q_t(X(t), t) + X'(t)q_x(X(t), t)$$

$$= q_t + u(X(t))q_x$$

$$= q_t + (u(X(t))q)_x - u'(X(t))q$$

$$= -u'(X(t))q(X(t), t). \tag{2.18}$$

Note that when u is not constant, the curves are no longer straight lines and the solution q is no longer constant along the curves, but still the original partial differential equation has been reduced to solving sets of ordinary differential equations.

The operator $\partial_t + u\partial_x$ is often called the *material derivative*, since it represents differentiation along the characteristic curve, and hence computes the rate of change observed by a material particle moving with the fluid.

The equation (2.16) is an advection equation in conservation form. In some applications it is more natural to derive a nonconservative advection equation of the form

$$q_t + u(x)q_x = 0. \tag{2.19}$$

Again the characteristic curves satisfy (2.17) and track the motion of material points. For this equation the second line of the right-hand side of (2.18) reduces to zero, so that q is now constant along characteristic curves. Which form (2.16) or (2.19) arises often depends simply on what units are used to measure physical quantities, e.g., whether we measure concentration in grams per meter as was assumed above (giving (2.16)), or whether we use grams per cubic meter, as might seem to be a more reasonable definition of concentration in a physical fluid. The latter choice leads to (2.19), as is discussed in detail in Chapter 9, and further treatment of variable-coefficient problems is deferred until that point.

2.2 Diffusion and the Advection–Diffusion Equation

Now suppose that the fluid in the pipe is not flowing, and has zero velocity. Then according to the advection equation, $q_t = 0$ and the initial profile $\overset{\circ}{q}(x)$ does not change with time. However, if $\overset{\circ}{q}$ is not constant in space, then in fact it should still tend to slowly change due to molecular diffusion. The velocity $\bar{u}$ should really be thought of as a *mean velocity*, the average velocity that the roughly 10^{23} molecules in a given drop of water have. But individual molecules are bouncing around in different directions, and so molecules of the substance we are tracking will tend to get spread around in the water, as a drop of ink spreads. There will tend to be a net motion from regions where the density is large to regions where it is smaller. *Fick's law of diffusion* states that the net flux is proportional to the *gradient* of q, which in one space dimension is simply the derivative q_x. The flux at a point x now depends on the value of q_x at this point, rather than on the value of q, so we write

$$\text{flux of } q = f(q_x) = -\beta q_x, \tag{2.20}$$

where β is the *diffusion coefficient*. Using this flux in (2.10) gives

$$q_t = \beta q_{xx}, \tag{2.21}$$

which is known as the *diffusion equation*.

In some problems the diffusion coefficient may vary with x. Then $f = -\beta(x)q_x$ and the equation becomes

$$q_t = (\beta(x)q_x)_x. \tag{2.22}$$

Returning to the example of fluid flow, more generally there would be both advection and diffusion occurring simultaneously. Then the flux is $f(q, q_x) = \bar{u}q - \beta q_x$, giving the *advection–diffusion* equation

$$q_t + \bar{u}q_x = \beta q_{xx}. \tag{2.23}$$

The diffusion and advection–diffusion equations are examples of the general class of PDEs called *parabolic*.

2.3 The Heat Equation

The equation (2.21) (or more generally (2.22)) is often called the *heat equation*, for heat diffuses in much the same way as a chemical concentration. In the case of heat, there may be no net motion of the material, but thermal vibration of molecules causes neighboring molecules to vibrate and this internal energy diffuses through the material. Let $q(x, t)$ now be the temperature of the material at point x (e.g., a metal rod, since we are in one space dimension). The density of internal energy at point x is then given by

$$E(x, t) = \kappa(x)q(x, t),$$

where $\kappa(x)$ is the *heat capacity* of the material at this point. It is this energy that is conserved, and hence varies in a test section $[x_1, x_2]$ only due to the flux of energy past the endpoints. The heat flux is given by *Fourier's law of heat conduction*,

$$\text{flux} = -\beta q_x,$$

where β is the *coefficient of thermal conductivity*. This looks identical to Fick's law for diffusion, but note that Fourier's law says that the *energy* flux is proportional to the *temperature* gradient. If the heat capacity is identically constant, say $\kappa \equiv 1$, then this is identical to Fick's law, but there is a fundamental difference if κ varies. Equation (2.22) is the heat equation when $\kappa \equiv 1$. More generally the heat equation is derived from the conservation law

$$\frac{d}{dt} \int_{x_1}^{x_2} \kappa(x)q(x, t)\, dx = -\beta(x)q_x(x, t)\Big|_{x_1}^{x_2}, \tag{2.24}$$

and has the differential form

$$(\kappa q)_t = (\beta q_x)_x. \tag{2.25}$$

Typically κ does not vary with time and so this can be written as

$$\kappa q_t = (\beta q_x)_x. \tag{2.26}$$

2.4 Capacity Functions

In the previous section we saw how the heat capacity comes into the conservation law for heat conduction. There are also other situations where a "capacity" function naturally arises in the derivation of a conservation law, where again the flux of a quantity is naturally defined in terms of one variable q, whereas it is a different quantity κq that is conserved. If the flux function is $f(q)$, then the obvious generalization of (2.24) yields the conservation law

$$\kappa q_t + f(q)_x = 0. \tag{2.27}$$

While it may be possible to incorporate κ into the definition of $f(q)$, it is often preferable numerically to work directly with the form (2.27). This is discussed in Section 6.16 and is useful in many applications. In fluid flow problems, κ might represent the capacity of the medium to hold fluid. For flow through a pipe with a varying diameter, $\kappa(x)$ might be the cross-sectional area, for example (see Section 9.1). For flow in porous media, κ would be the porosity, the fraction of the medium available to fluid. On a nonuniform grid a capacity κ appears in the numerical method that is related to the size of a physical grid cell; see Section 6.17 and Chapter 23.

2.5 Source Terms

In some situations $\int_{x_1}^{x_2} q(x, t)\,dx$ changes due to effects other than flux through the endpoints of the section, if there is some source or sink of the substance within the section. Denote the density function for such a source by $\psi(q, x, t)$. (Negative values of ψ correspond to a sink rather than a source.) Then the equation becomes

$$\frac{d}{dt} \int_{x_1}^{x_2} q(x, t)\,dx = \int_{x_1}^{x_2} \frac{\partial}{\partial x} f(q(x, t))\,dx + \int_{x_1}^{x_2} \psi(q(x, t), x, t)\,dx.$$

This leads to the PDE

$$q_t(x, t) + f(q(x, t))_x = \psi(q(x, t), x, t). \tag{2.28}$$

In this section we mention only a few effects that lead to source terms. Conservation laws with source terms are more fully discussed in Chapter 17.

2.5.1 External Heat Sources

As one example, consider heat conduction in a rod as in Section 2.3, with $\kappa \equiv 1$ and $\beta \equiv$ constant, but now suppose there is also an external energy source distributed along the rod with density ψ. Then we obtain the equation

$$q_t(x, t) = \beta q_{xx}(x, t) + \psi(x, t).$$

This assumes the heat source is independent of the current temperature. In some cases the strength of the source may depend on the value of q. For example, if the rod is immersed in a liquid that is held at constant temperature q_0, then the flux of heat into the rod at the point (x, t) is proportional to $q_0 - q(x, t)$ and the equation becomes

$$q_t(x, t) = \beta q_{xx}(x, t) + D(q_0 - q(x, t)),$$

where D is the conductivity coefficient between the rod and the bath.

2.5.2 Reacting Flow

As another example, consider a fluid flowing through a pipe at constant velocity as in Section 2.1, but now suppose there are several different chemical species being advected in this flow (in minute quantities compared to the bulk fluid). If these chemicals react with one another, then the mass of each species individually will not be conserved, since it is used up or produced by the chemical reactions. We will have an advection equation for each species, but these will include source terms arising from the chemical kinetics.

As an extremely simple example, consider the advection of a radioactive isotope with concentration measured by q^1, which decays spontaneously at some rate α into a different isotope with concentration q^2. If this decay is taking place in a fluid moving with velocity $\bar{u}$, then we have a system of two advection equations with source terms:

$$q_t^1 + \bar{u} q_x^1 = -\alpha q^1,$$
$$q_t^2 + \bar{u} q_x^2 = +\alpha q^1. \tag{2.29}$$

This has the form $q_t + Aq_x = \psi(q)$, in which the coefficient matrix A is diagonal with both diagonal elements equal to $\bar{u}$. This is a hyperbolic system, with a source term. More generally we might have m species with various chemical reactions occurring simultaneously between them. Then we would have a system of m advection equations (with diagonal coefficient matrix $A = \bar{u}I$) and source terms given by the standard kinetics equations of mass action.

If there are spatial variations in concentrations, then these equations may be augmented with diffusion terms for each species. This would lead to a system of reaction–advection–diffusion equations of the form

$$q_t + Aq_x = \beta q_{xx} + \psi(q). \tag{2.30}$$

The diffusion coefficient could be different for each species, in which case β would be a diagonal matrix instead of a scalar.

Other types of source terms arise from external forces such as gravity or from geometric transformations used to simplify the equations. See Chapter 17 for some other examples.

2.6 Nonlinear Equations in Fluid Dynamics

In the pipe-flow model discussed above, the function $q(x, t)$ represented the density of some tracer that was carried along with the fluid but was present in such small quantities that the distribution of q has no effect on the fluid velocity. Now let's consider the density of the fluid itself, again in grams per meter, say, for this one-dimensional problem. We will denote the fluid density by the standard symbol $\rho(x, t)$. If the fluid is incompressible (as most liquids can be assumed to be for most purposes), then $\rho(x, t)$ is constant and this one-dimensional problem is not very interesting. If we consider a gas, however, then the molecules are far enough apart that compression or expansion is possible and the density may vary from point to point.

If we again assume that the velocity $\bar{u}$ is constant, then the density ρ will satisfy the same advection equation as before (since the flux is simply $\bar{u}\rho$ and $\bar{u}$ is constant),

$$\rho_t + \bar{u}\rho_x = 0, \tag{2.31}$$

and any initial variation in density will simply translate at speed $\bar{u}$. However, this is not what we would expect to happen physically. If the gas is compressed in some region (i.e., the density is higher here than nearby) then we would expect that the gas would tend to push into the neighboring gas, spreading out, and lowering the density in this region while raising the density nearby. (This does in fact happen provided that the *pressure* is also higher in this region; see below.) In order for the gas to spread out it must move relative to the neighboring gas, and hence we expect the velocity to change as a result of the variation in density.

While previously we assumed the tracer density q had no effect on the velocity, this is no longer the case. Instead we must view the velocity $u(x, t)$ as another unknown to be determined along with $\rho(x, t)$. The density flux still takes the form (2.3), and so the conservation law for ρ has the form

$$\rho_t + (\rho u)_x = 0, \tag{2.32}$$

which agrees with (2.31) only if u is constant. This equation is generally called the *continuity equation* in fluid dynamics, and models the *conservation of mass*.

In addition to this equation we now need a second equation for the velocity. The velocity itself is not a conserved quantity, but the momentum is. The product $\rho(x, t)u(x, t)$ gives the density of momentum, in the sense that the integral of ρu between any two points x_1 and x_2 yields the total momentum in this interval, and this can change only due to the flux of momentum through the endpoints of the interval. The momentum flux past any point x consists of two parts. First there is momentum carried past this point along with the moving fluid. For any density function q this flux has the form qu, as we have already seen at the beginning of this chapter, and so for the momentum $q = \rho u$ this contribution to the flux is $(\rho u)u = \rho u^2$. This is essentially an advective flux, although in the case where the quantity being advected is the velocity or momentum of the fluid itself, the phenomenon is often referred to as *convection* rather than advection.

In addition to this macroscopic convective flux, there is also a microscopic momentum flux due to the *pressure* of the fluid, as described in Section 14.1. This enters into the momentum flux, which now becomes

$$\text{momentum flux} = \rho u^2 + p.$$

The integral form of the conservation law (2.7) is then

$$\frac{d}{dt} \int_{x_1}^{x_2} \rho(x, t)u(x, t)\, dx = -[\rho u^2 + p]_{x_1}^{x_2}. \tag{2.33}$$

Note that it is only a *difference* in pressure between the two ends of the interval that will cause a change in the net momentum, as we would expect. We can think of this pressure difference as a net force that causes an acceleration of the fluid, though this isn't strictly correct and a better interpretation is given in Section 14.1.

If we assume that ρ, u, and p are all smooth, then we obtain the differential equation

$$(\rho u)_t + (\rho u^2 + p)_x = 0, \tag{2.34}$$

modeling *conservation of momentum*. Combining this with the continuity equation (2.32), we have a system of two conservation laws for the conservation of mass and momentum. These are coupled equations, since ρ and ρu appear in both. They are also clearly nonlinear, since products of the unknowns appear.

In developing the conservation law for ρu we have introduced a new unknown, the pressure $p(x, t)$. It appears that we need a third differential equation for this. Pressure is not a conserved quantity, however, and so instead we introduce a fourth variable, the *energy*, and an additional equation for the *conservation of energy*. The density of energy will be denoted by $E(x, t)$. This still does not determine the pressure, and to close the system we must add an *equation of state*, an algebraic equation that determines the pressure at any point in terms of the mass, momentum, and energy at the point. The energy equation and equations of state will be discussed in detail in Chapter 14, where we will derive the full system of three conservation laws.

For the time being we consider special types of flow where we can drop the conservation-of-energy equation and use a simpler equation of state that determines p from ρ alone. For example, if no shock waves are present, then it is often correct to assume that the *entropy* of the gas is constant. Such a flow is called *isentropic*. This is discussed further in Chapter 14. This assumption is reasonable in particular if we wish to derive the equations of *linear acoustics*, which we will do in the next section. In this case we look at very small-amplitude motions (sound waves) and the flow remains isentropic. In the isentropic case the equation of state is simply

$$p = \hat{\kappa} \rho^\gamma \equiv P(\rho), \tag{2.35}$$

where $\hat{\kappa}$ and γ are two constants (with $\gamma \approx 1.4$ for air).

More generally we could assume an equation of state of the form

$$p = P(\rho), \tag{2.36}$$

where $P(\rho)$ is a given function specifying the pressure in terms of density. To be physically realistic we can generally assume that

$$P'(\rho) > 0 \quad \text{for } \rho > 0. \tag{2.37}$$

This matches our intuition (already used above) that increasing the density of the gas will cause a corresponding increase in pressure. Note that the isentropic equation of state (2.35) has this property. We will see below that the assumption (2.37) is necessary in order to obtain a hyperbolic system.

Using the equation of state (2.36) in (2.34), together with the continuity equation (2.32), gives a closed system of two equations:

$$\rho_t + (\rho u)_x = 0,$$
$$(\rho u)_t + (\rho u^2 + P(\rho))_x = 0. \tag{2.38}$$

This is a coupled system of two nonlinear conservation laws, which we can write in the form

$$q_t + f(q)_x = 0 \tag{2.39}$$

if we define

$$q = \begin{bmatrix} \rho \\ \rho u \end{bmatrix} = \begin{bmatrix} q^1 \\ q^2 \end{bmatrix}, \qquad f(q) = \begin{bmatrix} \rho u \\ \rho u^2 + P(\rho) \end{bmatrix} = \begin{bmatrix} q^2 \\ (q^2)^2/q^1 + P(q^1) \end{bmatrix}. \tag{2.40}$$

More generally, a system of m conservation laws takes the form (2.39) with $q \in \mathbb{R}^m$ and $f : \mathbb{R}^m \to \mathbb{R}^m$. The components of f are the fluxes of the respective components of q, and in general each flux may depend on the values of any or all of the conserved quantities at that point.

Again it should be stressed that this differential form of the conservation law is derived under the assumption that q is smooth, from the more fundamental integral form. Note that when q is smooth, we can also rewrite (2.39) as

$$q_t + f'(q)q_x = 0, \tag{2.41}$$

where $f'(q)$ is the Jacobian matrix with (i, j) entry given by $\partial f_i/\partial q_j$. The form (2.41) is called the *quasilinear form* of the equation, because it resembles the linear system

$$q_t + A q_x = 0, \tag{2.42}$$

where A is a given $m \times m$ matrix. In the linear case this matrix does not depend on q, while in the quasilinear equation (2.41) it does. A thorough understanding of linear systems of the form (2.42) is required before tackling nonlinear systems, and the first 10 chapters concern only linear problems. There is a close connection between these theories, and the Jacobian matrix $f'(q)$ plays an important role in the nonlinear theory.

2.7 Linear Acoustics

In general one can always obtain a linear system from a nonlinear problem by *linearizing* about some state. This amounts to defining $A = f'(q_0)$ for some fixed state q_0 in the linear system (2.42), and gives a mathematically simpler problem that is useful in some situations, particularly when the interest is in studying small perturbations about some constant state.

To see how this comes about, suppose we wish to model the propagation of sound waves in a one-dimensional tube of gas. An acoustic wave is a very small pressure disturbance that propagates through the compressible gas, causing infinitesimal changes in the density and pressure of the gas via small motions of the gas with infinitesimal values of the velocity u.

Our eardrums are extremely sensitive to small changes in pressure and translate small oscillations in the pressure into nerve impulses that we interpret as sound. Consequently, most sound waves are essentially linear phenomena: the magnitudes of disturbances from the background state are so small that products or powers of the perturbation amplitude can be ignored. As linear phenomena, they also do not involve shock waves, and so a linearization of the isentropic equations introduced above is suitable. (An exception is the "sonic boom" caused by supersonic aircraft – this is a nonlinear shock wave, or at least originates as such.)

To perform the linearization of (2.40), let

$$q(x, t) = q_0 + \tilde{q}(x, t), \tag{2.43}$$

where $q_0 = (\rho_0, \rho_0 u_0)$ is the background state we are linearizing about and $\tilde{q}$ is the perturbation we wish to determine. Typically $u_0 = 0$, but it can be nonzero if we wish to study the propagation of sound in a constant-strength wind, for example. Using (2.43) in (2.11) and discarding any terms that involve powers or products of the $\tilde{q}$ variables, we obtain the linearized equations

$$\tilde{q}_t + f'(q_0)\tilde{q}_x = 0 \tag{2.44}$$

This is a constant-coefficient linear system modeling the evolution of small disturbances.

To obtain the acoustics equations, we compute the Jacobian matrix for the simplified system of gas dynamics (2.38). Differentiating the flux function from (2.40) gives

$$
\begin{aligned}
f'(q) &= \begin{bmatrix} \partial f^1/\partial q^1 & \partial f^1/\partial q^2 \\ \partial f^2/\partial q^1 & \partial f^2/\partial q^2 \end{bmatrix} \\[2mm]
&= \begin{bmatrix} 0 & 1 \\ -(q^2)^2/(q^1)^2 + P'(q^1) & 2q^2/q^1 \end{bmatrix} \\[2mm]
&= \begin{bmatrix} 0 & 1 \\ -u^2 + P'(\rho) & 2u \end{bmatrix}.
\end{aligned}
\tag{2.45}
$$

The equations of linear acoustics thus take the form of a constant-coefficient linear system (2.44) with

$$A = f'(q_0) = \begin{bmatrix} 0 & 1 \\ -u_0^2 + P'(\rho_0) & 2u_0 \end{bmatrix}. \tag{2.46}$$

Note that the vector $\tilde{q}$ in the system (2.44) has components $\tilde{\rho}$ and $\widetilde{\rho u}$, the perturbation of density and momentum. When written out in terms of its components, the system is

$$
\begin{aligned}
\tilde{\rho}_t + (\widetilde{\rho u})_x &= 0 \\
(\widetilde{\rho u})_t + \left(-u_0^2 + P'(\rho_0)\right)\tilde{\rho}_x + 2u_0(\widetilde{\rho u})_x &= 0.
\end{aligned}
\tag{2.47}
$$

Physically it is often more natural to model perturbations $\tilde{u}$ and $\tilde{p}$ in velocity and pressure, since these can often be measured directly. To obtain such equations, first note that pressure perturbations can be related to density perturbations through the equation of state,

$$p_0 + \tilde{p} = P(\rho_0 + \tilde{\rho}) = P(\rho_0) + P'(\rho_0)\tilde{\rho} + \cdots,$$

and since $p_0 = P(\rho_0)$, we obtain

$$\tilde{p} \approx P'(\rho_0)\tilde{\rho}.$$

Also we have

$$\rho u = (\rho_0 + \tilde{\rho})(u_0 + \tilde{u}) = \rho_0 u_0 + \tilde{\rho}u_0 + \rho_0\tilde{u} + \tilde{\rho}\tilde{u},$$

and so

$$\widetilde{\rho u} \approx u_0\tilde{\rho} + \rho_0\tilde{u}.$$

Using these expressions in the equations (2.47) and performing some manipulations (Exercise 2.1) leads to the alternative form of the linear acoustics equations

$$\begin{aligned}
\tilde{p}_t + u_0\tilde{p}_x + K_0\tilde{u}_x &= 0, \\
\rho_0\tilde{u}_t + \tilde{p}_x + \rho_0 u_0\tilde{u}_x &= 0,
\end{aligned} \tag{2.48}$$

where

$$K_0 = \rho_0 P'(\rho_0). \tag{2.49}$$

The equations (2.48) can be written as a linear system

$$\begin{bmatrix} p \\ u \end{bmatrix}_t + \begin{bmatrix} u_0 & K_0 \\ 1/\rho_0 & u_0 \end{bmatrix} \begin{bmatrix} p \\ u \end{bmatrix}_x = 0. \tag{2.50}$$

Here and from now on we will generally drop the tilde on p and u and use

$$q(x, t) = \begin{bmatrix} p(x, t) \\ u(x, t) \end{bmatrix}$$

to denote the pressure and velocity perturbations in acoustics.

The system (2.50) can also be derived by first rewriting the conservation laws (2.38) as a nonconservative set of equations for u and p, which is valid only for smooth solutions, and then linearizing this system; see Exercise 2.2.

An important special case of these equations is obtained by setting $u_0 = 0$, so that we are linearizing about the motionless state. In this case the coefficient matrix A appearing in the system (2.50) is

$$A = \begin{bmatrix} 0 & K_0 \\ 1/\rho_0 & 0 \end{bmatrix} \tag{2.51}$$

and the equations reduce to

$$\begin{aligned}
p_t + K_0 u_x &= 0, \\
\rho_0 u_t + p_x &= 0.
\end{aligned} \tag{2.52}$$

In Section 2.12 we will see that essentially the same set of equations can be derived for one-dimensional acoustics in an elastic solid. The parameter K_0 is called the *bulk modulus of compressibility* of the material; see Section 22.1.2 for more about this parameter.

2.8 Sound Waves

If we solve the equations just obtained for linear acoustics in a stationary gas, we expect the solution to consist of sound waves propagating to the left and right. Since the equations are linear, we should expect that the general solution consists of a linear superposition of waves moving in each direction, and that each wave propagates at constant speed (the speed of sound) with its shape unchanged. This suggests looking for solutions to the system (2.52) of the form

$$q(x, t) = \bar{q}(x - st)$$

for some speed s, where $\bar{q}(\xi)$ is some function of one variable. With this *Ansatz* we compute that

$$q_t(x, t) = -s\bar{q}'(x - st), \qquad q_x(x, t) = \bar{q}'(x - st),$$

and so the equation $q_t + Aq_x = 0$ reduces to

$$A\bar{q}'(x - st) = s\bar{q}'(x - st). \tag{2.53}$$

Since s is a scalar while A is a matrix, this is only possible if s is an eigenvalue of the matrix A, and $\bar{q}'(\xi)$ must also be a corresponding eigenvector of A for each value of ξ. Make sure you understand why this is so, as this is a key concept in understanding the structure of hyperbolic systems.

For the matrix A in (2.51) we easily compute that the eigenvalues are

$$\lambda^1 = -c_0 \quad \text{and} \quad \lambda^2 = +c_0, \tag{2.54}$$

where

$$c_0 = \sqrt{K_0/\rho_0}, \tag{2.55}$$

which must be the *speed of sound* in the gas. As expected, waves can propagate in either direction with this speed. Recalling (2.49), we see that

$$c_0 = \sqrt{P'(\rho_0)}. \tag{2.56}$$

The intuitively obvious assumption (2.37) (that pressure increases with density so that $P'(\rho) > 0$) turns out to be important mathematically in order for the speed of sound c_0 to be a real number.

For the more general coefficient matrix A of (2.50) with $u_0 \neq 0$, the eigenvalues are found to be

$$\lambda^1 = u_0 - c_0 \quad \text{and} \quad \lambda^2 = u_0 + c_0. \tag{2.57}$$

When the fluid is moving with velocity u_0, sound waves still propagate at speed $\pm c_0$ relative to the fluid, and at velocities λ^1 and λ^2 relative to a fixed observer. (See Figure 3.7.)

Regardless of the value of u_0, the eigenvectors of the coefficient matrix are

$$r^1 = \begin{bmatrix} -\rho_0 c_0 \\ 1 \end{bmatrix}, \qquad r^2 = \begin{bmatrix} \rho_0 c_0 \\ 1 \end{bmatrix}. \tag{2.58}$$

Any scalar multiple of each vector would still be an eigenvector. We choose the particular normalization of (2.58) because the quantity

$$Z_0 \equiv \rho_0 c_0 \tag{2.59}$$

is an important parameter in acoustics, called the *impedance* of the medium.

A sound wave propagating to the left with velocity $-c_0$ must have the general form

$$q(x, t) = \bar{w}^1(x + c_0 t) r^1 \tag{2.60}$$

for some scalar function $\bar{w}^1(\xi)$, so that

$$q(x, t) = \bar{w}^1(x + c_0 t) r^1 \equiv \bar{q}(x + c_0 t)$$

and hence $\bar{q}'(\xi)$ is a scalar multiple of r^1 as required by (2.53) for $s = -c_0$. In terms of the components of q this means that

$$\begin{aligned} p(x, t) &= -Z_0 \bar{w}^1(x + c_0 t), \\ u(x, t) &= \bar{w}^1(x + c_0 t). \end{aligned} \tag{2.61}$$

We see that in a left-going sound wave the pressure and velocity perturbations are always related by $p = -Z_0 u$. Analogously, in a right-going sound wave $p = +Z_0 u$ everywhere and $q(x, t) = \bar{w}^2(x - c_0 t) r^2$ for some scalar function $\bar{w}^2(\xi)$. (See Figure 3.1.)

The general solution to the acoustic equations consists of a superposition of left-going and right-going waves, and has

$$q(x, t) = \bar{w}^1(x + c_0 t) r^1 + \bar{w}^2(x - c_0 t) r^2 \tag{2.62}$$

for some scalar functions $\bar{w}^1(\xi)$ and $\bar{w}^2(\xi)$. Exactly what these functions are will depend on the *initial data* given for the problem. Let

$$q(x, 0) = \overset{\circ}{q}(x) = \begin{bmatrix} \overset{\circ}{p}(x) \\ \overset{\circ}{u}(x) \end{bmatrix}$$

be the pressure and velocity perturbation at time $t = 0$. To compute the resulting solution $q(x, t)$ we need to determine the scalar functions $\bar{w}^1$ and $\bar{w}^2$ in (2.62). To do so we can evaluate (2.62) at time $t = 0$ and set this equal to the given data $\overset{\circ}{q}$, obtaining

$$\bar{w}^1(x) r^1 + \bar{w}^2(x) r^2 = \overset{\circ}{q}(x).$$

At each point x this gives a 2×2 linear system of equations to solve for $\bar{w}^1(x)$ and $\bar{w}^2(x)$ at this particular point (since the vectors r^1, r^2, and $\overset{\circ}{q}(x)$ are all known). Let

$$R = [r^1 | r^2] \tag{2.63}$$

be the 2×2 matrix with columns r^1 and r^2. Then this system of equations can be written as

$$R\bar{w}(x) = \overset{\circ}{q}(x), \tag{2.64}$$

where $\bar{w}(x)$ is the vector with components $\bar{w}^1(x)$ and $\bar{w}^2(x)$. For acoustics the matrix R is

$$R = \begin{bmatrix} -Z_0 & Z_0 \\ 1 & 1 \end{bmatrix}, \tag{2.65}$$

which is a nonsingular matrix provided $Z_0 > 0$ as it will be in practice. The solution to (2.64) can be found in terms of the inverse matrix

$$R^{-1} = \frac{1}{2Z_0} \begin{bmatrix} -1 & Z_0 \\ 1 & Z_0 \end{bmatrix}. \tag{2.66}$$

We find that

$$\bar{w}^1(x) = \frac{1}{2Z_0}[-\overset{\circ}{p}(x) + Z_0\overset{\circ}{u}(x)],$$

$$\bar{w}^2(x) = \frac{1}{2Z_0}[\overset{\circ}{p}(x) + Z_0\overset{\circ}{u}(x)]. \tag{2.67}$$

The solution (2.62) then becomes

$$p(x, t) = \frac{1}{2}[\overset{\circ}{p}(x + c_0 t) + \overset{\circ}{p}(x - c_0 t)] - \frac{Z_0}{2}[\overset{\circ}{u}(x + c_0 t) - \overset{\circ}{u}(x - c_0 t)],$$

$$u(x, t) = -\frac{1}{2Z_0}[\overset{\circ}{p}(x + c_0 t) - \overset{\circ}{p}(x - c_0 t)] + \frac{1}{2}[\overset{\circ}{u}(x + c_0 t) + \overset{\circ}{u}(x - c_0 t)]. \tag{2.68}$$

2.9 Hyperbolicity of Linear Systems

The process we have just gone through to solve the acoustics equations motivates the definition of a first-order *hyperbolic system* of partial differential equations. This process generalizes to solve any linear constant-coefficient hyperbolic system.

Definition 2.1. *A linear system of the form*

$$q_t + Aq_x = 0 \tag{2.69}$$

is called hyperbolic *if the $m \times m$ matrix A is diagonalizable with real eigenvalues.*

We denote the eigenvalues by

$$\lambda^1 \le \lambda^2 \le \cdots \le \lambda^m.$$

The matrix is diagonalizable if there is a *complete* set of eigenvectors, i.e., if there are nonzero vectors $r^1, r^2, \ldots, r^m \in \mathbb{R}^m$ such that

$$Ar^p = \lambda^p r^p \quad \text{for } p = 1, 2, \ldots, m, \tag{2.70}$$

and these vectors are linearly independent. In this case the matrix

$$R = [r^1|r^2|\cdots|r^m], \tag{2.71}$$

formed by collecting the vectors $r^1, r^2, \ldots, r^m$ together, is nonsingular and has an inverse R^{-1}. We then have

$$R^{-1}AR = \Lambda \quad \text{and} \quad A = R\Lambda R^{-1}, \tag{2.72}$$

where

$$\Lambda = \begin{bmatrix} \lambda^1 & & & \\ & \lambda^2 & & \\ & & \ddots & \\ & & & \lambda^m \end{bmatrix} \equiv \text{diag}(\lambda^1, \lambda^2, \ldots, \lambda^m).$$

Hence we can bring A to diagonal form by a similarity transformation, as displayed in (2.72). The importance of this from the standpoint of the PDE is that we can then rewrite the linear system (2.69) as

$$R^{-1}q_t + R^{-1}ARR^{-1}q_x = 0. \tag{2.73}$$

If we define $w(x, t) \equiv R^{-1}q(x, t)$, then this takes the form

$$w_t + \Lambda w_x = 0. \tag{2.74}$$

Since Λ is diagonal, this system *decouples* into m independent advection equations for the components w^p of w:

$$w_t^p + \lambda^p w_x^p = 0 \quad \text{for } p = 1, 2, \ldots, m. \tag{2.75}$$

Since each λ^p is real, these advection equations make sense physically and can be used to solve the original system of equations (2.69). Complete details are given in the next chapter, but clearly the solution will consist of a linear combination of m "waves" traveling at the *characteristic speeds* $\lambda^1, \lambda^2, \ldots, \lambda^m$. (Recall that eigenvalues are also sometimes called "characteristic values.") These values define the *characteristic curves* $X(t) = x_0 + \lambda^p t$ along which information propagates in the decoupled advection equations. The functions $w^p(x, t)$ are called the *characteristic variables*; see Section 3.2.

There are some special classes of matrices A for which the system is certainly hyperbolic. If A is a *symmetric* matrix ($A = A^T$), then A is always diagonalizable with real eigenvalues and the system is said to be *symmetric hyperbolic*. Also, if A has *distinct* real eigenvalues $\lambda^1 < \lambda^2 < \cdots < \lambda^m$, then the eigenvectors must be linearly independent and the system is hyperbolic. Such a system is called *strictly hyperbolic*. The equations of linear acoustics are strictly hyperbolic, for example. The homogeneous part of the system (2.29) (i.e., setting $\alpha = 0$) is symmetric hyperbolic but not strictly hyperbolic. Difficulties can arise in studying certain nonstrictly hyperbolic equations, as discussed briefly in Section 16.2. If A has real eigenvalues but is not diagonalizable, then the system is *weakly hyperbolic*; see Section 16.3.

2.9.1 Second-Order Wave Equations

From the acoustics equations (2.52) we can eliminate the velocity u and obtain a second-order equation for the pressure. Differentiating the pressure equation with respect to t and the velocity equation with respect to x and then combining the results gives

$$p_{tt} = -K_0 u_{xt} = -K_0 u_{tx} = K_0 \left(\frac{1}{\rho_0} p_x \right)_x = c_0^2 p_{xx}.$$

This yields the *second-order wave equation* of the classical form

$$p_{tt} = c_0^2 p_{xx} \qquad (c_0 \equiv \text{constant}). \tag{2.76}$$

This is also a *hyperbolic* equation according to the standard classification of second-order differential equations. In this book, however, we concentrate almost entirely on first-order hyperbolic systems as defined at the start of Section 2.9. There is a certain equivalence as suggested by the above transformation for acoustics. Conversely, given a second-order equation of the type (2.76), we can derive a first-order hyperbolic system by defining new variables

$$q^1 = p_t, \qquad q^2 = -p_x,$$

so that (2.76) becomes $q_t^1 + c_0^2 q_x^2 = 0$, while the equality of mixed partial derivatives gives $q_t^2 + q_x^1 = 0$. These two equations taken together give a system $q_t + \tilde{A} q_x = 0$, with the coefficient matrix

$$\tilde{A} = \begin{bmatrix} 0 & c_0^2 \\ 1 & 0 \end{bmatrix}. \tag{2.77}$$

This matrix is *similar* to the matrix A of (2.51), meaning that there is a similarity transformation $\tilde{A} = SAS^{-1}$ relating the two matrices. The matrix S relates the two sets of variables and leads to a corresponding change in the eigenvector matrix, while the eigenvalues of the two matrices are the same, $\pm c_0$.

Many books take the viewpoint that the equation (2.76) is the fundamental wave equation and a first-order system can be derived from it by introducing "artificial" variables such as p_t and p_x. In fact, however, it is the first-order system that follows directly from the physics, as we have seen. Since effective numerical methods are more easily derived for the first-order system than for the second-order scalar equation, there is no need for us to consider the second-order equation further.

2.10 Variable-Coefficient Hyperbolic Systems

A variable-coefficient linear system of PDEs might take the form

$$q_t + A(x) q_x = 0. \tag{2.78}$$

This system is hyperbolic at any point x where the coefficient matrix satisfies the conditions laid out in Section 2.9. In Section 9.6, for example, we will see that the equations of acoustics in a heterogeneous medium (where the density and bulk modulus vary with x) can be written as such a system which is hyperbolic everywhere, with eigenvalues given by $\pm c(x)$, where the sound speed $c(x)$ varies with position depending on the material parameters.

In some cases we might have a conservative system of linear equations of the form

$$q_t + (A(x)q)_x = 0, \tag{2.79}$$

in which the flux function $f(q, x) = A(x)q$ depends explicitly on x. This system could be rewritten as

$$q_t + A(x)q_x = -A'(x)q \tag{2.80}$$

as a system of the form (2.78) with the addition of a *source term*. Again the problem is hyperbolic at any point where $A(x)$ is diagonalizable with real eigenvalues. Such problems are discussed further in Chapter 9.

2.11 Hyperbolicity of Quasilinear and Nonlinear Systems

A *quasilinear* system

$$q_t + A(q, x, t)q_x = 0 \tag{2.81}$$

is said to be *hyperbolic* at a point (q, x, t) if the matrix $A(q, x, t)$ satisfies the hyperbolicity condition (diagonalizable with real eigenvalues) at this point.

The nonlinear conservation law (2.11) is *hyperbolic* if the Jacobian matrix $f'(q)$ appearing in the quasilinear form (2.41) satisfies the hyperbolicity condition for each physically relevant value of q.

Example 2.1. The nonlinear equations of isentropic gas dynamics (2.38) have the Jacobian matrix (2.45). The eigenvalues are

$$\lambda^1 = u - c, \qquad \lambda^2 = u + c,$$

where the velocity u may now vary from point to point, as does the sound speed

$$c = \sqrt{P'(\rho)}. \tag{2.82}$$

However, since $P'(\rho) > 0$ at all points in the gas, this nonlinear system is strictly hyperbolic. (Provided we stay away from the "vacuum state" where ρ and p go to zero. For the equation of state (2.35), $c \to 0$ as well in this case, and the nonstrict hyperbolicity at this point causes additional difficulties in the nonlinear analysis.)

Solutions to nonlinear hyperbolic systems also involve wave propagation, and for a system of m equations we will often be able to find m independent waves at each point. However, since the wave speeds depend on the value of the solution q, wave shapes will typically deform, and the solution procedure is greatly complicated by this nonlinear structure. Nonlinear conservation laws are discussed starting in Chapter 11.

In the remainder of this chapter some other hyperbolic systems are introduced. These sections can be skipped at this point without loss of continuity.

2.12 Solid Mechanics and Elastic Waves

The equations of linear acoustics were derived in Section 2.8 by linearizing the equations of isentropic gas dynamics. Essentially the same system of equations can be derived from elasticity theory to obtain the equations modeling a one-dimensional acoustic wave in a solid, which again is a small-amplitude compressional disturbance in which the material moves back and forth in the same direction as the wave propagates, leading to small-scale changes in density and pressure. Unlike a gas or liquid, however, a solid also supports a second distinct type of small-amplitude waves called *shear waves*, in which the motion of the material is orthogonal to the direction of wave propagation. These two types of waves travel at distinct speeds, as illustrated in Figure 2.2. In general these two types of waves are coupled together and the equations of linear elasticity are a single set of hyperbolic equations that must be solved for all motions of the solid, which are coupled together. However, if we restrict our attention to one-dimensional plane waves, in which all quantities vary only in one direction, then these equations decouple into two independent hyperbolic systems of two equations each. Mathematically these linear systems are not very interesting, since each has the same structure as the acoustics equations we have already studied in detail. Because of this, however, some of the basic concepts of wave propagation in solids can be most easily introduced in this context, and this foundation will be useful when we develop the multidimensional equations.

Figure 2.2 shows the two distinct types of plane-wave motion in an elastic solid. Other types of waves can also be observed in solids, such as *surface waves* at a free surface or interface between different solids, but these have a truly multidimensional structure. The strips shown in Figure 2.2 should be viewed as taken from an infinite three-dimensional solid in which all quantities vary only with x, so the motion shown extends infinitely far in the

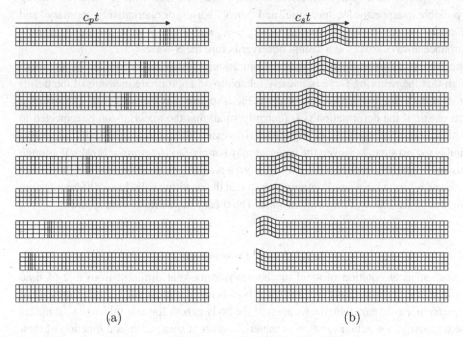

(a)   (b)

Fig. 2.2. Illustration of (a) P-waves and (b) S-waves in elastic solids. Time advances going upwards.

y (upward) direction and in z (normal to the page). Related one-dimensional equations can also be used to study elastic motion in a finite elastic bar, but then additional complications arise in that a compression in the x-direction will typically result in some spreading in the y- and z-directions, whereas in the infinite plane-wave case there is no opportunity for such spreading; see Chapter 22.

In Figure 2.2(a) the material is compressed at the left edge by a motion in the x-direction confined to a small region in x. This compressional wave moves in the x-direction at some speed c_p and is analogous to an acoustic wave in a gas. Compressing the material leads to an increase in stress and hence to acceleration in the x-direction. These changes in stress and velocity are coupled together to result in wave motion.

Figure 2.2(b) shows a different type of wave, a shear wave in which the material is displaced in the y-direction over a small region in x. In a gas or liquid, a shear displacement of this type would not result in any restoring force or wave motion. There is no compression or expansion of the material, and hence no stress results. Molecules of a gas or liquid may be freely rearranged as long as there is no change in the pressure and there will be no restoring force. Of course frictional (viscous) forces will arise during a rearrangement as the molecules move past one another, but once rearranged they are not pulled back towards their original locations. A solid is fundamentally different in that the constituent molecules are attached to one another by chemical bonds that resist any deformations. The bonds will stretch slightly to allow small elastic deformations, but like tiny springs they exert a restoring force that typically grows with the magnitude of any deformation. This operates like pressure in the case of compressional waves, but these bonds also resist shear deformations, and the restoring forces result in shear waves as illustrated in Figure 2.2(b). These waves move at a speed c_s that we will see is always smaller then the speed c_p of compressional waves. The two types of waves are often called *P-waves* and *S-waves*, with "P" and "S" having two possible interpretations: "pressure" and "shear" waves, or alternatively "primary" and "secondary" waves in view of the fact that $c_p > c_s$ and so the P-wave arising from some disturbance always arrives at a distant observer before the S-wave.

The theory of *linear elasticity* results from assuming that the deformations are small enough that the restoring force is linearly related to an appropriate measure of the deformation of the solid. For larger deformations the response may be nonlinear. The material is still elastic if the deformation is sufficiently small that the material can be expected to eventually return to its original configuration if all external forces are removed. If the deformation is too extreme, however, the material may simply fail (fracture or break), if enough bonds are irreparably broken, or it may undergo a *plastic deformation*, in which bonds are broken and reformed in a new configuration so that the resulting solid has a different resting configuration from that of the original solid. The theory of *plasticity* then applies.

2.12.1 Elastic Deformations

The mathematical notation of solid mechanics is somewhat different from that of fluid dynamics. For an elastic body we are typically concerned with small displacements about some *reference configuration*, the location of the body at rest, for example, and so it makes sense to consider the actual location of material points at some time as a function of their reference location. For example, in two space dimensions we can let $(X(x, y), Y(x, y))$

represent the location at time t of the material whose reference location is (x, y). The *displacement vector* $\vec{\delta}$ is then defined to be

$$\vec{\delta}(x, y, t) = \begin{bmatrix} \delta^1(x, y, t) \\ \delta^2(x, y, t) \end{bmatrix} = \begin{bmatrix} X(x, y, t) \\ Y(x, y, t) \end{bmatrix} - \begin{bmatrix} x \\ y \end{bmatrix}. \tag{2.83}$$

The symbol $\vec{u}$ is often used for the displacement vector, but we reserve this for the velocity vector, which is the time derivative of the displacement,

$$\vec{u}(x, y, t) = \begin{bmatrix} u(x, y, t) \\ v(x, y, t) \end{bmatrix} = \begin{bmatrix} \delta^1_t(x, y, t) \\ \delta^2_t(x, y, t) \end{bmatrix}. \tag{2.84}$$

Displacements of the body often lead to *strains* within the body. A strain is a deformation that results in changes of length or shape within the body. These strains in turn lead to *stress*, the interior forces due to the stretching or compression of atomic bonds. These forces result in acceleration of the material, affecting the motion and hence the evolution of the strains. The equations of elasticity consist of Newton's law relating force to acceleration together with *stress–strain relations* describing the force that results from a given strain. This *constitutive relation* depends on the particular material (similarly to the *equation of state* for a gas). For sufficiently small strains the stress may be assumed to vary linearly with strain, resulting in the equations of linear elasticity.

2.12.2 Strain

Not all deformations result in a strain. Rigid-body motions (translations and rotations) in which the body is simply moved as a rigid entity do not lead to any internal strain or stress. Rigid translations correspond to a displacement vector $\vec{\delta}(x, y, t)$ that varies only with t and is independent of spatial position. Clearly there will be a strain in the material only if $\vec{\delta}$ varies in space, so that some points are displaced relative to other points in the body. Hence the strain depends only on the *displacement gradient*

$$\nabla\vec{\delta} = \begin{bmatrix} \delta^1_x & \delta^1_y \\ \delta^2_x & \delta^2_y \end{bmatrix} = \begin{bmatrix} X_x - 1 & X_y \\ Y_x & Y_y - 1 \end{bmatrix}, \tag{2.85}$$

where the subscripts denote partial derivatives. Note that for a rigid translation $\nabla\vec{\delta} = 0$.

We still need to eliminate solid-body rotations, which can be done by splitting $\nabla\vec{\delta}$ into the sum of a symmetric and a skew-symmetric matrix,

$$\nabla\vec{\delta} = \epsilon + \Omega, \tag{2.86}$$

with

$$\epsilon = \frac{1}{2}[\nabla\vec{\delta} + (\nabla\vec{\delta})^T] = \begin{bmatrix} \delta^1_x & \frac{1}{2}(\delta^1_y + \delta^2_x) \\ \frac{1}{2}(\delta^1_y + \delta^2_x) & \delta^2_y \end{bmatrix} \tag{2.87}$$

and

$$\Omega = \frac{1}{2}[\nabla\vec{\delta} - (\nabla\vec{\delta})^T] = \begin{bmatrix} 0 & \frac{1}{2}(\delta^1_y - \delta^2_x) \\ -\frac{1}{2}(\delta^1_y - \delta^2_x) & 0 \end{bmatrix}. \tag{2.88}$$

The *rotation matrix* Ω measures rigid rotations, whereas the symmetric matrix ϵ is the desired *strain matrix*, which will also be written as

$$\epsilon = \begin{bmatrix} \epsilon^{11} & \epsilon^{12} \\ \epsilon^{21} & \epsilon^{22} \end{bmatrix}.$$

The diagonal elements ϵ^{11} and ϵ^{22} measure *extensional strains* in the x- and y-directions, whereas $\epsilon^{12} = \epsilon^{21}$ is the *shear strain*.

Example 2.2. The P-wave shown in Figure 2.2 has a displacement of the form

$$\vec{\delta}(x, y, t) = \begin{bmatrix} w(x - c_p t) \\ 0 \end{bmatrix}$$

for some wave form w, and hence

$$\epsilon = \begin{bmatrix} w'(x - c_p t) & 0 \\ 0 & 0 \end{bmatrix}$$

with only ϵ^{11} nonzero.

The S-wave shown in Figure 2.2 has a displacement of the form

$$\vec{\delta}(x, y, t) = \begin{bmatrix} 0 \\ w(x - c_s t) \end{bmatrix}$$

for some waveform w, and hence

$$\epsilon = \begin{bmatrix} 0 & \frac{1}{2} w'(x - c_s t) \\ \frac{1}{2} w'(x - c_s t) & 0 \end{bmatrix}$$

with only the shear strain nonzero.

To study one-dimensional elastic waves of the sort shown in Figure 2.2, we need only consider the components ϵ^{11} and ϵ^{12} of the strain and must assume that these are functions of (x, t) alone, independent of y and z. For two-dimensional elasticity we must consider ϵ^{22} as well, with all three variables being functions of (x, y, t). For full three-dimensional elasticity the displacement vector and strain matrix must be extended to three dimensions. The formula (2.87) still holds, and ϵ is now a 3×3 symmetric matrix with six independent elements, three extensional strains on the diagonal, and three shear strains off the diagonal. See Chapter 22 for more discussion of these equations and their proper relation to three-dimensional elasticity.

2.12.3 Stress

A strain in an elastic body typically results in a restoring force called the *stress*. For one-dimensional elasticity as described above, we need only be concerned with two components of the stress: $\sigma^{11}(x, t)$, the force in the x-direction (the *normal stress*), and $\sigma^{12}(x, t)$, the force in the y-direction (the *shear stress*).

In one-dimensional linear elasticity there is a complete decoupling of compressional and shear effects. The normal stress σ^{11} depends only on the strain ϵ^{11}, while the shear stress σ^{12} depends only on the shear strain ϵ^{12}, and these constitutive relations are linear:

$$\sigma^{11} = (\lambda + 2\mu)\epsilon^{11} \qquad \text{with } \lambda + 2\mu > 0, \tag{2.89}$$

$$\sigma^{12} = 2\mu\,\epsilon^{12} \qquad \text{with } \mu > 0. \tag{2.90}$$

Here λ and μ are the *Lamé parameters* characterizing the material. The parameter μ is also called the *shear modulus*. The parameter λ does not have a simple physical meaning, but is related to other properties of the material in Section 22.1. It is unfortunate that the symbol λ is standard for this parameter, which should not be confused with an eigenvalue.

2.12.4 The Equations of Motion

We are now ready to write down the equations of motion for one-dimensional elastic waves. P-waves are governed by the system of equations

$$\begin{aligned} \epsilon_t^{11} - u_x &= 0, \\ \rho u_t - \sigma_x^{11} &= 0, \end{aligned} \tag{2.91}$$

where $\rho > 0$ is the density of the material. The first equation follows from the equality $X_{xt} = X_{tx}$, since

$$\begin{aligned} \epsilon^{11}(x, t) = X_x(x, t) - 1 &\Longrightarrow \epsilon_t^{11} = X_{xt}, \\ u(x, t) = X_t(x, t) &\Longrightarrow u_x = X_{tx}. \end{aligned} \tag{2.92}$$

The second equation of (2.91) is Newton's second law since u_t is the acceleration.

The system (2.91) involves both ϵ^{11} and σ^{11}, and one of these must be eliminated using the constitutive relation (2.89). If we eliminate σ^{11}, we obtain

$$\begin{bmatrix} \epsilon^{11} \\ u \end{bmatrix}_t + \begin{bmatrix} 0 & -1 \\ -(\lambda + 2\mu)/\rho & 0 \end{bmatrix} \begin{bmatrix} \epsilon^{11} \\ u \end{bmatrix}_x = 0. \tag{2.93}$$

This is a hyperbolic system, since the matrix has eigenvalues $\lambda = \pm c_p$ with

$$c_p = \sqrt{(\lambda + 2\mu)/\rho}. \tag{2.94}$$

If we instead eliminate ϵ^{11} we obtain

$$\begin{bmatrix} \sigma^{11} \\ u \end{bmatrix}_t + \begin{bmatrix} 0 & -(\lambda + 2\mu) \\ -1/\rho & 0 \end{bmatrix} \begin{bmatrix} \sigma^{11} \\ u \end{bmatrix}_x = 0. \tag{2.95}$$

Again the coefficient matrix has eigenvalues $\pm c_p$. Note that this form is essentially equivalent to the acoustic equations derived in Section 2.8 if we identify

$$p(x, t) = -\sigma^{11}(x, t). \tag{2.96}$$

Since ϵ^{11} measures the extensional stress (positive when the material is stretched, negative when compressed), a positive pressure corresponds to a negative value of σ^{11}. Note that

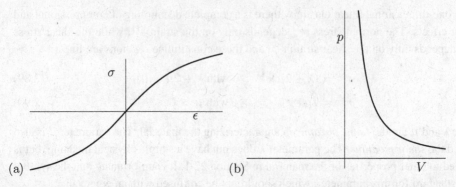

Fig. 2.3. (a) A typical stress–strain relation $\sigma = \sigma(\epsilon)$ for the nonlinear elasticity equation (2.97). (b) The equation of state $p = p(V)$ for isentropic gas dynamics in a Lagrangian frame using the p-system (2.108).

the stress σ^{11} can have either sign, depending on whether the material is compressed or stretched, while the pressure in a gas can only be positive. A gas that is "stretched" by allowing it to expand to a larger volume will not attempt to contract back to its original volume the way a solid will. This is another consequence of the fact that there are no intermolecular bonds between the gas molecules.

One-dimensional *nonlinear P-waves* can be modeled by the more general form of (2.91) given by

$$\begin{aligned} \epsilon_t - u_x &= 0, \\ \rho u_t - \sigma(\epsilon)_x &= 0, \end{aligned} \tag{2.97}$$

where ϵ is the extensional strain ϵ^{11} and $\sigma^{11} = \sigma(\epsilon)$ is a more general nonlinear constitutive relation between stress and strain. A typical stress–strain relation might look something like what is shown in Figure 2.3(a). In the case shown the derivative of the stress with respect to strain decreases as the magnitude of the strain is increased. This is shown for small values of $|\epsilon|$, in particular for $-\epsilon \ll 1$, since $\epsilon = -1$ corresponds to a state of complete compression, $X_x = 0$. Elasticity theory typically breaks down long before this. For very small deformations ϵ, this nonlinear function can generally be replaced by a linearization $\sigma = (\lambda + 2\mu)\epsilon$, where $(\lambda + 2\mu) \equiv \sigma'(0)$. This is the relation (2.90) used in linear elasticity.

The equations for a linear S-wave are essentially identical to (2.91) but involve the shear strain, shear stress, and vertical velocity:

$$\begin{aligned} \epsilon_t^{12} - \frac{1}{2}v_x &= 0, \\ \rho v_t - \sigma_x^{12} &= 0. \end{aligned} \tag{2.98}$$

The relationship (2.90) is now used to eliminate either ϵ^{12} or σ^{12}, resulting in a closed system of two equations, either

$$\begin{bmatrix} \epsilon^{12} \\ v \end{bmatrix}_t + \begin{bmatrix} 0 & -1/2 \\ -2\mu/\rho & 0 \end{bmatrix} \begin{bmatrix} \epsilon^{12} \\ v \end{bmatrix}_x = 0 \tag{2.99}$$

if ϵ^{12} is used, or

$$\begin{bmatrix} \sigma^{12} \\ v \end{bmatrix}_t + \begin{bmatrix} 0 & -\mu \\ -1/\rho & 0 \end{bmatrix} \begin{bmatrix} \sigma^{12} \\ v \end{bmatrix}_x = 0 \qquad (2.100)$$

if σ^{12} is used. In either case, the eigenvalues of the coefficient matrix are $\lambda = \pm c_s$, with the wave speed

$$c_s = \sqrt{\mu/\rho}. \qquad (2.101)$$

In general, $\mu < \lambda + 2\mu$ and so $c_s < c_p$.

We have assumed that shear-wave motion is in the y-direction. In a three-dimensional body one could also observe a plane shear wave propagating in the x-direction for which the shear motion is in the z-direction. These are governed by a set of equations identical to (2.100) but involving ϵ^{13} and σ^{13} in place of ϵ^{12} and σ^{12}, and the z-component of velocity w in place of v. Shear motion need not be aligned with either the y- or the z-axis, but can occur in any direction perpendicular to x. Motion in any other direction is simply a linear combination of these two, however, so that there are really two decoupled systems of equations for S-waves, along with the system of equations of P-waves, needed to describe all *plane waves* in x. Note that the systems (2.95) and (2.100) both have the same mathematical structure as the acoustics equations studied previously.

For a general two- or three-dimensional motion of an elastic solid it is not possible to decompose the resulting equations into independent sets of equations for P-waves and S-waves. Instead one obtains a single coupled hyperbolic system. For motions that are fully two-dimensional (but independent of the third direction), one obtains a system of five equations for the velocities u, v and the components of the stress tensor σ^{11}, σ^{12}, and σ^{22} (or alternatively the three components of the strain tensor). Only in the case of purely one-dimensional motions do these equations decouple into independent sets. These decoupled systems are related to the full three-dimensional equations in Chapter 22.

2.13 Lagrangian Gas Dynamics and the p-System

The fluid dynamics equations derived in Section 2.6 are in *Eulerian form*, meaning that x represents a fixed location in space, and quantities such as the velocity $u(x, t)$ refer to the velocity of whatever fluid particle happens to be at the point x at time t. Alternatively, the equations can be written in *Lagrangian form*, where fixing the coordinate ξ corresponds to tracking a particular fluid particle. The Lagrangian velocity $U(\xi, t)$ then gives the velocity of this particle at time t. We must then determine the mapping $X(\xi, t)$ that gives the physical location of the particle labeled ξ at time t. This is more like the approach used in elasticity, as described in Section 2.12, and in one dimension a system of equations very similar to (2.97) results. (The term *fluid particle* refers to an infinitesimally small volume of fluid, but one that still contains a huge number of molecules so that the small-scale random variations in velocity can be ignored.)

To set up the labeling of points initially, we take an arbitrary physical location x_0 (say $x_0 = 0$) and then at each point x assign the label

$$\xi = \int_{x_0}^{x} \overset{\circ}{\rho}(s) \, ds \tag{2.102}$$

to the particle initially located at x, where $\overset{\circ}{\rho}$ is the initial data for the density. If the density is positive everywhere, then this gives a one–one map. Note that ξ has units of mass and the label ξ gives the total mass between x_0 and $X(\xi, t)$. Moreover $\xi_2 - \xi_1$ is the total mass of all particles between those labeled ξ_1 and ξ_2 (at any time t, since particles cannot cross in this one-dimensional model).

The Lagrangian velocity is related to the Eulerian velocity by

$$U(\xi, t) = u(X(\xi, t), t).$$

Since $X(\xi, t)$ tracks the location of this particle, we must have

$$X_t(\xi, t) = U(\xi, t).$$

We could define a Lagrangian density function similarly, but the conservation of mass equation in the Lagrangian framework is more naturally written in terms of the *specific volume*

$$V(\xi, t) = \frac{1}{\rho(X(\xi, t), t)}.$$

This has units of volume/mass (which is just length/mass in one dimension), so it makes sense to integrate this over ξ. Since integrating the specific volume over a fixed set of particles gives the volume occupied by these particles at time t, we must have

$$\int_{\xi_1}^{\xi_2} V(\xi, t) \, d\xi = X(\xi_2, t) - X(\xi_1, t). \tag{2.103}$$

Differentiating this with respect to t gives

$$\frac{d}{dt} \int_{\xi_1}^{\xi_2} V(\xi, t) \, d\xi = U(\xi_2, t) - U(\xi_1, t)$$

$$= \int_{\xi_1}^{\xi_2} \frac{\partial}{\partial \xi} U(\xi, t) \, d\xi. \tag{2.104}$$

Rearranging this and using the fact that it must hold for all choices of ξ_1 and ξ_2 gives the differential form of the conservation law,

$$V_t - U_\xi = 0. \tag{2.105}$$

Now consider the conservation of momentum. In Eulerian form ρu is the density of momentum in units of momentum/volume. In Lagrangian form we instead consider $U(\xi, t)$, which can be interpreted as the momentum per unit mass, with

$$\int_{\xi_1}^{\xi_2} U(\xi, t) \, d\xi$$

being the total momentum of all particles between ξ_1 and ξ_2. By conservation of momentum, this integral changes only due to flux at the endpoints. Since in the Lagrangian framework the endpoints are moving with the fluid, there is no "advective flux" and the only change in momentum comes from the pressure difference between the two endpoints, so

$$\frac{d}{dt} \int_{\xi_1}^{\xi_2} U(\xi, t)\, d\xi = p(\xi_1, t) - p(\xi_2, t),$$

which leads to the conservation law

$$U_t + p_\xi = 0. \tag{2.106}$$

If we consider isentropic or isothermal flow, then we have only these two conservation laws and the equation of state gives p in terms of V alone. Then (2.105) and (2.106) give the system of conservation laws known as the *p-system*,

$$\begin{aligned} V_t - U_\xi &= 0, \\ U_t + p(V)_\xi &= 0. \end{aligned} \tag{2.107}$$

This is another simple system of two equations that is useful in understanding conservation laws. It is slightly simpler than the corresponding Eulerian equations (2.38) in that the only nonlinearity is in the function $p(V)$. This system if hyperbolic if $p'(V) < 0$ (see Exercise 2.7). Note that for isentropic flow we have $p(V) = \hat{k} V^{-\gamma}$, corresponding to the equation of state (2.35), with the shape shown in Figure 2.3(b).

Frequently the p-system is written using lowercase symbols as

$$\begin{aligned} v_t - u_x &= 0, \\ u_t + p(v)_x &= 0, \end{aligned} \tag{2.108}$$

and we will generally use this notation when the p-system is used as a generic example of a hyperbolic system. To relate this system to the Eulerian gas dynamics equations, however, it is important to use distinct notation as derived above.

The p-system (2.108) has a very similar structure to the nonlinear elasticity equation (2.97) if we equate p with the negative stress $-\sigma$ as discussed in Section 2.12. Note, however, that in the gas dynamics case we must have $V > 0$ and $p > 0$, whereas in elasticity the stress and strain can each be either positive or negative, corresponding to extension and compression respectively (recall Figure 2.3(a)).

2.14 Electromagnetic Waves

Electromagnetic waves are governed by Maxwell's equations. In the simplest case this is a hyperbolic system of equations, though in materials where waves are attenuated due to induced electric currents these are modified by additional source terms. If we assume there is no net electric charge or current in the material through which the wave is propagating, then Maxwell's equations reduce to

$$\vec{D}_t - \nabla \times \vec{H} = 0, \tag{2.109}$$

$$\vec{B}_t + \nabla \times \vec{E} = 0, \tag{2.110}$$

$$\nabla \cdot \vec{D} = 0, \tag{2.111}$$

$$\nabla \cdot \vec{B} = 0. \tag{2.112}$$

Here $\vec{E}$, $\vec{D}$, $\vec{B}$, and $\vec{H}$ are all vectors with three spatial components. The electric field $\vec{E}$ and the magnetic field $\vec{B}$ are related to the two other fields $\vec{D}$ and $\vec{H}$ via *constitutive relations* that characterize the medium in which the wave is propagating. These are similar to the stress–strain relations needed in elasticity theory. In general they take the form

$$\vec{D} = \epsilon \vec{E}, \tag{2.113}$$

$$\vec{B} = \mu \vec{H}, \tag{2.114}$$

where ϵ is the *permittivity* and μ is the *magnetic permeability* of the medium. In a homogeneous isotropic material these are both scalar constants. More generally they could be 3×3 matrices and also vary in space.

If the initial data satisfies the divergence-free conditions (2.111) and (2.112), then it can be shown that these will hold for all time, and so equations (2.109) and (2.110) for $\vec{D}_t$ and $\vec{B}_t$ can be taken as the time evolution equations for electromagnetic waves.

If ϵ and μ are scalar constants, then we can eliminate $\vec{D}$ and $\vec{H}$ and rewrite the wave-propagation equations as

$$\vec{E}_t - \frac{1}{\epsilon \mu} \nabla \times \vec{B} = 0,$$
$$\vec{B}_t + \nabla \times \vec{E} = 0. \tag{2.115}$$

This is a linear hyperbolic system of equations in three dimensions.

In this chapter we consider only the simplest case of a plane wave propagating in the x-direction. The B- and E-fields then oscillate in the y–z plane, so that electromagnetic waves are somewhat like shear waves, with the oscillations orthogonal to the direction of propagation. However, there are now two fields, and there is an additional relationship that the B-field oscillations for a given wave are orthogonal to the E-field oscillations. Figure 2.4 illustrates a wave in which the E-field oscillates in y while the B-field oscillates in z. In this case only the y-component of $\vec{E}$ and the z-component of $\vec{B}$ are nonzero, and both vary only with x and t. Then $\vec{E}$ and $\vec{B}$ have the form

$$\vec{E} = \begin{bmatrix} 0 \\ E^2(x,t) \\ 0 \end{bmatrix}, \qquad \vec{B} = \begin{bmatrix} 0 \\ 0 \\ B^3(x,t) \end{bmatrix}. \tag{2.116}$$

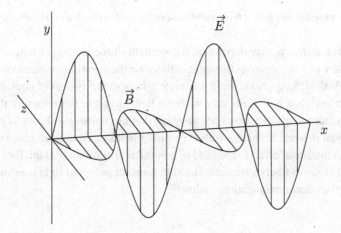

Fig. 2.4. The $\vec{E}$ and $\vec{B}$ fields for an electromagnetic plane wave propagating in the x-direction.

Maxwell's equations (2.115) then reduce to

$$E_t^2 + \frac{1}{\epsilon\mu} B_x^3 = 0,$$

$$B_t^3 + E_x^2 = 0.$$

(2.117)

This has exactly the same structure as the one-dimensional linear acoustics and elasticity equations considered previously, with the coefficient matrix

$$A = \begin{bmatrix} 0 & 1/\epsilon\mu \\ 1 & 0 \end{bmatrix}.$$

(2.118)

The eigenvalues are $\lambda^{1,2} = \pm c$, where

$$c = \frac{1}{\sqrt{\epsilon\mu}}$$

(2.119)

is the *speed of light* in the medium. In a vacuum the parameters ϵ and μ take particular constant values ϵ_0 and μ_0 known as the *permittivity and permeability of free space*, and

$$c_0 = \frac{1}{\sqrt{\epsilon_0\mu_0}}$$

(2.120)

is the *speed of light in a vacuum*. For any other medium we have $c < c_0$.

In a heterogeneous medium that consists of one-dimensional layers of isotropic material, ϵ and μ would be scalar but vary with x and we would obtain the variable-coefficient hyperbolic system

$$\epsilon(x)E_t^2(x,t) + \left(\frac{1}{\mu(x)} B^3(x,t)\right)_x = 0,$$

$$B_t^3(x,t) + E_x^2(x,t) = 0.$$

(2.121)

The methods discussed in Section 9.6 for variable-coefficient acoustics could also be applied to this system.

In some media ϵ and/or μ may depend on the strength of the electric or magnetic field, and hence will vary as a wave passes through. In this case the constitutive relations become nonlinear, and Maxwell's equations yield a nonlinear hyperbolic system of equations. Actually, in most materials ϵ and μ do vary with the field strength, but normally the fields associated with electromagnetic waves are so weak that the linearized theory is perfectly adequate. However, in some problems with very strong fields or special materials it is necessary to consider nonlinear effects. The field of *nonlinear optics* is important, for example, in the design and study of fiber-optic cables used to transmit pulses of light over thousands of kilometers in the telecommunications industry.

Exercises

2.1. Derive the equations (2.48) of linear acoustics from the linearized system (2.47).

2.2. (a) Show that for smooth solutions the conservation laws (2.38) can be manipulated into the following set of nonconservative nonlinear equations for the pressure and velocity:

$$
\begin{aligned}
p_t + u p_x + \rho P'(\rho) u_x &= 0, \\
u_t + (1/\rho) p_x + u u_x &= 0,
\end{aligned}
\tag{2.122}
$$

where we assume that the equation of state can be inverted to define ρ as a function of p to complete this system. Note that linearizing this nonlinear system about some state $(\rho_0, u_0, p_0 = P(\rho_0))$ again gives the acoustics system (2.47).

   (b) Show that the nonlinear system (2.122) is hyperbolic provided $P'(\rho) > 0$, and has the same characteristic speeds as the conservative version (2.38).

2.3. Determine the eigenvalues and eigenvectors the matrix $\tilde{A}$ in (2.77) and also the similarity transformation relating this to A from (2.51) when $u_0 = 0$.

2.4. Determine the eigenvalues and eigenvectors the matrix A from (2.46), and show that these agree with (2.57). Determine the similarity transformation relating this matrix to A from (2.51).

2.5. Determine the condition on the function $\sigma(\epsilon)$ that is required in order for the nonlinear elasticity equation (2.91) to be hyperbolic.

2.6. Show that $X_\xi(\xi, t) = V(\xi, t)$ and hence (2.105) is simply the statement that $X_{\xi t} = X_{t\xi}$.

2.7. Show that the p-system (2.108) is hyperbolic provided the function $p(V)$ satisfies $p'(V) < 0$ for all V.

2.8. Isothermal flow is modeled by the system (2.38) with $P(\rho) = a^2 \rho$, where a is constant; see Section 14.6.

   (a) Determine the wave speeds of the linearized equations (2.50) in this case.

   (b) The Lagrangian form of the isothermal equations have $p(V) = a^2/V$. Linearize the p-system (2.107) in this case about V_0, U_0, and compute the wave speeds for Lagrangian acoustics. Verify that these are what you expect in relation to the Eulerian acoustic wave speeds.

3

Characteristics and Riemann Problems for Linear Hyperbolic Equations

In this chapter we will further explore the characteristic structure of linear hyperbolic systems of equations. In particular, we will study solutions to the *Riemann problem*, which is simply the given equation together with very special initial data consisting of a piecewise constant function with a single jump discontinuity. This problem and its solution are discussed starting in Section 3.8, after laying some more groundwork. This simple problem plays a very important role in understanding the structure of more general solutions. It is also a fundamental building block for the finite volume methods discussed in this book.

Linear hyperbolic systems of the form

$$q_t + Aq_x = 0 \tag{3.1}$$

were introduced in the last chapter. Recall that the problem is hyperbolic if $A \in \mathbb{R}^{m \times m}$ is diagonalizable with real eigenvalues, so that we can write

$$A = R \Lambda R^{-1}, \tag{3.2}$$

where R is the matrix of right eigenvectors. Then introducing the new variables

$$w = R^{-1} q$$

allows us to reduce the system (3.1) to

$$w_t + \Lambda w_x = 0, \tag{3.3}$$

which is a set of m decoupled advection equations. Note that this assumes A is constant. If A varies with x and/or t, then the problem is still linear, but R and Λ will typically depend on x and t as well and the manipulations used to obtain (3.3) are no longer valid. See Chapter 9 for discussion of variable-coefficient problems.

3.1 Solution to the Cauchy Problem

Consider the Cauchy problem for the constant-coefficient system (3.1), in which we are given data

$$q(x, 0) = \overset{\circ}{q}(x) \quad \text{for } -\infty < x < \infty.$$

From this data we can compute data

$$\mathring{w}(x) \equiv R^{-1}\mathring{q}(x)$$

for the system (3.3). The pth equation of (3.3) is the advection equation

$$w_t^p + \lambda^p w_x^p = 0 \tag{3.4}$$

with solution

$$w^p(x, t) = w^p(x - \lambda^p t, 0) = \mathring{w}^p(x - \lambda^p t).$$

Having computed all components $w^p(x, t)$ we can combine these into the vector $w(x, t)$, and then

$$q(x, t) = Rw(x, t) \tag{3.5}$$

gives the solution to the original problem. This is exactly the process we used to obtain the solution (2.68) to the acoustics equations in the previous chapter.

3.2 Superposition of Waves and Characteristic Variables

Note that we can write (3.5) as

$$q(x, t) = \sum_{p=1}^m w^p(x, t) r^p, \tag{3.6}$$

so that we can view the vector $q(x, t)$ as being some linear combination of the right eigen-vectors $r^1, \ldots, r^m$ at each point in space–time, and hence as a superposition of waves propagating at different velocities λ^p. The scalar values $w^p(x, t)$ for $p = 1, \ldots, m$ give the coefficients of these eigenvectors at each point, and hence the *strength* of each wave. The requirements of hyperbolicity insure that these m vectors are linearly independent and hence every vector q has a unique representation in this form. The manipulations resulting in (3.4) show that the eigencoefficient $\mathring{w}^p(x) = w^p(x, 0)$ is simply advected at constant speed λ^p as time evolves, i.e., $w^p(x, t) \equiv \mathring{w}^p(x_0)$ all along the curve $X(t) = x_0 + \lambda^p t$. These curves are called *characteristics of the pth family*, or simply *p-characteristics*. These are straight lines in the case of a constant-coefficient system. Note that for a strictly hyperbolic system, m distinct characteristic curves pass through each point in the x–t plane.

The coefficient $w^p(x, t)$ of the eigenvector r^p in the eigenvector expansion (3.6) of $q(x, t)$ is constant along any p-characteristic. The functions $w^p(x, t)$ are called the *characteristic variables*.

As an example, for the acoustics equations with A given by (2.51), we found in Section 2.8 that the characteristic variables are $-p + Z_0 u$ and $p + Z_0 u$ (or any scalar multiples of these functions), where Z_0 is the impedance; see (2.67).

3.3 Left Eigenvectors

Let $L = R^{-1}$, and denote the rows of the matrix L by $\ell^1, \ell^2, \ldots, \ell^m$. These row vectors are the left eigenvectors of the matrix A,

$$\ell^p A = \lambda^p \ell^p,$$

whereas the r^p are the right eigenvectors. For example, the left eigenvectors for acoustics are given by the rows of the matrix R^{-1} in (2.66).

We can write the characteristic variable $w^p(x, t)$, which is the pth component of $R^{-1}q(x, t)$ $= Lq(x, t)$, simply as

$$w^p(x, t) = \ell^p q(x, t). \tag{3.7}$$

We can then rewrite the solution $q(x, t)$ from (3.6) succinctly in terms of the initial data $\overset{\circ}{q}$ as

$$q(x, t) = \sum_{p=1}^{m} [\ell^p \overset{\circ}{q}(x - \lambda^p t)]r^p. \tag{3.8}$$

3.4 Simple Waves

We can view the solution $q(x, t)$ as being the superposition of m waves, each of which is advected independently with no change in shape. The pth wave has shape $\overset{\circ}{w}^p(x)r^p$ and propagates with speed λ^p. This solution has a particularly simple form if $w^p(x, 0)$ is constant in x for all but one value of p, say $\overset{\circ}{w}^p(x) \equiv \bar{w}^p$ for $p \neq i$. Then the solution has the form

$$q(x, t) = \overset{\circ}{w}^i(x - \lambda^i t)r^i + \sum_{p \neq i} \bar{w}^p r^p \tag{3.9}$$

$$= \overset{\circ}{q}(x - \lambda^i t)$$

and the initial data simply propagates with speed λ^i. Since $m - 1$ of the characteristic variables are constant, the equation essentially reduces to $q_t + \lambda^i q_x = 0$, which governs the behavior of the ith family. Nonlinear equations have analogous solutions, called *simple waves*, in which variations occur only in one characteristic family; see Section 13.8.

3.5 Acoustics

An arbitrary solution to the acoustics equations, as derived in Section 2.8, can be decomposed as in (3.6),

$$\begin{bmatrix} p(x, t) \\ u(x, t) \end{bmatrix} = w^1(x, t) \begin{bmatrix} -Z_0 \\ 1 \end{bmatrix} + w^2(x, t) \begin{bmatrix} Z_0 \\ 1 \end{bmatrix}, \tag{3.10}$$

where $w^1 = [-p + Z_0 u]/2Z_0$ is the strength of the left-going 1-wave, and $w^2 = [p + Z_0 u]/2Z_0$ is the strength of the right-going 2-wave. The functions $w^1(x, t)$ and $w^2(x, t)$

satisfy scalar advection equations,

$$w_t^1 - c_0 w_x^1 = 0 \quad \text{and} \quad w_t^2 + c_0 w_x^2 = 0, \tag{3.11}$$

so from arbitrary initial data we can compute

$$w^1(x, t) = w^1(x + c_0 t, 0) = \mathring{w}^1(x + c_0 t),$$
$$w^2(x, t) = w^2(x - c_0 t, 0) = \mathring{w}^2(x - c_0 t), \tag{3.12}$$

where $\mathring{w}(x) = R^{-1}\mathring{q}(x)$ is the initial data for w, and (3.6) agrees with (2.68).

The advection equations (3.11) are often called the *one-way wave equations*, since each one models the strength of an acoustic wave going in only one direction.

If one of the characteristic variables w^1 or w^2 is identically constant, then the solution (3.10) is a *simple wave* as defined in Section 3.4. Suppose, for example, that $w^1 \equiv \bar{w}^1 =$ constant, in which case

$$q(x, t) = \bar{w}^1 r^1 + \mathring{w}^2(x - c_0 t) r^2.$$

In this case it is also easy to check that the full solution q satisfies the one-way wave equation $q_t + c_0 q_x = 0$.

Simple waves often arise in physical problems. Suppose for example that we take initial data in which $p = u = 0$ everywhere except in some small region near the origin. If we choose p and u as arbitrary functions in this region, unrelated to one another, then the solution will typically involve a superposition of a left-going and a right-going wave. Figure 3.1 shows the time evolution in a case where

$$p(x, 0) = \frac{1}{2} \exp(-80x^2) + S(x),$$
$$u(x, 0) = 0, \tag{3.13}$$

with

$$S(x) = \begin{cases} 0.5 & \text{if } -0.3 < x < -0.1, \\ 0 & \text{otherwise.} \end{cases}$$

For small time the solution changes in a seemingly haphazard way as the left-going and right-going waves superpose. But observe that eventually the two waves separate and for larger t their individual forms are easy to distinguish. Once they have separated, each wave is a simple wave that propagates at constant velocity with its shape unchanged. In this example $\rho_0 = 1$ and $K_0 = 0.25$, so that $c_0 = Z_0 = 1/2$. Notice that the left-going wave has $p = -u/2$ while the right-going wave has $p = u/2$, as expected from the form of the eigenvectors.

3.6 Domain of Dependence and Range of Influence

Let (X, T) be some fixed point in space–time. We see from (3.8) that the solution $q(X, T)$ depends only on the data $\mathring{q}$ at m particular points $X - \lambda^p T$ for $p = 1, 2, \ldots, m$. This set

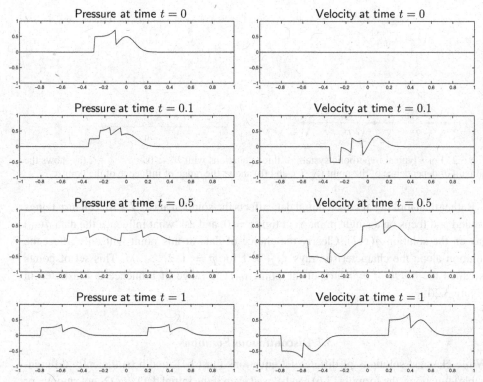

Fig. 3.1. Evolution of an initial pressure perturbation, concentrated near the origin, into distinct simple waves propagating with velocities $-c_0$ and c_0. The left column shows the pressure perturbation $q^1 = p$, and the right column shows the velocity $q^2 = u$. (Time increases going downwards.) [claw/book/chap3/acousimple]

of points,

$$\mathcal{D}(X, T) = \{X - \lambda^p T: p = 1, 2, \ldots, m\}, \tag{3.14}$$

is called the *domain of dependence* of the point (X, T). See Figure 3.2(a). The value of the initial data at other points has no influence on the value of q at (X, T).

For hyperbolic equations more generally, the domain of dependence is always a *bounded set*, though for nonlinear equations the solution may depend on data over a whole interval rather than at only a finite number of distinct points. The bounded domain of dependence results from the fact that information propagates at finite speed in a hyperbolic equation, as we expect from wave motion or advection. This has important consequences in the design of numerical methods, and means that explicit methods can often be efficiently used.

By contrast, for the heat equation $q_t = \beta q_{xx}$, the domain of dependence of any point (X, T) is the entire real line. Changing the data anywhere would in principle change the value of the solution at (X, T), though the contribution dies away exponentially fast, so data at points far away may have little effect. Nonetheless, this means that *implicit* numerical methods are often needed in solving parabolic equations. This is discussed further in Section 4.4 in relation to the *CFL condition*.

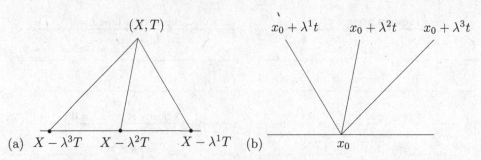

Fig. 3.2. For a typical hyperbolic system of three equations with $\lambda^1 < 0 < \lambda^2 < \lambda^3$, (a) shows the domain of dependence of the point (X, T), and (b) shows the range of influence of the point x_0.

Rather than looking at which initial data affects the solution at (X, T), we can turn things around and focus on a single point x_0 at time $t = 0$, and ask what influence the data $\mathring{q}(x_0)$ has on the solution $q(x, t)$. Clearly the choice of data at this point will only affect the solution along the characteristic rays $x_0 + \lambda^p t$ for $p = 1, 2, \ldots, m$. This set of points is called the *range of influence* of the point x_0. The range of influence is illustrated in Figure 3.2(b).

3.7 Discontinuous Solutions

While classical solutions of differential equations must be smooth (sufficiently differentiable) functions, the formula (3.6) can be used even if the initial data $\mathring{q}(x)$ is not smooth, or is even discontinuous, at some points. If the data has a singularity (a discontinuity in some derivative) at some point x_0, then one or more of the characteristic variables $w^p(x, 0)$ will also have a singularity at this point. Such singularities in the initial data can then propagate along the characteristics and lead to singularities in the solution $q(x, t)$ at some or all of the points $x_0 + \lambda^p t$.

Conversely, if the initial data is smooth in a neighborhood of all the points $\bar{x} - \lambda^p \bar{t}$, then the solution $q(x, t)$ must be smooth in a neighborhood of the point $(\bar{x}, \bar{t})$. This means that singularities can *only* propagate along characteristics for a linear system.

3.8 The Riemann Problem for a Linear System

The *Riemann problem* consists of the hyperbolic equation together with special initial data that is piecewise constant with a single jump discontinuity,

$$\mathring{q}(x) = \begin{cases} q_l & \text{if } x < 0, \\ q_r & \text{if } x > 0. \end{cases}$$

By the remarks in Section 3.7, we expect this discontinuity to propagate along the characteristic curves.

For the scalar advection equation $q_t + \bar{u} q_x = 0$, the coefficient "matrix" is the 1×1 scalar value $\bar{u}$. The single eigenvalue is $\lambda^1 = \bar{u}$, and we can choose the eigenvector to be $r^1 = 1$. The solution to the Riemann problem consists of the discontinuity $q_r - q_l$ propagating at speed $\bar{u}$, along the characteristic, and the solution is $q(x, t) = \mathring{q}(x - \bar{u}t)$.

For a general $m \times m$ linear system we can solve the Riemann problem explicitly using the information we have obtained above. It is very important to understand the structure of this solution, since we will see later that Riemann solutions for nonlinear conservation laws have a similar structure. Moreover, many of the numerical methods we will discuss (beginning in Chapter 4) are based on using solutions to the Riemann problem to construct approximate solutions with more general data.

For the Riemann problem we can simplify the notation if we decompose q_l and q_r as

$$q_l = \sum_{p=1}^{m} w_l^p r^p \quad \text{and} \quad q_r = \sum_{p=1}^{m} w_r^p r^p. \tag{3.15}$$

Then the pth advection equation (3.4) has Riemann data

$$\overset{\circ}{w}^p(x) = \begin{cases} w_l^p & \text{if } x < 0, \\ w_r^p & \text{if } x > 0, \end{cases} \tag{3.16}$$

and this discontinuity simply propagates with speed λ^p, so

$$w^p(x, t) = \begin{cases} w_l^p & \text{if } x - \lambda^p t < 0, \\ w_r^p & \text{if } x - \lambda^p t > 0. \end{cases} \tag{3.17}$$

If we let $P(x, t)$ be the maximum value of p for which $x - \lambda^p t > 0$, then

$$q(x, t) = \sum_{p=1}^{P(x,t)} w_r^p r^p + \sum_{p=P(x,t)+1}^{m} w_l^p r^p, \tag{3.18}$$

which we will write more concisely as

$$q(x, t) = \sum_{p:\lambda^p < x/t} w_r^p r^p + \sum_{p:\lambda^p > x/t} w_l^p r^p. \tag{3.19}$$

The determination of $q(x, t)$ at a given point (X, T) is illustrated in Figure 3.3. In the case shown, $w^1 = w_r^1$ while $w^2 = w_l^2$ and $w^3 = w_l^3$. The solution at the point illustrated is thus

$$q(X, T) = w_r^1 r^1 + w_l^2 r^2 + w_l^3 r^3. \tag{3.20}$$

Note that the solution is the same at any point in the wedge between the $x = \lambda^1 t$ and $x = \lambda^2 t$ characteristics. As we cross the pth characteristic, the value of $x - \lambda^p t$ passes through 0 and the corresponding w^p jumps from w_l^p to w_r^p. The other coefficients w^i ($i \neq p$) remain constant.

The solution is constant in each of the wedges as shown in Figure 3.3. Across the pth characteristic the solution jumps with the jump in q given by

$$\left(w_r^p - w_l^p\right) r^p \equiv \alpha^p r^p. \tag{3.21}$$

Note that this jump in q is an eigenvector of the matrix A (being a scalar multiple of r^p). This is an extremely important fact, and a generalization of this statement is what will allow us to solve the Riemann problem for nonlinear systems of equations. This condition,

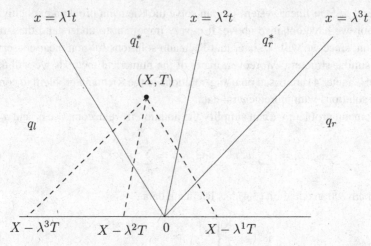

Fig. 3.3. Construction of the solution to the Riemann problem at (X, T). We trace back along the pth characteristic to determine the value of w^p from the initial data. The value of q is constant in each wedge of the x–t plane: $q_l = w_l^1 r^1 + w_l^2 r^2 + w_l^3 r^3$ $q_l^* = w_r^1 r^1 + w_l^2 r^2 + w_l^3 r^3$ $q_r^* = w_r^1 r^1 + w_r^2 r^2 + w_l^3 r^3$ $q_r = w_r^1 r^1 + w_r^2 r^2 + w_r^3 r^3$. Note that the jump across each discontinuity in the solution is an eigenvector of A.

called the *Rankine–Hugoniot jump condition*, will be derived from the integral form of the conservation law and seen to hold across any propagating discontinuity; see Section 11.8. Typically the given data (q_l, q_r) will not satisfy this condition, and the process of solving the Riemann problem can be viewed as an attempt to split up the jump $q_r - q_l$ into a series of jumps, defining the different waves, each of which does satisfy this condition.

For the case of a linear system, solving the Riemann problem consists of taking the initial data (q_l, q_r) and decomposing the jump $q_r - q_l$ into eigenvectors of A:

$$q_r - q_l = \alpha^1 r^1 + \cdots + \alpha^m r^m. \tag{3.22}$$

This requires solving the linear system of equations

$$R\alpha = q_r - q_l \tag{3.23}$$

for the vector α, and so $\alpha = R^{-1}(q_r - q_l)$. The vector α has components $\alpha^p = \ell^p (q_r - q_l)$, where ℓ^p is the left eigenvector defined in Section 3.3, and $\alpha^p = w_r^p - w_l^p$. Since $\alpha^p r^p$ is the jump in q across the pth wave in the solution to the Riemann problem, we introduce the notation

$$\mathcal{W}^p = \alpha^p r^p \tag{3.24}$$

for these waves.

The solution $q(x, t)$ from (3.8) can be written in terms of the waves in two different forms:

$$q(x, t) = q_l + \sum_{p:\lambda^p < x/t} \mathcal{W}^p \tag{3.25}$$

$$= q_r - \sum_{p:\lambda^p \geq x/t} \mathcal{W}^p. \tag{3.26}$$

This can also be written as

$$q(x, t) = q_l + \sum_{p=1}^{m} H(x - \lambda^p t) \mathcal{W}^p, \tag{3.27}$$

where $H(x)$ is the *Heaviside function*

$$H(x) = \begin{cases} 0 & \text{if } x < 0, \\ 1 & \text{if } x > 0. \end{cases} \tag{3.28}$$

3.9 The Phase Plane for Systems of Two Equations

It is illuminating to view the splitting of $q_r - q_l$ in *state space*, often called the *phase plane* for systems of two equations. This is simply the q^1–q^2 plane, where $q = (q^1, q^2)$. Each vector $q(x, t)$ is represented by a point in this plane. In particular, q_l and q_r are points in this plane, and a discontinuity with left and right states q_l and q_r can propagate as a single discontinuity only if $q_r - q_l$ is an eigenvector of A, which means that the line segment from q_l to q_r must be parallel to the eigenvector r^1 or r^2. Figure 3.4 shows an example. For the state q_l illustrated there, the jump from q_l to q_r can propagate as a single discontinuity if and only if q_r lies on one of the two lines drawn through q_l in the directions r^1 and r^2. These lines give the locus of all points that can be connected to q_l by a 1-wave or a 2-wave. This set of states is called the *Hugoniot locus*. We will see that there is a direct generalization of this to nonlinear systems in Chapter 13.

Similarly, there is a Hugoniot locus through any point q_r that gives the set of all points q_l that can be connected to q_r by an elementary p-wave. These curves are again in the directions r^1 and r^2.

For a general Riemann problem with arbitrary q_l and q_r, the solution consists of two discontinuities traveling with speeds λ^1 and λ^2, with a new constant state in between that we will call q_m. By the discussion above,

$$q_m = w_r^1 r^1 + w_l^2 r^2, \tag{3.29}$$

so that $q_m - q_l = (w_r^1 - w_l^1)r^1$ and $q_r - q_m = (w_r^2 - w_l^2)r^2$. The location of q_m in the phase

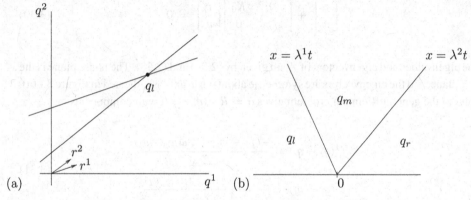

Fig. 3.4. (a) The Hugoniot locus of the state q_l consists of all states that differ from q_l by a scalar multiple of r^1 or r^2. (b) Solution to the Riemann problem in the x–t plane.

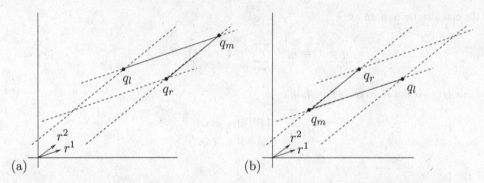

Fig. 3.5. The new state q_m arising in the solution to the Riemann problem for two different choices of q_l and q_r. In each case the jump from q_l to q_m lies in the direction of the eigenvector r^1 corresponding to the lower speed, while the jump from q_m to q_r lies in the direction of the eigenvector r^2.

plane must be where the 1-wave locus through q_l intersects the 2-wave locus through q_r. This is illustrated in Figure 3.5(a).

Note that if we interchange q_r and q_l in this picture, the location of q_m changes as illustrated in Figure 3.5(b). In each case we travel from q_l to q_r by first going in the direction r^1 and then in the direction r^2. This is required by the fact that $\lambda^1 < \lambda^2$, since clearly the jump between q_l and q_m must travel slower than the jump between q_m and q_r (see Figure 3.4(b)) if we are to obtain a single-valued solution.

For systems with more than two equations, the same interpretation is possible but becomes harder to draw, since the state space is now m-dimensional. Since the m eigenvectors r^p are linearly independent, we can decompose any jump $q_r - q_l$ into the sum of jumps in these directions via (3.22), obtaining a piecewise linear path from q_l to q_r in m-dimensional space.

3.9.1 Acoustics

As a specific example, consider the acoustics equations discussed in Sections 2.7–2.8 with $u_0 = 0$,

$$\begin{bmatrix} p \\ u \end{bmatrix}_t + \begin{bmatrix} 0 & K_0 \\ 1/\rho_0 & 0 \end{bmatrix} \begin{bmatrix} p \\ u \end{bmatrix}_x = 0. \tag{3.30}$$

The eigenvalues and eigenvectors of A are given by (2.54) and (2.58). The phase plane is the p–u plane, and the eigenvectors are symmetric about the u-axis as indicated in Figure 3.6(a). Solving the general Riemann problem gives $\alpha = R^{-1}(q_r - q_l)$ with components

$$\alpha^1 = \ell^1(q_r - q_l) = \frac{-(p_r - p_l) + Z_0(u_r - u_l)}{2Z_0},$$

$$\alpha^2 = \ell^2(q_r - q_l) = \frac{(p_r - p_l) + Z_0(u_r - u_l)}{2Z_0}, \tag{3.31}$$

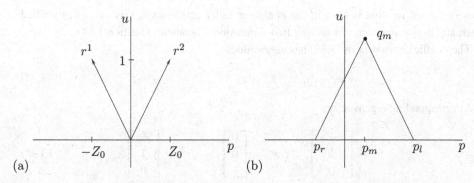

Fig. 3.6. (a) Eigenvectors for the acoustics equations in the p–u phase plane, where Z_0 is the impedance. (b) Solution to a Riemann problem in which $u_l = u_r = 0$ and $p_r < p_l$.

and the waves are $\mathcal{W}^1 = \alpha^1 r^1$ and $\mathcal{W}^2 = \alpha^2 r^2$. The intermediate state is

$$q_m = q_l + \alpha^1 r^1 = \frac{1}{2} \begin{bmatrix} (p_l + p_r) - Z_0(u_r - u_l) \\ (u_l + u_r) - (p_r - p_l)/Z_0 \end{bmatrix}. \tag{3.32}$$

Example 3.1. Consider a Riemann problem in which $u_l = u_r = 0$ and there is only a jump in pressure with $p_r < p_l$. The phase-plane solution to the Riemann problem is sketched in Figure 3.6(b), and we compute that

$$\alpha^1 = \frac{p_l - p_r}{2Z_0}, \qquad \alpha^2 = \frac{p_r - p_l}{2Z_0},$$

so that the intermediate state is

$$q_m = q_l + \alpha^1 r^1 = q_r - \alpha^2 r^2 = \frac{1}{2} \begin{bmatrix} p_l + p_r \\ -(p_r - p_l)/Z_0 \end{bmatrix}.$$

(Recall that p represents the perturbation of pressure from the constant state p_0, so it is fine for it to be negative.)

3.10 Coupled Acoustics and Advection

Now consider acoustics in a fluid moving at constant speed $u_0 > 0$, and to make the problem more interesting suppose that there is also a passive tracer being advected in this fluid, with density denoted by $\phi(x, t)$. Then we can solve the acoustics and advection equation together as a system of three equations,

$$\begin{bmatrix} p \\ u \\ \phi \end{bmatrix}_t + \begin{bmatrix} u_0 & K_0 & 0 \\ 1/\rho_0 & u_0 & 0 \\ 0 & 0 & u_0 \end{bmatrix} \begin{bmatrix} p \\ u \\ \phi \end{bmatrix}_x. \tag{3.33}$$

Of course the acoustics and advection could be decoupled into two separate problems, but it is illuminating to solve the Riemann problem for this full system, since its structure is closely related to what is seen in the nonlinear Euler equations of gas dynamics studied later, and is also important in solving two-dimensional acoustics (Section 18.4).

The coefficient matrix in (3.33) has eigenvalues

$$\lambda^1 = u_0 - c_0, \qquad \lambda^2 = u_0, \qquad \lambda^3 = u_0 + c_0, \tag{3.34}$$

and corresponding eigenvectors

$$r^1 = \begin{bmatrix} -Z_0 \\ 1 \\ 0 \end{bmatrix}, \qquad r^2 = \begin{bmatrix} 0 \\ 0 \\ 1 \end{bmatrix}, \qquad r^1 = \begin{bmatrix} Z_0 \\ 1 \\ 0 \end{bmatrix}. \tag{3.35}$$

The solution to the Riemann problem is easily determined:

$$\begin{bmatrix} p_r - p_l \\ u_r - u_l \\ \phi_r - \phi_l \end{bmatrix} = \alpha^1 \begin{bmatrix} -Z_0 \\ 1 \\ 0 \end{bmatrix} + \alpha^2 \begin{bmatrix} 0 \\ 0 \\ 1 \end{bmatrix} + \alpha^3 \begin{bmatrix} Z_0 \\ 1 \\ 0 \end{bmatrix},$$

where

$$\alpha^1 = \frac{1}{2Z_0}[-(p_r - p_l) + Z_0(u_r - u_l)],$$

$$\alpha^2 = \phi_r - \phi_l, \tag{3.36}$$

$$\alpha^3 = \frac{1}{2Z_0}[(p_r - p_l) + Z_0(u_r - u_l)].$$

Note that the 1-wave and 3-wave are standard acoustic waves independent of ϕ, while the 2-wave gives the advection of ϕ.

Suppose ϕ measures the concentration of a dye in the fluid and that $\phi_r > \phi_l$, so that at time $t = 0$ the fluid to the left is dark while the fluid to the right is light. Then the 2-wave marks the interface between the dark and light fluids as time evolves, as indicated in Figure 3.7. The two fluids remain in contact across this discontinuity in ϕ, which has no dynamic effect, since this tracer does not affect the fluid dynamics and the pressure and velocity are both constant across the 2-wave. This wave is called a *contact discontinuity*.

Within each of the two fluids there is an acoustic wave moving at speed c_0 (relative to the fluid) away from the origin. The jump in pressure and/or velocity in the original Riemann data creates a "noise," that moves through the fluids at the speed of sound.

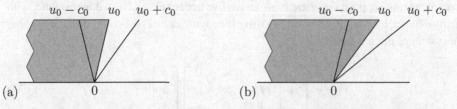

Fig. 3.7. Solution to the Riemann problem for the coupled acoustics and advection problem. The interface between dark and light fluid advects at the fluid velocity u_0, and acoustic waves move at speed c_0 relative to the fluid. The speed of each wave is indicated. (a) A subsonic case. (b) A supersonic case.

Figure 3.7 shows two different situations. In Figure 3.7(a) the fluid velocity u_0 is positive but *subsonic* ($u_0 < c_0$), and so the left-going acoustic wave (the 1-wave) has a negative velocity $u_0 - c_0 < 0$ relative to a fixed observer. Figure 3.7(b) illustrates a *supersonic* flow, where $u_0 > c_0$ and so $u_0 - c_0 > 0$. In this case all three waves propagate to the right and no information can propagate upstream from the observer. This distinction is not very important in this linear example. In nonlinear gas dynamics the distinction can be very important. The ratio $M = |u_0|/c_0$ is called the *Mach number* of the flow.

3.11 Initial–Boundary-Value Problems

Now consider a hyperbolic system on a bounded interval $a \leq x \leq b$. This is called the *initial–boundary-value problem*, or IBVP for short, since it is a time-dependent problem for which we need both initial data and boundary data. For a system of m equations we need a total of m boundary conditions. Typically some conditions must be prescribed at the left boundary $x = a$ and some at the right boundary $x = b$. How many are required at each boundary depends on the number of eigenvalues of A that are positive and negative, respectively.

We considered the IBVP for the advection equation in Section 2.1 and saw that we need a boundary condition only at $x = a$ if $\bar{u} > 0$ and only at $x = b$ if $\bar{u} < 0$. So if we diagonalize a general linear system to obtain a decoupled set of advection equations

$$w_t^p + \lambda^p w_x^p = 0,$$

then we need to specify boundary data on $w^p(x, t)$ at $x = a$ if $\lambda^p > 0$ and at $x = b$ if $\lambda^p < 0$. (For now assume all eigenvalues are nonzero, i.e., that the boundary is *noncharacteristic*.)

So if the system of m equations has $n \leq m$ negative eigenvalues and $m - n$ positive eigenvalues, i.e.,

$$\lambda^1 \leq \lambda^2 \leq \cdots \leq \lambda^n < 0 < \lambda^{n+1} \leq \cdots \leq \lambda^m,$$

then we need to specify $m - n$ boundary conditions at $x = a$ and n boundary conditions at $x = b$. What sort of boundary data should we impose? Partition the vector w as

$$w = \begin{bmatrix} w^{\mathrm{I}} \\ w^{\mathrm{II}} \end{bmatrix}, \tag{3.37}$$

where $w^{\mathrm{I}} \in \mathbb{R}^n$ and $w^{\mathrm{II}} \in \mathbb{R}^{m-n}$. Then at the left boundary $x = a$, for example, we must specify the components of w^{II}, while w^{I} are outflow variables. It is valid to specify w^{II} in terms of w^{I}. For example, we might use a linear boundary condition of the form

$$w^{\mathrm{II}}(a, t) = B_1 w^{\mathrm{I}}(a, t) + g_1(t), \tag{3.38}$$

where $B_1 \in \mathbb{R}^{(m-n)\times n}$ and $g_1 \in \mathbb{R}^{m-n}$. If $B_1 = 0$, then we are simply specifying given values for the inflow variables. But at a physical boundary there is often some *reflection* of outgoing waves, and this requires a nonzero B_1.

Boundary conditions should be specified as part of the problem and are determined by the physical setup – generally not in terms of the characteristic variables, unfortunately. It is not always easy to see what the correct conditions are to impose on the mathematical equation. We may have several pieces of information about what is happening at the boundary. Which are the correct ones to specify at the boundary? If we specify too few or too many conditions, or inappropriate conditions (such as trying to specify the value of an outflow characteristic variable), then the mathematical problem is ill posed and will have no solution, or perhaps many solutions. It often helps greatly to know what the characteristic structure is, which reveals how many boundary conditions we need and allows us to check that we are imposing appropriate conditions for a well-posed problem. In Chapter 7 boundary conditions are discussed further, and we will see how to impose such boundary conditions numerically.

Example 3.2. Consider the acoustics problem (2.50) in a closed tube of gas, $a \leq x \leq b$. We expect an acoustic wave hitting either closed end to be reflected. Since the system has eigenvalues $-c_0$ and $+c_0$, we need to specify one condition at each end ($n = m - n = 1$ and $w^{\mathrm{I}} = w^1$, $w^{\mathrm{II}} = w^2$). We do not have any information on values of the pressure at the boundary *a priori*, but we do know that the velocity must be zero at each end at all times, since the gas cannot flow through the solid walls (and shouldn't flow away from the walls or a vacuum would appear). This suggests that we should set

$$u(a, t) = u(b, t) = 0 \qquad (3.39)$$

as our two boundary conditions, and this is correct. Note that we are specifying the same thing at each end, although the ingoing characteristic variable is different at the two ends. From Section 2.8 we know that the characteristic variables are

$$w^1 = -p + Z_0 u, \qquad w^2 = p + Z_0 u. \qquad (3.40)$$

We can combine w^1 and w^2 to see that specifying $u = 0$ amounts to requiring that $w^1 + w^2 = 0$ at each end. At $x = a$ we can write this as

$$w^2(a, t) = -w^1(a, t),$$

which has the form (3.38) with $B_1 = 1$ and $g_1 = 0$. The outgoing wave is completely reflected and feeds back into the incoming wave. Conversely, at $x = b$ we can interpret the boundary condition $u = 0$ as

$$w^1(b, t) = -w^2(b, t),$$

which sets the incoming variable at this boundary, again by complete reflection of the outgoing variable.

Example 3.3. Suppose we set $B_1 = 0$ and $g_1 = 0$ in (3.38), so that this becomes $w^{\mathrm{II}}(a, t) = 0$. Then there is nothing flowing into the domain at the left boundary, and any left-going waves will simply leave the domain with no reflection. These are called *outflow boundary conditions*.

Figure 3.8 shows a continuation of the example shown in Figure 3.1 to later times, with a solid wall at the left and outflow boundary conditions imposed at the right, which amount to setting $w^1(b, t) = 0$ and hence $p(b, t) = Z_0 u(b, t)$.

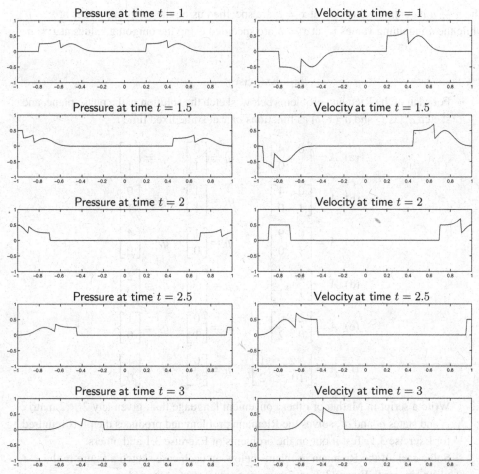

Fig. 3.8. Continuation of the example shown in Figure 3.1 with a solid wall at the left and outflow boundary conditions imposed at the right. Note that the wave that strikes the left boundary is a 1-wave with $p = -u/2$, while the reflected wave is a 2-wave with $p = u/2$. [claw/book/chap3/acousimple]

Example 3.4. A set of boundary conditions that is often useful mathematically is the *periodic boundary conditions*

$$q(a, t) = q(b, t). \tag{3.41}$$

This set of boundary conditions couples information at the two boundaries, and the idea is that waves going out one end should reenter at the other end. Solving the IBVP with periodic boundary conditions is equivalent to solving a Cauchy problem with periodic initial data, where the data given in $a \leq x \leq b$ is periodically extended to the whole real line.

We are specifying m coupled boundary conditions rather than $m - n$ at one end and n at the other, but we can reinterpret (3.41) in terms of the characteristic variables as

$$w^{\mathrm{II}}(a, t) = w^{\mathrm{II}}(b, t),$$
$$w^{\mathrm{I}}(b, t) = w^{\mathrm{I}}(a, t). \tag{3.42}$$

The $m - n$ incoming values w^{II} at $x = a$ are specified using the outgoing values at $x = b$, while the n incoming values w^{I} at $x = b$ are specified using the outgoing values at $x = a$.

Exercises

3.1. For each of the Riemann problems below, sketch the solution in the phase plane, and sketch $q^1(x, t)$ and $q^2(x, t)$ as functions of x at some fixed time t:

$$\text{(a) } A = \begin{bmatrix} 0 & 4 \\ 1 & 0 \end{bmatrix}, \quad q_l = \begin{bmatrix} 0 \\ 1 \end{bmatrix}, \quad q_r = \begin{bmatrix} 1 \\ 1 \end{bmatrix},$$

$$\text{(b) } A = \begin{bmatrix} 0 & 4 \\ 1 & 0 \end{bmatrix}, \quad q_l = \begin{bmatrix} 1 \\ 1 \end{bmatrix}, \quad q_r = \begin{bmatrix} 0 \\ 1 \end{bmatrix}.$$

$$\text{(c) } A = \begin{bmatrix} 0 & 9 \\ 1 & 0 \end{bmatrix}, \quad q_l = \begin{bmatrix} 1 \\ 0 \end{bmatrix}, \quad q_r = \begin{bmatrix} 4 \\ 0 \end{bmatrix}.$$

$$\text{(d) } A = \begin{bmatrix} 1 & 1 \\ 1 & 1 \end{bmatrix}, \quad q_l = \begin{bmatrix} 1 \\ 0 \end{bmatrix}, \quad q_r = \begin{bmatrix} 2 \\ 0 \end{bmatrix}.$$

$$\text{(e) } A = \begin{bmatrix} 2 & 0 \\ 0 & 2 \end{bmatrix}, \quad q_l = \begin{bmatrix} 0 \\ 1 \end{bmatrix}, \quad q_r = \begin{bmatrix} 1 \\ 0 \end{bmatrix}.$$

$$\text{(f) } A = \begin{bmatrix} 2 & 1 \\ 10^{-4} & 2 \end{bmatrix}, \quad q_l = \begin{bmatrix} 0 \\ 1 \end{bmatrix}, \quad q_r = \begin{bmatrix} 1 \\ 0 \end{bmatrix}.$$

3.2. Write a script in Matlab or other convenient language that, given any 2×2 matrix A and states q_l and q_r, solves the Riemann problem and produces the plots required for Exercise 3.1. Test it out on the problems of Exercise 3.1 and others.

3.3. Solve each of the Riemann problems below. In each case sketch a figure in the x–t plane similar to Figure 3.3, indicating the solution in each wedge.

$$\text{(a) } A = \begin{bmatrix} 0 & 0 & 4 \\ 0 & 1 & 0 \\ 1 & 0 & 0 \end{bmatrix}, \quad q_l = \begin{bmatrix} 1 \\ 2 \\ 0 \end{bmatrix}, \quad q_r = \begin{bmatrix} 1 \\ 5 \\ 1 \end{bmatrix}.$$

$$\text{(b) } A = \begin{bmatrix} 1 & 0 & 2 \\ 0 & 2 & 0 \\ 0 & 0 & 3 \end{bmatrix}, \quad q_l = \begin{bmatrix} 1 \\ 1 \\ 1 \end{bmatrix}, \quad q_r = \begin{bmatrix} 3 \\ 3 \\ 3 \end{bmatrix}.$$

3.4. Consider the acoustics equations (3.30) with

$$A = \begin{bmatrix} 0 & K_0 \\ 1/\rho_0 & 0 \end{bmatrix}, \quad \overset{\circ}{p}(x) = \begin{cases} 1 & \text{if } 1 \le x \le 2, \\ 0 & \text{otherwise}, \end{cases} \quad \overset{\circ}{u}(x) \equiv 0.$$

Find the solution for $t > 0$. This might model a popping balloon, for example (in one dimension).

3.5. Solve the IBVP for the acoustics equations from Exercise 3.4 on the finite domain $0 \le x \le 4$ with boundary conditions $u(0, t) = u(4, t) = 0$ (solid walls). Sketch the solution (u and p as functions of x) at times $t = 0, 0.5, 1, 1.5, 2, 3$.

3.6. In the problem of Exercise 3.5, what is the domain of dependence of the point $X = 1$, $T = 10$? In this case the domain of dependence should be defined to include not only the set of points x where the initial data affects the solution, but also the set of times on each boundary where the boundary conditions can affect the solution at the point (X, T).

3.7. Suppose a tube of gas is bounded by a piston at $x = 0$ and a solid wall at $x = 1$, and that the piston is very slowly pushed into the tube with constant speed $\epsilon \ll c$, where c is the speed of sound. Then we might expect the gas in the tube to be simply compressed slowly with the pressure essentially uniform through the tube and increasing in time like $p = p_0 + \epsilon t K_0$, where K_0 is the bulk modulus. The velocity should be roughly linear in x, varying from $u = \epsilon$ at the piston to $u = 0$ at the solid wall. For very small ϵ we can model this using linear acoustics on the fixed interval $0 \leq x \leq 1$ with initial data

$$\overset{\circ}{u}(x) = 0, \qquad \overset{\circ}{p}(x) = p_0,$$

and boundary conditions

$$u(0, t) = \epsilon, \qquad u(1, t) = 0.$$

The solution consists of a single acoustic wave bouncing back and forth between the piston and solid wall (very rapidly relative to the wall motion), with p and u piecewise constant. Determine this solution, and show that by appropriately averaging this rapidly varying solution one observes the expected behavior described above. This illustrates the fact that slow-scale motion is sometimes mediated by high-speed waves.

3.8. Consider a general hyperbolic system $q_t + A q_x = 0$ in which $\lambda = 0$ is a simple or multiple eigenvalue. How many boundary conditions do we need to impose at each boundary in this case? As a specific example consider the system (3.33) in the case $u_0 = 0$.

4

Finite Volume Methods

In this chapter we begin to study finite volume methods for the solution of conservation laws and hyperbolic systems. The fundamental concepts will be introduced, and then we will focus on first-order accurate methods for linear equations, in particular the *upwind method* for advection and for hyperbolic systems. This is the linear version of *Godunov's method*, which is the fundamental starting point for methods for nonlinear conservation laws, discussed beginning in Chapter 15. These methods are based on the solution to Riemann problems as discussed in the previous chapter for linear systems.

Finite volume methods are closely related to finite difference methods, and a finite volume method can often be interpreted directly as a finite difference approximation to the differential equation. However, finite volume methods are derived on the basis of the integral form of the conservation law, a starting point that turns out to have many advantages.

4.1 General Formulation for Conservation Laws

In one space dimension, a finite volume method is based on subdividing the spatial domain into intervals (the "finite volumes," also called *grid cells*) and keeping track of an approximation to the integral of q over each of these volumes. In each time step we update these values using approximations to the flux through the endpoints of the intervals.

Denote the ith grid cell by

$$\mathcal{C}_i = \left(x_{i-1/2}, x_{i+1/2}\right),$$

as shown in Figure 4.1. The value Q_i^n will approximate the average value over the ith interval at time t_n:

$$Q_i^n \approx \frac{1}{\Delta x} \int_{x_{i-1/2}}^{x_{i+1/2}} q(x, t_n)\,dx \equiv \frac{1}{\Delta x} \int_{\mathcal{C}_i} q(x, t_n)\,dx, \tag{4.1}$$

where $\Delta x = x_{i+1/2} - x_{i-1/2}$ is the length of the cell. For simplicity we will generally assume a uniform grid, but this is not required. (Nonuniform grids are discussed in Section 6.17.)

If $q(x, t)$ is a smooth function, then the integral in (4.1) agrees with the value of q at the midpoint of the interval to $\mathcal{O}(\Delta x^2)$. By working with cell averages, however, it is easier to use important properties of the conservation law in deriving numerical methods. In particular, we can insure that the numerical method is *conservative* in a way that mimics the

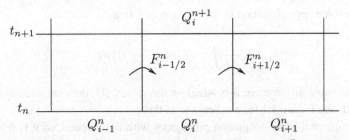

Fig. 4.1. Illustration of a finite volume method for updating the cell average Q_i^n by fluxes at the cell edges. Shown in x–t space.

true solution, and this is extremely important in accurately calculating shock waves, as we will see in Section 12.9. This is because $\sum_{i=1}^{N} Q_i^n \, \Delta x$ approximates the integral of q over the entire interval $[a, b]$, and if we use a method that is in *conservation form* (as described below), then this discrete sum will change only due to fluxes at the boundaries $x = a$ and $x = b$. The total mass within the computational domain will be preserved, or at least will vary correctly provided the boundary conditions are properly imposed.

The integral form of the conservation law (2.2) gives

$$\frac{d}{dt} \int_{C_i} q(x, t)\, dx = f\big(q(x_{i-1/2}, t)\big) - f\big(q(x_{i+1/2}, t)\big). \tag{4.2}$$

We can use this expression to develop an explicit time-marching algorithm. Given Q_i^n, the cell averages at time t_n, we want to approximate Q_i^{n+1}, the cell averages at the next time t_{n+1} after a time step of length $\Delta t = t_{n+1} - t_n$. Integrating (4.2) in time from t_n to t_{n+1} yields

$$\int_{C_i} q(x, t_{n+1})\, dx - \int_{C_i} q(x, t_n)\, dx = \int_{t_n}^{t_{n+1}} f\big(q(x_{i-1/2}, t)\big)\, dt - \int_{t_n}^{t_{n+1}} f\big(q(x_{i+1/2}, t)\big)\, dt.$$

Rearranging this and dividing by Δx gives

$$\frac{1}{\Delta x} \int_{C_i} q(x, t_{n+1})\, dx = \frac{1}{\Delta x} \int_{C_i} q(x, t_n)\, dx$$
$$- \frac{1}{\Delta x} \left[\int_{t_n}^{t_{n+1}} f\big(q(x_{i+1/2}, t)\big)\, dt - \int_{t_n}^{t_{n+1}} f\big(q(x_{i-1/2}, t)\big)\, dt \right]. \tag{4.3}$$

This tells us exactly how the cell average of q from (4.1) should be updated in one time step. In general, however, we cannot evaluate the time integrals on the right-hand side of (4.3) exactly, since $q(x_{i\pm1/2}, t)$ varies with time along each edge of the cell, and we don't have the exact solution to work with. But this does suggest that we should study numerical methods of the form

$$Q_i^{n+1} = Q_i^n - \frac{\Delta t}{\Delta x}\big(F_{i+1/2}^n - F_{i-1/2}^n\big), \tag{4.4}$$

where $F_{i-1/2}^n$ is some approximation to the average flux along $x = x_{i-1/2}$:

$$F_{i-1/2}^n \approx \frac{1}{\Delta t} \int_{t_n}^{t_{n+1}} f\big(q\big(x_{i-1/2}, t\big)\big)\, dt. \tag{4.5}$$

If we can approximate this average flux based on the values Q^n, then we will have a fully discrete method. See Figure 4.1 for a schematic of this process.

For a hyperbolic problem information propagates with finite speed, so it is reasonable to first suppose that we can obtain $F_{i-1/2}^n$ based only on the values Q_{i-1}^n and Q_i^n, the cell averages on either side of this interface (see Section 4.4 for some discussion of this). Then we might use a formula of the form

$$F_{i-1/2}^n = \mathcal{F}\big(Q_{i-1}^n, Q_i^n\big) \tag{4.6}$$

where $\mathcal{F}$ is some *numerical flux function*. The method (4.4) then becomes

$$Q_i^{n+1} = Q_i^n - \frac{\Delta t}{\Delta x}\big[\mathcal{F}\big(Q_i^n, Q_{i+1}^n\big) - \mathcal{F}\big(Q_{i-1}^n, Q_i^n\big)\big]. \tag{4.7}$$

The specific method obtained depends on how we choose the formula $\mathcal{F}$, but in general any method of this type is an explicit method with a *three-point stencil*, meaning that the value Q_i^{n+1} will depend on the three values Q_{i-1}^n, Q_i^n, and Q_{i+1}^n at the previous time level. Moreover, it is said to be in *conservation form*, since it mimics the property (4.3) of the exact solution. Note that if we sum $\Delta x\, Q_i^{n+1}$ from (4.4) over any set of cells, we obtain

$$\Delta x \sum_{i=I}^{J} Q_i^{n+1} = \Delta x \sum_{i=I}^{J} Q_i^n - \frac{\Delta t}{\Delta x}\big(F_{J+1/2}^n - F_{I-1/2}^n\big). \tag{4.8}$$

The sum of the flux differences cancels out except for the fluxes at the extreme edges. Over the full domain we have exact conservation except for fluxes at the boundaries. (Numerical boundary conditions are discussed later.)

The method (4.7) can be viewed as a direct finite difference approximation to the conservation law $q_t + f(q)_x = 0$, since rearranging it gives

$$\frac{Q_i^{n+1} - Q_i^n}{\Delta t} + \frac{F_{i+1/2}^n - F_{i-1/2}^n}{\Delta x} = 0. \tag{4.9}$$

Many methods can be equally well viewed as finite difference approximations to this equation or as finite volume methods.

4.2 A Numerical Flux for the Diffusion Equation

The above derivation was presented for a conservation law in which the flux $f(q)$ depends only on the state q. The same derivation works more generally, however, for example if the flux depends explicitly on x or if it depends on derivatives of the solution such as q_x. As an example consider the diffusion equation (2.22), where the flux (2.20) is

$$f(q_x, x) = -\beta(x)q_x.$$

Given two cell averages Q_{i-1} and Q_i, the numerical flux $\mathcal{F}(Q_{i-1}, Q_i)$ at the cell interface between can very naturally be defined as

$$\mathcal{F}(Q_{i-1}, Q_i) = -\beta_{i-1/2}\left(\frac{Q_i - Q_{i-1}}{\Delta x}\right), \tag{4.10}$$

where $\beta_{i-1/2} \approx \beta(x_{i-1/2})$. This numerical flux has the natural physical interpretation that the conserved quantity measured by q flows from one grid cell to its neighbor at a rate proportional to the difference in Q-values in the two cells, with $\beta_{i-1/2}$ measuring the conductivity of the interface between these cells. This is a macroscopic version of Fick's law or Fourier's law (or Newton's law of cooling).

Using (4.10) in (4.7) gives a standard finite difference discretization of the diffusion equation,

$$Q_i^{n+1} = Q_i^n + \frac{\Delta t}{\Delta x^2}[\beta_{i+1/2}(Q_{i+1}^n - Q_i^n) - \beta_{i-1/2}(Q_i^n - Q_{i-1}^n)]. \tag{4.11}$$

If $\beta \equiv$ constant, then this takes the simpler form

$$Q_i^{n+1} = Q_i^n + \frac{\Delta t}{\Delta x^2}\beta(Q_{i-1}^n - 2Q_i^n + Q_{i+1}^n) \tag{4.12}$$

and we recognize the centered approximation to q_{xx}.

For parabolic equations, explicit methods of this type are generally not used, since they are only stable if $\Delta t = \mathcal{O}(\Delta x^2)$. Instead an implicit method is preferable, such as the standard *Crank–Nicolson method*,

$$Q_i^{n+1} = Q_i^n + \frac{\Delta t}{2\,\Delta x^2}[\beta_{i+1/2}(Q_{i+1}^n - Q_i^n) - \beta_{i-1/2}(Q_i^n - Q_{i-1}^n)$$
$$+ \beta_{i+1/2}(Q_{i+1}^{n+1} - Q_i^{n+1}) - \beta_{i-1/2}(Q_i^{n+1} - Q_{i-1}^{n+1})]. \tag{4.13}$$

This can also be viewed as a finite volume method, with the flux

$$F_{i-1/2}^n = -\frac{1}{2\,\Delta x}[\beta_{i-1/2}(Q_i^n - Q_{i-1}^n) + \beta_{i-1/2}(Q_i^{n+1} - Q_{i-1}^{n+1})].$$

This is a natural approximation to the time-averaged flux (4.5), and in fact has the advantage of being a second-order accurate approximation (since it is centered in both space and time) as well as giving an unconditionally stable method.

The stability difficulty with explicit methods for the diffusion equation arises from the fact that the flux (4.10) contains Δx in the denominator, leading to stability restrictions involving $\Delta t/(\Delta x)^2$ after multiplying by $\Delta t/\Delta x$ in (4.4). For first-order hyperbolic equations the flux function involves only q and not q_x, and *explicit* methods are generally more efficient. However, some care must be taken to obtain stable methods in the hyperbolic case as well.

4.3 Necessary Components for Convergence

Later in this chapter we will introduce various ways to define the numerical flux function of (4.6) for hyperbolic equations, leading to various different finite volume methods. There are several considerations that go into judging how good a particular flux function is for numerical computation. One essential requirement is that the resulting method should be *convergent*, i.e., the numerical solution should converge to the true solution of the differential equation as the grid is refined (as $\Delta x, \Delta t \to 0$). This generally requires two conditions:

- The method must be *consistent* with the differential equation, meaning that it approximates it well locally.
- The method must be *stable* in some appropriate sense, meaning that the small errors made in each time step do not grow too fast in later time steps.

Stability and convergence theory are discussed in more detail in Chapter 8. At this stage we simply introduce some essential ideas that are useful in discussing the basic methods.

4.3.1 Consistency

The numerical flux should approximate the integral in (4.5). In particular, if the function $q(x, t) \equiv \bar{q}$ is constant in x, then q will not change with time and the integral in (4.5) simply reduces to $f(\bar{q})$. As a result, if $Q_{i-1}^n = Q_i^n = \bar{q}$, then we expect the numerical flux function $\mathcal{F}$ of (4.6) to reduce to $f(\bar{q})$, so we require

$$\mathcal{F}(\bar{q}, \bar{q}) = f(\bar{q}) \tag{4.14}$$

for any value $\bar{q}$. This is part of the basic consistency condition. We generally also expect continuity in this function as Q_{i-1} and Q_i vary, so that $\mathcal{F}(Q_{i-1}, Q_i) \to f(\bar{q})$ as $Q_{i-1}, Q_i \to \bar{q}$. Typically some requirement of *Lipschitz continuity* is made, e.g., there exists a constant L so that

$$|\mathcal{F}(Q_{i-1}, Q_i) - f(\bar{q})| \leq L \max(|Q_i - \bar{q}|, |Q_{i-1} - \bar{q}|). \tag{4.15}$$

4.4 The CFL Condition

Stability analysis is considered in detail in Chapter 8. Here we mention only the CFL condition, which is a *necessary* condition that must be satisfied by any finite volume or finite difference method if we expect it to be stable and converge to the solution of the differential equation as the grid is refined. It simply states that the method must be used in such a way that information has a chance to propagate at the correct physical speeds, as determined by the eigenvalues of the flux Jacobian $f'(q)$.

With the explicit method (4.7) the value Q_i^{n+1} depends only on three values Q_{i-1}^n, Q_i^n, and Q_{i+1}^n at the previous time step. Suppose we apply such a method to the advection equation $q_t + \bar{u} q_x = 0$ with $\bar{u} > 0$ so that the exact solution simply translates at speed $\bar{u}$ and propagates a distance $\bar{u} \Delta t$ over one time step. Figure 4.2(a) shows a situation where $\bar{u} \Delta t < \Delta x$, so that information propagates less than one grid cell in a single time step. In this case it makes sense to define the flux at $x_{i-1/2}$ in terms of Q_{i-1}^n and Q_i^n alone. In Figure 4.2(b), on the other hand, a larger time step is used with $\bar{u} \Delta t > \Delta x$. In this case the true flux at $x_{i-1/2}$ clearly depends on the value of Q_{i-2}^n, and so should the new cell average Q_i^{n+1}. The method (4.7) would certainly be unstable when applied with such a large time

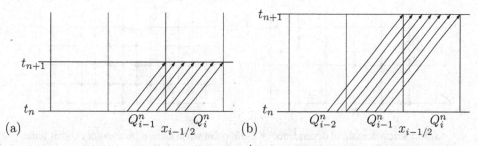

Fig. 4.2. Characteristics for the advection equation, showing the information that flows into cell $\mathcal{C}_i$ during a single time step. (a) For a small enough time step, the flux at $x_{i-1/2}$ depends only on the values in the neighboring cells – only on Q_{i-1}^n in this case where $\bar{u} > 0$. (b) For a larger time step, the flux should depend on values farther away.

step, no matter how the flux (4.6) was specified, if this numerical flux depended only on Q_{i-1}^n and Q_i^n.

This is a consequence of the *CFL condition*, named after Courant, Friedrichs, and Lewy. They wrote one of the first papers on finite difference methods for partial differential equations [93] in 1928. (There is an English translation in [94].) They used finite difference methods as an analytic tool for proving the existence of solutions of certain PDEs. The idea is to define a sequence of approximate solutions (via finite difference equations), prove that they converge as the grid is refined, and then show that the limit function must satisfy the PDE, giving the existence of a solution. In the course of proving convergence of this sequence (which is precisely what we are interested in numerically), they recognized the following necessary stability condition for any numerical method:

> **CFL Condition:** *A numerical method can be convergent only if its numerical domain of dependence contains the true domain of dependence of the PDE, at least in the limit as Δt and Δx go to zero.*

It is very important to note that the CFL condition is only a *necessary* condition for stability (and hence convergence). It is not always *sufficient* to guarantee stability. In the next section we will see a numerical flux function yielding a method that is unstable even when the CFL condition is satisfied.

The domain of dependence $\mathcal{D}(X, T)$ for a PDE has been defined in Section 3.6. The numerical domain of dependence of a method can be defined in a similar manner as the set of points where the initial data can possibly affect the numerical solution at the point (X, T). This is easiest to illustrate for a finite difference method where pointwise values of Q are used, as shown in Figure 4.3 for a three-point method. In Figure 4.3(a) we see that Q_i^2 depends on Q_{i-1}^1, Q_i^1, Q_{i+1}^1 and hence on $Q_{i-2}^0, \ldots, Q_{i+2}^0$. Only initial data in the interval $X - 2\,\Delta x^a \le x \le X + 2\,\Delta x^a$ can affect the numerical solution at $(X, T) = (x_i, t_2)$. If we now refine the grid by a factor of 2 in both space and time ($\Delta x^b = \Delta x^a / 2$), but continue to focus on the same physical point (X, T), then we see in Figure 4.3(b) that the numerical approximation at this point now depends on initial data at more points in the interval $X - 4\,\Delta x^b \le x \le X + 4\,\Delta x^b$. But this is the same interval as before. If we continue to refine the grid with the ratio $\Delta t / \Delta x \equiv r$ fixed, then the numerical domain of dependence of a general point (X, T) is $X - T/r \le x \le X + T/r$.

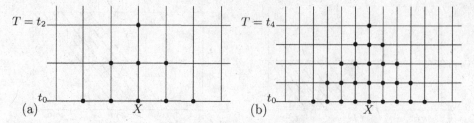

Fig. 4.3. (a) Numerical domain of dependence of a grid point when using a three-point explicit finite difference method, with mesh spacing Δx^a. (b) On a finer grid with mesh spacing $\Delta x^b = \frac{1}{2}\Delta x^a$. Similar figures can be drawn for finite volume methods.

In order for the CFL condition to be satisfied, the domain of dependence of the true solution must lie within this interval. For the advection equation $q_t + \bar{u}q_x = 0$, for example, $\mathcal{D}(X, T)$ is the single point $X - \bar{u}T$, since $q(X, T) = \overset{\circ}{q}(X - \bar{u}T)$. The CFL condition then requires

$$X - T/r \le X - \bar{u}T \le X + T/r$$

and hence

$$\nu \equiv \left| \frac{\bar{u}\,\Delta t}{\Delta x} \right| \le 1. \tag{4.16}$$

If this condition is not satisfied, then a change in the initial data $\overset{\circ}{q}$ at $X - \bar{u}T$ would change the true solution at (X, T) but could have no effect on the numerical solution at this point. Clearly the method cannot converge to the proper solution for all choices of initial data under these circumstances.

The ratio ν in (4.16) is sometimes called the CFL number, or more frequently the *Courant number*. Returning to the finite volume method illustrated in Figure 4.2, note that the Courant number measures the fraction of a grid cell that information propagates through in one time step. For a hyperbolic system of equations there are generally a set of m wave speeds $\lambda^1, \ldots, \lambda^m$ as described in Chapter 3, and the true domain of dependence is given by (3.14). In this case we define the Courant number by

$$\nu = \frac{\Delta t}{\Delta x} \max_p |\lambda^p|. \tag{4.17}$$

For a three-point method the CFL condition again leads to a necessary condition $\nu \le 1$.

Note that if the method has a wider stencil, then the CFL condition will lead to a more lenient condition on the time step. For a centered five-point stencil in which Q_i^{n+1} depends also on Q_{i-2}^n and Q_{i+2}^n, the CFL condition gives $\nu \le 2$. Again this will only be a necessary condition, and a more detailed analysis of stability would be required to determine the actual stability constraint needed to guarantee convergence.

For hyperbolic equations we typically use explicit methods and grids for which the Courant number is somewhat smaller than 1. This allows keeping $\Delta t/\Delta x$ fixed as the grid is refined, which is sensible in that generally we wish to add more resolution at the same rate in both space and in time in order to improve the solution.

For a parabolic equation such as the diffusion equation, on the other hand, the CFL condition places more severe constraints on an explicit method. The domain of dependence of any point (X, T) for $T > 0$ is now the whole real line, $\mathcal{D}(X, T) = (-\infty, \infty)$, and data at every point can affect the solution everywhere. Because of this infinite propagation speed, the CFL condition requires that the numerical domain of dependence must include the whole real line, at least in the limit as $\Delta t, \Delta x \rightarrow 0$. For an explicit method this can be accomplished by letting Δt approach zero more rapidly than Δx as we refine the grid, e.g., by taking $\Delta t = \mathcal{O}(\Delta x^2)$ as required for the method (4.11). A better way to satisfy the CFL condition in this case is to use an implicit method. In this case the numerical domain of dependence is the entire domain, since all grid points are coupled together.

4.5 An Unstable Flux

We now return to the general finite volume method (4.4) for a hyperbolic system and consider various ways in which the numerical flux might be defined. In particular we consider flux functions $\mathcal{F}$ as in (4.6). We wish to define the average flux at $x_{i-1/2}$ based on the data Q_{i-1}^n and Q_i^n to the left and right of this point. A first attempt might be the simple arithmetic average

$$F_{i-1/2}^n = \mathcal{F}(Q_{i-1}^n, Q_i^n) = \frac{1}{2}[f(Q_{i-1}^n) + f(Q_i^n)]. \tag{4.18}$$

Using this in (4.4) would give

$$Q_i^{n+1} = Q_i^n - \frac{\Delta t}{2 \Delta x}[f(Q_{i+1}^n) - f(Q_{i-1}^n)]. \tag{4.19}$$

Unfortunately, this method is generally unstable for hyperbolic problems and cannot be used, even if the time step is small enough that the CFL condition is satisfied. (See Exercise 8.1.)

4.6 The Lax–Friedrichs Method

The classical Lax–Friedrichs (LxF) method has the form

$$Q_i^{n+1} = \frac{1}{2}(Q_{i-1}^n + Q_{i+1}^n) - \frac{\Delta t}{2 \Delta x}[f(Q_{i+1}^n) - f(Q_{i-1}^n)]. \tag{4.20}$$

This is very similar to the unstable method (4.19), but the value Q_i^n is replaced by the average $\frac{1}{2}(Q_{i-1}^n + Q_{i+1}^n)$. For a linear hyperbolic equation this method is stable provided $\nu \leq 1$, where the Courant number ν is defined in (4.17).

At first glance the method (4.20) does not appear to be of the form (4.4). However, it can be put into this form by defining the numerical flux as

$$\mathcal{F}(Q_{i-1}^n, Q_i^n) = \frac{1}{2}[f(Q_{i-1}^n) + f(Q_i^n)] - \frac{\Delta x}{2 \Delta t}(Q_i^n - Q_{i-1}^n). \tag{4.21}$$

Note that this flux looks like the unstable centered flux (4.18) with the addition of another term similar to the flux (4.10) of the diffusion equation. By using this flux we appear to be modeling the advection–diffusion equation $q_t + f(q)_x = \beta q_{xx}$ with $\beta = \frac{1}{2}(\Delta x)^2/\Delta t$. But if we fix $\Delta t / \Delta x$, then we see that this coefficient vanishes as the grid is refined, so in the limit the method is still consistent with the original hyperbolic equation. This additional term can be interpreted as *numerical diffusion* that damps the instabilities arising in (4.19) and gives a method that can be shown to be stable for Courant number up to 1 (which is also the CFL limit for this three-point method). However, the Lax–Friedrichs method introduces much more diffusion than is actually required, and gives numerical results that are typically badly smeared unless a very fine grid is used.

4.7 The Richtmyer Two-Step Lax–Wendroff Method

The Lax–Friedrichs method is only first-order accurate. Second-order accuracy can be achieved by using a better approximation to the integral in (4.5). One approach is to first approximate q at the midpoint in time, $t_{n+1/2} = t_n + \frac{1}{2}\Delta t$, and evaluate the flux at this point. The *Richtmyer method* is of this form with

$$F_{i-1/2}^n = f\left(Q_{i-1/2}^{n+1/2}\right), \tag{4.22}$$

where

$$Q_{i-1/2}^{n+1/2} = \frac{1}{2}\left(Q_{i-1}^n + Q_i^n\right) - \frac{\Delta t}{2\,\Delta x}\left[f(Q_i^n) - f(Q_{i-1}^n)\right]. \tag{4.23}$$

Note that $Q_{i-1/2}^{n+1/2}$ is obtained by applying the Lax–Friedrichs method at the cell interface with Δx and Δt replaced by $\frac{1}{2}\Delta x$ and $\frac{1}{2}\Delta t$ respectively.

For a linear system of equations, $f(q) = Aq$, the Richtmyer method reduces to the standard Lax–Wendroff method, discussed further in Section 6.1. As we will see, these methods often lead to spurious oscillations in solutions, particularly when solving problems with discontinuous solutions. Additional numerical diffusion (or *artificial viscosity*) can be added to eliminate these oscillations, as first proposed by von Neumann and Richtmyer [477]. In Chapter 6 we will study a different approach to obtaining better accuracy that allows us to avoid these oscillations more effectively.

4.8 Upwind Methods

The methods considered above have all been centered methods, symmetric about the point where we are updating the solution. For hyperbolic problems, however, we expect information to propagate as waves moving along characteristics. For a system of equations we have several waves propagating at different speeds and perhaps in different directions. It makes sense to try to use our knowledge of the structure of the solution to determine better numerical flux functions. This idea gives rise to *upwind* methods in which the information for each characteristic variable is obtained by looking in the direction from which this information should be coming.

For the scalar advection equation there is only one speed, which is either positive or negative, and so an upwind method is typically also a *one-sided* method, with Q_i^{n+1} determined based on values only to the left or only to the right. This is discussed in the next section.

For a system of equations there may be waves traveling in both directions, so an upwind method must still use information from both sides, but typically uses characteristic decomposition (often via the solution of Riemann problems) to select *which* information to use from each side. Upwind methods for systems of equations are discussed beginning in Section 4.10.

4.9 The Upwind Method for Advection

For the constant-coefficient advection equation $q_t + \bar{u}q_x = 0$, Figure 4.2(a) indicates that the flux through the left edge of the cell is entirely determined by the value Q_{i-1}^n in the cell to the left of this cell. This suggests defining the numerical flux as

$$F_{i-1/2}^n = \bar{u} Q_{i-1}^n. \tag{4.24}$$

This leads to the standard *first-order upwind method* for the advection equation,

$$Q_i^{n+1} = Q_i^n - \frac{\bar{u}\,\Delta t}{\Delta x}(Q_i^n - Q_{i-1}^n). \tag{4.25}$$

Note that this can be rewritten as

$$\frac{Q_i^{n+1} - Q_i^n}{\Delta t} + \bar{u}\left(\frac{Q_i^n - Q_{i-1}^n}{\Delta x}\right) = 0,$$

whereas the unstable centered method (4.19) applied to the advection equation is

$$\frac{Q_i^{n+1} - Q_i^n}{\Delta t} + \bar{u}\left(\frac{Q_{i+1}^n - Q_{i-1}^n}{2\,\Delta x}\right) = 0.$$

The upwind method uses a one-sided approximation to the derivative q_x in place of the centered approximation.

Another interpretation of the upwind method is suggested by Figure 4.4(a). If we think of the Q_i^n as being values at grid points, $Q_i^n \approx q(x_i, t_n)$, as is standard in a finite difference method, then since $q(x, t)$ is constant along characteristics we expect

$$Q_i^{n+1} \approx q(x_i, t_{n+1}) = q(x_i - \bar{u}\,\Delta t, t_n).$$

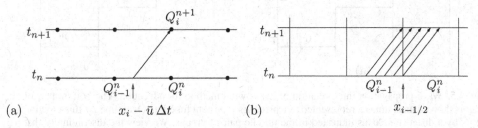

Fig. 4.4. Two interpretations of the upwind method for advection. (a) If Q_i^n represents the value at a grid point, then we can trace the characteristic back and interpolate. (b) If Q_i^n represents the cell average, then the flux at the interface is determined by the cell value on the upwind side.

If we approximate the value on the right by a linear interpolation between the grid values Q_{i-1}^n and Q_i^n, we obtain the method

$$Q_i^{n+1} = \frac{\bar{u}\,\Delta t}{\Delta x} Q_{i-1}^n + \left(1 - \frac{\bar{u}\,\Delta t}{\Delta x}\right) Q_i^n. \tag{4.26}$$

This is simply the upwind method, since a rearrangement gives (4.25).

Note that we must have

$$0 \le \frac{\bar{u}\,\Delta t}{\Delta x} \le 1 \tag{4.27}$$

in order for the characteristic to fall between the neighboring points so that this interpolation is sensible. In fact, (4.27) must be satisfied in order for the upwind method to be stable, and also follows from the CFL condition. Note that if (4.27) is satisfied then (4.26) expresses Q_i^{n+1} as a *convex combination* of Q_i^n and Q_{i-1}^n (i.e., the weights are both nonnegative and sum to 1). This is a key fact in proving stability of the method. (See Section 8.3.4.)

We are primarily interested in finite volume methods, and so other interpretations of the upwind method will be more valuable. Figure 4.4(b) and Figure 4.5 show the finite volume viewpoint, in which the value Q_i^n is now seen as a cell average of q over the ith grid cell C_i. We think of mixing up the tracer within this cell so that it has this average value at every point in the cell, at time t_n. This defines a piecewise constant function at time t_n with the

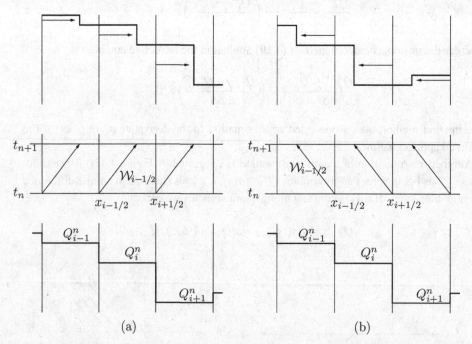

(a)   (b)

Fig. 4.5. Wave-propagation interpretation of the upwind method for advection. The bottom pair of graphs shows data at time t_n, represented as a piecewise constant function. Over time Δt this function shifts by a distance $\bar{u}\,\Delta t$ as indicated in the middle pair of graphs. We view the discontinuity that originates at $x_{i-1/2}$ as a wave $\mathcal{W}_{i-1/2}$. The top pair shows the piecewise constant function at the end of the time step after advecting. The new cell averages Q_i^{n+1} in each cell are then computed by averaging this function over each cell. (a) shows a case with $\bar{u} > 0$, while (b) shows $\bar{u} < 0$.

value Q_i^n in cell C_i. As time evolves, this piecewise constant function advects to the right with velocity $\bar{u}$, and the jump between states Q_{i-1}^n and Q_i^n shifts a distance $\bar{u}\,\Delta t$ into cell C_i. At the end of the time step we can compute a new cell average Q_i^{n+1} in order to repeat this process. To compute Q_i^{n+1} we must average the piecewise constant function shown in the top of Figure 4.5 over the cell. Computing this average results in the same convex combination (4.26) as was motivated by the characteristic-based approach of Figure 4.4(a), as the reader should verify.

We can also take a wave-propagation viewpoint, which will prove useful in extending and implementing the upwind method. The jump $\mathcal{W}_{i-1/2} \equiv Q_i^n - Q_{i-1}^n$ can be viewed as a wave that is moving into cell C_i at velocity $\bar{u}$. This wave modifies the value of q by $-\mathcal{W}_{i-1/2}$ at each point it passes. Over the time step it moves a distance $\bar{u}\,\Delta t$ and passes through a fraction $\bar{u}\,\Delta t/\Delta x$ of the grid cell, and hence the cell average is modified by this fraction of $-\mathcal{W}_{i-1/2}$:

$$Q_i^{n+1} = Q_i^n + \frac{\bar{u}\,\Delta t}{\Delta x}(-\mathcal{W}_{i-1/2}).\tag{4.28}$$

This again results in the upwind method (4.25).

In the above discussion we have assumed that $\bar{u} > 0$. On the other hand if $\bar{u} < 0$ then the upwind direction is to the right and so the numerical flux at $x_{i-1/2}$ is

$$F_{i-1/2}^n = \bar{u} Q_i^n.\tag{4.29}$$

The upwind method then has the form

$$Q_i^{n+1} = Q_i^n - \frac{\bar{u}\,\Delta t}{\Delta x}(Q_{i+1}^n - Q_i^n).\tag{4.30}$$

This can also be written in wave-propagation form as

$$Q_i^{n+1} = Q_i^n - \frac{\bar{u}\,\Delta t}{\Delta x}\mathcal{W}_{i+1/2},\tag{4.31}$$

with $\mathcal{W}_{i+1/2} = Q_{i+1}^n - Q_i^n$. All the interpretations presented above carry over to this case $\bar{u} < 0$, with the direction of flow reversed. The method (4.31) is stable provided that

$$-1 \le \frac{\bar{u}\,\Delta t}{\Delta x} \le 0.\tag{4.32}$$

The two formulas (4.24) and (4.29) can be combined into a single upwind formula that is valid for $\bar{u}$ of either sign,

$$F_{i-1/2}^n = \bar{u}^- Q_i^n + \bar{u}^+ Q_{i-1}^n,\tag{4.33}$$

where

$$\bar{u}^+ = \max(\bar{u}, 0), \qquad \bar{u}^- = \min(\bar{u}, 0).\tag{4.34}$$

The wave-propagation versions of the upwind method in (4.28) and (4.31) can also be combined to give the more general formula

$$Q_i^{n+1} = Q_i^n - \frac{\Delta t}{\Delta x}\left(\bar{u}^+ \mathcal{W}_{i-1/2} + \bar{u}^- \mathcal{W}_{i+1/2}\right). \qquad (4.35)$$

This formulation will be useful in extending this method to more general hyperbolic problems. Not all hyperbolic equations are in conservation form; consider for example the variable-coefficient linear equation (2.78) or the quasilinear system (2.81) with suitable coefficient matrix. Such equations do not have a flux function, and so numerical methods of the form (4.7) cannot be applied. However, these hyperbolic problems can still be solved using finite volume methods that result from a simple generalization of the high-resolution methods developed for hyperbolic conservation laws. The unifying feature of all hyperbolic equations is that they model waves that travel at finite speeds. In particular, the solution to a Riemann problem with piecewise constant initial data (as discussed in Chapter 3) consists of waves traveling at constant speeds away from the location of the jump discontinuity in the initial data.

In Section 4.12 we will generalize (4.35) to obtain an approach to solving hyperbolic systems that is more general than the flux-differencing form (4.4). First, however, we see how the upwind method can be extended to systems of equations.

4.10 Godunov's Method for Linear Systems

The upwind method for the advection equation can be derived as a special case of the following approach, which can also be applied to systems of equations. This will be referred to as the *REA algorithm*, for reconstruct–evolve–average. These are one-word summaries of the three steps involved.

Algorithm 4.1 (REA).

1. Reconstruct *a piecewise polynomial function* $\tilde{q}^n(x, t_n)$ *defined for all x, from the cell averages* Q_i^n. *In the simplest case this is a piecewise constant function that takes the value* Q_i^n *in the ith grid cell, i.e.,*

$$\tilde{q}^n(x, t_n) = Q_i^n \qquad \text{for all } x \in \mathcal{C}_i.$$

2. Evolve *the hyperbolic equation exactly (or approximately) with this initial data to obtain* $\tilde{q}^n(x, t_{n+1})$ *a time Δt later.*
3. Average *this function over each grid cell to obtain new cell averages*

$$Q_i^{n+1} = \frac{1}{\Delta x}\int_{\mathcal{C}_i} \tilde{q}^n(x, t_{n+1})\, dx.$$

This whole process is then repeated in the next time step.

In order to implement this procedure, we must be able to solve the hyperbolic equation in step 2. Because we are starting with piecewise constant data, this can be done using the theory of Riemann problems as introduced for linear problems in Chapter 3. When applied to the advection equation, this leads to the upwind algorithm, as illustrated in Figure 4.5.

The general approach of Algorithm 4.1 was originally proposed by Godunov [157] as a method for solving the nonlinear Euler equations of gas dynamics. Application in that context hinges on the fact that, even for this nonlinear system, the Riemann problem with piecewise constant initial data can be solved and the solution consists of a finite set of waves traveling at constant speeds, as we will see in Chapter 13.

Godunov's method for gas dynamics revolutionized the field of computational fluid dynamics, by overcoming many of the difficulties that had plagued earlier numerical methods for compressible flow. Using the wave structure determined by the Riemann solution allows shock waves to be handled in a properly "upwinded" manner even for systems of equations where information propagates in both directions. We will explore this for linear systems in the remainder of this chapter.

In step 1 we reconstruct a function $\tilde{q}^n(x, t_n)$ from the discrete cell averages. In Godunov's original approach this reconstruction is a simple piecewise constant function, and for now we concentrate on this form of reconstruction. This leads most naturally to Riemann problems, but gives only a first-order accurate method, as we will see. To obtain better accuracy one might consider using a better reconstruction, for example a piecewise linear function that is allowed to have a nonzero slope σ_i^n in the ith grid cell. This idea forms the basis for the *high-resolution* methods that are considered starting in Chapter 6.

Clearly the exact solution at time t_{n+1} can be constructed by piecing together the Riemann solutions, provided that the time step Δt is short enough that the waves from two adjacent Riemann problems have not yet started to interact. Figure 4.6 shows a schematic diagram of this process for the equations of linear acoustics with constant sound speed c, in which case this requires that

$$c\,\Delta t \le \frac{1}{2}\Delta x,$$

so that each wave goes at most halfway through the grid cell. Rearranging gives

$$\frac{c\,\Delta t}{\Delta x} \le \frac{1}{2}. \tag{4.36}$$

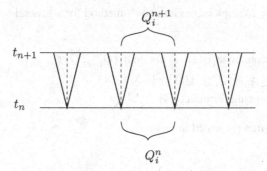

Fig. 4.6. An illustration of the process of Algorithm 4.1 for the case of linear acoustics. The Riemann problem is solved at each cell interface, and the wave structure is used to determine the exact solution time Δt later. This solution is averaged over the grid cell to determine Q_i^{n+1}.

The quantity $c\,\Delta t/\Delta x$ is simply the Courant number, so it appears that we are limited in (4.36) to a Courant number less than $1/2$. But we will see below that this method is easily extended to Courant numbers up to 1.

4.11 The Numerical Flux Function for Godunov's Method

We now develop a finite volume method based on Algorithm 4.1 that can be easily implemented in practice. As presented, the algorithm seems cumbersome to implement. The exact solution $\tilde{q}^n(x, t_{n+1})$ will typically contain several discontinuities and we must compute its integral over each grid cell in order to determine the new cell averages Q_i^{n+1}. However, it turns out to be easy to determine the numerical flux function $\mathcal{F}$ that corresponds to Godunov's method.

Recall the formula (4.5), which states that the numerical flux $F_{i-1/2}^n$ should approximate the time average of the flux at $x_{i-1/2}$ over the time step,

$$F_{i-1/2}^n \approx \frac{1}{\Delta t} \int_{t_n}^{t_{n+1}} f(q(x_{i-1}, t))\, dt.$$

In general the function $q(x_{i-1/2}, t)$ varies with t, and we certainly don't know this variation of the exact solution. However, we can compute this integral *exactly* if we replace $q(x, t)$ by the function $\tilde{q}^n(x, t)$ defined in Algorithm 4.1 using Godunov's piecewise constant reconstruction. The structure of this function is shown in Figure 3.3, for example, and so clearly $\tilde{q}^n(x_{i-1/2}, t)$ is constant over the time interval $t_n < t < t_{n+1}$. The Riemann problem centered at $x_{i-1/2}$ has a similarity solution that is constant along rays $(x - x_{i-1/2})/(t - t_n) =$ constant, and looking at the value along $(x - x_{i-1/2})/t = 0$ gives the value of $\tilde{q}^n(x_{i-1/2}, t)$. Denote this value by $Q_{i-1/2}^\vee = q^\vee(Q_{i-1}^n, Q_i^n)$. This suggests defining the numerical flux $F_{i-1/2}^n$ by

$$F_{i-1/2}^n = \frac{1}{\Delta t} \int_{t_n}^{t_{n+1}} f\big(q^\vee(Q_{i-1}^n, Q_i^n)\big)\, dt$$

$$= f\big(q^\vee(Q_{i-1}^n, Q_i^n)\big). \tag{4.37}$$

This gives a simple way to implement Godunov's method for a general system of conservation laws:

- Solve the Riemann problem at $x_{i-1/2}$ to obtain $q^\vee(Q_{i-1}^n, Q_i^n)$.
- Define the flux $F_{i-1/2}^n = \mathcal{F}(Q_{i-1}^n, Q_i^n)$ by (4.37).
- Apply the flux-differencing formula (4.4).

Godunov's method is often presented in this form.

4.12 The Wave-Propagation Form of Godunov's Method

By taking a slightly different viewpoint, we can also develop simple formulas for Godunov's method on linear systems of equations that are analogous to the form (4.35) for the upwind

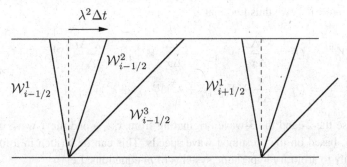

Fig. 4.7. An illustration of the process of Algorithm 4.1 for the case of a linear system of three equations. The Riemann problem is solved at each cell interface, and the wave structure is used to determine the exact solution time Δt later. The wave $\mathcal{W}^2_{i-1/2}$, for example, has moved a distance $\lambda^2 \Delta t$ into the cell.

method on the advection equation. This viewpoint is particularly useful in extending Godunov's method to hyperbolic systems that are not in conservation form.

Figure 4.7 shows a more complicated version of Figure 4.6, in which a linear system of three equations is solved assuming $\lambda^1 < 0 < \lambda^2 < \lambda^3$. The function $\tilde{q}^n(x, t_{n+1})$ will typically have three discontinuities in the grid cell C_i, at the points $x_{i-1/2} + \lambda^2 \Delta t$, $x_{i-1/2} + \lambda^3 \Delta t$, and $x_{i+1/2} + \lambda^1 \Delta t$.

Instead of trying to work with this function directly to compute the new cell average, recall from Section 3.8 that for a linear system the solution to the Riemann problem can be expressed as a set of waves,

$$Q_i - Q_{i-1} = \sum_{p=1}^{m} \alpha^p_{i-1/2} r^p \equiv \sum_{p=1}^{m} \mathcal{W}^p_{i-1/2}. \tag{4.38}$$

Let's investigate what effect each wave has on the cell average. Consider the wave denoted by $\mathcal{W}^2_{i-1/2}$ in Figure 4.7, for example. It consists of a jump in q given by

$$\mathcal{W}^2_{i-1/2} = \alpha^2_{i-1/2} r^2,$$

propagating at speed λ^2, and hence after time Δt it has moved a distance $\lambda^2 \Delta t$. This wave modifies the value of q over a fraction of the grid cell given by $\lambda^2 \Delta t / \Delta x$. It follows that the effect of this wave on the *cell average* of q is to change the average value by the amount

$$-\frac{\lambda^2 \Delta t}{\Delta x} \mathcal{W}^2_{i-1/2}.$$

The minus sign arises because the value $\mathcal{W}^2_{i-1/2}$ measures the jump from right to left, and is analogous to the minus sign in (4.28).

Each of the waves entering the grid cell has an analogous effect on the cell average, and the new cell average can be found by simply adding up these independent effects. For the case shown in Figure 4.7, we thus find that

$$Q_i^{n+1} = Q_i^n - \frac{\lambda^2 \Delta t}{\Delta x} \mathcal{W}_{i-1/2}^2 - \frac{\lambda^3 \Delta t}{\Delta x} \mathcal{W}_{i-1/2}^3 - \frac{\lambda^1 \Delta t}{\Delta x} \mathcal{W}_{i+1/2}^1$$

$$= Q_i^n - \frac{\Delta t}{\Delta x} \left(\lambda^2 \mathcal{W}_{i-1/2}^2 + \lambda^3 \mathcal{W}_{i-1/2}^3 + \lambda^1 \mathcal{W}_{i+1/2}^1 \right). \tag{4.39}$$

Note that we use the 2- and the 3-wave originating from $x_{i-1/2}$ and the 1-wave originating from $x_{i+1/2}$, based on the presumed wave speeds. This can be written in a form that generalizes easily to arbitrary hyperbolic systems of m equations. Let

$$\lambda^+ = \max(\lambda, 0), \qquad \lambda^- = \min(\lambda, 0), \tag{4.40}$$

and suppose the solution of the Riemann problem consists of m waves $\mathcal{W}^p$ traveling at speeds λ^p, each of which may be positive or negative. Then the cell average is updated by

$$Q_i^{n+1} = Q_i^n - \frac{\Delta t}{\Delta x} \left[\sum_{p=1}^m (\lambda^p)^+ \mathcal{W}_{i-1/2}^p + \sum_{p=1}^m (\lambda^p)^- \mathcal{W}_{i+1/2}^p \right]. \tag{4.41}$$

The cell average is affected by all right-going waves from $x_{i-1/2}$ and by all left-going waves from $x_{i+1/2}$. This is a generalization of (4.35). Understanding this formulation of Godunov's method is crucial to understanding many of the other algorithms presented in this book.

As a shorthand notation, we will also introduce the following symbols:

$$\mathcal{A}^- \Delta Q_{i-1/2} = \sum_{p=1}^m (\lambda^p)^- \mathcal{W}_{i-1/2}^p,$$

$$\tag{4.42}$$

$$\mathcal{A}^+ \Delta Q_{i-1/2} = \sum_{p=1}^m (\lambda^p)^+ \mathcal{W}_{i-1/2}^p,$$

so that (4.41) can be rewritten as

$$Q_i^{n+1} = Q_i^n - \frac{\Delta t}{\Delta x} \left(\mathcal{A}^+ \Delta Q_{i-1/2} + \mathcal{A}^- \Delta Q_{i+1/2} \right). \tag{4.43}$$

The symbol $\mathcal{A}^+ \Delta Q_{i-1/2}$ should be interpreted as a single entity that measures the net effect of all right-going waves from $x_{i-1/2}$, while $\mathcal{A}^- \Delta Q_{i-1/2}$ measures the net effect of all left-going waves from this same interface. These net effects will also sometimes be called *fluctuations*. Note that within cell $\mathcal{C}_i$, it is the right-going fluctuation from the left edge, $\mathcal{A}^+ \Delta Q_{i-1/2}$, and the left-going fluctuation from the right edge, $\mathcal{A}^- \Delta Q_{i+1/2}$, that affect the cell average.

The notation introduced in (4.42) is motivated by the following observation. For the constant-coefficient linear system $q_t + A q_x = 0$, we have

$$\mathcal{W}_{i-1/2}^p = \alpha_{i-1/2}^p r^p,$$

where r^p is the pth eigenvector of A, and the propagation speed is the corresponding eigenvalue λ^p. Define the matrices

$$\Lambda^+ = \begin{bmatrix} (\lambda^1)^+ & & & \\ & (\lambda^2)^+ & & \\ & & \ddots & \\ & & & (\lambda^m)^+ \end{bmatrix}, \quad \Lambda^- = \begin{bmatrix} (\lambda^1)^- & & & \\ & (\lambda^2)^- & & \\ & & \ddots & \\ & & & (\lambda^m)^- \end{bmatrix}.$$

$$(4.44)$$

Thus Λ^+ has only the positive eigenvalues on the diagonal, with negative ones replaced by zero, and conversely for Λ^-. Now define

$$A^+ = R\Lambda^+ R^{-1} \quad \text{and} \quad A^- = R\Lambda^- R^{-1},$$

$$(4.45)$$

and note that

$$A^+ + A^- = R(\Lambda^+ + \Lambda^-)R^{-1} = R\Lambda R^{-1} = A.$$

$$(4.46)$$

This gives a useful splitting of the coefficient matrix A into pieces essential for right-going and left-going propagation. Now if we let $\Delta Q_{i-1/2} = Q_i - Q_{i-1}$ and multiply this vector by A^+, we obtain

$$\begin{aligned} A^+ \Delta Q_{i-1/2} &= R\Lambda^+ R^{-1}(Q_i - Q_{i-1}) \\ &= R\Lambda^+ \alpha_{i-1/2} \\ &= \sum_{p=1}^{m} (\lambda^p)^+ \alpha_{i-1/2}^p r^p \\ &= \mathcal{A}^+ \Delta Q_{i-1/2}. \end{aligned}$$

$$(4.47)$$

Similarly, we compute that

$$\begin{aligned} A^- \Delta Q_{i-1/2} &= \sum_{p=1}^{m} (\lambda^p)^- \alpha_{i-1/2}^p r^p \\ &= \mathcal{A}^- \Delta Q_{i-1/2}. \end{aligned}$$

$$(4.48)$$

So in the linear constant-coefficient case, each of the fluctuations $\mathcal{A}^+ \Delta Q_{i-1/2}$ and $\mathcal{A}^- \Delta Q_{i-1/2}$ can be computed by simply multiplying the matrix A^+ or A^- by the jump in Q. For variable-coefficient or nonlinear problems the situation is not quite so simple, and hence we introduce the more general notation (4.42) for the fluctuations, which can still be computed by solving Riemann problems and combining the appropriate waves. We will see that the form (4.43) of Godunov's method can still be used.

For the constant-coefficient linear problem, the wave-propagation form (4.43) of Godunov's method can be related directly to the numerical flux function (4.37). Note that the value of q in the Riemann solution along $x = x_{i-1/2}$ is

$$Q_{i-1/2}^{\vee} = q^{\vee}(Q_{i-1}, Q_i) = Q_{i-1} + \sum_{p:\lambda^p < 0} \mathcal{W}_{i-1/2}^p,$$

using the summation notation introduced in (3.19).

In the linear case $f(Q_{i-1/2}^{\downarrow}) = A Q_{i-1/2}^{\downarrow}$ and so (4.37) gives

$$F_{i-1/2}^n = A Q_{i-1} + \sum_{p:\lambda^p < 0} A \mathcal{W}_{i-1/2}^p.$$

Since $\mathcal{W}_{i-1/2}^p$ is an eigenvector of A with eigenvalue λ^p, this can be rewritten as

$$F_{i-1/2}^n = A Q_{i-1} + \sum_{p=1}^m (\lambda^p)^- \mathcal{W}_{i-1/2}^p. \qquad (4.49)$$

Alternatively, we could start with the formula

$$Q_{i-1/2}^{\downarrow} = Q_i - \sum_{p:\lambda^p > 0} \mathcal{W}_{i-1/2}^p$$

and obtain

$$F_{i-1/2}^n = A Q_i - \sum_{p=1}^m (\lambda^p)^+ \mathcal{W}_{i-1/2}^p. \qquad (4.50)$$

Similarly, there are two ways to express $F_{i+1/2}^n$. Choosing the form

$$F_{i+1/2}^n = A Q_i + \sum_{p=1}^m (\lambda^p)^- \mathcal{W}_{i+1/2}^p$$

and combining this with (4.50) in the flux-differencing formula (4.4) gives

$$
\begin{aligned}
Q_i^{n+1} &= Q_i^n - \frac{\Delta t}{\Delta x}\left(F_{i+1/2}^n - F_{i-1/2}^n\right) \\
&= Q_i^n - \frac{\Delta t}{\Delta x}\left[\sum_{p=1}^m (\lambda^p)^- \mathcal{W}_{i+1/2}^p + \sum_{p=1}^m (\lambda^p)^+ \mathcal{W}_{i-1/2}^p\right],
\end{aligned} \qquad (4.51)
$$

since the $A Q_i$ terms cancel out. This is exactly the same expression obtained in (4.41).

For a more general conservation law $q_t + f(q)_x = 0$, we can define

$$F_{i-1/2}^n = f(Q_{i-1}) + \sum_{p=1}^m (\lambda^p)^- \mathcal{W}_{i-1/2}^p \equiv f(Q_{i-1}) + \mathcal{A}^- \Delta Q_{i-1/2} \qquad (4.52)$$

or

$$F_{i-1/2}^n = f(Q_i) - \sum_{p=1}^m (\lambda^p)^+ \mathcal{W}_{i-1/2}^p \equiv f(Q_i) - \mathcal{A}^+ \Delta Q_{i-1/2}, \qquad (4.53)$$

corresponding to (4.49) and (4.50) respectively, where the speeds λ^p and waves $\mathcal{W}^p$ come out of the solution to the Riemann problem.

4.13 Flux-Difference vs. Flux-Vector Splitting

Note that if we subtract (4.52) from (4.53) and rearrange, we obtain

$$f(Q_i) - f(Q_{i-1}) = \mathcal{A}^- \Delta Q_{i-1/2} + \mathcal{A}^+ \Delta Q_{i-1/2}. \tag{4.54}$$

This indicates that the terms on the right-hand side correspond to a so-called *flux-difference splitting*. The difference between the fluxes computed based on each of the cell averages Q_{i-1} and Q_i is split into a left-going fluctuation that updates Q_{i-1} and a right-going fluctuation that updates Q_i.

We can define a more general class of flux-difference splitting methods containing any method based on some splitting of the flux difference as in (4.54), followed by application of the formula (4.43). Such a method is guaranteed to be conservative, and corresponds to a flux-differencing method with numerical fluxes

$$F^n_{i-1/2} = f(Q_i) - \mathcal{A}^+ \Delta Q_{i-1/2} = f(Q_{i-1}) + \mathcal{A}^- \Delta Q_{i-1/2}. \tag{4.55}$$

For a linear system, there are other ways to rewrite the numerical flux $F^n_{i-1/2}$ that give additional insight. Using (4.47) in (4.49), we obtain

$$\begin{aligned} F^n_{i-1/2} &= (A^+ + A^-)Q_{i-1} + A^-(Q_i - Q_{i-1}) \\ &= A^+ Q_{i-1} + A^- Q_i. \end{aligned} \tag{4.56}$$

Since $A^+ + A^- = A$, the formula (4.56) gives a flux that is *consistent* with the correct flux in the sense of (4.14): If $Q_{i-1} = Q_i = \bar{q}$, then (4.56) reduces to $F_{i-1/2} = A\bar{q} = f(q)$. This has a very natural interpretation as a *flux-vector splitting*. The flux function is $f(q) = Aq$, and so AQ_{i-1} and AQ_i give two possible approximations to the flux at $x_{i-1/2}$. In Section 4.5 we considered the possibility of simply averaging these to obtain $F^n_{i-1/2}$ and rejected this because it gives an unstable method. The formula (4.56) suggests instead a more sophisticated average in which we take the part of AQ_{i-1} corresponding to right-going waves and combine it with the part of AQ_i corresponding to left-going waves in order to obtain the flux in between. This is the proper generalization to systems of equations of the upwind flux for the scalar advection equation given in (4.33).

This is philosophically a different approach from the flux-difference splitting discussed in relation to (4.54). What we have observed is that for a constant-coefficient linear system, the two viewpoints lead to exactly the same method. This is not typically the case for nonlinear problems, and in this book we concentrate primarily on methods that correspond to flux-difference splittings. However, note that given *any* flux-vector splitting, one can define a corresponding splitting of the flux difference in the form (4.54). If we have split

$$f(Q_{i-1}) = f_{i-1}^{(-)} + f_{i-1}^{(+)} \quad \text{and} \quad f(Q_i) = f_i^{(-)} + f_i^{(+)},$$

and wish to define the numerical flux as

$$F^n_{i-1/2} = f_{i-1}^{(+)} + f_i^{(-)}, \tag{4.57}$$

then we can define the corresponding fluctuations as

$$\mathcal{A}^- \Delta Q_{i-1/2} = f_i^{(-)} - f_{i-1}^{(-)},$$
$$\mathcal{A}^+ \Delta Q_{i-1/2} = f_i^{(+)} - f_{i-1}^{(+)},$$

(4.58)

to obtain a flux-difference splitting that satisfies (4.54) and again yields $F_{i-1/2}^n$ via the formula (4.55). Flux-vector splittings for nonlinear problems are discussed in Section 15.7.

4.14 Roe's Method

For a constant-coefficient linear problem there is yet another way to rewrite the flux $F_{i-1/2}^n$ appearing in (4.49), (4.50), and (4.56), which relates it directly to the unstable naive averaging of AQ_{i-1} and AQ_i given in (4.18). Averaging the expressions (4.49) and (4.50) gives

$$F_{i-1/2}^n = \frac{1}{2}\left[(AQ_{i-1} + AQ_i) - \sum_{p=1}^{m}[(\lambda^p)^+ - (\lambda^p)^-]\mathcal{W}_{i-1/2}^p \right].$$

(4.59)

Notice that $\lambda^+ - \lambda^- = |\lambda|$. Define the matrix $|A|$ by

$$|A| = R|\Lambda|R^{-1}, \qquad \text{where } |\Lambda| = \text{diag}(|\lambda^p|).$$

(4.60)

Then (4.59) becomes

$$F_{i-1/2}^n = \frac{1}{2}(AQ_{i-1} + AQ_i) - \frac{1}{2}|A|(Q_i - Q_{i-1})$$
$$= \frac{1}{2}[f(Q_{i-1}) + f(Q_i)] - \frac{1}{2}|A|(Q_i - Q_{i-1}).$$

(4.61)

This can be viewed as the arithmetic average plus a correction term that stabilizes the method.

For the constant-coefficient linear problem this is simply another way to rewrite the Godunov or upwind flux, but this form is often seen in extensions to nonlinear problems based on approximate Riemann solvers, as discussed in Section 15.3. This form of the flux is often called *Roe's method* in this connection. This formulation is also useful in studying the numerical dissipation of the upwind method.

Using the flux (4.61) in the flux-differencing formula (4.4) gives the following updating formula for Roe's method on a linear system:

$$Q_i^{n+1} = Q_i^n - \frac{1}{2}\frac{\Delta t}{\Delta x} A\left(Q_{i+1}^n - Q_{i-1}^n\right)$$
$$-\frac{1}{2}\frac{\Delta t}{\Delta x}\sum_{p=1}^{m}\left(|\lambda^p|\mathcal{W}_{i+1/2}^p - |\lambda^p|\mathcal{W}_{i-1/2}^p\right).$$

(4.62)

This can also be derived directly from (4.41) by noting that another way to express (4.40) is

$$\lambda^+ = \frac{1}{2}(\lambda + |\lambda|), \qquad \lambda^- = \frac{1}{2}(\lambda - |\lambda|). \tag{4.63}$$

We will see in Section 12.3 that for nonlinear problems it is sometimes useful to modify these definitions of $\lambda^{\pm}$.

Exercises

4.1. (a) Determine the matrices A^+ and A^- as defined in (4.45) for the acoustics equations (2.50).

(b) Determine the waves $W^1_{i-1/2}$ and $W^2_{i-1/2}$ that result from arbitrary data Q_{i-1} and Q_i for this system.

4.2. If we apply the upwind method (4.25) to the advection equation $q_t + \bar{u}q_x = 0$ with $\bar{u} > 0$, and choose the time step so that $\bar{u}\,\Delta t = \Delta x$, then the method reduces to

$$Q_i^{n+1} = Q_{i-1}^n.$$

The initial data simply shifts one grid cell each time step and the exact solution is obtained, up to the accuracy of the initial data. (If the data Q_i^0 is the exact cell average of $\overset{\circ}{q}(x)$, then the numerical solution will be the exact cell average for every step.) This is a nice property for a numerical method to have and is sometimes called the *unit CFL condition*.

(a) Sketch figures analogous to Figure 4.5(a) for this case to illustrate the wave-propagation interpretation of this result.

(b) Does the Lax–Friedrichs method (4.20) satisfy the unit CFL condition? Does the two-step Lax–Wendroff method of Section 4.7?

(c) Show that the exact solution (in the same sense as above) is also obtained for the constant-coefficient acoustics equations (2.50) with $u_0 = 0$ if we choose the time step so that $c\,\Delta t = \Delta x$ and apply Godunov's method. Determine the formulas for p_i^{n+1} and u_i^{n+1} that result in this case, and show how they are related to the solution obtained from characteristic theory.

(d) Is it possible to obtain a similar exact result by a suitable choice of Δt in the case where $u_0 \neq 0$ in acoustics?

4.3. Consider the following method for the advection equation with $\bar{u} > 0$:

$$Q_i^{n+1} = Q_i^n - (Q_i^n - Q_{i-1}^n) - \left(\frac{\bar{u}\,\Delta t - \Delta x}{\Delta x}\right)(Q_{i-1}^n - Q_{i-2}^n)$$

$$= Q_{i-1}^n - \left(\frac{\bar{u}\,\Delta t}{\Delta x} - 1\right)(Q_{i-1}^n - Q_{i-2}^n). \tag{4.64}$$

(a) Show that this method results from a wave-propagation algorithm of the sort illustrated in Figure 4.5(a) in the case where $\Delta x \leq \bar{u}\,\Delta t \leq 2\,\Delta x$, so that each wave propagates all the way through the adjacent cell and part way through the next.

(b) Give an interpretation of this method based on linear interpolation, similar to what is illustrated in Figure 4.4(a).

(c) Show that this method is exact if $\bar{u} \, \Delta t / \Delta x = 1$ or $\bar{u} \, \Delta t / \Delta x = 2$.

(d) For what range of Courant numbers is the CFL condition satisfied for this method? (See also Exercise 8.6.)

(e) Determine a method of this same type that works if each wave propagates through more than two cells but less than three, i.e., if $2 \, \Delta x \leq \bar{u} \, \Delta t \leq 3 \, \Delta x$.

Large-time-step methods of this type can also be applied, with limited success, to nonlinear problems, e.g., [42], [181], [274], [275], [279], [316].

Introduction to the CLAWPACK Software

The basic class of finite volume methods developed in this book has been implemented in the software package CLAWPACK. This allows these algorithms to be applied to a wide variety of hyperbolic systems simply by providing the appropriate Riemann solver, along with initial data and boundary conditions. The high-resolution methods introduced in Chapter 6 are implemented, but the simple first-order Godunov method of Chapter 4 is obtained as a special case by setting the input parameters appropriately. (Specifically, set method(2)=1 as described below.) In this chapter an overview of the software is given along with examples of its application to simple problems of advection and acoustics.

The software includes more advanced features that will be introduced later in the book, and can solve linear and nonlinear problems in one, two, and three space dimensions, as well as allowing the specification of capacity functions introduced in Section 2.4 (see Section 6.16) and source terms (see Chapter 17). CLAWPACK is used throughout the book to illustrate the implementation and behavior of various algorithms and their application on different physical systems. Nearly all the computational results presented have been obtained using CLAWPACK with programs that can be downloaded to reproduce these results or investigate the problems further. These samples also provide templates that can be adapted to solve other problems. See Section 1.5 for details on how to access webpages for each example.

Only the one-dimensional software is introduced here. More extensive documentation, including discussion of the multidimensional software and adaptive mesh refinement capabilities, can be downloaded from the webpage

        http://www.amath.washington.edu/~claw/

See also the papers [283], [257] for more details about the multidimensional algorithms, and [32] for a discussion of some features of the adaptive mesh refinement code.

5.1 Basic Framework

In one space dimension, the CLAWPACK routine claw1 (or the simplified version claw1ez) can be used to solve a system of equations of the form

$$\kappa(x)q_t + f(q)_x = \psi(q, x, t), \qquad (5.1)$$

where $q = q(x, t) \in \mathbb{R}^m$. The standard case of a homogeneous conservation law has $\kappa \equiv 1$ and $\psi \equiv 0$,

$$q_t + f(q)_x = 0. \tag{5.2}$$

The flux function $f(q)$ can also depend explicitly on x and t as well as on q. Hyperbolic systems that are not in conservation form, e.g.,

$$q_t + A(x, t)q_x = 0, \tag{5.3}$$

can also be solved.

The basic requirement on the homogeneous system is that it be hyperbolic in the sense that a Riemann solver can be specified that, for any two states Q_{i-1} and Q_i, returns a set of M_w waves $\mathcal{W}_{i-1/2}^p$ and speeds $s_{i-1/2}^p$ satisfying

$$\sum_{p=1}^{M_w} \mathcal{W}_{i-1/2}^p = Q_i - Q_{i-1} \equiv \Delta Q_{i-1/2}.$$

The Riemann solver must also return a left-going fluctuation $\mathcal{A}^- \Delta Q_{i-1/2}$ and a right-going fluctuation $\mathcal{A}^+ \Delta Q_{i-1/2}$. In the standard conservative case (5.2) these should satisfy

$$\mathcal{A}^- \Delta Q_{i-1/2} + \mathcal{A}^+ \Delta Q_{i-1/2} = f(Q_i) - f(Q_{i-1}) \tag{5.4}$$

and the fluctuations then define a *flux-difference splitting* as described in Section 4.13. Typically

$$\mathcal{A}^- \Delta Q_{i-1/2} = \sum_p (s_{i-1/2}^p)^- \mathcal{W}_{i-1/2}^p, \qquad \mathcal{A}^+ \Delta Q_{i-1/2} = \sum_p (s_{i-1/2}^p)^+ \mathcal{W}_{i-1/2}^p, \tag{5.5}$$

where $s^- = \min(s, 0)$ and $s^+ = \max(s, 0)$. In the nonconservative case (5.3), there is no flux function $f(q)$, and the constraint (5.4) need not be satisfied.

Only the fluctuations are used for the first-order Godunov method, which is implemented in the form introduced in Section 4.12,

$$Q_i^{n+1} = Q_i^n - \frac{\Delta t}{\Delta x} (\mathcal{A}^+ \Delta Q_{i-1/2} + \mathcal{A}^- \Delta Q_{i+1/2}), \tag{5.6}$$

assuming $\kappa \equiv 1$.

The Riemann solver must be supplied by the user in the form of a subroutine rp1, as described below. Typically the Riemann solver first computes waves and speeds and then uses these to compute $\mathcal{A}^+ \Delta Q_{i-1/2}$ and $\mathcal{A}^- \Delta Q_{i-1/2}$ internally in the Riemann solver. The waves and speeds must also be returned by the Riemann solver in order to use the high-resolution methods described in Chapter 6. These methods take the form

$$Q_i^{n+1} = Q_i^n - \frac{\Delta t}{\Delta x} (\mathcal{A}^+ \Delta Q_{i-1/2} + \mathcal{A}^- \Delta Q_{i+1/2}) - \frac{\Delta t}{\Delta x} (\tilde{F}_{i+1/2} - \tilde{F}_{i-1/2}), \tag{5.7}$$

where

$$\tilde{F}_{i-1/2} = \frac{1}{2} \sum_{p=1}^m |s_{i-1/2}^p| \left(1 - \frac{\Delta t}{\Delta x} |s_{i-1/2}^p| \right) \tilde{\mathcal{W}}_{i-1/2}^p. \tag{5.8}$$

Here $\widetilde{\mathcal{W}}_{i-1/2}^p$ represents a limited version of the wave $\mathcal{W}_{i-1/2}^p$, obtained by comparing $\mathcal{W}_{i-1/2}^p$ to $\mathcal{W}_{i-3/2}^p$ if $s^p > 0$ or to $\mathcal{W}_{i+1/2}^p$ if $s^p < 0$.

When a capacity function $\kappa(x)$ is present, the Godunov method becomes

$$Q_i^{n+1} = Q_i^n - \frac{\Delta t}{\kappa_i \, \Delta x}(\mathcal{A}^+\Delta Q_{i-1/2} + \mathcal{A}^-\Delta Q_{i+1/2}), \qquad (5.9)$$

See Section 6.16 for discussion of this algorithm and its extension to the high-resolution method.

If the equation has a source term, a routine src1 must also be supplied that solves the source-term equation $q_t = \psi(q, \kappa)$ over a time step. A fractional-step method is used to couple this with the homogeneous solution, as described in Chapter 17. Boundary conditions are imposed by setting values in ghost cells each time step, as described in Chapter 7. A few standard boundary conditions are implemented in a library routine, but this can be modified to impose other conditions; see Section 5.4.4.

5.2 Obtaining CLAWPACK

The latest version of CLAWPACK can be downloaded from the web, at

    http://www.amath.washington.edu/~claw/

Go to "download software" and select the portion you wish to obtain. At a minimum, you will need

    claw/clawpack

If you plan to use Matlab to plot results, some useful scripts are in

    claw/matlab

Other plotting packages can also be used, but you will have to figure out how to properly read in the solution produced by CLAWPACK.

The basic CLAWPACK directories 1d, 2d, and 3d each contain one or two examples in directories such as

    claw/clawpack/1d/example1

that illustrate the basic use of CLAWPACK. The directory

    claw/book

contains drivers and data for all the examples presented in this book. You can download this entire directory or selectively download specific examples as you need them. Some other applications of CLAWPACK can be found in

    claw/applications

5.3 Getting Started

The discussion here assumes you are using the Unix (or Linux) operating system. The Unix prompt is denoted by unix>.

5.3.1 Creating the Directories

The files you download will be gzipped tar files. Before installing any of CLAWPACK, you should create a directory named <path>/claw where the pathname <path> depends on where you want these files to reside and the local naming conventions on your computer. You should download any CLAWPACK or claw/book files to this directory. After downloading any file of the form name.tar.gz, execute the following commands:

```
unix> gunzip name.tar.gz
unix> tar -xvf name.tar
```

This will create the appropriate subdirectories within <path>/claw.

5.3.2 Environment variables for the path

You should now set the environment variable CLAW in Unix so that the proper files can be found:

```
unix> setenv CLAW <path>/claw
```

You might want to put this line in your .cshrc file so it will automatically be executed when you log in or create a new window. Now you can refer to $CLAW/clawpack/1d, for example, and reach the correct directory.

5.3.3 Compiling the code

Go to the directory claw/clawpack/1d/example1. There is a file in this directory named compile, which should be executable so that you can type

```
unix> compile
```

This should invoke f77 to compile all the necessary files and create an executable called xclaw. To run the program, type

```
unix> xclaw
```

and the program should run, producing output files that start with fort. In particular, fort.q0000 contains the initial data, and fort.q0001 the solution at the first output time. The file fort.info has some information about the performance of CLAWPACK.

5.3.4 Makefiles

The compile file simply compiles all of the routines needed to run CLAWPACK on this example. This is simple, but if you make one small change in one routine, then everything has to be recompiled. Instead it is generally easier to use a Makefile, which specifies what set of object files (ending with .o) are needed to make the executable, and which Fortran files (ending with .f) are needed to make the object files. If a Fortran file is changed, then it is only necessary to recompile this one rather than everything. This is done simply by typing

```
unix> make
```

A complication arises in that the example1 directory only contains a few of the necessary Fortran files, the ones specific to this particular problem. All the standard CLAWPACK files are in the directory claw/clawpack/1d/lib. You should first go into that directory and type make to create the object files for these library routines. This only needs to be done once if these files are never changed. Now go to the example1 directory and also type make. Again an executable named xclaw should be created. See the comments at the start of the Makefile for some other options.

5.3.5 *Matlab Graphics*

If you wish to use Matlab to view the results, you should download the directory claw/matlab and then set the environment variable

```
unix> setenv MATLABPATH ".:\$CLAW/matlab"
```

before starting Matlab, in order to add this directory to your Matlab search path. This directory contains the plotting routines plotclaw1.m and plotclaw2.m for plotting results in one and two dimensions respectively.

With Matlab running in the example1 directory, type

```
Matlab> plotclaw1
```

to see the results of this computation. You should see a pulse advecting to the right with velocity 1, and wrapping around due to the periodic boundary conditions applied in this example.

5.4 Using CLAWPACK – a Guide through example1

The program in claw/clawpack/1d/example1 solves the advection equation

$$q_t + uq_x = 0$$

with constant velocity $u = 1$ and initial data consisting of a Gaussian hump

$$q(x, 0) = \exp(-\beta(x - 0.3)^2). \tag{5.10}$$

The parameters $u = 1$ and $\beta = 200$ are specified in the file setprob.data. These values are read in by the routine setprob.f described in Section 5.5.

5.4.1 *The Main Program* (driver.f)

The main program is located in the file driver.f. It simply allocates storage for the arrays needed in CLAWPACK and then calls claw1ez, described below. Several parameters are set and used to declare these arrays. The proper values of these parameters depends on the particular problem. They are:

maxmx: The maximum number of grid cells to be used. (The actual number mx is later read in from the input file claw1ez.data and must satisfy mx ≤ maxmx.)

meqn: The number of equations in the hyperbolic system, e.g., meqn = 1 for a scalar equation, meqn = 2 for the acoustics equations (2.50), etc.

mwaves: The number of waves produced in each Riemann solution, called M_w in the text. Often mwaves = meqn, but not always.

mbc: The number of ghost cells used for implementing boundary conditions, as described in Chapter 7. Setting mbc = 2 is sufficient unless changes have been made to the CLAWPACK software resulting in a larger stencil.

mwork: A work array of dimension mwork is used internally by CLAWPACK for various purposes. How much space is required depends on the other parameters:

$$mwork \geq (maxmx + 2^*mbc) * (2 + 4^*meqn + mwaves + meqn^*mwaves)$$

If the value of mwork is set too small, CLAWPACK will halt with an error message telling how much space is required.

maux: The number of "auxiliary" variables needed for information specifying the problem, which is used in declaring the dimensions of the array aux (see below).

Three arrays are declared in driver.f:

q(1-mbc:maxmx+mbc, meqn): This array holds the approximation Q_i^n (a vector with meqn components) at each time t_n. The value of i ranges from 1 to mx where mx <= maxmx is set at run time from the input file. The additional ghost cells numbered (1-mbc):0 and (mx+1):(mx+mbc) are used in setting boundary conditions.

work(mwork): Used as work space.

aux(1-mbc:maxmx+mbc, maux): Used for auxiliary variables if maux > 0. For example, in a variable-coefficient advection problem the velocity in the ith cell might be stored in aux(i,1). See Section 5.6 for an example and more discussion. If maux = 0, then there are no auxiliary variables, and aux can simply be declared as a scalar or not declared at all, since this array will not be referenced.

5.4.2 *The Initial Conditions* (qinit.f)

The subroutine qinit.f sets the initial data in the array q. For a system with meqn components, q(i,m) should be initialized to a cell average of the mth component in the ith grid cell. If the data is given by a smooth function, then it may be simplest to evaluate this function just at the center of the cell, which agrees with the cell average to $\mathcal{O}((\Delta x)^2)$. The left edge of the cell is at xlower + (i-1)*dx, and the right edge is at xlower + i*dx. It is only necessary to set values in cells i = 1:mx, not in the ghost cells. The values of xlower, dx, and mx are passed into qinit.f, having been set in claw1ez.

5.4.3 *The* claw1ez *Routine*

The main program driver.f sets up array storage and then calls the subroutine claw1ez, which is located in claw/clawpack/1d/lib, along with other standard CLAWPACK subroutines described below. The claw1ez routine provides an easy way to use CLAWPACK that should suffice for many applications. It reads input data from a file claw1ez.data, which is assumed to be in a standard form described below. It also makes other assumptions about what the user is providing and what type of output is desired. After checking the inputs for consistency, claw1ez calls the CLAWPACK routine claw1 repeatedly to produce the solution at each desired output time.

The claw1 routine (located in claw/clawpack/1d/lib/claw1.f) is much more general and can be called directly by the user if more flexibility is needed. See the documentation in the source code for this routine.

5.4.4 Boundary Conditions

Boundary conditions must be set in every time step, and claw1 calls a subroutine bc1 in every step to accomplish this. The manner in which this is done is described in detail in Chapter 7. For many problems the choice of boundary conditions provided in the default routine claw/clawpack/1d/lib/bc1.f will be sufficient. For other boundary conditions the user must provide an appropriate routine. This can be done by copying the bc1.f routine to the application directory and modifying it to insert the appropriate boundary conditions at the points indicated.

When using claw1ez, the claw1ez.data file contains parameters specifying what boundary condition is to be used at each boundary (see Section 5.4.6, where the mthbc array is described).

5.4.5 The Riemann Solver

The file claw/clawpack/1d/example1/rp1ad.f contains the Riemann solver, a subroutine that should be named rp1 if claw1ez is used. (More generally the name of the subroutine can be passed as an argument to claw1.) The Riemann solver is the crucial user-supplied routine that specifies the hyperbolic equation being solved. The input data consists of two arrays ql and qr. The value ql(i,:) is the value Q_i^L at the left edge of the ith cell, while qr(i,:) is the value Q_i^R at the right edge of the ith cell, as indicated in Figure 5.1. Normally ql = qr and both values agree with Q_i^n, the cell average. More flexibility is allowed because in some applications, or in adapting CLAWPACK to implement different algorithms, it is useful to allow different values at each edge. For example, we might want to define a piecewise linear function within the grid cell as illustrated in Figure 5.1 and

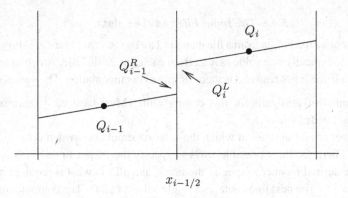

$x_{i-1/2}$

Fig. 5.1. The states used in solving the Riemann problem at the interface $x_{i-1/2}$.

then solve the Riemann problems between these values. This approach to high-resolution methods is discussed in Section 10.1.

Note that the Riemann problem at the interface $x_{i-1/2}$ between cells $i-1$ and i has data

$$\text{left state:} \qquad Q_{i-1}^R = \texttt{qr(i - 1, :)},$$
$$\text{right state:} \qquad Q_i^L = \texttt{ql(i, :)}. \tag{5.11}$$

This notation is potentially confusing in that normally we use q_l to denote the left state and q_r to denote the right state in specifying Riemann data. The routine rp1 must solve the Riemann problem for each value of i, and return the following:

amdq(i,1:meqn), the vector $\mathcal{A}^-\Delta Q_{i-1/2}$ containing the left-going fluctuation as described in Section 4.12.

apdq(i,1:meqn), the vector $\mathcal{A}^+\Delta Q_{i-1/2}$ containing the right-going fluctuation as described in Section 4.12.

wave(i,1:meqn,p), the vector $\mathcal{W}_{i-1/2}^p$ representing the jump in q across the pth wave in the Riemann solution at $x_{i-1/2}$, for $p = 1, 2, \ldots,$ mwaves. (In the code mw is typically used in place of p.)

s(i,p), the wave speed $s_{i-1/2}^p$ for each wave.

For Godunov's method, only the fluctuations amdq and apdq are actually used, and the update formula (5.6) is employed. The waves and speeds are only used for high-resolution correction terms as described in Chapter 6.

For the advection equation, the Riemann solver in example1 returns

$$\text{wave}(i, 1, 1) = \texttt{ql}(i) - \texttt{qr}(i - 1),$$
$$\text{s}(i, 1) = u,$$
$$\text{amdq}(i, 1) = \min(u, 0)*\text{wave}(i, 1, 1),$$
$$\text{apdq}(i, 1) = \max(u, 0)*\text{wave}(i, 1, 1).$$

Sample Riemann solvers for a variety of other applications can be found in claw/book and claw/applications. Often these can be used directly rather than writing a new Riemann solver.

5.4.6 The Input File claw1ez.data

The claw1ez routine reads data from a file named claw1ez.data. Figure 5.2 shows the file from example1. Typically one value is read from each line of this file. Any text following this value on each line is not read and is there simply as documentation. The values read are:

mx: The number of grid cells for this computation. (Must have mx $\leq$ maxmx, where maxmx is set in driver.f.)

nout: Number of output times at which the solution should be written out.

outstyle: There are three possible ways to specify the output times. This parameter selects the desired manner to specify the times, and affects what is required next.

outstyle = 1: The next line contains a single value tfinal. The computation should proceed to this time, and the nout outputs will be at times t0 + (tfinal − t0)/nout, where the initial time t0 is set below.

```
50              mx              = cells in x direction

11              nout            = number of output times to print results
1               outstyle        = style of specifying output times
2.2d0           tfinal          = final time

0.1d0           dtv(1)          = initial dt (used in all steps if method(1)=0)
1.0d99          dtv(2)          = max allowable dt
1.0d0           cflv(1)         = max allowable Courant number
0.9d0           cflv(2)         = desired Courant number
500             nv(1)           = max number of time steps per call to claw1

1               method(1)       = 1 for variable dt
2               method(2)       = order
0               method(3)       = not used in one dimension
1               method(4)       = verbosity of output
0               method(5)       = source term splitting
0               method(6)       = mcapa
0               method(7)       = maux (should agree with parameter in driver)

1               meqn            = number of equations in hyperbolic system
1               mwaves          = number of waves in each Riemann solution
3               mthlim(mw)      = limiter for each wave  (mw=1,mwaves)

0.d0            t0              = initial time
0.0d0           xlower          = left edge of computational domain
1.0d0           xupper          = right edge of computational domain

2               mbc             = number of ghost cells at each boundary
2               mthbc(1)        = type of boundary conditions at left
2               mthbc(2)        = type of boundary conditions at right
```

Fig. 5.2. A typical claw1ez.data file, from claw/clawpack/1d/example1 for advection.

outstyle = 2: The next line(s) contain a list of nout times at which the outputs are desired. The computation will end when the last of these times is reached.

outstyle = 3: The next line contains two values

nstepout, nsteps

A total of nsteps time steps will be taken, with output after every nstepout time steps. The value of nout is ignored. This is most useful if you want to insure that time steps of maximum length are always taken with a desired Courant number. With the other output options, the time steps are adjusted to hit the desired times exactly. This option is also useful for debugging if you want to force the solution to be output every time step, by setting nstepout = 1.

dtv(1): The initial value of Δt used in the first time step. If method(1) = 0 below, then fixed-size time steps are used and this is the value of Δt in all steps. In this case Δt must divide the time increment between all requested outputs an integer number of times.

dtv(2): The maximum time step Δt to be allowed in any step (in the case where method(1) = 1 and variable Δt is used). Variable time steps are normally chosen based on the Courant number, and this parameter can then simply be set to some very large value so that it has no effect. For some problems, however, it may be necessary to restrict the time step to a smaller value based on other considerations, e.g., the behavior of source terms in the equations.

cflv(1): The maximum Courant number to be allowed. The Courant number is calculated after all the Riemann problems have been solved by determining the maximum wave speed seen. If the Courant number is no larger than cflv(1), then this step is accepted. If the Courant number is larger, then:

method(1)=0: (fixed time steps), the calculation aborts.

method(1)=1: (variable time steps), the step is rejected and a smaller time step is taken.

Usually cflv(1) = 1 can be used.

cflv(2): The desired Courant number for this computation. Used only if method(1)=1 (variable time steps). In each time step, the next time increment Δt is based on the maximum wave speed found in solving all Riemann problems in the *previous* time step. If the wave speeds do not change very much, then this will lead to roughly the desired Courant number. It's typically best to take cflv(2) to be slightly smaller than cflv(1), say cflv(2) = 0.9.

nv(1): The maximum number of time steps allowed in any single call to claw1. This is provided as a precaution to avoid too lengthy runs.

method(1): Tells whether fixed or variable size time steps are to be used.

method(1) = 0: A fixed time step of size dtv(1) will be used in all steps.

method(1) = 1: CLAWPACK will automatically select the time step as described above, based on the desired Courant number.

method(2): The order of the method.

method(2) = 1: The first-order Godunov's method described in Chapter 4 is used.

method(2) = 2: High-resolution correction terms are also used, as described in Chapter 6.

method(3): This parameter is not used in one space dimension. In two and three dimensions it is used to further specify which high-order correction terms are applied.

method(4): This controls the amount of output printed by claw1 on the screen as CLAWPACK progresses.

method(4) = 0: Information is printed only when output files are created.

method(4) = 1: Every time step the value Δt and Courant number are reported.

method(5): Tells whether there is a source term in the equation. If so, then a fractional-step method is used as described in Chapter 17. Time steps on the homogeneous hyperbolic equation are alternated with time steps on the source term. The solution operator for the source terms must be provided by the user in the routine src1.f.

method(5) = 0: There is no source term. In this case the default routine claw/clawpack/1d/lib/src1.f can be used, which does nothing, and in fact this routine will never be called.

method(5) = 1: A source term is specified in src1.f, and the first-order (Godunov) fractional-step method should be used.

method(5) = 2: A source term is specified in src1.f, and a Strang splitting is used. The Godunov splitting is generally recommended rather than the Strang splitting, for reasons discussed in Chapter 17.

`method(6)`: Tells whether there is a "capacity function" in the equation, as introduced in Section 2.4.

`method(6)` = 0: No capacity function; $\kappa \equiv 1$ in (2.27).

`method(6)` = `mcapa` > 0: There is a capacity function, and the value of κ in the ith cell is given by `aux(i,mcapa)`, i.e., the `mcapa` component of the `aux` array is used to store this function. In this case *capacity-form differencing* is used, as described in Section 6.16.

`method(7)`: Tells whether there are any auxiliary variables stored in an `aux` array.

`method(7)` = 0: No auxiliary variables. In this case the array `aux` is not referenced and can be a dummy variable.

`method(7)` = `maux` > 0: There is an `aux` array with `maux` components. In this case the array must be properly declared in `driver.f`.

Note that we must always have `maux` $\geq$ `mcapa`. The value of `method(7)` specified here must agree with the value of `maux` set in `driver.f`.

`meqn`: The number of equations in the hyperbolic system. This is also set in `driver.f` and the two should agree.

`mwaves`: The number of waves in each Riemann solution. This is often equal to `meqn` but need not be. This is also set in `driver.f`, and the two should agree.

`mthlim(1:mwaves)`: The limiter to be applied in each wave family as described in Chapter 6. Several different limiters are provided in CLAWPACK [see (6.39)]:

`mthlim(mw)` = 0: No limiter (Lax–Wendroff)

`mthlim(mw)` = 1: Minmod

`mthlim(mw)` = 2: Superbee

`mthlim(mw)` = 3: van Leer

`mthlim(mw)` = 4: MC (monotonized centered)

Other limiters can be added by modifying the routine `claw/clawpack/1d/lib/philim.f`, which is called by `claw/clawpack/1d/lib/limiter.f`.

`t0`: The initial time.

`xlower`: The left edge of the computational domain.

`xupper`: The right edge of the computational domain.

`mbc`: The number of ghost cells used for setting boundary conditions. Usually `mbc` = 2 is used. See Chapter 7.

`mthbc(1)`: The type of boundary condition to be imposed at the left boundary. See Chapter 7 for more description of these and how they are implemented. The following values are recognized:

`mthbc(1)` = 0: The user will specify a boundary condition. In this case you must copy the file `claw/clawpack/1d/lib/bc1.f` to your application directory and modify it to insert the proper boundary conditions in the location indicated.

`mthbc(1)` = 1: Zero-order extrapolation.

`mthbc(1)` = 2: Periodic boundary conditions. In this case you must also set `mthbc(2)` = 2.

`mthbc(1)` = 3: Solid wall boundary conditions. This set of boundary conditions only makes sense for certain systems of equations; see Section 7.3.3.

`mthbc(2)`: The type of boundary condition to be imposed at the right boundary. The same values are recognized as described above.

5.5 Other User-Supplied Routines and Files

Several other routines may be provided by the user but are not required. In each case there is a default version provided in the library `claw/clawpack/1d/lib` that does nothing but `return`. If you wish to provide a version, copy the library version to the application directory, modify it as required, and also modify the `Makefile` to point to the modified version rather than to the library version.

setprob.f The `claw1ez` routine always calls setprob at the beginning of execution. The user can provide a subroutine that sets any problem-specific parameters or does other initialization. For the advection problem solved in `example1`, this is used to set the advection velocity u. This value is stored in a common block so that it can be passed into the Riemann solver, where it is required. The parameter `beta` is also set and passed into the routine `qinit.f` for use in setting the initial data according to (5.10). When `claw1ez` is used, a `setprob` subroutine must always be provided. If there is nothing to be done, the default subroutine `claw/clawpack/1d/lib/setprob.f` can be used, which does nothing but `return`.

setaux.f The `claw1ez` routine calls a subroutine `setaux` before the first call to `claw1`. This routine should set the array `aux` to contain any necessary data used in specifying the problem. For the example in `example1` no aux array is used (`maux = 0` in `driver.f`) and the default subroutine `claw/clawpack/1d/lib/setaux.f` is specified in the `Makefile`.

b4step1.f Within `claw1` there is a call to a routine b4step1 before each call to step1 (the CLAWPACK routine that actually takes a single time step). The user can supply a routine b4step1 in place of the default routine `claw/clawpack/1d/lib/b4step1.f` in order to perform additional tasks that might be required each time step. One example might be to modify the `aux` array values each time step, as described in Section 5.6.

src1.f If the equation includes a source term ψ as in (5.1), then a routine src1 must be provided in place of the default routine `claw/clawpack/1d/lib/src1.f`. This routine must solve the equation $q_t = \psi$ over one time step. Often this requires solving an ordinary differential equation in each grid cell. In some cases a partial differential equation must be solved, for example if diffusive terms are included with $\psi = q_{xx}$, then the diffusion equation must be solved over one time step.

5.6 Auxiliary Arrays and `setaux.f`

The array `q(i,1:meqn)` contains the finite-volume solution in the ith grid cell. Often other arrays defined over the grid are required to specify the problem in the first place. For example, in a variable-coefficient advection problem

$$q_t + u(x)q_x = 0$$

the Riemann solution at any cell interface $x_{i-1/2}$ depends on the velocities u_{i-1} and u_i. The aux array can be used to store these values and pass them into the Riemann solver. In the advection example we need only one auxiliary variable, so `maux = 1` and we store the velocity u_i in `aux(i,1)`. See Chapter 9 for more discussion of variable-coefficient problems.

Of course one could hard-wire the specific function $u(x)$ into the Riemann solver or pass it in using a common block, but the use of the auxiliary arrays gives a uniform treatment of such data arrays. This is useful in particular when adaptive mesh refinement is applied, in which case there are many different q grids covering different portions of the computational domain and it is very convenient to have an associated aux array corresponding to each.

The clawlez routine always calls a subroutine setaux before beginning the computation. This routine, normally stored in setaux.f, should set the values of all auxiliary arrays. If maux = 0 then the default routine claw/clawpack/1d/lib/setaux.f can be used, which does nothing. See claw/applications/acoustics/1d/varying/interface for an example of the use of auxiliary arrays.

In some problems the values stored in the aux arrays must be time-dependent, for example in an advection equation of the form $q_t + u(x, t)q_x = 0$. The routine setaux is called only once at the beginning of the computation and cannot be used to modify values later. The user can supply a routine b4step1 in place of the default routine claw/clawpack/1d/lib/b4step1.f in order to modify the aux array values each time step.

5.7 An Acoustics Example

The directory claw/clawpack/1d/example2 contains a sample code for the constant-coefficient acoustics equations (2.50). The values of the density and bulk modulus are set in setprob.f (where they are read in from a data file setprob.data). In this routine the sound speed and impedance are also computed and passed to the Riemann solver in a common block. The Riemann solver uses the formulas (3.31) to obtain α^1 and α^2, and then the waves are $\mathcal{W}^p = \alpha^p r^p$. The boundary conditions are set for a reflecting wall at $x = -1$ and nonreflecting outflow at $x = 1$.

Exercises

The best way to do these exercises, or more generally to use CLAWPACK on a new problem, is to copy an existing directory to a new directory with a unique name and modify the routines in that new directory.

5.1. The example in claw/clawpack/1d/example2 has method(2)=2 set in clawlez.data, and hence uses a high-resolution method. Set method(2)=1 to use the upwind method instead, and compare the computed results.

5.2. Modify the data from Exercise 5.1 to take time steps for which the Courant number is 1.1. (You must change both cflv(1) and cflv(2) in clawlez.data.) Observe that the upwind method is unstable in this case.

5.3. The initial data in claw/clawpack/1d/example2/qinit.f has $q^2(x, 0) = 0$ and hence the initial pressure pulse splits into left going and right going pulses. Modify $q^2(x, 0)$ so that the initial pulse is purely left going.

6

High-Resolution Methods

In Chapter 4 we developed the basic ideas of Godunov's method, an upwind finite volume method for hyperbolic systems, in the context of constant-coefficient linear systems. Godunov's method is only first-order accurate and introduces a great deal of numerical diffusion, yielding poor accuracy and smeared results, as can be seen in Figure 6.1(a), for example. In this chapter we will see how this method can be greatly improved by introducing correction terms into (4.43), to obtain a method of the form

$$Q_i^{n+1} = Q_i - \frac{\Delta t}{\Delta x}\left(A^+\Delta Q_{i-1/2} + A^-\Delta Q_{i+1/2}\right) - \frac{\Delta t}{\Delta x}\left(\tilde{F}_{i+1/2} - \tilde{F}_{i-1/2}\right). \quad (6.1)$$

The fluxes $\tilde{F}_{i-1/2}$ are based on the waves resulting from the Riemann solution, which have already been computed in the process of determining the fluctuations $\mathcal{A}^\pm\Delta Q_{i-1/2}$. The basic form of these correction terms is motivated by the *Lax–Wendroff method*, a standard second-order accurate method described in the next section. The addition of crucial *limiters* leads to great improvement, as discussed later in this chapter.

6.1 The Lax–Wendroff Method

The Lax–Wendroff method for the linear system $q_t + Aq_x = 0$ is based on the Taylor series expansion

$$q(x, t_{n+1}) = q(x, t_n) + \Delta t\, q_t(x, t_n) + \frac{1}{2}(\Delta t)^2 q_{tt}(x, t_n) + \cdots. \quad (6.2)$$

From the differential equation we have that $q_t = -Aq_x$, and differentiating this gives

$$q_{tt} = -Aq_{xt} = A^2 q_{xx},$$

where we have used $q_{xt} = q_{tx} = (-Aq_x)_x$. Using these expressions for q_t and q_{tt} in (6.2) gives

$$q(x, t_{n+1}) = q(x, t_n) - \Delta t\, Aq_x(x, t_n) + \frac{1}{2}(\Delta t)^2 A^2 q_{xx}(x, t_n) + \cdots. \quad (6.3)$$

100

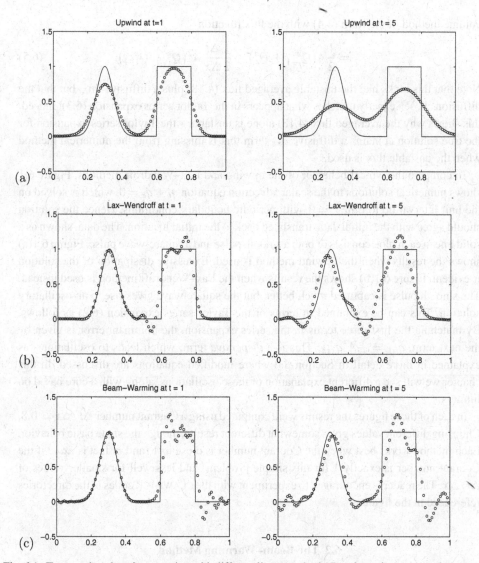

Fig. 6.1. Tests on the advection equation with different linear methods. Results at time $t = 1$ and $t = 5$ are shown, corresponding to 1 and 5 revolutions through the domain in which the equation $q_t + q_x = 0$ is solved with periodic boundary conditions: (a) upwind, (b) Lax–Wendroff, (c) Beam–Warming. [claw/book/chap6/compareadv]

Keeping only the first three terms on the right-hand side and replacing the spatial derivatives by central finite difference approximations gives the *Lax–Wendroff method*,

$$Q_i^{n+1} = Q_i^n - \frac{\Delta t}{2\,\Delta x} A\big(Q_{i+1}^n - Q_{i-1}^n\big) + \frac{1}{2}\left(\frac{\Delta t}{\Delta x}\right)^2 A^2\big(Q_{i-1}^n - 2Q_i^n + Q_{i+1}^n\big). \quad (6.4)$$

By matching three terms in the Taylor series and using centered approximations, we obtain a second-order accurate method.

This derivation of the method is based on a finite difference interpretation, with Q_i^n approximating the pointwise value $q(x_i, t_n)$. However, we can reinterpret (6.4) as a finite volume method of the form (4.4) with the flux function

$$F_{i-1/2}^n = \frac{1}{2}A(Q_{i-1}^n + Q_i^n) - \frac{1}{2}\frac{\Delta t}{\Delta x}A^2(Q_i^n - Q_{i-1}^n). \tag{6.5}$$

Note that this looks like the unstable averaged flux (4.18) plus a diffusive flux, but that the diffusion chosen exactly matches what appears in the Taylor series expansion (6.3). Indeed, this shows why the averaged flux (4.18) alone is unstable – the Taylor series expansion for the true solution contains a diffusive q_{xx} term that is missing from the numerical method when the unstable flux is used.

To compare the typical behavior of the upwind and Lax–Wendroff methods, Figure 6.1 shows numerical solutions to the scalar advection equation $q_t + q_x = 0$, which is solved on the unit interval up to time $t = 1$ with periodic boundary conditions. Hence the solution should agree with the initial data, translated back to the initial location. The data, shown as a solid line in each plot, consists of both a smooth pulse and a square-wave pulse. Figure 6.1(a) shows the results when the upwind method is used. Excessive dissipation of the solution is evident. Figure 6.1(b) shows the results when the Lax–Wendroff method is used instead. The smooth pulse is captured much better, but the square wave gives rise to an oscillatory solution. This can be explained in terms of the Taylor series expansion (6.2) as follows. By matching the first three terms in the series expansion, the dominant error is given by the next term, $q_{ttt} = -A^3 q_{xxx}$. This is a *dispersive* term, which leads to oscillations, as explained in more detail in Section 8.6 where modified equations are discussed. In this chapter we will see a different explanation of these oscillations, along with a cure based on limiters.

In each of these figures the results were computed using a Courant number $\Delta t / \Delta x = 0.8$. Choosing different values gives somewhat different results, though the same basic behavior. Each method works best when the Courant number is close to 1 (and in fact is *exact* if the Courant number is exactly 1 for this simple problem) and less well for smaller values of $\Delta t / \Delta x$. The reader is encouraged to experiment with the CLAWPACK codes in the directories referenced in the figures.

6.2 The Beam–Warming Method

The Lax–Wendroff method (6.4) is a centered three-point method. If we have a system for which all the eigenvalues of A are positive (e.g., the scalar advection equation with $\bar{u} > 0$), then we might think it is preferable to use a one-sided formula. In place of the centered formula for q_x and q_{xx}, we might use

$$q_x(x_i, t_n) = \frac{1}{2\,\Delta x}[3q(x_i, t_n) - 4q(x_{i-1}, t_n) + q(x_{i-2}, t_n)] + \mathcal{O}(\Delta x^2),$$

$$q_{xx}(x_i, t_n) = \frac{1}{(\Delta x)^2}[q(x_i, t_n) - 2q(x_{i-1}, t_n) + q(x_{i-2}, t_n)] + \mathcal{O}(\Delta x). \tag{6.6}$$

Using these in (6.3) gives a method that is again second-order accurate,

$$Q_i^{n+1} = Q_i^n - \frac{\Delta t}{2\,\Delta x}A(3Q_i^n - 4Q_{i-1}^n + Q_{i-2}^n) + \frac{1}{2}\left(\frac{\Delta t}{\Delta x}\right)^2 A^2(Q_i^n - 2Q_{i-1}^n + Q_{i-2}^n). \tag{6.7}$$

This is known as the *Beam–Warming method*, and was originally introduced in [481]. It can be written as a flux-differencing finite volume method with

$$F^n_{i-1/2} = A Q^n_{i-1} + \frac{1}{2} A \left(1 - \frac{\Delta t}{\Delta x} A \right) (Q^n_{i-1} - Q^n_{i-2}). \tag{6.8}$$

Figure 6.1(c) shows the results of the previous advection test using the Beam–Warming method. The behavior is similar to that of the Lax–Wendroff method in that oscillations appear, though the oscillations are now ahead of the discontinuities rather than behind.

6.3 Preview of Limiters

Second-order accurate methods such as the Lax–Wendroff or Beam–Warming give much better accuracy on smooth solutions than the upwind method, as seen in Figure 6.1, but fail near discontinuities, where oscillations are generated. In fact, even when the solution is smooth, oscillations may appear due to the dispersive nature of these methods, as evident in Figure 6.1. Upwind methods have the advantage of keeping the solution monotonically varying in regions where the solution should be monotone, even though the accuracy is not very good. The idea with *high-resolution* methods is to combine the best features of both methods. Second-order accuracy is obtained where possible, but we don't insist on it in regions where the solution is not behaving smoothly (and the Taylor series expansion is not even valid). With this approach we can achieve results like those shown in Figure 6.2.

The dispersive nature of the Lax–Wendroff method also causes a slight shift in the location of the smooth hump, a *phase error*, that is visible in Figure 6.1, particularly at the later time $t = 5$. Another advantage of using limiters is that this phase error can be essentially eliminated. Figure 6.3 shows a computational example where the initial data consists of a *wave packet*, a high-frequency signal modulated by a Gaussian. With a dispersive method such a packet will typically propagate at an incorrect speed corresponding to the numerical *group velocity* of the method. The Lax–Wendroff method is clearly quite dispersive. The high-resolution method shown in Figure 6.3(c) performs much better. There is some dissipation of the wave, but much less than with the upwind method. The main goal of this chapter is to develop the class of high-resolution methods used to obtain these results.

A hint of how this can be done is seen by rewriting the Lax-Wendroff flux (6.5) as

$$F^n_{i-1/2} = \left(A^- Q^n_i + A^+ Q^n_{i-1}\right) + \frac{1}{2} |A| \left(I - \frac{\Delta t}{\Delta x} |A| \right) (Q^n_i - Q^n_{i-1}), \tag{6.9}$$

using the notation A^-, A^+, $|A|$ defined in Section 4.12. This now has the form of the upwind flux (4.56) with a correction term. Using this in the flux-differencing method (4.4) gives a method of the form (6.1). Note that the correction term in (6.9) looks like a diffusive flux, since it depends on $Q^n_i - Q^n_{i-1}$ and has the form of (4.10), but the coefficient is positive if the CFL condition is satisfied. Hence it corresponds to an *antidiffusive flux* that has the effect of sharpening up the overly diffusive upwind approximation.

The idea is now to modify the final term in (6.9) by applying some form of *limiter* that changes the magnitude of the correction actually used, depending on how the solution is behaving. The limiting process is complicated by the fact that the solution to a hyperbolic system typically consists of a superposition of waves in several different families. At a given

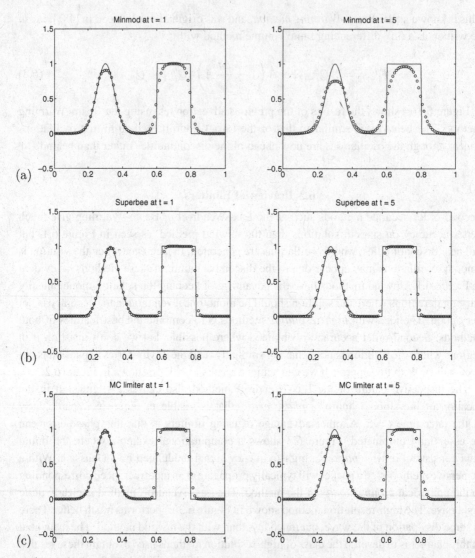

Fig. 6.2. Tests on the advection equation with different high-resolution methods, as in Figure 6.1: (a) minmod limiter, (b) superbee limiter, (c) MC limiter. `[claw/book/chap6/compareadv]`

point and time, some of the waves passing by may be smooth while others are discontinuous. Ideally we would like to apply the limiters in such a way that the discontinuous portion of the solution remains nonoscillatory while the smooth portion remains accurate. To do so we must use the characteristic structure of the solution. We will see that this is easily accomplished once we have solved the Riemann problem necessary to implement the upwind Godunov method. The second-order correction terms can be computed based on the waves arising in that Riemann solution, with each wave limited independently from the others. This process is fully described later in this chapter.

More generally, one can consider combining any low-order flux formula $\mathcal{F}_L(Q_{i-1}, Q_i)$ (such as the upwind flux) and any higher-order formula $\mathcal{F}_H(Q_{i-1}, Q_i)$ (such as the

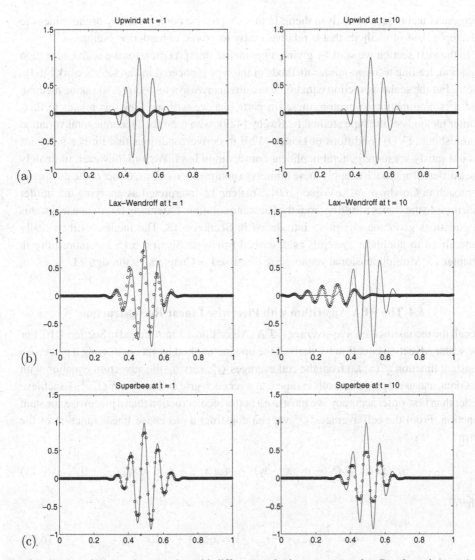

Fig. 6.3. Tests on the advection equation with different methods on a wave packet. Results at time $t = 1$ and $t = 10$ are shown, corresponding to 1 and 10 revolutions through the domain in which the equation $q_t + q_x = 0$ is solved with periodic boundary conditions. `[claw/book/chap6/wavepacket]`

Lax–Wendroff) to obtain a *flux-limiter* method with

$$F_{i-1/2}^n = \mathcal{F}_L(Q_{i-1}, Q_i) + \Phi_{i-1/2}^n [\mathcal{F}_H(Q_{i-1}, Q_i) - \mathcal{F}_L(Q_{i-1}, Q_i)]. \tag{6.10}$$

If $\Phi_{i-1/2}^n = 0$, then this reduces to the low-order method, while if $\Phi_{i-1/2}^n = 1$, we obtain the high-order method. This idea of applying limiters to improve the behavior of high-order methods appeared in the early 1970s in the *hybrid method* of Harten and Zwas [190] and the *flux-corrected transport* (FCT) method of Boris and Book [38] (see also [348], [496]). A wide variety of other methods of this form have since been developed, along with better theoretical techniques to analyze them. In this chapter we combine many of these ideas to develop a class of methods that is relatively easy to extend to harder problems.

In the next section we start by giving a geometric interpretation for the scalar advection equation, leading to *slope-limiter* methods of the type pioneered in van Leer's work [464]–[468]. For the scalar advection equation there are many ways to interpret the same method, and it is illuminating to explore these. In particular we will see how this relates to flux-limiter methods of the type studied by Sweby [429], who used the algebraic total variation diminishing (TVD) conditions of Harten [179] to derive conditions that limiter functions should satisfy for more general nonlinear conservation laws. We will, however, ultimately use a different approach to apply these limiters to nonlinear problems, closer to the geometric approach in Goodman & LeVeque [160]. This can be interpreted as applying the limiter functions to the waves resulting from the Riemann solution. Extending this to linear systems of equations gives the algorithm introduced in Section 6.13. The method is then easily generalized to nonlinear systems, as described briefly in Section 6.15 and more fully in Chapter 15. Multidimensional versions are discussed in Chapters 19 through 23.

6.4 The REA Algorithm with Piecewise Linear Reconstruction

Recall the reconstruct–evolve–average (REA) Algorithm 4.1 introduced in Section 4.10. For the scalar advection equation we derived the upwind method by reconstructing a piecewise constant function $\tilde{q}^n(x, t_n)$ from the cell averages Q_i^n, solving the advection equation with this data, and averaging the result at time t_{n+1} over each grid cell to obtain Q_i^{n+1}. To achieve better than first-order accuracy, we must use a better reconstruction than a piecewise constant function. From the cell averages Q_i^n we can construct a piecewise linear function of the form

$$\tilde{q}^n(x, t_n) = Q_i^n + \sigma_i^n(x - x_i) \qquad \text{for } x_{i-1/2} \leq x < x_{i+1/2}, \tag{6.11}$$

where

$$x_i = \frac{1}{2}\left(x_{i-1/2} + x_{i+1/2}\right) = x_{i-1/2} + \frac{1}{2}\Delta x \tag{6.12}$$

is the center of the ith grid cell and σ_i^n is the slope on the ith cell. The linear function (6.11) on the ith cell is defined in such a way that its value at the cell center x_i is Q_i^n. More importantly, the average value of $\tilde{q}^n(x, t_n)$ over cell C_i is Q_i^n (regardless of the slope σ_i^n), so that the reconstructed function has the cell average Q_i^n. This is crucial in developing conservative methods for conservation laws. Note that steps 2 and 3 are conservative in general, and so Algorithm 4.1 is conservative provided we use a *conservative reconstruction* in step 1, as we have in (6.11). Later we will see how to write such methods in the standard conservation form (4.4).

For the scalar advection equation $q_t + \bar{u}q_x = 0$, we can easily solve the equation with this data, and compute the new cell averages as required in step 3 of Algorithm 4.1. We have

$$\tilde{q}^n(x, t_{n+1}) = \tilde{q}^n(x - \bar{u}\,\Delta t, t_n).$$

Until further notice we will assume that $\bar{u} > 0$ and present the formulas for this particular case. The corresponding formulas for $\bar{u} < 0$ should be easy to derive, and in Section 6.10 we will see a better way to formulate the methods in the general case.

Suppose also that $|\bar{u} \, \Delta t / \Delta x| \leq 1$, as is required by the CFL condition. Then it is straightforward to compute that

$$Q_i^{n+1} = \frac{\bar{u} \, \Delta t}{\Delta x} \left(Q_{i-1}^n + \frac{1}{2}(\Delta x - \bar{u} \, \Delta t)\sigma_{i-1}^n \right) + \left(1 - \frac{\bar{u} \, \Delta t}{\Delta x} \right) \left(Q_i^n - \frac{1}{2}\bar{u} \, \Delta t \, \sigma_i^n \right)$$

$$= Q_i^n - \frac{\bar{u} \, \Delta t}{\Delta x} (Q_i^n - Q_{i-1}^n) - \frac{1}{2} \frac{\bar{u} \, \Delta t}{\Delta x} (\Delta x - \bar{u} \, \Delta t) \left(\sigma_i^n - \sigma_{i-1}^n \right). \qquad (6.13)$$

Again note that this is the upwind method with a correction term that depends on the slopes.

6.5 Choice of Slopes

Choosing slopes $\sigma_i^n \equiv 0$ gives Godunov's method (the upwind method for the advection equation), since the final term in (6.13) drops out. To obtain a second-order accurate method we want to choose nonzero slopes in such a way that σ_i^n approximates the derivative q_x over the ith grid cell. Three obvious possibilities are

$$\text{Centered slope:} \qquad \sigma_i^n = \frac{Q_{i+1}^n - Q_{i-1}^n}{2 \, \Delta x} \qquad \text{(Fromm)}, \qquad (6.14)$$

$$\text{Upwind slope:} \qquad \sigma_i^n = \frac{Q_i^n - Q_{i-1}^n}{\Delta x} \qquad \text{(Beam–Warming)}, \qquad (6.15)$$

$$\text{Downwind slope:} \qquad \sigma_i^n = \frac{Q_{i+1}^n - Q_i^n}{\Delta x} \qquad \text{(Lax–Wendroff)}. \qquad (6.16)$$

The centered slope might seem like the most natural choice to obtain second-order accuracy, but in fact all three choices give the same formal order of accuracy, and it is the other two choices that give methods we have already derived using the Taylor series expansion. Only the downwind slope results in a centered three-point method, and this choice gives the Lax–Wendroff method. The upwind slope gives a fully-upwinded 3-point method, which is simply the Beam–Warming method.

The centered slope (6.14) may seem the most symmetric choice at first glance, but because the reconstructed function is then advected in the positive direction, the final updating formula turns out to be a nonsymmetric four-point formula,

$$Q_i^{n+1} = Q_i^n - \frac{\bar{u} \, \Delta t}{4 \, \Delta x} \left(Q_{i+1}^n + 3Q_i^n - 5Q_{i-1}^n + Q_{i-2}^n \right)$$

$$- \frac{\bar{u}^2 \, \Delta t^2}{4 \, \Delta x^2} \left(Q_{i+1}^n - Q_i^n - Q_{i-1}^n + Q_{i-2}^n \right). \qquad (6.17)$$

This method is known as *Fromm's method*.

6.6 Oscillations

As we have seen in Figure 6.1, second-order methods such as the Lax–Wendroff or Beam–Warming (and also Fromm's method) give oscillatory approximations to discontinuous solutions. This can be easily understood using the interpretation of Algorithm 4.1.

Consider the Lax–Wendroff method, for example, applied to piecewise constant data with values

$$Q_i^n = \begin{cases} 1 & \text{if } i \leq J, \\ 0 & \text{if } i > J. \end{cases}$$

Choosing slopes in each grid cell based on the Lax–Wendroff prescription (6.16) gives the piecewise linear function shown in Figure 6.4(a). The slope σ_i^n is nonzero only for $i = J$.

The function $\tilde{q}^n(x, t_n)$ has an *overshoot* with a maximum value of 1.5 regardless of Δx. When we advect this profile a distance $\bar{u} \, \Delta t$, and then compute the average over the Jth cell, we will get a value that is greater than 1 for any Δt with $0 < \bar{u} \, \Delta t < \Delta x$. The worst case is when $\bar{u} \, \Delta t = \Delta x/2$, in which case $\tilde{q}^n(x, t_{n+1})$ is shown in Figure 6.4(b) and $Q_J^{n+1} = 1.125$. In the next time step this overshoot will be accentuated, while in cell $J - 1$ we will now have a positive slope, leading to a value Q_{J-1}^{n+1} that is less than 1. This oscillation then grows with time.

The slopes proposed in the previous section were based on the assumption that the solution is smooth. Near a discontinuity there is no reason to believe that introducing this slope will improve the accuracy. On the contrary, if one of our goals is to avoid nonphysical oscillations, then in the above example we must set the slope to zero in the Jth cell. Any $\sigma_J^n < 0$ will lead to $Q_J^{n+1} > 1$, while a positive slope wouldn't make much sense. On the other hand we don't want to set all slopes to zero all the time, or we simply have the first-order upwind method. Where the solution is smooth we want second-order accuracy. Moreover, we will see below that even near a discontinuity, once the solution is somewhat smeared out over more than one cell, introducing nonzero slopes can help keep the solution from smearing out too far, and hence will significantly increase the resolution and keep discontinuities fairly sharp, as long as care is taken to avoid oscillations.

This suggests that we must pay attention to *how the solution is behaving* near the ith cell in choosing our formula for σ_i^n. (And hence the resulting updating formula will be nonlinear even for the linear advection equation). Where the solution is smooth, we want to choose something like the Lax–Wendroff slope. Near a discontinuity we may want to limit this slope, using a value that is smaller in magnitude in order to avoid oscillations. Methods based on this idea are known as *slope-limiter* methods. This approach was introduced by van

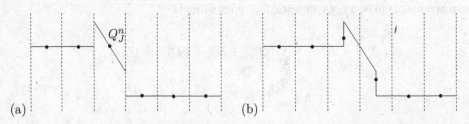

(a)   (b)

Fig. 6.4. (a) Grid values Q^n and reconstructed $\tilde{q}^n(\cdot, t_n)$ using Lax–Wendroff slopes. (b) After advection with $\bar{u} \, \Delta t = \Delta x/2$. The dots show the new cell averages Q^{n+1}. Note the overshoot.

Leer in a series of papers, [464] through [468], where he developed the approach known as *MUSCL* (monotonic upstream-centered scheme for conservation laws) for nonlinear conservation laws.

6.7 Total Variation

How much should we limit the slope? Ideally we would like to have a mathematical prescription that will allow us to use the Lax–Wendroff slope whenever possible, for second-order accuracy, while guaranteeing that no nonphysical oscillations will arise. To achieve this we need a way to measure oscillations in the solution. This is provided by the notion of the *total variation* of a function. For a grid function Q we define

$$TV(Q) = \sum_{i=-\infty}^{\infty} |Q_i - Q_{i-1}|. \tag{6.18}$$

For an arbitrary function $q(x)$ we can define

$$TV(q) = \sup \sum_{j=1}^{N} |q(\xi_j) - q(\xi_{j-1})|, \tag{6.19}$$

where the supremum is taken over all subdivisions of the real line $-\infty = \xi_0 < \xi_1 < \cdots < \xi_N = \infty$. Note that for the total variation to be finite, Q or q must approach constant values $q^{\pm}$ as $x \rightarrow \pm\infty$.

Another possible definition for functions is

$$TV(q) = \limsup_{\epsilon \to 0} \frac{1}{\epsilon} \int_{-\infty}^{\infty} |q(x) - q(x - \epsilon)| \, dx. \tag{6.20}$$

If q is differentiable, then this reduces to

$$TV(q) = \int_{-\infty}^{\infty} |q'(x)| \, dx. \tag{6.21}$$

We can use (6.21) also for nondifferentiable functions (distributions) if we interpret $q'(x)$ as the distribution derivative (which includes delta functions at points where q is discontinuous). Note that if we define a function $\tilde{q}(x)$ from a grid function Q using a piecewise constant approximation, then $TV(\tilde{q}) = TV(Q)$.

The true solution to the advection equation simply propagates at speed $\bar{u}$ with unchanged shape, so that the total variation $TV(q(\cdot, t))$ must be constant in time. A numerical solution to the advection equation might not have constant total variation, however. If the method introduces oscillations, then we would expect the total variation of Q^n to *increase* with time. We can thus attempt to avoid oscillations by requiring that the method not increase the total variation:

Definition 6.1. *A two-level method is called* total variation diminishing (TVD) *if, for any set of data Q^n, the values Q^{n+1} computed by the method satisfy*

$$TV(Q^{n+1}) \leq TV(Q^n). \tag{6.22}$$

Note that the total variation need not actually diminish in the sense of decreasing; it may remain constant in time. A better term might be *total variation nonincreasing*. In fact this term (and the abbreviation TVNI) was used in the original work of Harten [179], who introduced the use of this tool in developing and analyzing numerical methods for conservation laws. It was later changed to TVD as a less cumbersome term.

If a method is TVD, then in particular data that is initially monotone, say

$$Q_i^n \geq Q_{i+1}^n \qquad \text{for all } i,$$

will remain monotone in all future time steps. Hence if we discretize a single propagating discontinuity (as in Figure 6.4), the discontinuity may become smeared in future time steps but cannot become oscillatory. This property is especially useful, and we make the following definition.

Definition 6.2. *A method is called* monotonicity-preserving *if*

$$Q_i^n \geq Q_{i+1}^n \qquad \text{for all } i$$

implies that

$$Q_i^{n+1} \geq Q_{i+1}^{n+1} \qquad \text{for all } i.$$

Any TVD method is monotonicity-preserving; see Exercise 6.3.

6.8 TVD Methods Based on the REA Algorithm

How can we derive a method that is TVD? One easy way follows from the reconstruct–evolve–average approach to deriving methods described by Algorithm 4.1. Suppose that we perform the reconstruction in such a way that

$$\text{TV}(\tilde{q}^n(\cdot, t_n)) \leq \text{TV}(Q^n). \tag{6.23}$$

Then the method will be TVD. The reason is that the evolving and averaging steps cannot possibly increase the total variation, and so it is only the reconstruction that we need to worry about.

In the evolve step we clearly have

$$\text{TV}(\tilde{q}^n(\cdot, t_{n+1})) = \text{TV}(\tilde{q}^n(\cdot, t_n)) \tag{6.24}$$

for the advection equation, since $\tilde{q}^n$ simply advects without changing shape. The total variation turns out to be a very useful concept in studying nonlinear problems as well; for we will see later that a wide class of nonlinear scalar conservation laws also have this property, that the true solution has a nonincreasing total variation.

It is a simple exercise (Exercise 6.4) to show that the averaging step gives

$$\text{TV}(Q^{n+1}) \leq \text{TV}(\tilde{q}^n(\cdot, t_{n+1})). \tag{6.25}$$

Combining (6.23), (6.24), and (6.25) then shows that the method is TVD.

6.9 Slope-Limiter Methods

We now return to the derivation of numerical methods based on piecewise linear reconstruction, and consider how to limit the slopes so that (6.23) is satisfied. Note that setting $\sigma_i^n \equiv 0$ works, since the piecewise constant function has the same TV as the discrete data. Hence *the first-order upwind method is TVD* for the advection equation. The upwind method may smear solutions but cannot introduce oscillations.

One choice of slope that gives second-order accuracy for smooth solutions while still satisfying the TVD property is the *minmod slope*

$$\sigma_i^n = \text{minmod}\left(\frac{Q_i^n - Q_{i-1}^n}{\Delta x}, \frac{Q_{i+1}^n - Q_i^n}{\Delta x}\right), \tag{6.26}$$

where the minmod function of two arguments is defined by

$$\text{minmod}(a, b) = \begin{cases} a & \text{if } |a| < |b| \text{ and } ab > 0, \\ b & \text{if } |b| < |a| \text{ and } ab > 0, \\ 0 & \text{if } ab \leq 0. \end{cases} \tag{6.27}$$

If a and b have the same sign, then this selects the one that is smaller in modulus, else it returns zero.

Rather than defining the slope on the ith cell by always using the downwind difference (which would give the Lax–Wendroff method), or by always using the upwind difference (which would give the Beam–Warming method), the minmod method compares the two slopes and chooses the one that is smaller in magnitude. If the two slopes have different sign, then the value Q_i^n must be a local maximum or minimum, and it is easy to check in this case that we must set $\sigma_i^n = 0$ in order to satisfy (6.23).

Figure 6.2(a) shows results using the minmod method for the advection problem considered previously. We see that the minmod method does a fairly good job of maintaining good accuracy in the smooth hump and also sharp discontinuities in the square wave, with no oscillations.

Sharper resolution of discontinuities can be achieved with other limiters that do not reduce the slope as severely as minmod near a discontinuity. Figure 6.5(a) shows some sample data representing a discontinuity smeared over two cells, along with the minmod slopes. Figure 6.5(b) shows that we can increase the slopes in these two cells to *twice* the value of the minmod slopes and still have (6.23) satisfied. This sharper reconstruction will lead to sharper resolution of the discontinuity in the next time step than we would obtain with the minmod slopes.

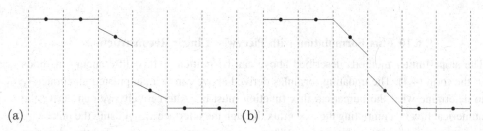

Fig. 6.5. Grid values Q^n and reconstructed $\tilde{q}^n(\cdot, t_n)$ using (a) minmod slopes, (b) superbee or MC slopes. Note that these steeper slopes can be used and still have the TVD property.

One choice of limiter that gives the reconstruction of Figure 6.5(b), while still giving second order accuracy for smooth solutions, is the so-called *superbee limiter* introduced by Roe [378]:

$$\sigma_i^n = \text{maxmod}(\sigma_i^{(1)}, \sigma_i^{(2)}), \tag{6.28}$$

where

$$\sigma_i^{(1)} = \text{minmod}\left(\left(\frac{Q_{i+1}^n - Q_i^n}{\Delta x}\right), \ 2\left(\frac{Q_i^n - Q_{i-1}^n}{\Delta x}\right)\right),$$

$$\sigma_i^{(2)} = \text{minmod}\left(2\left(\frac{Q_{i+1}^n - Q_i^n}{\Delta x}\right), \ \left(\frac{Q_i^n - Q_{i-1}^n}{\Delta x}\right)\right).$$

Each one-sided slope is compared with *twice* the opposite one-sided slope. Then the maxmod function in (6.28) selects the argument with *larger* modulus. In regions where the solution is smooth this will tend to return the *larger* of the two one-sided slopes, but will still be giving an approximation to q_x, and hence we expect second-order accuracy. We will see later that the superbee limiter is also TVD in general.

Figure 6.2(b) shows the same test problem as before but with the superbee method. The discontinuity stays considerably sharper. On the other hand, we see a tendency of the smooth hump to become steeper and squared off. This is sometimes a problem with superbee – by choosing the larger of the neighboring slopes it tends to steepen smooth transitions near inflection points.

Another popular choice is the *monotonized central-difference limiter* (MC limiter), which was proposed by van Leer [467]:

$$\sigma_i^n = \text{minmod}\left(\left(\frac{Q_{i+1}^n - Q_{i-1}^n}{2\Delta x}\right), \ 2\left(\frac{Q_i^n - Q_{i-1}^n}{\Delta x}\right), \ 2\left(\frac{Q_{i+1}^n - Q_i^n}{\Delta x}\right)\right). \tag{6.29}$$

This compares the central difference of Fromm's method with *twice* the one-sided slope to either side. In smooth regions this reduces to the centered slope of Fromm's method and hence does not tend to artificially steepen smooth slopes to the extent that superbee does. Numerical results with this limiter are shown in Figure 6.2(c). The MC limiter appears to be a good default choice for a wide class of problems.

6.10 Flux Formulation with Piecewise Linear Reconstruction

The slope-limiter methods described above can be written as flux-differencing methods of the form (4.4). The updating formulas derived above can be manipulated algebraically to determine what the numerical flux function must be. Alternatively, we can derive the numerical flux by computing the exact flux through the interface $x_{i-1/2}$ using the piecewise linear solution $\tilde{q}^n(x, t)$, by integrating $\bar{u}\tilde{q}^n(x_{i-1/2}, t)$ in time from t_n to t_{n+1}. For the advection equation this is easy to do and we find that

$$F_{i-1/2}^n = \frac{1}{\Delta t} \int_{t_n}^{t_{n+1}} \bar{u}\tilde{q}^n(x_{i-1/2}, t)\, dt$$

$$= \frac{1}{\Delta t} \int_{t_n}^{t_{n+1}} \bar{u}\tilde{q}^n\left(x_{i-1/2} - \bar{u}(t - t_n), t_n\right) dt$$

$$= \frac{1}{\Delta t} \int_{t_n}^{t_{n+1}} \bar{u}\left[Q_{i-1}^n + \left(x_{i-1/2} - \bar{u}(t - t_n) - x_{i-1}\right)\sigma_{i-1}^n\right] dt$$

$$= \bar{u} Q_{i-1}^n + \frac{1}{2}\bar{u}(\Delta x - \bar{u}\,\Delta t)\sigma_{i-1}^n.$$

Using this in the flux-differencing formula (4.4) gives

$$Q_i^{n+1} = Q_i^n - \frac{\bar{u}\,\Delta t}{\Delta x}\left(Q_i^n - Q_{i-1}^n\right) - \frac{1}{2}\frac{\bar{u}\,\Delta t}{\Delta x}(\Delta x - \bar{u}\,\Delta t)\left(\sigma_i^n - \sigma_{i-1}^n\right),$$

which agrees with (6.13).

If we also consider the case $\bar{u} < 0$, then we will find that in general the numerical flux for a slope-limiter method is

$$F_{i-1/2}^n = \begin{cases} \bar{u} Q_{i-1}^n + \frac{1}{2}\bar{u}(\Delta x - \bar{u}\,\Delta t)\sigma_{i-1}^n, & \text{if } \bar{u} \ge 0, \\ \bar{u} Q_i^n - \frac{1}{2}\bar{u}(\Delta x + \bar{u}\,\Delta t)\sigma_i^n & \text{if } \bar{u} \le 0, \end{cases} \qquad (6.30)$$

where σ_i^n is the slope in the ith cell C_i, chosen by one of the formulas discussed previously.

Rather than associating a slope σ_i^n with the ith cell, the idea of writing the method in terms of fluxes between cells suggests that we should instead associate our approximation to q_x with the cell interface at $x_{i-1/2}$ where $F_{i-1/2}^n$ is defined. Across the interface $x_{i-1/2}$ we have a jump

$$\Delta Q_{i-1/2}^n = Q_i^n - Q_{i-1}^n, \qquad (6.31)$$

and this jump divided by Δx gives an approximation to q_x. This suggests that we write the flux (6.30) as

$$F_{i-1/2}^n = \bar{u}^- Q_i^n + \bar{u}^+ Q_{i-1}^n + \frac{1}{2}|\bar{u}|\left(1 - \left|\frac{\bar{u}\,\Delta t}{\Delta x}\right|\right)\delta_{i-1/2}^n, \qquad (6.32)$$

where

$$\delta_{i-1/2}^n = \text{a limited version of } \Delta Q_{i-1/2}^n. \qquad (6.33)$$

If $\delta_{i-1/2}^n$ is the jump $\Delta Q_{i-1/2}^n$ itself, then (6.32) gives the Lax–Wendroff method (see Exercise 6.6). From the form (6.32), we see that the Lax–Wendroff flux can be interpreted as a modification to the upwind flux (4.33). By limiting this modification we obtain a different form of the high-resolution methods, as explored in the next section.

6.11 Flux Limiters

From the above discussion it is natural to view the Lax–Wendroff method as the basic second-order method based on piecewise linear reconstruction, since defining the jump $\delta^n_{i-1/2}$ in (6.33) in the most obvious way as $\Delta Q^n_{i-1/2}$ at the interface $x_{i-1/2}$ results in that method. Other second-order methods have fluxes of the form (6.32) with different choices of $\delta^n_{i-1/2}$. The slope-limiter methods can then be reinterpreted as *flux-limiter methods* by choosing $\delta^n_{i-1/2}$ to be a limited version of (6.31). In general we will set

$$\delta^n_{i-1/2} = \phi(\theta^n_{i-1/2})\Delta Q^n_{i-1/2}, \tag{6.34}$$

where

$$\theta^n_{i-1/2} = \frac{\Delta Q^n_{I-1/2}}{\Delta Q^n_{i-1/2}}. \tag{6.35}$$

The index I here is used to represent the interface on the *upwind* side of $x_{i-1/2}$:

$$I = \begin{cases} i-1 & \text{if } \bar{u} > 0, \\ i+1 & \text{if } \bar{u} < 0. \end{cases} \tag{6.36}$$

The ratio $\theta^n_{i-1/2}$ can be thought of as a measure of the smoothness of the data near $x_{i-1/2}$. Where the data is smooth we expect $\theta^n_{i-1/2} \approx 1$ (except at extrema). Near a discontinuity we expect that $\theta^n_{i-1/2}$ may be far from 1.

The function $\phi(\theta)$ is the *flux-limiter function*, whose value depends on the smoothness. Setting $\phi(\theta) \equiv 1$ for all θ gives the Lax–Wendroff method, while setting $\phi(\theta) \equiv 0$ gives upwind. More generally we might want to devise a limiter function ϕ that has values near 1 for $\theta \approx 1$, but that reduces (or perhaps increases) the slope where the data is not smooth.

There are many other ways one might choose to measure the smoothness of the data besides the variable θ defined in (6.35). However, the framework proposed above results in very simple formulas for the function ϕ corresponding to many standard methods, including all the methods discussed so far.

In particular, note the nice feature that choosing

$$\phi(\theta) = \theta \tag{6.37}$$

results in (6.34) becoming

$$\delta^n_{i-1/2} = \left(\frac{\Delta Q^n_{I-1/2}}{\Delta Q^n_{i-1/2}}\right)\Delta Q^n_{i-1/2} = \Delta Q^n_{I-1/2}.$$

Hence this choice results in the jump at the interface *upwind* from $x_{i-1/2}$ being used to define $\delta^n_{i-1/2}$ instead of the jump at this interface. As a result, the method (6.32) with the choice of limiter (6.37) reduces to the Beam–Warming method.

Since the centered difference (6.14) is the average of the one-sided slopes (6.15) and (6.16), we also find that Fromm's method can be obtained by choosing

$$\phi(\theta) = \frac{1}{2}(1 + \theta). \tag{6.38}$$

Also note that $\phi(\theta) = 2$ corresponds to using $\delta_{i-1/2}^n = 2\,\Delta Q_{i-1/2}^n$, i.e., twice the jump at this interface, while $\phi(\theta) = 2\theta$ results in using twice the jump at the upwind interface. Recall that these are necessary ingredients in some of the slope limiters discussed in Section 6.9. Translating the various slope limiters into flux-limiter functions, we obtain the functions found below for the methods previously introduced.

Linear methods:

$$\begin{aligned}
\text{upwind}: \quad & \phi(\theta) = 0, \\
\text{Lax--Wendroff}: \quad & \phi(\theta) = 1, \\
\text{Beam--Warming}: \quad & \phi(\theta) = \theta, \\
\text{Fromm}: \quad & \phi(\theta) = \frac{1}{2}(1 + \theta).
\end{aligned}$$
(6.39a)

High-resolution limiters:

$$\begin{aligned}
\text{minmod}: \quad & \phi(\theta) = \text{minmod}(1, \theta), \\
\text{superbee}: \quad & \phi(\theta) = \max(0, \min(1, 2\theta), \min(2, \theta)), \\
\text{MC}: \quad & \phi(\theta) = \max(0, \min((1 + \theta)/2, 2, 2\theta)) \\
\text{van Leer}: \quad & \phi(\theta) = \frac{\theta + |\theta|}{1 + |\theta|}.
\end{aligned}$$
(6.39b)

The van Leer limiter listed here was proposed in [465]. A wide variety of other limiters have also been proposed in the literature. Many of these limiters are built into CLAWPACK. The parameter `mthlim` in `claw1ez` (see Section 5.4.6) determines which limiter is used. Other limiters are easily added to the code by modifying the file `claw/clawpack/1d/lib/philim.f`.

The flux-limiter method has the flux (6.32) with $\delta_{i-1/2}^n$ given by (6.34). Let $\nu = \bar{u}\,\Delta t / \Delta x$ be the Courant number. Then the flux-limiter method takes the form

$$\begin{aligned}
Q_i^{n+1} = Q_i^n &- \nu\big(Q_i^n - Q_{i-1}^n\big) \\
&- \frac{1}{2}\nu(1 - \nu)\big[\phi\big(\theta_{i+1/2}^n\big)\big(Q_{i+1}^n - Q_i^n\big) - \phi\big(\theta_{i-1/2}^n\big)\big(Q_i^n - Q_{i-1}^n\big)\big]
\end{aligned}$$
(6.40)

if $\bar{u} > 0$, or

$$\begin{aligned}
Q_i^{n+1} = Q_i^n &- \nu\big(Q_{i+1}^n - Q_i^n\big) \\
&+ \frac{1}{2}\nu(1 + \nu)\big[\phi\big(\theta_{i+1/2}^n\big)\big(Q_{i+1}^n - Q_i^n\big) - \phi\big(\theta_{i-1/2}^n\big)\big(Q_i^n - Q_{i-1}^n\big)\big]
\end{aligned}$$
(6.41)

if $\bar{u} < 0$.

6.12 TVD Limiters

For simple limiters such as minmod, it is clear from the derivation as a slope limiter (Section 6.9) that the resulting method is TVD, since it is easy to check that (6.23) is satisfied. For more complicated limiters we would like to have an algebraic proof that the resulting method is TVD. A fundamental tool in this direction is the following theorem of Harten [179], which can be used to derive explicit algebraic conditions on the function ϕ required for a TVD method. For some other discussions of TVD conditions, see [180], [349], [429], [435], [465].

Theorem 6.1 (Harten). *Consider a general method of the form*

$$Q_i^{n+1} = Q_i^n - C_{i-1}^n(Q_i^n - Q_{i-1}^n) + D_i^n(Q_{i+1}^n - Q_i^n) \qquad (6.42)$$

over one time step, where the coefficients C_{i-1}^n and D_i^n are arbitrary values (which in particular may depend on values of Q^n in some way, i.e., the method may be nonlinear). Then

$$\mathrm{TV}(Q^{n+1}) \le \mathrm{TV}(Q^n)$$

provided the following conditions are satisfied:

$$C_{i-1}^n \ge 0 \quad \forall i,$$
$$D_i^n \ge 0 \quad \forall i, \qquad (6.43)$$
$$C_i^n + D_i^n \le 1 \quad \forall i.$$

Note: the updating formula for Q_i^{n+1} uses C_{i-1}^n and D_i^n, but the last condition involves C_i^n and D_i^n.

For the proof see Exercise 8.5. We can apply this theorem to the flux-limiter method for $q_t + \bar{u}q_x = 0$. We consider the case $\bar{u} > 0$ here (see Exercise 6.7 for the case $\bar{u} < 0$), so that the method has the form (6.40). There are many ways to rewrite this in the form (6.42), since C_{i-1}^n and D_i^n are allowed to depend on Q^n. The obvious choice is

$$C_{i-1}^n = \nu - \frac{1}{2}\nu(1 - \nu)\phi(\theta_{i-1/2}^n),$$
$$D_i^n = -\frac{1}{2}\nu(1 - \nu)\phi(\theta_{i+1/2}^n),$$

but this can't be effectively used to prove the method is TVD, as there is no hope of satisfying the condition (6.43) using this. If $0 \le \nu \le 1$ then $D_i^n < 0$ when ϕ is near 1.

Instead note that

$$Q_{i+1}^n - Q_i^n = (Q_i^n - Q_{i-1}^n)/\theta_{i+1/2}^n,$$

and so the formula (6.40) can be put into the form (6.42) with

$$C_{i-1}^n = \nu + \frac{1}{2}\nu(1 - \nu)\left(\frac{\phi(\theta_{i+1/2}^n)}{\theta_{i+1/2}^n} - \phi(\theta_{i-1/2}^n)\right),$$
$$D_i^n = 0.$$

The conditions (6.43) are then satisfied provided that

$$0 \le C_{i-1}^n \le 1.$$

This in turn holds provided that the CFL condition $0 \le \nu \le 1$ holds, along with the bound

$$\left| \frac{\phi(\theta_1)}{\theta_1} - \phi(\theta_2) \right| \le 2 \qquad \text{for all values of } \theta_1, \theta_2. \tag{6.44}$$

If $\theta \le 0$, then we are at an extremum, and we know from our previous discussion that we should take $\phi(\theta) = 0$ in this case to achieve a TVD method. Also, when $\theta > 0$ we want $\phi(\theta) > 0$, since it generally doesn't make sense to negate the sign of the slope in applying the limiter. Since θ_1 and θ_2 in (6.44) are independent, we then see that we must require

$$0 \le \frac{\phi(\theta)}{\theta} \le 2 \quad \text{and} \quad 0 \le \phi(\theta) \le 2 \tag{6.45}$$

for all values of $\theta \ge 0$ in order to guarantee that condition (6.44) is satisfied (along with $\phi(\theta) = 0$ for $\theta < 0$). These constraints can be rewritten concisely as

$$0 \le \phi(\theta) \le \mathrm{minmod}(2, 2\theta). \tag{6.46}$$

This defines the *TVD region* in the θ–ϕ plane: the curve $\phi(\theta)$ must lie in this region, which is shown as the shaded region in Figure 6.6(a). This figure also shows the functions $\phi(\theta)$

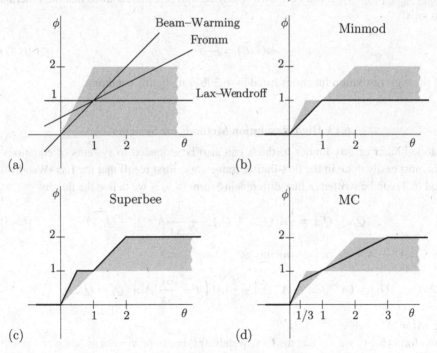

Fig. 6.6. Limiter functions $\phi(\theta)$. (a) The shaded regions shows where function values must lie for the method to be TVD. The second-order linear methods have functions $\phi(\theta)$ that leave this region. (b) The shaded region is the Sweby region of second-order TVD methods. The minmod limiter lies along the lower boundary. (c) The superbee limiter lies along the upper boundary. (d) The MC limiter is smooth at $\phi = 1$.

from (6.39a) for the Lax–Wendroff, Beam–Warming, and Fromm methods. All of these functions lie outside the shaded region for some values of θ, and indeed these methods are not TVD. This graphical analysis of ϕ was first presented by Sweby [429], who analyzed a wide class of flux-limiter methods (for nonlinear conservation laws as well as the advection equation).

Note that for any second-order accurate method we must have $\phi(1) = 1$. Sweby found, moreover, that it is best to take ϕ to be a convex combination of $\phi = 1$ (Lax–Wendroff) and $\phi = \theta$ (Beam–Warming). Other choices apparently give too much compression, and smooth data such as a sine wave tends to turn into a square wave as time evolves, as is already seen to happen with the superbee limiter. Imposing this additional restriction gives the second-order TVD region of Sweby, which is shown in Figure 6.6(b).

The high-resolution limiter functions from (6.39b) are all seen to satisfy the constraints (6.46), and these limiters all give TVD methods. The functions ϕ are graphed in Figure 6.6. Note that minmod lies along the lower boundary of the Sweby region, while superbee lies along the upper boundary. The fact that these functions are not smooth at $\theta = 1$ corresponds to the fact that there is a switch in the choice of one-sided approximation used as θ crosses this point. For full second-order accuracy we would like the function ϕ to be smooth near $\theta = 1$, as for the MC limiter. The van Leer limiter is an even smoother version of this.

We also generally want to impose a symmetry condition on the function $\phi(\theta)$. If the data Q^n is symmetric in x, then we might expect the reconstructed piecewise linear function to have this same property. It can be shown (Exercise 6.8) that this requires that the function ϕ also satisfy

$$\phi(1/\theta) = \frac{\phi(\theta)}{\theta}. \qquad (6.47)$$

All of the high-resolution functions listed in (6.39b) satisfy this condition.

6.13 High-Resolution Methods for Systems

The slope-limiter or flux-limiter methods can also be extended to systems of equations. This is most easily done in the flux-limiter framework. First recall that the Lax–Wendroff method (6.4) can be written in flux-differencing form (4.4) if we define the flux by

$$\mathcal{F}(Q_{i-1}, Q_i) = \frac{1}{2} A(Q_{i-1} + Q_i) - \frac{1}{2} \frac{\Delta t}{\Delta x} A^2 (Q_i - Q_{i-1}). \qquad (6.48)$$

Since $A = A^+ + A^-$, we can rewrite this as

$$\mathcal{F}(Q_{i-1}, Q_i) = (A^+ Q_{i-1} + A^- Q_i) + \frac{1}{2} |A| \left(I - \frac{\Delta t}{\Delta x} |A| \right) (Q_i - Q_{i-1}), \qquad (6.49)$$

where $|A| = A^+ - A^-$.

In the form (6.49), we see that the Lax–Wendroff flux can be viewed as being composed of the upwind flux (4.56) plus a correction term, just as for the scalar advection equation. To define a flux-limiter method we must limit the magnitude of this correction term according to how the data is varying. But for a system of equations, $\Delta Q_{i-1/2} = Q_i - Q_{i-1}$ is a vector, and it is not so clear how to compare this vector with the neighboring jump vector $\Delta Q_{i-3/2}$

or $\Delta Q_{i+1/2}$ to generalize (6.34). It is also not clear which neighboring jump to consider, since the "upwind" direction is different for each eigencomponent. The solution, of course, is that we must decompose the correction term in (6.49) into eigencomponents and limit each scalar eigencoefficient separately, based on the algorithm for scalar advection.

We can rewrite the correction term as

$$\frac{1}{2}|A|\left(I - \frac{\Delta t}{\Delta x}|A|\right)(Q_i - Q_{i-1}) = \frac{1}{2}|A|\left(I - \frac{\Delta t}{\Delta x}|A|\right)\sum_{p=1}^{m}\alpha_{i-1/2}^p r^p,$$

where r^p are the eigenvectors of A and the coefficients $\alpha_{i-1/2}^p$ are defined by (4.38). The flux-limiter method is defined by replacing the scalar coefficient $\alpha_{i-1/2}^p$ by a limited version, based on the scalar formulas of Section 6.11. We set

$$\tilde{\alpha}_{i-1/2}^p = \alpha_{i-1/2}^p \phi(\theta_{i-1/2}^p), \tag{6.50}$$

where

$$\theta_{i-1/2}^p = \frac{\alpha_{I-1/2}^p}{\alpha_{i-1/2}^p} \quad \text{with } I = \begin{cases} i-1 & \text{if } \lambda^p > 0, \\ i+1 & \text{if } \lambda^p < 0, \end{cases} \tag{6.51}$$

and ϕ is one of the limiter functions of Section 6.11. The flux function for the flux-limiter method is then

$$F_{i-1/2} = A^+ Q_{i-1} + A^- Q_i + \tilde{F}_{i-1/2}, \tag{6.52}$$

where the first term is the upwind flux and the correction flux $\tilde{F}_{i-1/2}$ is defined by

$$\tilde{F}_{i-1/2} = \frac{1}{2}|A|\left(1 - \frac{\Delta t}{\Delta x}|A|\right)\sum_{p=1}^{m}\tilde{\alpha}_{i-1/2}^p r^p. \tag{6.53}$$

Note that in the case of a scalar equation, we can take $r^1 = 1$ as the eigenvector of $A = \bar{u}$, so that $\Delta Q_{i-1/2} = \alpha_{i-1/2}^1$, which is what we called $\delta_{i-1/2}$ in Section 6.11. The formula (6.52) then reduces to (6.32). Also note that the flux $\tilde{F}_{i-1/2}$ (and hence $F_{i-1/2}$) depends not only on Q_{i-1} and Q_i, but also on Q_{i-2} and Q_{i+1} in general, because neighboring jumps are used in defining the limited values $\tilde{\alpha}_{i-1/2}^p$ in (6.50). The flux-limiter method thus has a five-point stencil rather than the three-point stencil of the Lax–Wendroff. This is particularly important in specifying boundary conditions (see Chapter 7). Note that this widening of the stencil gives a relaxation of the CFL restriction on the time step. For a five-point method the CFL condition requires only that the Courant number be less than 2. However, the CFL condition gives only a necessary condition on stability, and in fact these high-resolution methods are generally *not* stable for Courant numbers between 1 and 2. The larger stencil does not lead to a greater stability limit because the additional information is used only to limit the second-order correction terms.

Note that $|A|r^p = |\lambda^p|r^p$, so that (6.53) may be rewritten as

$$\tilde{F}_{i-1/2} = \frac{1}{2}\sum_{p=1}^{m}|\lambda^p|\left(1 - \frac{\Delta t}{\Delta x}|\lambda^p|\right)\tilde{\alpha}_{i-1/2}^p r^p. \tag{6.54}$$

This method for a system of equations can also be viewed as a *slope-limiter* method, if we think of $(Q_i - Q_{i-1})/\Delta x$ as approximating the slope vector, with each element giving an approximation to the slope for the corresponding element of q in a piecewise linear reconstruction. One might be tempted to limit the slope for each element of the vector separately, but in fact it is much better to proceed as above and limit the eigencoefficients obtained when the slope vector is expanded in eigenvectors of A. As a result, each wave is limited independently of other families, and the accuracy of smooth waves is not adversely affected by limiters being applied to other waves that may not be smooth.

For these wave-propagation algorithms it is perhaps most natural to view the limiter function as being a *wave limiter* rather than a slope limiter or flux limiter, since it is really the individual waves that are being limited after solving the Riemann problem. This viewpoint carries over naturally to nonlinear systems as introduced in Section 6.15 and explored more fully in Chapter 15. In the more general context, it is natural to use the notation

$$\widetilde{\mathcal{W}}^p_{i-1/2} = \tilde{\alpha}^p_{i-1/2} r^p \tag{6.55}$$

to denote the limited wave appearing in (6.54).

6.14 Implementation

For the constant-coefficient linear system, we could compute the matrices A^+, A^-, and $|A|$ once and for all and compute the fluxes directly from the formulas given above. However, with limiters we must solve the Riemann problem at each interface to obtain a decomposition of $\Delta Q_{i-1/2}$ into waves $\alpha^p_{i-1/2} r^p$ and wave speeds λ^p, and these can be used directly in the computation of Q^{n+1}_i without ever forming the matrices. This approach also generalizes directly to nonlinear systems of conservation laws, where we do not have a single matrix A, but can still solve a Riemann problem at each interface for waves and wave speeds. This generalization is discussed briefly in the next section.

To accomplish this most easily, note that if we use the flux (6.52) in the flux-differencing formula (4.4) and then rearrange the upwind terms, we can write the formula for Q^{n+1}_i as

$$Q^{n+1}_i = Q_i - \frac{\Delta t}{\Delta x}\left(A^+\Delta Q_{i-1/2} + A^-\Delta Q_{i+1/2}\right) - \frac{\Delta t}{\Delta x}\left(\tilde{F}_{i+1/2} - \tilde{F}_{i-1/2}\right),$$

where we drop the superscript n from the current time step because we will need to use superscript p below to denote the wave family. Each of the terms in this expression can be written in terms of the waves $\mathcal{W}^p_{i-1/2} = \alpha^p_{i-1/2} r^p$ and wave speeds λ^p:

$$A^+\Delta Q_{i-1/2} = \sum_{p=1}^{m}(\lambda^p)^+\mathcal{W}^p_{i-1/2},$$

$$A^-\Delta Q_{i-1/2} = \sum_{p=1}^{m}(\lambda^p)^-\mathcal{W}^p_{i-1/2}, \tag{6.56}$$

$$\tilde{F}_{i-1/2} = \frac{1}{2}\sum_{p=1}^{m}|\lambda^p|\left(1 - \frac{\Delta t}{\Delta x}|\lambda^p|\right)\widetilde{\mathcal{W}}^p_{i-1/2}.$$

6.15 Extension to Nonlinear Systems

The full extension of these methods to nonlinear problems will be discussed in Chapter 15 after developing the theory of nonlinear equations. However, the main idea is easily explained as a simple extension of what has been done for linear systems. Given states Q_{i-1} and Q_i, the solution to the Riemann problem will be seen to yield a set of waves $\mathcal{W}_{i-1/2}^p \in \mathbb{R}^m$ and speeds $s_{i-1/2}^p \in \mathbb{R}$, analogous to the linear problem, though now the speeds will vary with i, and so will the directions of the vectors $\mathcal{W}_{i-1/2}^p$ in phase space; they will no longer all be scalar multiples of a single set of eigenvectors r^p.

The quantities $A^+ \Delta Q_{i-1/2}$ and $A^- \Delta Q_{i-1/2}$ have been generalized to *fluctuations* in Chapter 4, denoted by $\mathcal{A}^+ \Delta Q_{i-1/2}$ and $\mathcal{A}^- \Delta Q_{i-1/2}$, with the property that

$$\mathcal{A}^- \Delta Q_{i-1/2} + \mathcal{A}^+ \Delta Q_{i-1/2} = f(Q_i) - f(Q_{i-1}). \tag{6.57}$$

Note that for the linear case $f(Q_i) - f(Q_{i-1}) = A \Delta Q_{i-1/2}$, and this property is satisfied. In general we can think of setting

$$\mathcal{A}^- \Delta Q_{i-1/2} = \sum_{p=1}^m (s_{i-1/2}^p)^- \mathcal{W}_{i-1/2}^p,$$
$$\mathcal{A}^+ \Delta Q_{i-1/2} = \sum_{p=1}^m (s_{i-1/2}^p)^+ \mathcal{W}_{i-1/2}^p, \tag{6.58}$$

a direct extension of (6.56). There are, however, some issues concerned with rarefaction waves and entropy conditions that make the nonlinear case more subtle, and the proper specification of these flux differences will be discussed later. Once the waves, speeds, and flux differences have been suitably defined, the algorithm is virtually identical with what has already been defined in the linear case. We set

$$Q_i^{n+1} = Q_i^n - \frac{\Delta t}{\Delta x}\left(\mathcal{A}^- \Delta Q_{i+1/2} + \mathcal{A}^+ \Delta Q_{i-1/2}\right) - \frac{\Delta t}{\Delta x}\left(\tilde{F}_{i+1/2} - \tilde{F}_{i-1/2}\right), \tag{6.59}$$

where

$$\tilde{F}_{i-1/2} = \frac{1}{2}\sum_{p=1}^m |s_{i-1/2}^p|\left(1 - \frac{\Delta t}{\Delta x}|s_{i-1/2}^p|\right)\widetilde{\mathcal{W}}_{i-1/2}^p. \tag{6.60}$$

Here $\widetilde{\mathcal{W}}_{i-1/2}^p$ represents a limited version of the wave $\mathcal{W}_{i-1/2}^p$, obtained by comparing this jump with the jump $\mathcal{W}_{I-1/2}^p$ in the same family at the neighboring Riemann problem in the upwind direction, so

$$I = \begin{cases} i - 1 & \text{if } s_{i-1/2}^p > 0, \\ i + 1 & \text{if } s_{i-1/2}^p < 0. \end{cases} \tag{6.61}$$

This limiting procedure is slightly more complicated than in the constant-coefficient case, in that $\mathcal{W}_{i-1/2}^p$ and $\mathcal{W}_{I-1/2}^p$ are in general no longer scalar multiples of the same vector r^p. So we cannot simply apply the scalar limiter function $\phi(\theta)$ to the ratio of these scalar coefficients as in the constant-coefficient linear case. Instead we can, for example, project the vector $\mathcal{W}^p_{I-1/2}$ onto $\mathcal{W}^p_{i-1/2}$ and compare the length of this projection with the length of $\mathcal{W}^p_{i-1/2}$. This same issue arises for variable-coefficient linear systems and is discussed in Section 9.13.

When no limiting is used, this method is formally second-order accurate provided certain conditions are satisfied by the Riemann solution used; see Section 15.6. The methods generally perform much better, however, when limiters are applied.

6.16 Capacity-Form Differencing

In many applications the system of conservation laws to be solved most naturally takes the form

$$\kappa(x)q_t + f(q)_x = 0, \tag{6.62}$$

where $\kappa(x)$ is a spatially varying capacity function as introduced in Section 2.4. In the remainder of this chapter we will see how to extend the high-resolution methods defined above to this situation. In particular, this allows us to extend the methods to nonuniform grids as we do in Section 6.17. This material can be skipped without loss of continuity.

The equation (6.62) generally arises from an integral conservation law of the form

$$\frac{d}{dt} \int_{x_1}^{x_2} \kappa(x)q(x,t)\,dx = f(q(x_1,t)) - f(q(x_2,t)), \tag{6.63}$$

in which $\kappa(x)q(x,t)$ is the conserved quantity while the flux at a point depends most naturally on the value of q itself. It may then be most natural to solve Riemann problems corresponding to the flux function $f(q)$, i.e., by solving the equation

$$q_t + f(q)_x = 0 \tag{6.64}$$

locally at each cell interface, and then use the resulting wave structure to update the solution to (6.62). This suggests a method of the form

$$\kappa_i Q_i^{n+1} = \kappa_i Q_i^n - \frac{\Delta t}{\Delta x}\left[\mathcal{F}(Q_i^n, Q_{i+1}^n) - \mathcal{F}(Q_{i-1}^n, Q_i^n)\right]. \tag{6.65}$$

Dividing by κ_i gives

$$Q_i^{n+1} = Q_i^n - \frac{\Delta t}{\kappa_i \Delta x}\left[\mathcal{F}(Q_i^n, Q_{i+1}^n) - \mathcal{F}(Q_{i-1}^n, Q_i^n)\right]. \tag{6.66}$$

The form of equation (6.66) will be called *capacity-form differencing*. It is a generalization of conservation-form differencing to include the capacity function. The factor $\kappa_i \Delta x$ appearing in (6.66) has a natural interpretation as the effective volume of the ith grid cell if we think of κ_i as a fraction of the volume available to the fluid (as in porous-media flow, for example). Note that the updating formula (6.66) has the advantage of being conservative in the proper manner for the conservation law (6.63), since computing

$$\Delta x \sum_i \kappa_i Q_i^{n+1} = \Delta x \sum_i \kappa_i Q_i^n + \text{boundary fluxes}$$

from (6.65) shows that all the fluxes cancel except at the boundaries, where the boundary conditions must come into play.

We can rewrite the method (6.66) in the framework of the high-resolution wave-propagation algorithm as

$$Q_i^{n+1} = Q_i^n - \frac{\Delta t}{\kappa_i \Delta x}\left(\mathcal{A}^+\Delta Q_{i-1/2}^n + \mathcal{A}^-\Delta Q_{i+1/2}^n\right) - \frac{\Delta t}{\kappa_i \Delta x}\left(\tilde{F}_{i+1/2}^n - \tilde{F}_{i-1/2}^n\right), \quad (6.67)$$

where as usual $\mathcal{A}^\pm\Delta Q_{i-1/2}$ are the fluctuations and $\tilde{F}_{i-1/2}$ is the correction flux based on the Riemann solution for (6.64) at the interface $x_{i-1/2}$. We now use the correction flux

$$\tilde{F}_{i-1/2} = \frac{1}{2}\sum_{p=1}^{M_w}\left(1 - \frac{\Delta t}{\kappa_{i-1/2}\Delta x}|s_{i-1/2}^p|\right)|s_{i-1/2}^p|\tilde{\mathcal{W}}_{i-1/2}^p. \quad (6.68)$$

This is the general form for a system of m equations with M_w waves, as introduced in Section 6.15. The wave $\tilde{\mathcal{W}}_{i-1/2}^p$ is a limited version of the wave $\mathcal{W}_{i-1/2}^p$, just as before, and these waves again are obtained by solving the Riemann problem for (6.64), ignoring the capacity function. The only modification from the formula for $\tilde{F}_{i-1/2}$ given in (6.60) is that Δx is replaced by $\kappa_{i-1/2}\Delta x$, where

$$\kappa_{i-1/2} = \frac{1}{2}(\kappa_{i-1} + \kappa_i). \quad (6.69)$$

Clearly, replacing Δx by some form of $\kappa\Delta x$ is necessary on the basis of dimensional arguments. If $\kappa(x)$ is smoothly varying, then using either $\kappa_{i-1}\Delta x$ or $\kappa_i\Delta x$ would also work, and the resulting method is second-order accurate for smooth solutions with any of these choices.

The choice $\kappa_{i-1/2}$ seems most reasonable, however, since the flux $\tilde{F}_{i-1/2}$ is associated with the interface between cells $\mathcal{C}_{i-1}$ and $\mathcal{C}_i$, rather than with either of the two cells. This term in the correction flux gives an approximation to the $\frac{1}{2}\Delta t^2 q_{tt}$ term in the Taylor series expansion once it is inserted in the flux-differencing term of (6.67). We compute, for smooth solutions, that

$$q_{tt} = \frac{1}{\kappa}\left(\frac{1}{\kappa}[f'(q)]^2 q_x\right)_x,$$

and centering of these terms again suggests that the inner $1/\kappa$ should be evaluated at $x_{i\pm1/2}$. In Section 6.17.1, where nonuniform grids are discussed, we will see that this choice makes clear physical sense in that case, and is correct even if κ is not smoothly varying.

6.17 Nonuniform Grids

One natural example of a capacity function arises if we solve a standard conservation law $q_t + f(q)_x = 0$ on a nonuniform grid such as shown on the right of Figure 6.7. There are two philosophically different approaches that could be taken to derive a numerical method on this grid:

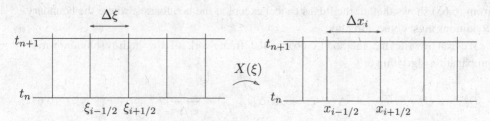

Fig. 6.7. The grid mapping $X(\xi)$ maps a uniform computational grid in ξ–t space (on the left) to the nonuniform x–t grid in physical space (on the right).

1. Work directly in the *physical space* (the right side of Figure 6.7), and derive a finite volume method on the nonuniform gird for the integral form of the conservation law on the physical-grid cells.
2. View the nonuniform grid as resulting from some coordinate mapping applied to a uniform grid in *computational space*. Such a grid is illustrated on the left side of Figure 6.7, where $\Delta\xi$ is constant and the mapping function $X(\xi)$ defines $x_{i-1/2} = X(\xi_{i-1/2})$. If we can transform the equation in x and t to an equivalent equation in ξ and t, then we can solve the transformed equation on the uniform grid.

The first approach is generally easier to use in developing finite volume methods for conservation laws, for several reasons:

- The transformed equation in computational space includes *metric terms* involving $X'(\xi)$, as we will see below. The mapping must be smooth in order to achieve good accuracy if we attempt to discretize the transformed equations directly. In practice we might want to use a highly nonuniform grid corresponding to a nonsmooth function $X(\xi)$.
- Even if $X(\xi)$ is smooth, it may not be easy to discretize the transformed equation in ξ in a way that exactly conserves the correct physical quantities in x. By contrast, a finite volume method in the physical space automatically achieves this if we write it in terms of fluxes at the cell edges.
- Using a Godunov-type method requires solving a Riemann problem at each cell interface, and the transformed equations lead to a transformed Riemann problem. It is often simpler to solve the original, physically meaningful Riemann problem in x.

For these reasons we will derive finite volume methods in the physical domain. However, once the method has been derived, we will see that we can then view it as a discretization of the transformed equation. This viewpoint is useful in implementing the methods, as we can then express the method as a finite volume method with a capacity function applied on the uniform ξ-grid. As we will see, the capacity κ_i of the ith grid cell $[\xi_{i-1/2}, \xi_{i+1/2}]$ is then the ratio of the physical cell length $\Delta x_i \equiv (x_{i+1/2} - x_{i-1/2})$ to $\Delta\xi$, which is a natural measure of the capacity of the computational cell. For a smooth mapping $X(\xi)$ this can be viewed as an approximation to the capacity function $\kappa(x) \equiv X'(\xi)$, the Jacobian of the mapping, but the finite volume method remains valid and accurate even if the mapping is not smooth.

In Chapter 23, we will see that this approach extends naturally to quadrilateral grids in more than one space dimension. Again we will derive finite volume methods in the irregular physical grid cells, but implement the methods using capacity-form differencing on a uniform rectangular grid.

We now derive a finite volume method in the physical domain. We have

$$
\int_{x_{i-1/2}}^{x_{i+1/2}} q(x, t_{n+1}) \, dx = \int_{x_{i-1/2}}^{x_{i+1/2}} q(x, t_n) \, dx
$$
$$
- \left(\int_{t_n}^{t_{n+1}} f\left(q\left(x_{i+1/2}, t\right)\right) dt - \int_{t_n}^{t_{n+1}} f\left(q\left(x_{i-1/2}, t\right)\right) dt \right),
$$

(6.70)

which suggests the finite volume method

$$
Q_i^{n+1} = Q_i^n - \frac{\Delta t}{\Delta x_i} \left[\mathcal{F}(Q_i^n, Q_{i+1}^n) - \mathcal{F}(Q_{i-1}^n, Q_i^n) \right],
$$

(6.71)

where $\Delta x_i = x_{i+1/2} - x_{i-1/2}$ is the width of the ith cell and

$$
Q_i^n \approx \frac{1}{\Delta x_i} \int_{x_{i-1/2}}^{x_{i+1/2}} q(x, t_n) \, dx.
$$

The numerical flux $\mathcal{F}(Q_{i-1}^n, Q_i^n)$ should be, as usual, an approximation to

$$
\frac{1}{\Delta t} \int_{t_n}^{t_{n+1}} f\left(q\left(x_{i-1/2}, t\right)\right) dt.
$$

For Godunov's method, this is determined simply by solving the Riemann problem and evaluating the flux along $x/t = 0$. The fact that the grid is nonuniform is immaterial in computing the Godunov flux. The nonuniformity of the grid does come into the second-order correction terms, however, since approximating the slopes q_x with cell values requires paying attention to the grid spacing. This is discussed in Section 6.17.1.

Next we will see that the method (6.71) can be reinterpreted as a method on the uniform computational grid. Let

$$
\kappa_i = \Delta x_i / \Delta \xi,
$$

where $\Delta \xi$ is the uniform cell size in the computational grid shown on the left in Figure 6.7. Then the method (6.71) can be written as

$$
Q_i^{n+1} = Q_i^n - \frac{\Delta t}{\kappa_i \Delta \xi} \left[\mathcal{F}(Q_i^n, Q_{i+1}^n) - \mathcal{F}(Q_{i-1}^n, Q_i^n) \right].
$$

(6.72)

Let

$$
\bar{q}(\xi, t) = q(X(\xi), t).
$$

If we now assume that the coordinate mapping $X(\xi)$ is differentiable, then the change of variables $x = X(\xi)$, for which $dx = X'(\xi) \, d\xi$, gives

$$
\int_{x_{i-1/2}}^{x_{i+1/2}} q(x, t) \, dx = \int_{\xi_{i-1/2}}^{\xi_{i+1/2}} q(X(\xi), t) \, X'(\xi) \, d\xi
$$
$$
= \int_{\xi_{i-1/2}}^{\xi_{i+1/2}} \bar{q}(\xi, t) \kappa(\xi) \, d\xi,
$$

where the capacity function $\kappa(\xi)$ is the Jacobian $X'(\xi)$, as expected. Hence the conservation law

$$\frac{d}{dt} \int_{x_{i-1/2}}^{x_{i+1/2}} q(x,t)\,dx = f\big(q\big(x_{i-1/2},t\big)\big) - f\big(q\big(x_{i+1/2},t\big)\big)$$

is transformed into

$$\frac{d}{dt} \int_{\xi_{i-1/2}}^{\xi_{i+1/2}} \bar{q}(\xi,t)\kappa(\xi)\,d\xi = f\big(\bar{q}\big(\xi_{i-1/2},t\big)\big) - f\big(\bar{q}\big(\xi_{i+1/2},t\big)\big).$$

This has the form of the integral conservation law (6.63) in the computational domain. The finite volume method (6.71), when rewritten as (6.72), can be viewed as a capacity-form differencing method on the uniform ξ-grid.

We can also interpret Q_i^n as an approximation to a cell average of $\bar{q}$ over the computational domain:

$$Q_i^n \approx \frac{1}{\Delta x_i} \int_{x_{i-1/2}}^{x_{i+1/2}} q(x,t_n)\,dx$$

$$= \frac{1}{\kappa_i\,\Delta\xi} \int_{\xi_{i-1/2}}^{\xi_{i+1/2}} \bar{q}(\xi,t_n)\kappa(\xi)\,d\xi$$

$$\approx \frac{1}{\Delta\xi} \int_{\xi_{i-1/2}}^{\xi_{i+1/2}} \bar{q}(\xi,t_n)\,d\xi. \tag{6.73}$$

6.17.1 High-Resolution Corrections

One way to derive high-resolution methods for the advection equation $q_t + \bar{u}q_x = 0$ (with $\bar{u} > 0$, say) on a nonuniform grid is to use the REA algorithm approach from Section 6.4, which is easily extended to nonuniform grids. Given cell averages, we reconstruct a piecewise linear function on each of the grid cells, evolve the advection equation exactly by shifting this function over a distance $\bar{u}\,\Delta t$, and average onto the grid cells to obtain new cell averages. If the slopes are all chosen to be $\sigma_i^n = 0$, then this reduces to the upwind method

$$Q_i^{n+1} = Q_i^n - \frac{\bar{u}\,\Delta t}{\kappa_i\,\Delta\xi}\big(Q_i^n - Q_{i-1}^n\big),$$

which has the form (6.72) with $\mathcal{F}(Q_{i-1}^n, Q_i^n) = \bar{u}Q_{i-1}^n$. With nonzero slopes we obtain

$$Q_i^{n+1} = Q_i^n - \frac{\bar{u}\,\Delta t}{\kappa_i\,\Delta\xi}\big(Q_i^n - Q_{i-1}^n\big)$$

$$- \frac{1}{2}\frac{\bar{u}\,\Delta t}{\kappa_i\,\Delta\xi}\Big[(\kappa_i\,\Delta\xi - \bar{u}\,\Delta t)\sigma_i^n - (\kappa_{i-1}\,\Delta\xi - \bar{u}\,\Delta t)\sigma_{i-1}^n\Big]. \tag{6.74}$$

Recall that on a uniform grid the Lax–Wendroff method is obtained by taking slopes $\sigma_{i-1}^n = (Q_i^n - Q_{i-1}^n)/\Delta x$. On a nonuniform grid the distance between cell centers is

$\kappa_{i-1/2} \Delta\xi$, where $\kappa_{i-1/2} = \frac{1}{2}(\kappa_{i-1} + \kappa_i)$, and the natural generalization of the Lax–Wendroff method is obtained by setting

$$\sigma_{i-1}^n = \frac{Q_i^n - Q_{i-1}^n}{\kappa_{i-1/2}\,\Delta\xi}. \tag{6.75}$$

This corresponds to a correction flux

$$\tilde{F}_{i-1/2} = \frac{1}{2}\left(\frac{\kappa_{i-1}}{\kappa_{i-1/2}} - \frac{\bar{u}\,\Delta t}{\kappa_{i-1/2}\,\Delta\xi}\right)\bar{u}\left(Q_i^n - Q_{i-1}^n\right). \tag{6.76}$$

The natural generalization of this to systems of equations gives a correction flux similar to (6.68) but with the 1 replaced by $\kappa_i/\kappa_{i-1/2}$:

$$\tilde{F}_{i-1/2} = \frac{1}{2}\sum_{p=1}^{M_w}\left(\frac{\kappa_{i-1}}{\kappa_{i-1/2}} - \frac{\Delta t}{\kappa_{i-1/2}\,\Delta\xi}\left|s_{i-1/2}^p\right|\right)\left|s_{i-1/2}^p\right|\widetilde{\mathcal{W}}_{i-1/2}^p. \tag{6.77}$$

For smooth κ this is essentially the same as (6.68), but for nonsmooth κ (6.77) might give better accuracy.

Exercises

6.1. Verify that (6.13) results from integrating the piecewise linear solution $\tilde{q}^n(x, t_{n+1})$.

6.2. Compute the total variation of the functions

(a)

$$q(x) = \begin{cases} 1 & \text{if } x < 0, \\ \sin(\pi x) & \text{if } 0 \le x \le 3, \\ 2 & \text{if } x > 3, \end{cases}$$

(b)

$$q(x) = \begin{cases} 1 & \text{if } x < 0 \text{ or } x = 3, \\ 1 & \text{if } 0 \le x \le 1 \text{ or } 2 \le x < 3, \\ -1 & \text{if } 1 < x < 2, \\ 2 & \text{if } x > 3. \end{cases}$$

6.3. Show that any TVD method is monotonicity-preserving. (But note that the converse is not necessarily true: a monotonicity-preserving method may not be TVD on more general data.)

6.4. Show that (6.25) is valid by showing that, for any function $q(x)$, if we define discrete values Q_i by averaging $q(x)$ over grid cells, then $\text{TV}(Q) \le \text{TV}(q)$. Hint: Use the definition (6.19) and the fact that the average value of q lies between the maximum and minimum values on each grid cell.

6.5. Show that the *minmod* slope guarantees that (6.23) will be satisfied in general, and hence the *minmod* method is TVD.

6.6.   Show that taking

$$\delta_{i-1/2}^n = Q_i^n - Q_{i-1}^n$$

in (6.32) corresponds to using the downwind slope for σ in both cases $\bar{u} > 0$ and $\bar{u} < 0$, and that the resulting flux gives the Lax–Wendroff method.

6.7.   Show that if $\bar{u} < 0$ we can apply Theorem 6.1 to the flux-limiter method (6.41) by choosing

$$C_{i-1}^n = 0,$$

$$D_i^n = -v + \frac{1}{2}v(1 + v)\left(\phi(\theta_{i-1/2}^n) - \frac{\phi(\theta_{i+1/2}^n)}{\theta_{i+1/2}^n}\right)$$

in order to show that the method is TVD provided $-1 \le v \le 0$ and the bound (6.44) holds. Hence the same restrictions on limiter functions are found in this case as discussed in Section 6.12.

6.8.   Verify that (6.47) is required for symmetry.

6.9.   Verify that (6.48) and (6.49) are equivalent and yield the Lax–Wendroff method.

6.10.   Plot the van Leer limiter from (6.39b), and verify that it lies in the Sweby region shown in Figure 6.6.

Boundary Conditions and Ghost Cells

So far we have only studied methods for updating the cell average Q_i^n assuming that we have neighboring cell values Q_{i-1}^n and Q_{i+1}^n and perhaps values further away as needed in order to compute the fluxes $F_{i-1/2}^n$ and $F_{i+1/2}^n$. In practice we must always compute on some finite set of grid cells covering a bounded domain, and in the first and last cells we will not have the required neighboring information. Instead we have some set of physical boundary conditions, as discussed in Section 3.11, that must be used in updating these cell values.

One approach is to develop special formulas for use near the boundaries, which will depend both on what type of boundary conditions are specified and on what sort of method we are trying to match. However, in general it is much easier to think of extending the computational domain to include a few additional cells on either end, called *ghost cells*, whose values are set at the beginning of each time step in some manner that depends on the boundary conditions and perhaps the interior solution.

Figure 7.1 shows a grid with two ghost cells at each boundary. These values provide the neighboring-cell values needed in updating the cells near the physical domain. The updating formula is then exactly the same in all cells, and there is no need to develop a special flux-limiter method, say, that works with boundary data instead of initial data. Instead the boundary conditions must be used in deciding how to set the values of the ghost cells, but this can generally be done in a way that depends only on the boundary conditions and is decoupled entirely from the choice of numerical method that is then applied.

Suppose the problem is on the physical domain $[a, b]$, which is subdivided into cells $C_1, C_2, \ldots, C_N$ with $x_1 = a$ and $x_{N+1} = b$, so that $\Delta x = (b - a)/N$. If we use a method for which $F_{i-1/2}^n$ depends only on Q_{i-1}^n and Q_i^n, then we need only one ghost cell on either end. The ghost cell $C_0 = (a - \Delta x, a)$ allows us to calculate the flux $F_{1/2}$ at the left boundary while the ghost cell $C_{N+1} = (b, b + \Delta x)$ is used to calculate $F_{N+1/2}^n$ at $x = b$. With a flux-limiter method of the type developed in Chapter 6, we will generally need two ghost cells at each boundary, since, for example, the jump $Q_0 - Q_{-1}$ will be needed in limiting the flux correction in $F_{1/2}$. For a method with an even wider stencil, additional ghost cells would be needed. In general we might have m_{BC} ghost cells at each side.

We will refer to the solution in the original domain $[a, b]$ as the *interior solution*; it is computed in each time step by the numerical method. At the start of each time step we have the interior values $Q_1^n, \ldots, Q_N^n$ obtained from the previous time step (or from the initial conditions if $n = 0$), and we apply a *boundary-condition procedure* to fill the ghost cells with values Q_i^n for $i = 1 - m_{BC}, \ldots, 0$ and $i = N + 1, \ldots, N + m_{BC}$ before applying the

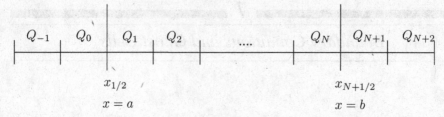

Fig. 7.1. The computational domain $[a, b]$ is extended to a set of ghost cells for specifying boundary conditions.

method on the next time step. In CLAWPACK this is done in a subroutine bc1.f that is called at the beginning of each time step. The default routine claw/clawpack/1d/lib/bc1 implements most of the particular boundary conditions discussed in this chapter. We will look at several examples to see how the ghost-cell values might be set in order to implement various physical boundary conditions. In general we will discuss the boundary conditions for the case $m_{\mathrm{BC}} = 2$, but they can easily be extended to larger values if necessary.

7.1 Periodic Boundary Conditions

Periodic boundary conditions $q(a, t) = q(b, t)$ are very easy to impose with any numerical method. In updating Q_1 we need values Q_0 to the left and Q_2 to the right (for a three-point method). By periodicity the value Q_0 should agree with the value Q_N in the last cell. One could code the formula for updating Q_1 specially to use Q_N in place of the value Q_{i-1} that would normally be used for $i > 1$, but it is simpler to use the ghost-cell approach and simply set $Q_0^n = Q_N^n$ before computing fluxes and updating the cell values, so that the same formula can then be used everywhere. With a five-point stencil we need to fill two ghost cells at each boundary, and we set

$$Q_{-1}^n = Q_{N-1}^n, \quad Q_0^n = Q_N^n, \quad Q_{N+1}^n = Q_1^n, \quad Q_{N+2}^n = Q_2^n \qquad (7.1)$$

at the start of each time step. These boundary conditions are implemented in the CLAWPACK routine claw/clawpack/1d/lib/bc1.f and invoked by setting mthbc(1) = 2 and mthbc(2) = 2.

7.2 Advection

Consider the advection equation on $[a, b]$ with $\bar{u} > 0$ and the boundary condition

$$q(a, t) = g_0(t), \qquad (7.2)$$

where $g_0(t)$ is a given function. Recall from Section 3.11 that we cannot specify a boundary condition at $x = b$. We may need to specify a boundary condition at $x = b$ for the numerical method, however, if the stencil for computing Q_i^{n+1} involves points to the right of x_i.

7.2.1 Outflow Boundaries

First we consider the outflow boundary at $x = b$. If we use a one-sided method such as upwind or Beam–Warming, then we do not need any numerical boundary condition. If we implement the method using ghost cells, then we can assign arbitrary values to the ghost cells on the right with no effect on the interior solution, since these values will never be used.

If we use a three-point method such as Lax–Wendroff that does use values to the right, then some numerical boundary conditions must be specified. We must in general take some care in how we specify these to obtain a stable and accurate method.

One possibility is to use a fully upwind method at the rightmost point, together with the Lax–Wendroff method at all other points. This works quite well. Note, however, that the Lax–Wendroff method allows information to propagate from right to left, even though the exact solution to the advection equation does not. So there is the possibility that noise generated at the right boundary by this switch in method will propagate back into the domain and contaminate the solution elsewhere, and perhaps even cause an instability. The Lax–Wendroff method is highly dissipative for left-going waves, and so this does not cause a noticeable problem, and one can prove that the resulting method is stable. Indeed, a similar change in method occurs frequently with flux-limiter methods, with only slight loss in accuracy.

In general, the theory of stability for the IBVP is much more subtle and difficult than for the Cauchy problem. See, for example, [427], [459] for an introduction to this topic.

Rather than explicitly switching to a different formula at the right boundary, we can achieve the same effect by setting ghost cell values by *extrapolation* from the interior solution. If the ghost-cell value Q^n_{N+1} is set based on $Q^n_N, Q^n_{N-1}, \ldots$, then the new value Q^{n+1}_N will effectively be computed on the basis of values to the left alone, even if the formula depends on Q^n_{N+1}, and hence this reduces to some sort of upwind method. The simplest approach is to use a *zero-order* extrapolation, meaning extrapolation by a constant function. We simply set

$$Q^n_{N+1} = Q^n_N, \qquad Q^n_{N+2} = Q^n_N \tag{7.3}$$

at the start of each time step. Then we have $\Delta Q^n_{N+1} = 0$ as the value $\delta^n_{N+1/2}$ used in the flux modification to $F^n_{N+1/2}$ in (6.32). This flux then reduces to the upwind flux,

$$F^n_{N+1/2} = \bar{u} Q^n_N,$$

since $\bar{u} > 0$. Note that the method may not reduce to the standard first-order upwind method in the last cell, however, since $\delta^n_{N-1/2}$ ($= \Delta Q^n_{N-1/2} = Q^n_N - Q^n_{N-1}$ for the Lax–Wendroff method) may be nonzero. But the resulting method behaves in roughly the same manner.

One might also consider *first-order* extrapolation, based on fitting a linear function through the interior solution. This gives

$$Q^n_{N+1} = Q^n_N + (Q^n_N - Q^n_{N-1}) = 2Q^n_N - Q^n_{N-1}. \tag{7.4}$$

In this case $\Delta Q^n_{N+1/2} = \Delta Q^n_{N-1/2}$ and the correction terms in $F^n_{N-1/2}$ and $F^n_{N+1/2}$ will cancel out (assuming the limiter function satisfies $\phi(1) = 1$). Now the update for Q^{n+1}_n does reduce to the standard first-order upwind method.

The idea of extrapolation at outflow boundaries turns out to be extremely powerful more generally, even for systems of equations where there are both incoming and outgoing characteristics. Often we have artificial computational boundaries that arise simply because we can only solve the problem on a bounded domain. At such boundaries we often want to have no incoming signal, though there may be outgoing waves that should leave the domain cleanly without generating spurious reflections at the artificial boundary. Zero-order extrapolation, coupled with the methods we are studying that perform characteristic decomposition in the solution to each Riemann problem, is often a very effective way to accomplish this. This is discussed in Section 7.3.1 for linear acoustics and will be investigated later for nonlinear and multidimensional problems. First-order extrapolation might seem even better, but can lead to stability problems and is not recommended in general.

Zero-order extrapolation boundary conditions are implemented in CLAWPACK as one of the options that can be automatically invoked when using claw1ez as described in Chapter 5. The ghost-cell values are set in claw/clawpack/1d/lib/bc1.f and invoked by setting mthbc(i) = 1, where i=1 for the left boundary or i=2 for the right boundary. (See Section 5.4.6.)

7.2.2 Inflow Boundary Conditions

Now consider the inflow boundary at $x = a$ for the advection equation, where we specify the boundary condition (7.2). One approach would be to compute the exact flux $F_{1/2}^n$ by integrating along the boundary,

$$F_{1/2}^n = \frac{1}{\Delta t} \int_{t_n}^{t_{n+1}} \bar{u} q(a, t) \, dt$$

$$= \frac{\bar{u}}{\Delta t} \int_{t_n}^{t_{n+1}} g_0(t) \, dt, \tag{7.5}$$

and use this in the flux-differencing formula for Q_1^{n+1}. Alternatively, a second-order accurate approximation such as

$$F_{1/2}^n = \bar{u} g_0(t_n + \Delta t/2) \tag{7.6}$$

could be used instead.

Note that for a five-point method we also need a special formula for the flux $F_{3/2}^n$, which might be more difficult to compute by this approach.

Again, for generality it is often simpler instead to find a way to set the ghost-cell values Q_0^n (and perhaps Q_{-1}^n) so that the same interior method can be used at all points. For the advection equation we can easily compute sensible values using our knowledge of the exact solution. We would like to set

$$Q_0^n = \frac{1}{\Delta x} \int_{a-\Delta x}^{a} q(x, t_n) \, dx.$$

Of course the true solution isn't even defined for $x < a$, but we can easily extend the solution past the boundary using our knowledge of characteristics, setting

$$q(x, t_n) = q\left(a, t_n + \frac{a-x}{\bar{u}}\right)$$

$$= g_0\left(t_n + \frac{a-x}{\bar{u}}\right). \tag{7.7}$$

Then

$$Q_0^n = \frac{1}{\Delta x} \int_{a-\Delta x}^{a} g_0\left(t_n + \frac{a-x}{\bar{u}}\right) dx$$

$$= \frac{\bar{u}}{\Delta x} \int_{t_n}^{t_n + \Delta x/\bar{u}} g_0(\tau) \, d\tau. \tag{7.8}$$

Again we could approximate this integral by the second-order midpoint approximation, obtaining

$$Q_0^n = g_0\left(t_n + \frac{\Delta x}{2\bar{u}}\right), \tag{7.9}$$

which is also what we would get if we simply evaluated the cell-center value $q(a - \Delta x/2, t_n)$ using (7.7). Similarly, for the second ghost cell we could set

$$Q_{-1}^n = g_0\left(t_n + \frac{3\,\Delta x}{2\bar{u}}\right).$$

Specifying such boundary conditions in CLAWPACK requires modifying the bc1 routine to set the ghost-cell values appropriately. See [claw/book/chap7/advinflow] for an example.

7.3 Acoustics

Similar ideas can be applied to develop boundary-condition procedures for systems of equations. The situation may be complicated by the fact that the system can have both positive and negative eigenvalues, so that each boundary will typically have both incoming and outgoing characteristics. Other sorts of physical boundary conditions, such as a solid wall where waves are expected to reflect, must also be studied. Many of these issues can be illustrated for the simple case of linear acoustics as developed in Section 2.7, a linear system of two equations with one eigenvalue of each sign. The boundary-condition procedures developed here will later be seen to extend very easily to nonlinear systems such as the full Euler equations of compressible flow.

We consider the acoustics equations from (2.52),

$$p_t + K_0 u_x = 0$$
$$\rho_0 u_t + p_x = 0, \tag{7.10}$$

with a variety of different boundary conditions. The characteristic variables for this system are

$$w^1(x, t) = \frac{1}{2Z_0}(-p + Z_0 u),$$

$$w^2(x, t) = \frac{1}{2Z_0}(p + Z_0 u), \tag{7.11}$$

where Z_0 is the impedance, as derived in Section 2.8.

7.3.1 Nonreflecting Boundary Conditions

Suppose we wish to solve the *Cauchy problem* with initial data $\mathring{u}(x)$ and $\mathring{p}(x)$ that vary with x only in some region $a_1 < x < b_1$, and are constant outside this interval, say

$$(p, u) = \begin{cases} (p_L, u_L) & \text{if } x < a_1, \\ (p_R, u_R) & \text{if } x > b_1. \end{cases}$$

We know from Chapter 3 that the solution to the Cauchy problem will consist of two waves, each with fixed shape depending on $\mathring{u}$ and $\mathring{p}$. One propagates to the left with speed $-c_0$ and the other to the right with speed c_0. Eventually these waves will separate completely from one another (certainly for times $t > (b_1 - a_1)/c_0$) and leave a new constant state in between. Note that the characteristic variables (7.11) satisfy

$$w^2(a_1, t) = \frac{1}{2Z_0}(p_L + Z_0 u_L) \qquad \text{for all } t \geq 0,$$

$$w^1(b_1, t) = \frac{1}{2Z_0}(-p_R + Z_0 u_R) \quad \text{for all } t \geq 0. \tag{7.12}$$

If we now want to compute this solution numerically, we must choose some finite domain $a \leq x \leq b$ on which to perform the computation, and we will suppose this domain includes the interval where the data is specified, i.e., $a < a_1 < b_1 < b$. For short times the solution should remain constant at the boundaries a and b, but eventually the acoustic waves should pass out through these boundaries. If we want to compute over longer times, then we must specify the boundary conditions in such a way that the waves leave the region cleanly, without generating any signal in the incoming characteristic variable propagating back into the computational domain.

The points $x = a$ and $x = b$ are artificial boundaries, in the sense that they are only introduced to make the computational domain finite, and do not correspond to physical boundaries. Hence we want to impose *nonreflecting boundary conditions* at these boundaries that do not introduce spurious reflections of the outgoing acoustic waves. Such boundary conditions are also called *absorbing boundary conditions*, since they are supposed to completely absorb any wave that hits them. Such boundary conditions are extremely important in many practical applications. We often wish to model what is happening in a small portion of physical space that must be truncated artificially at some point. Consider the problem of modeling seismic waves in the earth, or flow around an airplane. We cannot hope to model the entire earth or atmosphere, nor should we need to in order to understand localized phenomena. We want to cut off the domain far enough away that we will have a good model, but as close in as possible to reduce the size of the computational domain and the computing time required. Specifying good absorbing boundary conditions is often crucial in obtaining useful results. A variety of sophisticated approaches have been developed for specifying appropriate numerical boundary conditions. See Section 21.8.5 for some references and further discussion.

With Godunov-type methods that involve solving the Riemann problem at each interface, it turns out that simply using zero-order extrapolation often gives a reasonable set of absorbing boundary conditions that is extremely simple to implement. For the simple one-dimensional acoustics problem, we can analyze this choice completely using the characteristic decomposition.

In fact, one approach to solving the problem for this linear system would be to diagonalize the equations, obtaining scalar advection equations for the characteristic variables w^1 and w^2 from (7.11). The numerical method can also be diagonalized and yields updating formulas for $W^1 = (-Q^1 + Z_0 Q^2)/2Z_0$ and $W^2 = (Q^1 + Z_0 Q^2)/2Z_0$. At the boundary $x = a$, for example, the variable w^1 is the outgoing variable and we already know that zero-order extrapolation can be used for this variable from Section 7.2.1. For the incoming variable w^2 we want to set the value as a function $g_0(t)$, following Section 7.2.2. Since we want to insure that no signal flows into the domain, regardless of what is flowing out in w^1, the correct value to set is

$$w^2(a, t) = \frac{1}{2Z_0}(p_L + Z_0 u_L)$$

for all t (recall (7.12)). Following Section 7.2.1 this would suggest setting both the ghost cell values to this value,

$$W_0^2 = \frac{1}{2Z_0}(p_L + Z_0 u_L), \qquad W_{-1}^2 = \frac{1}{2Z_0}(p_L + Z_0 u_L), \tag{7.13}$$

together with extrapolation of W^1. From these ghost-cell values for W we could then compute the required ghost-cell values for Q.

However, since W^2 is already constant near the point $x = a$, we expect that $W_1^2 = (p_L + Z_0 u_L)/2Z_0$, and so the boundary conditions (7.13) can also be obtained by simply using zero-order extrapolation for W^2 as well as for W^1. But now, if we extrapolate both of the characteristic variables and then compute Q_0 and Q_{-1}, we will find that these values are simply what would be obtained by zero-order extrapolation of Q. So we do not need to go through the diagonalization at all in setting the boundary conditions, but can simply set

$$Q_0^n = Q_1^n, \qquad Q_{-1}^n = Q_1^n$$

in each time step.

Note that by setting $Q_0 = Q_1$ we insure that the solution to the Riemann problem at the interface $x_{1/2}$ consists of no waves, or more properly that the wave strengths $\alpha_{1/2}^p$ are all zero. So in particular there are no waves generated at the boundary regardless of what is happening in the interior, as desired for nonreflecting boundary conditions.

There are also no outgoing waves generated at $x_{1/2}$, which may be incorrect if an acoustic wave in w^1 should in fact be leaving the domain. But any outgoing waves would only be used to update the solution in the ghost cell C_0, and it is not important that we update this cell properly. In fact, the value Q_0 is not updated at all by the interior algorithm. Instead this value is reset by extrapolation again at the start of the next time step.

7.3.2 Incoming Waves

In the previous section we assumed that we wanted incoming waves to have zero strength. In some problems we may need to specify some incoming signal. For example, we might wish to study what happens when a sinusoidal sound wave at some frequency ω hits a region with different material properties, in which case some of the energy will be reflected (see Section 9.6). Suppose the material properties vary only in a region $a_1 < x < b_1$ and we wish to compute on a larger region $a < x < b$. Then we want to impose the boundary condition

$$w^2(a, t) = \sin(\omega t) \tag{7.14}$$

as the incoming signal, together with nonreflection of any outgoing signal that has been generated within the domain. We must now apply some characteristic decomposition in the process of applying the boundary procedure in order to impose the correct boundary conditions. If we decompose Q_1 into $Q_1 = W_1^1 r^1 + W_1^2 r^2$, then the ghost cell value Q_0 should be set as

$$Q_0 = W_1^1 r^1 + \sin(\omega(t_n + \Delta x/2c_0)) \, r^2.$$

Alternatively, we could first extrapolate the whole vector Q and then reset the w^2 component, setting

$$Q_0 = Q_1 + \left(\sin(\omega(t_n + \Delta x/2c_0)) - W_1^1\right) r^2.$$

See [claw/book/chap7/acouinflow] for an example.

For the acoustics system with only two equations, the two approaches are essentially equivalent in terms of the work required. But if we had a larger system of m equations and wanted to impose an incoming wave in only one characteristic family, e.g.,

$$w^j(a, t) = g_0(t),$$

for some j (with zero-strength signal in other incoming characteristics and nonreflection of outgoing waves), then we could set

$$Q_0 = Q_1 + \left[g_0(t_n + \Delta x/2\lambda^j) - W_1^j\right] r^j,$$

where $W_1^j = \ell^j Q_1$ is the eigen-coefficient of r^j in Q_1 and λ^j is the corresponding eigenvalue.

7.3.3 Solid Walls

Consider a tube of gas with a solid wall at one end. In this case we expect acoustic waves to reflect at this boundary in a particular way – see Section 3.11 for the relation between w^1 and w^2. Rather than working with the characteristic variables, however, in this case it is easier to return to the physical boundary condition itself. For a solid wall at $x = a$ this is

$$u(a, t) = 0. \tag{7.15}$$

The key observation is the following: Suppose we take any data $(p^0(x), u^0(x))$ defined for all $x > a$ and extend it to $x < a$ by setting

$$p^0(a - \xi) = p^0(a + \xi),$$
$$u^0(a - \xi) = -u^0(a + \xi), \tag{7.16}$$

for $\xi > 0$. Then if we solve the Cauchy problem with this extended data, the solution we obtain for $x > a$ will agree exactly with the solution to the half-space problem on $x > a$ with a solid wall boundary condition (7.15) at $x = a$. This follows from symmetry: the conditions (7.16) will continue to hold for $t > 0$ and in particular $u(a) = -u(a)$ must be zero. See also Exercise 7.2.

This suggests the following formulas for ghost-cell values in each time step:

$$\begin{aligned} \text{for } Q_0: \quad & p_0 = p_1, \quad & u_0 = -u_1, \\ \text{for } Q_{-1}: \quad & p_{-1} = p_2, \quad & u_{-1} = -u_2. \end{aligned} \tag{7.17}$$

This imposes the necessary symmetry at the start of each time step.

Solid-wall boundary conditions are implemented in the CLAWPACK library routine `claw/clawpack/1d/lib/bc1.f` and invoked by setting `mthbc(i) = 3`, where `i = 1` for the left boundary or `i = 2` for the right boundary. This assumes that the solid-wall boundary condition can be set by reflecting all components of Q and then negating the second component, as in (7.17). This works for the acoustics equations and also for several other systems of equations that we will study in this book, including the shallow water equations and several forms of the gas dynamics equations. Related boundary conditions can also be used for elastic waves in solids with fixed or free boundaries; see Section 22.4.

7.3.4 Oscillating Walls

Now suppose that the solid wall at $x = a$ is oscillating with some very small amplitude, generating an acoustic wave in the gas. This is a common situation in acoustics: small-scale molecular vibrations of solid objects give rise to many familiar sounds. For very small-amplitude motions we can still use the linear acoustics equations on the fixed domain $a \le x \le b$ but with the boundary condition

$$u(a, t) = U(t) \tag{7.18}$$

to simulate the vibrating wall. For a pure-tone oscillation we might take

$$U(t) = \epsilon \sin(\omega t), \tag{7.19}$$

for example. We can implement this by setting the following ghost-cell values:

$$\begin{aligned} \text{for } Q_0: \quad & p_0 = p_1, \quad & u_0 = 2U(t_n) - u_1, \\ \text{for } Q_{-1}: \quad & p_{-1} = p_2, \quad & u_{-1} = 2U(t_n) - u_2. \end{aligned} \tag{7.20}$$

These reduce to (7.17) if $U(t) \equiv 0$. The rationale for this set of boundary conditions is explored in Exercise 7.2.

Exercises

7.1.   If we use the ghost-cell value Q_0^n from (7.9) in the Lax–Wendroff method, what flux $F_{1/2}^n$ will be computed, and how does it compare with (7.6)? Note that if the Courant number is near 1, then $\Delta x / \bar{u} \approx \Delta t$.

7.2.   (a)   For acoustics with a solid-wall boundary, we set the ghost-cell values (7.17) and then solve a Riemann problem at $x_{1/2} = a$ with data

$$Q_0 = \begin{bmatrix} p_1 \\ -u_1 \end{bmatrix}, \quad Q_1 = \begin{bmatrix} p_1 \\ u_1 \end{bmatrix}.$$

Show that the solution to this Riemann problem has an intermediate state q^* with $u^* = 0$ along the wall, another reason why this is the sensible boundary condition to impose.

(b)   Give a similar interpretation for the oscillating-wall boundary conditions (7.20).

7.3.   The directory [claw/book/chap7/standing] models a standing wave. The acoustics equations are solved with $\rho_0 = K_0 = 1$ in a closed tube of length 1 with with initial data $\overset{\circ}{p}(x) = \cos(2\pi x)$ and $\overset{\circ}{u}(x) = 0$. Solid-wall boundary conditions are used at each end. Modify the input data in claw1ez.data to instead use zero-order extrapolation at the right boundary $x = 1$. Explain the resulting solution using the theory of Section 3.11.

8

Convergence, Accuracy, and Stability

Whenever we use a numerical method to solve a differential equation, we should be concerned about the accuracy and convergence properties of the method. In practice we must apply the method on some particular discrete grid with a finite number of points, and we wish to ensure that the numerical solution obtained is a sufficiently good approximation to the true solution. For real problems we generally do not have the true solution to compare against, and we must rely on some combination of the following techniques to gain confidence in our numerical results:

- *Validation on test problems.* The method (and particular implementation) should be tested on simpler problems for which the true solution is known, or on problems for which a highly accurate comparison solution can be computed by other means. In some cases experimental results may also be available for comparison.
- *Theoretical analysis of convergence and accuracy.* Ideally one would like to prove that the method being used converges to the correct solution as the grid is refined, and also obtain reasonable error estimates for the numerical error that will be observed on any particular finite grid.

In this chapter we concentrate on the theoretical analysis. Here we consider only the Cauchy problem on the unbounded spatial domain, since the introduction of boundary conditions leads to a whole new set of difficulties in analyzing the methods. We will generally assume that the initial data has *compact support*, meaning that it is nonzero only over some bounded region. Then the solution to a hyperbolic problem (which has finite propagation speeds) will have compact support for all time, and so the integrals over the whole real line that appear below really reduce to finite intervals and we don't need to worry about issues concerning the behavior at infinity.

8.1 Convergence

In order to talk about accuracy or convergence, we first need a way to quantify the error. We are trying to approximate a function of space and time, and there are many possible ways to measure the magnitude of the error. In one space dimension we have an approximation Q_i^n at each point on space–time grid, or in each grid cell when using a finite volume method. For comparison we will let q_i^n represent the exact value we are hoping to approximate well.

For a finite difference method we would probably choose the pointwise value

$$q_i^n = q(x_i, t_n),\qquad(8.1)$$

while for a finite volume method we might instead want to compare Q_i^n with

$$q_i^n = \frac{1}{\Delta x} \int_{x_{i-1/2}}^{x_{i+1/2}} q(x, t_n)\, dx.\qquad(8.2)$$

If the function $q(x, t)$ is sufficiently smooth, then the pointwise value (8.1) evaluated at the cell center x_i agrees with the cell average (8.2) to $\mathcal{O}(\Delta x^2)$, and so for the methods considered in this book (which are generally at most second-order accurate), comparison with the pointwise value can be used even for finite volume methods and is often simpler.

To discuss convergence we must first pick some finite time T over which we wish to compute. We expect errors generally to grow with time, and so it would be unreasonable to expect that any finite grid would be capable of yielding good solutions at arbitrarily large times. Note that as we refine the grid, the number of time steps to reach time T will grow like $T/\Delta t$ and go to infinity (in the limit that must be considered in convergence theory), and so even in this case we must deal with an unbounded number of time steps. We will use N to indicate the time level corresponding to time $T = N\Delta t$. The *global error* at this time will be denoted by

$$E^N = Q^N - q^N,$$

and we wish to obtain bounds on this grid function as the grid is refined.

To simplify notation we will generally assume that Δt and Δx are related in a fixed manner as we refine the grid. For hyperbolic problems it is reasonable to assume that the ratio $\Delta t/\Delta x$ is fixed, for example. Then we can speak of letting $\Delta t \to 0$ to refine the grid, and speak of convergence with *order s* if the errors vanish like $\mathcal{O}(\Delta t^s)$ or as $\mathcal{O}(\Delta x^s)$, which are the same thing.

8.1.1 Choice of Norms

To quantify the error, we must choose some norm in which to measure the error at a fixed time. The standard set of norms most commonly used are the *p*-norms

$$\|E\|_p = \left(\Delta x \sum_{i=-\infty}^{\infty} |E_i|^p \right)^{1/p}.\qquad(8.3)$$

These are discrete analogues of the function-space norms

$$\|E\|_p = \left(\int_{-\infty}^{\infty} |E(x)|^p\, dx \right)^{1/p}.\qquad(8.4)$$

Note that the factor Δx in (8.3) is very important to give the correct scaling and order of accuracy as the grid is refined.

In particular, the 1-norm (with $p = 1$) is commonly used for conservation laws, since integrals of the solution itself are of particular importance. The 2-norm is often used for linear problems because of the utility of Fourier analysis in this case (the classical *von Neumann analysis* of linear finite difference methods; see Section 8.3.3). We will use $\|\cdot\|$ without any subscript when we don't wish to specify a particular norm. Note that for a system of m equations, $E \in \mathbb{R}^m$ and the absolute value in (8.3) and (8.4) represents some vector norm on $\mathbb{R}^m$.

We say that the method is convergent at time T in the norm $\|\cdot\|$ if

$$\lim_{\substack{\Delta t \to 0 \\ N \Delta t = T}} \|E^N\| = 0.$$

The method is said to be *accurate of order s* if

$$\|E^N\| = \mathcal{O}(\Delta t^s) \qquad \text{as } \Delta t \to 0. \tag{8.5}$$

Ideally we might hope to have *pointwise convergence* as the grid is refined. This amounts to using the max norm (or ∞-norm) to measure the error:

$$\|E\|_\infty = \max_{-\infty < i < \infty} |E_i|. \tag{8.6}$$

This is the limiting case $p \to \infty$ of (8.3).

If the solution $q(x, t)$ is smooth, then it may be reasonable to expect pointwise convergence. For problems with discontinuous solutions, on the other hand, there will typically always be some smearing at one or more grid points in the neighborhood of the discontinuity. In this case we cannot expect convergence in the max norm, no matter how good the method is. In such cases we generally don't care about pointwise convergence, however. Convergence in the 1-norm is more relevant physically, and we may still hope to obtain this. In general, the rate of convergence observed can depend on what norm is being used, and it is important to use an appropriate norm when measuring convergence. Differences are typically greatest between the max norm and other choices (see Section 8.5 for another example). Except in fairly rare cases, the 1-norm and 2-norm will give similar results, and the choice of norm may depend mostly on which yields an easier mathematical analysis, e.g., the 1-norm for conservation laws and the 2-norm for linear equations.

8.2 One-Step and Local Truncation Errors

It is generally impossible to obtain a simple closed-form expression for the global error after hundreds or thousands of time steps. Instead of trying to obtain the error directly, the approach that is widely used in studying numerical methods for differential equations consists of a two-pronged attack on the problem:

- Study the error introduced in a single time step, showing that the method is *consistent* with the differential equation and introduces a small error in any one step.

- Show that the method is *stable*, so that these local errors do not grow catastrophically and hence a bound on the global error can be obtained in terms of these local errors.

If we can get a bound on the local error in an appropriate sense, then stability can be used to convert this into a bound on the global error that can be used to prove convergence.

Moreover we can generally determine the rate of convergence and perhaps even obtain reasonable error bounds. The *fundamental theorem* of numerical methods for differential equations can then be summarized briefly as

$$\text{consistency} + \text{stability} \Longleftrightarrow \text{convergence}. \tag{8.7}$$

This theorem appears in various forms in different contexts, e.g., the Lax equivalence theorem for linear PDEs (Section 8.3.2) or Dahlquist's equivalence theorem for ODEs. The exact form of "stability" needed depends on the type of equation and method. In this section we will study the local error, and in Section 8.3 we turn to the question of stability.

A general explicit numerical method can be written as

$$Q^{n+1} = \mathcal{N}(Q^n),$$

where $\mathcal{N}(\cdot)$ represents the numerical operator mapping the approximate solution at one time step to the approximate solution at the next. The *one-step error* is defined by applying the numerical operator to the true solution (restricted to the grid) at some time and comparing this with the true solution at the next time:

$$\text{one-step error} = \mathcal{N}(q^n) - q^{n+1}. \tag{8.8}$$

Here q^n and q^{n+1} represent the true solution restricted to the grid by (8.1) or (8.2). This gives an indication of how much error is introduced in a single time step by the numerical method. The *local truncation error* is defined by dividing this by Δt:

$$\tau^n = \frac{1}{\Delta t}[\mathcal{N}(q^n) - q^{n+1}]. \tag{8.9}$$

As we will see in Section 8.3, the local truncation error typically gives an indication of the magnitude of the global error, and particularly the order of accuracy, in cases when the method is stable. If the local truncation error is $\mathcal{O}(\Delta x^s)$ as $s \to 0$, then we expect the global error to have this same behavior.

We say that the method is *consistent* with the differential equation if the local truncation error vanishes as $\Delta t \to 0$ for all smooth functions $q(x, t)$ satisfying the differential equation. In this case we expect the method to be convergent, provided it is stable.

The local truncation error is relatively easy to investigate, and for smooth solutions can be well approximated by simple Taylor series expansions. This is illustrated very briefly in the next example.

Example 8.1. Consider the first-order upwind method for the advection equation with $\bar{u} > 0$,

$$Q_i^{n+1} = Q_i^n - \frac{\Delta t}{\Delta x}\bar{u}(Q_i^n - Q_{i-1}^n).$$

Applying this method to the true solution gives the local truncation error

$$\tau^n = \frac{1}{\Delta t}\left(q(x_i, t_n) - \frac{\Delta t}{\Delta x}\bar{u}[q(x_i, t_n) - q(x_{i-1}, t_n)] - q(x_i, t_{n+1})\right). \tag{8.10}$$

We now expand $q(x_{i-1}, t_n)$ and $q(x_i, t_{n+1})$ in Taylor series about (x_i, t_n) and cancel common terms to obtain

$$\tau^n = -[q_t(x_i, t_n) + \bar{u}q_x(x_i, t_n)] + \frac{1}{2}\Delta x\, \bar{u}q_{xx}(x_i, t_n) - \frac{1}{2}\Delta t\, q_{tt}(x_i, t_n) + \cdots. \quad (8.11)$$

The first term in this expression is identically zero, because we assume that q is an exact solution to the advection equation and hence $q_t + \bar{u}q_x = 0$, so we find that

$$\text{Upwind:} \quad \tau^n = \frac{1}{2}\bar{u}\,\Delta x\, q_{xx}(x_i, t_n) - \frac{1}{2}\Delta t\, q_{tt}(x_i, t_n) + \mathcal{O}(\Delta t^2)$$

$$= \frac{1}{2}\bar{u}\,\Delta x\,(1 - \nu)\, q_{xx}(x_i, t_n) + \mathcal{O}(\Delta t^2), \quad (8.12)$$

where

$$\nu \equiv \bar{u}\,\Delta t / \Delta x \quad (8.13)$$

is the Courant number. The truncation error is dominated by an $\mathcal{O}(\Delta x)$ term, and so the method is first-order accurate.

Similarly, one can compute the local truncation errors for other methods, e.g.,

$$\text{Lax–Friedrichs:} \quad \tau^n = \frac{1}{2}\left(\frac{\Delta x^2}{\Delta t} - \bar{u}^2 \Delta t\right) q_{xx}(x_i, t_n) + \mathcal{O}(\Delta t^2) \quad (8.14)$$

$$= \frac{1}{2}\bar{u}\,\Delta x\,(1/\nu - \nu)q_{xx}(x_i, t_n) + \mathcal{O}(\Delta t^2), \quad (8.15)$$

$$\text{Lax–Wendroff:} \quad \tau^n = -\frac{1}{6}\bar{u}(\Delta x)^2(1 - \nu^2)q_{xxx}(x_i, t_n) + \mathcal{O}(\Delta t^3). \quad (8.16)$$

Note that the Lax–Wendroff method is second-order accurate and the dominant term in the truncation error depends on the third derivative of q, whereas the upwind and Lax–Friedrichs methods are both first-order accurate with the dominant error term depending on q_{xx}. The relation between these errors and the diffusive or dispersive nature of the methods is discussed in Section 8.6.

8.3 Stability Theory

In this section we review the basic ideas of stability theory and the derivation of global error bounds from information on the local truncation error. The form of stability bounds discussed here are particularly useful in analyzing linear methods. For nonlinear methods they may be hard to apply, and in Section 12.12 we discuss a different approach to stability theory for such methods.

The essential requirements and importance of stability can be easily seen from the following attempt to bound the global error using a recurrence relation. In time step n suppose we have an approximation Q^n with error E^n, so that

$$Q^n = q^n + E^n.$$

We apply the numerical method to obtain Q^{n+1}:

$$Q^{n+1} = \mathcal{N}(Q^n) = \mathcal{N}(q^n + E^n),$$

and the global error is now

$$
\begin{aligned}
E^{n+1} &= Q^{n+1} - q^{n+1} \\
&= \mathcal{N}(q^n + E^n) - q^{n+1} \\
&= \mathcal{N}(q^n + E^n) - \mathcal{N}(q^n) + \mathcal{N}(q^n) - q^{n+1} \\
&= [\mathcal{N}(q^n + E^n) - \mathcal{N}(q^n)] + \Delta t\, \tau^n.
\end{aligned}
\tag{8.17}
$$

By introducing $\mathcal{N}(q^n)$ we have written the new global error as the sum of two terms:

- $\mathcal{N}(q^n + E^n) - \mathcal{N}(q^n)$, which measures the effect of the numerical method on the *previous* global error E^n,
- $\Delta t\, \tau^n$, the new one-step error introduced in this time step.

The study of the local truncation error allows us to bound the new one-step error. Stability theory is required to bound the other term, $\mathcal{N}(q^n + E^n) - \mathcal{N}(q^n)$.

8.3.1 Contractive Operators

The numerical solution operator $\mathcal{N}(\cdot)$ is called *contractive* in some norm $\|\cdot\|$ if

$$\|\mathcal{N}(P) - \mathcal{N}(Q)\| \le \|P - Q\| \tag{8.18}$$

for any two grid functions P and Q. If the method is contractive, then it is stable in this norm and we can obtain a bound on the global error from (8.17) very simply using $P = q^n + E^n$ and $Q = q^n$:

$$
\begin{aligned}
\|E^{n+1}\| &\le \|\mathcal{N}(q^n + E^n) - \mathcal{N}(q^n)\| + \Delta t\, \|\tau^n\| \\
&\le \|E^n\| + \Delta t\, \|\tau^n\|.
\end{aligned}
\tag{8.19}
$$

Applying this recursively gives

$$\|E^N\| \le \|E^0\| + \Delta t \sum_{n=1}^{N-1} \|\tau^n\|.$$

Suppose the local truncation error is bounded by

$$\|\tau\| \equiv \max_{0 \le n \le N} \|\tau^n\|.$$

Then we have

$$
\begin{aligned}
\|E^N\| &\le \|E^0\| + N\Delta t\, \|\tau\| \\
&\le \|E^0\| + T\|\tau\| \qquad (\text{for } N\Delta t = T).
\end{aligned}
\tag{8.20}
$$

The term $\|E^0\|$ measures the error in the initial data on the grid, and we require that $\|E^0\| \to 0$ as $\Delta t \to 0$ in order to be solving the correct initial-value problem. If the method is consistent, then also $\|\tau\| \to 0$ as $\Delta t \to 0$ and we have proved convergence. Moreover, if $\|\tau\| = \mathcal{O}(\Delta t^s)$, then the global error will have this same behavior as $\Delta t \to 0$ (provided the initial data is sufficiently accurate), and the method has global order of accuracy s.

Actually a somewhat weaker requirement on the operator $\mathcal{N}$ is sufficient for stability. Rather than the contractive property (8.18), it is sufficient to have

$$\|\mathcal{N}(P) - \mathcal{N}(Q)\| \le (1 + \alpha \, \Delta t)\|P - Q\| \tag{8.21}$$

for all P and Q, where α is some constant independent of Δt as $\Delta t \to 0$. (Recall that the one-step operator $\mathcal{N}$ depends on Δt even though we haven't explicitly included this dependence in the notation.) If (8.21) holds then the above proof still goes through with a slight modification. We now have

$$\|E^{n+1}\| \le (1 + \alpha \, \Delta t)\|E^n\| + \Delta t \, \|\tau\|,$$

and so

$$\|E^N\| \le (1 + \alpha \, \Delta t)^N \|E^0\| + \Delta t \sum_{n=1}^{N-1} (1 + \alpha \, \Delta t)^{N-1-n} \|\tau\|$$

$$\le e^{\alpha T}(\|E^0\| + T\|\tau\|) \qquad \text{(for } N\Delta t = T). \tag{8.22}$$

In this case the error may grow exponentially in time, but the key fact is that this growth is bounded independently of the time step Δt. For fixed T we have a bound that goes to zero with Δt. This depends on the fact that any growth in error resulting from the operator $\mathcal{N}$ is at most order $\mathcal{O}(\Delta t)$ in one time step, which is what (8.21) guarantees.

8.3.2 Lax–Richtmyer Stability for Linear Methods

If the operator $\mathcal{N}(\cdot)$ is a *linear* operator, then $\mathcal{N}(q^n + E^n) = \mathcal{N}(q^n) + \mathcal{N}(E^n)$, and so $\mathcal{N}(q^n + E^n) - \mathcal{N}(q^n)$ reduces to simply $\mathcal{N}(E^n)$. In this case, the condition (8.21) reduces simply to requiring that

$$\|\mathcal{N}(E^n)\| \le (1 + \alpha \, \Delta t)\|E^n\| \tag{8.23}$$

for any grid function E^n, which is generally simpler to check. This can also be expressed as a bound on the norm of the linear operator $\mathcal{N}$,

$$\|\mathcal{N}\| \le 1 + \alpha \, \Delta t. \tag{8.24}$$

An even looser version of this stability requirement can be formulated in the linear case. We really only need that, for each time T, there is a constant C such that

$$\|\mathcal{N}^n\| \le C \tag{8.25}$$

for all $n \le N = T/\Delta t$, i.e., the nth power of the operator $\mathcal{N}$ is uniformly bounded up to this time, for then all the terms in (8.22) are uniformly bounded. For linear methods, this form of stability is generally referred to as *Lax–Richtmyer stability*. The result (8.7) is called the *Lax equivalence theorem* in this context. See [369] for a rigorous proof. Note that if (8.24) holds, then we can take $C = e^{\alpha T}$ in (8.25).

Classical methods such as the first-order upwind or the Lax–Wendroff method for the linear advection equation are all linear methods, and this form of the stability condition can be used. (See Section 8.3.4 for an example.) The high-resolution methods developed in Chapter 6 are *not* linear methods, however, since the limiter function introduces nonlinearity. Proving stability of these methods is more subtle and is discussed briefly in Section 8.3.5 and Chapter 15.

8.3.3 2-Norm Stability and von Neumann Analysis

For linear difference equations, stability analysis is often particularly easy in the 2-norm, since Fourier analysis can then be used to simplify the problem. This is the basis of *von Neumann stability analysis*, which is described more completely in many books on finite difference methods for partial differential equations (e.g., [333] or [427]).

Let Q_I^n ($-\infty < I < \infty$) represent an arbitrary grid function for the Cauchy problem. In this section we use I for the grid index ($x_I = I \Delta x$) so that $i = \sqrt{-1}$ can be used in the complex exponentials below. We suppose that Q_I^n has finite norm, so that it can be expressed as a Fourier series

$$Q_I^n = \frac{1}{\sqrt{2\pi}} \int_{-\infty}^{\infty} \hat{Q}(\xi)\, e^{i\xi I \Delta x}\, d\xi. \tag{8.26}$$

Applying a linear finite difference method to Q_I^n and manipulating the exponentials typically gives an expression of the form

$$Q_I^{n+1} = \frac{1}{\sqrt{2\pi}} \int_{-\infty}^{\infty} \hat{Q}(\xi)\, g(\xi, \Delta x, \Delta t)\, e^{i\xi I \Delta x}\, d\xi, \tag{8.27}$$

where $g(\xi, \Delta x, \Delta t)$ is called the *amplification factor* for wave number ξ, since

$$\hat{Q}^{n+1}(\xi) = g(\xi, \Delta x, \Delta t)\, \hat{Q}^n(\xi). \tag{8.28}$$

The 2-norm is now convenient because of *Parseval's relation*, which states that

$$\|Q^n\|_2 = \|\hat{Q}^n\|_2, \tag{8.29}$$

where

$$\|Q^n\|_2 = \left(\Delta x \sum_{I=-\infty}^{\infty} |Q_I^n|^2 \right)^{1/2} \quad \text{and} \quad \|\hat{Q}^n\|_2 = \left(\int_{-\infty}^{\infty} |\hat{Q}^n(\xi)|^2\, d\xi \right)^{1/2}.$$

To show that the 2-norm of Q^n remains bounded it suffices to show that the 2-norm of $\hat{Q}^n$ does. But whereas all elements of Q^n (as I varies) are coupled together via the difference equations, each element of $\hat{Q}^n$ (as ξ varies) satisfies an equation (8.28) that is decoupled from all other wave numbers. (The Fourier transform diagonalizes the linear difference operator.) So it suffices to consider an arbitrary single wave number ξ and data of the form

$$Q_l^n = e^{i\xi l \Delta x}. \tag{8.30}$$

From this we can compute $g(\xi, \Delta x, \Delta t)$. Then requiring that $|g(\xi, \Delta x, \Delta t)| \leq 1$ for all ξ gives a sufficient condition for stability. In fact it suffices to have $|g(\xi, \Delta x, \Delta t)| \leq 1 + \alpha \Delta t$ for some constant α independent of ξ.

Example 8.2. Consider the upwind method (4.25) for the advection equation $q_t + \bar{u} q_x = 0$ with $\bar{u} > 0$. Again use $\nu \equiv \bar{u} \Delta t / \Delta x$ as shorthand for the Courant number, and write the upwind method (4.25) as

$$Q_l^{n+1} = Q_l^n - \nu \left(Q_l^n - Q_{l-1}^n \right)$$
$$= (1 - \nu) Q_l^n + \nu Q_{l-1}^n. \tag{8.31}$$

We will use von Neumann analysis to demonstrate that this method is stable in the 2-norm provided that

$$0 \leq \nu \leq 1 \tag{8.32}$$

is satisfied, which agrees exactly with the CFL condition for this method (see Section 4.4). Using the data (8.30) yields

$$Q_l^{n+1} = (1 - \nu) e^{i\xi l \Delta x} + \nu e^{i\xi (l-1) \Delta x}$$
$$= [(1 - \nu) + \nu e^{-i\xi \Delta x}] e^{i\xi l \Delta x}$$
$$= g(\xi, \Delta x, \Delta t) Q_l^n \tag{8.33}$$

with

$$g(\xi, \Delta x, \Delta t) = (1 - \nu) + \nu e^{-i\xi \Delta x}. \tag{8.34}$$

As ξ varies, $g(\xi, \Delta x, \Delta t)$ lies on a circle of radius ν in the complex plane, centered on the real axis at $1 - \nu$. This circle lies entirely inside the unit circle (i.e., $|g| \leq 1$ for all ξ) if and only if $0 \leq \nu \leq 1$, giving the stability limit (8.32) for the upwind method.

8.3.4 1-Norm Stability of the Upwind Method

For conservation laws the 1-norm is often used, particularly for nonlinear problems. We will demonstrate that the upwind method (4.25) considered in the previous example is also stable in the 1-norm under the time-step restriction (8.32). We revert to the usual notation with i as the grid index and write the upwind method (8.31) as

$$Q_i^{n+1} = (1 - \nu) Q_i^n + \nu Q_{i-1}^n, \tag{8.35}$$

where v is again the Courant number (8.13). From this we compute

$$\|Q^{n+1}\|_1 = \Delta x \sum_i |Q_i^{n+1}|$$

$$= \Delta x \sum_i |(1-v)Q_i^n + vQ_{i-1}^n|$$

$$\leq \Delta x \sum_i [(1-v)|Q_i^n| + v|Q_{i-1}^n|]. \qquad (8.36)$$

In the final step we have used the triangle inequality and then pulled $1 - v$ and v outside the absolute values, since these are both positive if (8.32) is satisfied. The sum can be split up into two separate sums, each of which gives $\|Q^n\|_1$, obtaining

$$\|Q^{n+1}\|_1 \leq (1-v)\|Q^n\|_1 + v\|Q^n\|_1 = \|Q^n\|_1.$$

This proves stability in the 1-norm. Note that this only works if (8.32) is satisfied, since we need both $1 - v$ and v to be positive.

8.3.5 Total-Variation Stability for Nonlinear Methods

For a nonlinear numerical method, showing that (8.23) holds is generally not sufficient to prove convergence. The stronger contractivity property (8.21) would be sufficient, but is generally difficult to obtain. Even for the linear advection equation, the high-resolution methods of Chapter 6 are nonlinear (since the limiter function depends on the data), and so a different approach to stability must be adopted to prove convergence of these methods.

The total variation introduced in Section 6.7 turns out to be an effective tool for studying stability of nonlinear problems. We make the following definition.

Definition 8.1. *A numerical method is* total-variation bounded *(TVB) if, for any data Q^0 (with $\mathrm{TV}(Q^0) < \infty$) and time T, there is a constant $R > 0$ and a value $\Delta t_0 > 0$ such that*

$$\mathrm{TV}(Q^n) \leq R \qquad (8.37)$$

for all $n \, \Delta t \leq T$ whenever $\Delta t < \Delta t_0$.

This simply requires that we have a uniform bound on the total variation up to time T on all grids sufficiently fine (and hence as $\Delta t \to 0$).

In Section 12.12 we will see how this can be used to prove convergence of the numerical method (using a more subtle argument than the approach taken above for linear problems, based on the compactness of an appropriate function space). For now we just note that, in particular, a method that is TVD (see Section 6.7) is certainly TVB with $R = \mathrm{TV}(Q^0)$ in (8.37) for any T. So the notion of a TVD method, useful in insuring that no spurious oscillations are introduced, is also sufficient to prove convergence. In particular the high-resolution TVD methods introduced in Chapter 6 are all convergent provided the CFL condition is satisfied (since this is required in order to be TVD).

Of course a weaker condition than TVD is sufficient for convergence, since we only need a uniform bound of the form (8.37). For example, a method that satisfies

$$\text{TV}(Q^{n+1}) \le (1 + \alpha\,\Delta t)\text{TV}(Q^n) \tag{8.38}$$

for some constant α independent of Δt (at least for all Δt sufficiently small) will also be TVB.

8.4 Accuracy at Extrema

Examining the results of Figure 6.1 and Figure 6.3 shows that the high-resolution methods developed in Chapter 6 give rather poor accuracy at extrema (local maxima or minima in q), even though the solution is smooth. Figure 8.1(a) gives an indication of why this occurs. All of the limiters discussed in Section 6.9 will give slopes $\sigma = 0$ in cells $i - 2$ and $i - 1$ for this data. This is required in order to prove that the method is truly TVD, as any other choice will give a reconstructed function $\tilde{q}^n(x, t_n)$ with $\text{TV}(\tilde{q}^n(\cdot, t_n)) > \text{TV}(Q^n)$, allowing the possibility that $\text{TV}(Q^{n+1}) > \text{TV}(Q^n)$ with suitable choices of the data and time step. If the solution is smooth near the extremum, then the Lax–Wendroff slope should be close to zero anyway, so this is perhaps not a bad approximation. However, setting the slope to zero will lead to a *clipping* of the solution, and the extreme value will be diminished by $\mathcal{O}(\Delta x^2) = \mathcal{O}(\Delta t^2)$ in each time step. After $T/\Delta t$ time steps this can lead to a global error near extrema that is $\mathcal{O}(\Delta t)$, reducing the method to first-order accuracy in this region. This can be observed in Figure 6.2. Osher and Chakravarthy [351] prove that TVD methods must in fact degenerate to first-order accuracy at extremal points.

Using a better approximation to q_x near extrema, as indicated in Figure 8.1(b), would give a reconstruction that allows smooth peaks to be better represented over time, since the peaks are then reconstructed more accurately from the data. The cost is that the total variation will need to increase slightly at times in order to reconstruct such a peak. But as indicated in Section 8.3.5, the TVD property is not strictly needed for stability. The challenge is to find looser criteria that allow a small increase in the total variation near extremal points while still suppressing oscillations where necessary. This goal has led to several suggestions on

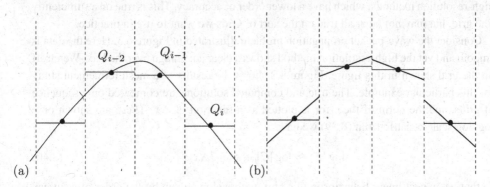

(a)   (b)

Fig. 8.1. (a) A smooth maximum and cell averages. A TVD slope reconstruction will give clipping of the peak. (b) Better accuracy can be obtained with a reconstruction that is not TVD relative to the cell averages.

other ways to choose the limiter function and criteria other than strictly requiring the TVD property. Shu [407] has developed the theory of *TVB* methods for which convergence can still be proved while better accuracy at extremal points may be obtained. The *essentially nonoscillatory* (ENO) methods are another approach to obtaining high-resolution that often give better than second-order accuracy in smooth regions, including extrema. These are briefly described in Section 10.4.4.

8.5 Order of Accuracy Isn't Everything

The quality of a numerical method is often summarized by a single number, its order of accuracy. This is indeed an important consideration, but it can be a mistake to put too much emphasis on this one attribute. It is not always true that a method with a higher order of accuracy is more accurate on a particular grid or for a particular problem.

Suppose a method has order of accuracy s. Then we expect the error to behave like

$$\| E^N \| = C(\Delta x)^s + \text{higher-order terms} \tag{8.39}$$

as the grid is refined and $\Delta x \to 0$. Here C is some constant that depends on the particular solution being computed (and the time T). The magnitude of the constant C is important as well as the value of s. Also, note that the "higher-order" terms, which depend on higher powers of Δx, are asymptotically negligible as $\Delta x \to 0$, but may in fact be larger than the "dominant" term $C(\Delta x)^s$ on the grid we wish to use in practice.

As a specific example, consider the high-resolution TVD methods developed in Chapter 6 for the scalar advection equation. Because of the nonlinear limiter function, these methods are formally not second-order accurate, even when applied to problems with smooth solutions. The limiter typically leads to a clipping of the solution near extrema, as discussed in Section 8.4.

For discontinuous solutions, as illustrated in Figure 6.1 and Figure 6.2, these methods have clear advantages over the "second-order" Lax–Wendroff or Beam–Warming methods, even though for discontinuous solutions none of these methods exhibit second-order convergence.

But suppose we compare these methods on a problem where the solution is smooth. At least in this case one might think the second-order methods should be better than the high-resolution methods, which have a lower order of accuracy. This is true on a sufficiently fine grid, but may not be at all true on the sort of grids we want to use in practice.

Consider the wave-packet propagation problem illustrated in Figure 6.3. Here the data is smooth and yet the high-resolution method shows a clear advantage over the Lax–Wendroff on the grid shown in this figure. Figure 8.2 shows the results of a mesh refinement study for this particular example. The true and computed solutions are compared on a sequence of grids, and the norm of the error is plotted as a function of Δx. These are shown on a log–log scale because from (8.39) we have

$$\log |E| \approx \log |C| + s \log |\Delta x|, \tag{8.40}$$

so that we expect linear behavior in this plot, with a slope given by the order of accuracy s. Figure 8.2(a) shows errors in the max norm, while Figure 8.2(b) shows the errors in the 1-norm. The Lax–Wendroff method is second-order accurate in both norms; the slope

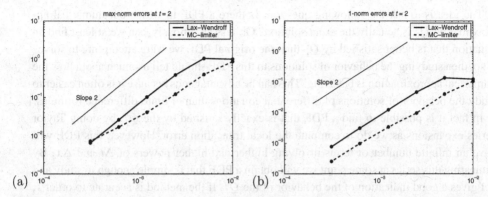

Fig. 8.2. Log–log plot of the error vs. grid size for the Lax–Wendroff and a high-resolution method on the wave-packet problem of Figure 6.3: (a) max-norm errors, (b) 1-norm errors. [claw/book/chap8/wavepacket]

of the corresponding curves is about 1.999. Due to the clipping of peaks observed with TVD methods, the error with the high-resolution method is dominated by errors near these few locations, and so the max norm shows larger errors than the 1-norm, which averages over the entire domain. In the max norm the observed order of accuracy is about 1.22, and on sufficiently fine grids the Lax–Wendroff method is superior. However, the crossover point for this particular example is at about $\Delta x = 0.00035$, meaning about 2800 grid cells in the unit interval. This is a much finer grid than one would normally want to use for this problem. Certainly for two- or three-dimensional analogues of this problem it would be unthinkable to use this many grid points in each direction. On the coarser grids one might use in practice, the high-resolution method is superior in spite of its lower order of accuracy.

In the 1-norm the high-resolution method looks even better. In this norm the observed order of accuracy is about 1.92, but the error constant C is about 5 times smaller than what is observed for the Lax–Wendroff method, so that on all the grids tested the error is essentially 5 times smaller. Of course, for very small Δx the Lax–Wendroff method would eventually prove superior, but extrapolating from the results seen in Figure 8.2(b) we find that the crossover point in the 1-norm is around $\Delta x = 10^{-33}$.

Later on we will see other examples where it is wise to look beyond order of accuracy in comparing different methods. For example, in Chapter 17 we will see that a fractional-step method for source terms that is often dismissed as being "only first-order accurate" is in fact essentially identical to second-order accurate methods for many practical purposes, and is often more efficient to use.

8.6 Modified Equations

As discussed in Section 8.2, the local truncation error of a method is determined by seeing how well the true solution of the differential equation satisfies the difference equation. Now we will study a slightly different approach that can be very illuminating in that it reveals much more about the structure and behavior of the numerical solution in addition to the order of the error.

The idea is to ask the following question: Is there a PDE to which our numerical approximation Q_i^n is actually the *exact* solution? Or, less ambitiously, can we at least find an equation that is better satisfied by Q_i^n than the original PDE we were attempting to solve? If so, then studying the behavior of solutions to this PDE should tell us much about how the numerical approximation is behaving. This can be advantageous because it is often easier to study the behavior of solutions of differential equations than of finite-difference formulas.

In fact it is possible to find a PDE that is exactly satisfied by the Q_i^n, by doing Taylor series expansions as we do to compute the local truncation error. However, this PDE will have an infinite number of terms involving higher and higher powers of Δt and Δx. By truncating this series at some point we will obtain a PDE that is simple enough to study and yet gives a good indication of the behavior of the Q_i^n. If the method is accurate to order s, then this equation is generally a modification of the original PDE with new terms of order s, and is called the *modified equation* for the method, or sometimes the *model equation*. Good descriptions of the theory and use of modified equations can be found in Hedstrom [193] or Warming & Hyett [480]. See [61], [114], [126], [170] for some further discussion and other applications of this approach.

8.6.1 The Upwind Method

The derivation of a modified equation is best illustrated with an example. Consider the first-order upwind method for the advection equation $q_t + \bar{u} q_x = 0$ in the case $\bar{u} > 0$,

$$Q_i^{n+1} = Q_i^n - \frac{\bar{u} \, \Delta t}{\Delta x} (Q_i^n - Q_{i-1}^n). \tag{8.41}$$

The process of deriving the modified equation is very similar to computing the local truncation error, only now we insert a function $v(x, t)$ into the numerical method instead of the true solution $q(x, t)$. Our goal is to determine a differential equation satisfied by v. We view the method as a finite difference method acting on grid-point values, and v is supposed to be a function that agrees exactly with Q_i^n at the grid points. So, unlike $q(x, t)$, the function $v(x, t)$ satisfies (8.41) exactly:

$$v(x, t + \Delta t) = v(x, t) - \frac{\bar{u} \, \Delta t}{\Delta x} [v(x, t) - v(x - \Delta x, t)].$$

Expanding these terms in Taylor series about (x, t) and simplifying gives

$$\left(v_t + \frac{1}{2} \Delta t \, v_{tt} + \frac{1}{6} (\Delta t)^2 v_{ttt} + \cdots \right) + \bar{u} \left(v_x - \frac{1}{2} \Delta x \, v_{xx} + \frac{1}{6} (\Delta x)^2 v_{xxx} + \cdots \right) = 0.$$

We can rewrite this as

$$v_t + \bar{u} v_x = \frac{1}{2} (\bar{u} \, \Delta x \, v_{xx} - \Delta t \, v_{tt}) - \frac{1}{6} [\bar{u} (\Delta x)^2 v_{xxx} + (\Delta t)^2 v_{ttt}] + \cdots. \tag{8.42}$$

This is the PDE that v satisfies. If we take $\Delta t / \Delta x$ fixed, then the terms on the right-hand side are $\mathcal{O}(\Delta t)$, $\mathcal{O}(\Delta t^2)$, etc., so that for small Δt we can truncate this series to get a PDE that is quite well satisfied by the Q_i^n.

If we drop all the terms on the right-hand side, we just recover the original advection equation. Since we have then dropped terms of $\mathcal{O}(\Delta t)$, we expect that Q_i^n satisfies this equation to $\mathcal{O}(\Delta t)$, as we know to be true, since this upwind method is first-order accurate.

If we keep the $\mathcal{O}(\Delta t)$ terms then we get something more interesting:

$$v_t + \bar{u} v_x = \frac{1}{2}(\bar{u} \, \Delta x \, v_{xx} - \Delta t \, v_{tt}). \tag{8.43}$$

This involves second derivatives in both x and t, but we can derive a slightly different modified equation with the same accuracy by differentiating (8.43) with respect to t to obtain

$$v_{tt} = -\bar{u} v_{xt} + \frac{1}{2}(\bar{u} \, \Delta x \, v_{xxt} - \Delta t \, v_{ttt})$$

and with respect to x to obtain

$$v_{tx} = -\bar{u} v_{xx} + \frac{1}{2}(\bar{u} \, \Delta x \, v_{xxx} - \Delta t \, v_{ttx}).$$

Combining these gives

$$v_{tt} = \bar{u}^2 v_{xx} + \mathcal{O}(\Delta t).$$

Inserting this in (8.43) gives

$$v_t + \bar{u} v_x = \frac{1}{2}(\bar{u} \, \Delta x \, v_{xx} - \bar{u}^2 \Delta t \, v_{xx}) + \mathcal{O}(\Delta t^2).$$

Since we have already decided to drop terms of $\mathcal{O}(\Delta t^2)$, we can drop these terms here also to obtain

$$v_t + \bar{u} v_x = \frac{1}{2}\bar{u} \, \Delta x \, (1 - v)v_{xx}, \tag{8.44}$$

where $v = \bar{u} \, \Delta t / \Delta x$ is the Courant number. This is now a familiar advection–diffusion equation. The grid values Q_i^n can be viewed as giving a *second-order accurate* approximation to the true solution of this equation (whereas they only give first-order accurate approximations to the true solution of the advection equation).

For higher-order methods this elimination of t-derivatives in terms of x-derivatives can also be done, but must be done carefully and is complicated by the need to include higher-order terms. Warming and Hyett [480] present a general procedure.

The fact that the modified equation for the upwind method is an advection–diffusion equation tells us a great deal about how the numerical solution behaves. Solutions to the advection–diffusion equation translate at the proper speed $\bar{u}$ but also diffuse and are smeared out. This was clearly visible in Figures 6.1 and 6.3, for example.

Note that the diffusion coefficient in (8.43) vanishes in the special case $\bar{u} \, \Delta t = \Delta x$. In this case we already know that the exact solution to the advection equation is recovered by the upwind method; see Figure 4.4 and Exercise 4.2.

Also note that the diffusion coefficient is positive only if $0 < \bar{u} \, \Delta t / \Delta x < 1$. This is precisely the stability limit of the upwind method. If it is violated, then the diffusion coefficient in the modified equation is negative, giving an ill-posed *backward heat equation* with exponentially growing solutions. Hence we see that some information about stability can also be extracted from the modified equation.

8.6.2 Lax–Wendroff Method

If the same procedure is followed for the Lax–Wendroff method, we find that all $\mathcal{O}(\Delta t)$ terms drop out of the modified equation, as is expected because this method is second-order accurate on the advection equation. The modified equation obtained by retaining the $\mathcal{O}(\Delta t^2)$ term and then replacing time derivatives by spatial derivatives is

$$v_t + \bar{u}v_x = -\frac{1}{6}\bar{u}(\Delta x)^2(1 - v^2)v_{xxx}. \tag{8.45}$$

The Lax–Wendroff method produces a *third-order* accurate solution to this equation. This equation has a very different character from (8.43). The v_{xxx} term leads to *dispersive* behavior rather than diffusion.

This dispersion is very clearly seen in the wave-packet computation of Figure 6.3, where the Q_i^n computed with the Lax–Wendroff method clearly travels at the wrong speed. Dispersive wave theory predicts that such a packet should travel at the *group velocity*, which for wavenumber ξ in the Lax–Wendroff method is

$$c_g = \bar{u} - \frac{1}{2}\bar{u}(\Delta x)^2(1 - v^2)\xi^2.$$

See for example [8], [50], [298], [427], [486] for discussions of dispersive equations and group velocities. The utility of this concept in the study of numerical methods has been stressed by Trefethen, in particular in relation to the stability of boundary conditions. A nice summary of some of this theory may be found in Trefethen [458].

The computation shown in Figure 6.3 has $\xi = 80$, $\bar{u} = 1$, $\Delta x = 1/200$, and $\bar{u}\,\Delta t/\Delta x = 0.8$, giving a group velocity of 0.9712 rather than the correct advection speed of 1. At time 10 this predicts the wave packet will be lagging the correct location by a distance of about 0.288, which agrees well with what is seen in the figure.

For data such as that used in Figure 6.1, dispersion means that the high-frequency components of data such as the discontinuity will travel substantially more slowly than the lower-frequency components, since the group velocity is less than $\bar{u}$ for all wave numbers and falls with increasing ξ. As a result the numerical result can be expected to develop a train of oscillations *behind* the peak, with the high wave numbers lagging farthest behind the correct location.

If we retain one more term in the modified equation for the Lax–Wendroff method, we find that the Q_i^n are fourth-order accurate solutions to an equation of the form

$$v_t + \bar{u}v_x = \frac{1}{6}\bar{u}(\Delta x)^2(v^2 - 1)v_{xxx} - \epsilon v_{xxxx}, \tag{8.46}$$

where the ϵ in the fourth-order dissipative term is $\mathcal{O}(\Delta x^3)$ and positive when the stability bound holds. This higher-order dissipation causes the highest wave numbers to be damped, so that there is a limit to the oscillations seen in practice.

Note that the dominant new term in the modified equation corresponds to the dominant term in the local truncation error for each of these methods. Compare (8.45) with (8.16), for example.

8.6.3 Beam–Warming Method

The second-order Beam–Warming method (6.7) has a modified equation similar to that of the Lax–Wendroff method,

$$v_t + \bar{u} v_x = \frac{1}{6}\bar{u}(\Delta x)^2(2 - 3v + v^2)v_{xxx}. \tag{8.47}$$

In this case the group velocity is greater than $\bar{u}$ for all wave numbers in the case $0 < v < 1$, so that the oscillations move ahead of the main hump. This can be observed in Figure 6.1, where $v = \bar{u}\,\Delta t/\Delta x = 0.8$ was used. If $1 < v < 2$, then the group velocity is less than $\bar{u}$ and the oscillations will fall behind.

8.7 Accuracy Near Discontinuities

In the previous section we derived the modified equation for various numerical methods, a PDE that models the behavior of the numerical solution. This equation was derived using Taylor series expansion, and hence is based on the assumption of smoothness, but it turns out that the modified equation is often a good model even when the true solution of the original hyperbolic problem contains discontinuities. This is because the modified equation typically contains diffusive terms that cause the discontinuities to be immediately smeared out, as also happens with the numerical solution, and so the solution we are studying is smooth and the Taylor series expansion is valid.

Figure 8.3 shows a simple example in which the upwind method has been applied to the scalar advection equation $q_t + \bar{u} q_x = 0$ with discontinuous data $\overset{\circ}{q}(x)$ having the value 2 for $x < 0$ and 0 for $x > 0$. The parameters $\bar{u} = 1$, $\Delta x = 0.05$, and $\Delta t = 0.04$ were used giving a Courant number $v = 0.8$. The dashed line is the true solution to this equation, $q(x, t) = \overset{\circ}{q}(x - \bar{u}t)$. The solid line is the exact solution to the modified equation (8.44). This advection–diffusion equation can be solved exactly to yield

$$v(x, t) = \text{erfc}\left(\frac{x - \bar{u}t}{\sqrt{4\beta t}}\right), \tag{8.48}$$

where

$$\beta = \frac{1}{2}\bar{u}\,\Delta x(1 - v) \tag{8.49}$$

is the diffusion coefficient from (8.44), and the *complementary error function* erfc is defined by

$$\text{erfc}(x) = \frac{2}{\sqrt{\pi}}\int_x^\infty e^{-z^2}\,dz. \tag{8.50}$$

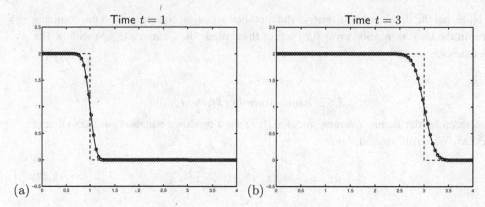

Fig. 8.3. Dashed line: exact solution to the advection equation. Points: numerical solution obtained with the upwind method. Solid line: exact solution to the modified equation (8.44). (a) At time $t = 1$. (b) At time $t = 3$. [book/chap8/modeqn]

The numerical solution to the advection equation obtained using the upwind method, marked by the symbols in Figure 8.3, is well approximated by the exact solution to the modified equation.

It follows that we can use the modified equation to give us some insight into the expected accuracy of the upwind method on this problem. Comparing $v(x, t)$ from (8.48) to the true solution $q(x, t) = 2H(\bar{u}t - x)$, it is possible to show that the 1-norm of the difference is

$$
\|q(\cdot, t) - v(\cdot, t)\|_1 = 2 \int_0^\infty \mathrm{erfc}\left(\frac{x}{\sqrt{4\beta t}}\right) dx
$$

$$
= 2\sqrt{4\beta t} \int_0^\infty \mathrm{erfc}(z)\, dz
$$

$$
= C_1\sqrt{\beta t} \tag{8.51}
$$

for some constant C_1 independent of β and t. Since β is given by (8.49), this gives

$$
\|q(\cdot, t) - v(\cdot, t)\|_1 \approx C_2\sqrt{\Delta x\, t} \tag{8.52}
$$

as $\Delta x \to 0$ with $\Delta t/\Delta x$ fixed. This indicates that the 1-norm of the error decays only like $(\Delta x)^{1/2}$ even though the method is formally "first-order accurate" based on the local truncation error, which is valid only for smooth solutions.

This informal analysis only gives an indication of the accuracy one might expect from a first-order method on a problem with a discontinuous solution. More detailed error analysis of numerical methods for discontinuous solutions (to nonlinear scalar equations) can be found, for example, in [251], [316], [339], [390], [436], [438].

Exercises

8.1.　Consider the centered method (4.19) for the scalar advection equation $q_t + \bar{u}q_x = 0$. Apply von Neumann analysis to show that this method is unstable in the 2-norm for any fixed $\Delta t/\Delta x$.

8.2. Following the proof of Section 8.3.4, show that the upwind method (4.30) is stable provided the CFL condition (4.32) is satisfied.

8.3. Consider the equation

$$q_t + \bar{u} q_x = aq, \qquad q(x, 0) = \overset{\circ}{q}(x),$$

with solution $q(x, t) = e^{at} \overset{\circ}{q}(x - \bar{u}t)$.

(a) Show that the method

$$Q_i^{n+1} = Q_i^n - \frac{\bar{u}\,\Delta t}{\Delta x}\left(Q_i^n - Q_{i-1}^n\right) + \Delta t\, a Q_i^n$$

is first-order accurate for this equation by computing the local truncation error.

(b) Show that this method is Lax–Richtmyer-stable in the 1-norm provided $|\bar{u}\,\Delta t/\Delta x| \le 1$, by showing that a bound of the form (8.23) holds. Note that when $a > 0$ the numerical solution is growing exponentially in time (as is the true solution) but the method is stable and convergent at any fixed time.

(c) Show that this method is TVB. Is it TVD?

8.4. Show that a method $Q^{n+1} = \mathcal{N}(Q^n)$ that is contractive in the 1-norm (L^1-contractive), so that $\|\mathcal{N}(P) - \mathcal{N}(Q)\|_1 \le \|P - Q\|_1$, must also be TVD, so $TV(\mathcal{N}(Q)) \le TV(Q)$. *Hint:* Define the grid function P by $P_i = Q_{i-1}$.

8.5. Prove Harten's Theorem 6.1. *Hint:* Note that

$$Q_{i+1}^{n+1} - Q_i^{n+1} = \left(1 - C_i^n - D_i^n\right)\left(Q_{i+1}^n - Q_i^n\right) + D_{i+1}^n\left(Q_{i+2}^n - Q_{i+1}^n\right)$$
$$+ C_{i-1}^n\left(Q_i^n - Q_{i-1}^n\right).$$

Sum $|Q_{i+1}^{n+1} - Q_i^{n+1}|$ over i and use the nonnegativity of each coefficient, as in the stability proof of Section 8.3.4.

8.6. Use the method of Section 8.3.4 to show that the method (4.64) is stable in the 1-norm for $\Delta x \le \bar{u}\,\Delta t \le 2\Delta x$.

8.7. View (8.41) as a numerical method for the equation (8.44). Compute the local truncation error, and verify that it is $\mathcal{O}(\Delta t^2)$.

8.8. Derive the modified equation (8.45) for the Lax–Wendroff method.

8.9. Determine the modified equation for the centered method (4.19), and show that the diffusion coefficient is always negative and this equation is hence ill posed. Recall that the method (4.19) is unstable for all fixed $\Delta t/\Delta x$.

9

Variable-Coefficient Linear Equations

In the preceding chapters we have primarily studied linear problems with constant co-efficients. Many interesting problems involve spatially-varying coefficients. This chapter is devoted to exploring some of the issues that arise, both analytically and numerically, in this case.

There are several distinct forms that a variable-coefficient hyperbolic system might take, each of which arises naturally in some contexts. One possibility is that the coefficient matrix A multiplying q_x varies with x, so the system is

$$q_t + A(x)q_x = 0. \tag{9.1}$$

This system is hyperbolic in some domain if $A(x)$ is diagonalizable with real eigenvalues at each x in the domain. Such problems still model wave propagation, and the finite volume methods developed in previous chapters can be applied fairly directly, in spite of the fact that this system (9.1) is not in conservation form and there is no flux function.

Another form of variable-coefficient problem that often arises is

$$q_t + (A(x)q)_x = 0, \tag{9.2}$$

in which the matrix $A(x)$ appears within the x-derivative. For the constant-coefficient problem with $A(x) \equiv A$, the two forms (9.1) and (9.2) are identical, but with variable coefficients they are distinct and have different solutions. The equation (9.2) is a conservation law, with the flux function

$$f(q, x) = A(x)q. \tag{9.3}$$

In this case the flux function depends explicitly on the location x as well as on the value of the conserved quantities q. Again this equation is hyperbolic whenever $A(x)$ is diagonalizable with real eigenvalues.

For a given physical problem it may be possible to derive an equation of either form (9.1) or (9.2), depending on how the vector q is chosen. For example, in Section 9.1 we will see how flow through a pipe can lead to an advection equation of either form, depending on how q is defined.

We can go back and forth between the forms at the expense of introducing source terms. By applying the product rule to the x-derivative in (9.2), we could rewrite this equation in the form

$$q_t + A(x)q_x = -A'(x)q,$$

which has the form of (9.1) with the addition of a source term. Conversely, we could rewrite (9.1) as

$$q_t + (A(x)q)_x = A'(x)q,$$

which now has the form of a conservation law with a source term. Normally we wish to avoid adding source terms to the equation if they can be avoided by a better choice of variables.

Another form of variable coefficients that often arises naturally is a *capacity function* $\kappa(x)$, as described in the context of heat capacity in Section 2.3, giving systems of the form

$$\kappa(x)q_t + (A(x)q)_x = 0, \qquad (9.4)$$

for example. Again this could be manipulated into other forms, e.g.,

$$q_t + \kappa^{-1}(x)A(x)q_x = -\kappa^{-1}(x)A'(x)q,$$

but for problems where κq is the proper conserved quantity it may be preferable to work directly with the form (9.4), using the capacity-form differencing algorithms introduced in Section 6.16.

9.1 Advection in a Pipe

Consider an incompressible fluid flowing through a pipe with variable cross-sectional area $\kappa(x)$, and suppose we want to model the advection of some tracer down the length of the pipe. We denote the area by κ, since we will see that this is a natural capacity function. There are several ways we might model this, depending on how we choose our variables, leading to different forms of the variable-coefficient advection equation.

We are assuming here that a one-dimensional formulation is adequate, i.e., that all quantities vary only with x (distance along the pipe) and t, and are essentially constant across any given cross section. In reality the flow through a pipe with varying diameter cannot be truly one-dimensional, since there must be a velocity component in the radial direction in regions where the walls converge or diverge. But we will assume this is sufficiently small that it can be safely ignored, which is true if the variation in κ is sufficiently smooth. The fluid velocity is then given by a single axial velocity $u(x)$ that varies only with x. (See Sections 9.4.1 and 9.4.2 for other contexts where this makes sense even if u is discontinuous.)

Note that if $\kappa(x)$ is measured in square meters and $u(x)$ in meters per second, say, then the product $\kappa(x)u(x)$ has units of cubic meters per second and measures the volume of fluid passing any point per unit time. Since the fluid is assumed to be incompressible, this must be the same at every point in the pipe, and we will denote the flow rate by U,

$$U = \kappa(x)u(x) \equiv \text{constant}. \qquad (9.5)$$

If we know U and the cross-sectional area $\kappa(x)$, then we can compute

$$u(x) = U/\kappa(x). \tag{9.6}$$

(In Section 9.4 we consider a formulation in which $\kappa(x)u(x)$ need not be constant.)

Now suppose that we introduce a tracer into the fluid (e.g., some food coloring or chemical contaminant into water, with very small concentration so that it does not affect the fluid dynamics). We wish to study how a given initial concentration profile advects downstream. There are two distinct ways we might choose to measure the concentration, leading to different forms of the advection equation.

One approach would be to measure the concentration in grams per cubic meter (mass per unit volume of fluid), which is what we would probably actually measure if we took a small sample of fluid from the pipe and determined the concentration. Call this variable $\bar{q}(x, t)$.

However, since we are solving a one-dimensional problem with quantities assumed to vary in only the x-direction, another natural possibility would be to measure the concentration in units of mass per unit length of the pipe (grams per meter). If we call this variable $q(x, t)$, then

$$\int_{x_1}^{x_2} q(x, t)\, dx \tag{9.7}$$

measures the total mass[1] of tracer in the section of pipe from x_1 to x_2. Since this mass can only change due to tracer moving past the endpoints, q is the proper conserved quantity for this problem. The one-dimensional velocity $u(x)$ measures the rate at which tracer moves past a point and the product $u(x)q(x, t)$ is the flux in grams per second. Hence with this choice of variables we obtain the conservative advection equation

$$q_t + (u(x)q)_x = 0. \tag{9.8}$$

There is a simple relation between q and $\bar{q}$, given by

$$q(x, t) = \kappa(x)\bar{q}(x, t), \tag{9.9}$$

since multiplying the mass per unit volume by the cross-sectional area gives the mass per unit length. Hence the total mass of tracer in $[x_1, x_2]$, given by (9.7), can be rewritten as

$$\int_{x_1}^{x_2} \kappa(x)\bar{q}(x, t)\, dx.$$

With this choice of variables we see that the cross-sectional area acts as a capacity function, which is quite natural in that this area clearly determines the fluid capacity of the pipe at each point.

Using the relation (9.9) in (9.8) gives an advection equation for $\bar{q}$,

$$\kappa(x)\bar{q}_t + (u(x)\kappa(x)\bar{q})_x = 0. \tag{9.10}$$

---

[1] In problems where chemical kinetics is involved, we should measure the "mass" in moles rather than grams, and the density in moles per meter or moles per cubic meter.

But now notice that applying (9.5) allows us to rewrite this as

$$\kappa(x)\bar{q}_t + U\bar{q}_x = 0. \tag{9.11}$$

Dividing by κ and using (9.6) gives

$$\bar{q}_t + u(x)\bar{q}_x = 0 \tag{9.12}$$

as another form of the advection equation.

Comparing the advection equations (9.8) and (9.12) shows that they have the form of (9.2) and (9.1), respectively. The velocity $u(x)$ is the same in either case, but depending on how we measure the concentration, we obtain either the nonconservative or the conservative form of the equation. The form (9.11) is a third distinct form, in which the capacity function and flow rate appear instead of the velocity.

The nonconservative form (9.12) of the advection equation is often called the *transport equation* or the *color equation*. If we think of food coloring in water, then it is $\bar{q}$, the mass per unit volume, that determines the color of the water. If we follow a parcel of water through the pipe, we expect the color to remain the same even as the area of the pipe and velocity of the water change. This is easily verified, since $\bar{q}(x, t)$ is constant along characteristic curves (recall the discussion of Section 2.1.1). By contrast the conserved quantity q is not constant along characteristic curves. The mass per unit length varies with the cross-sectional area even if the color is constant. From (9.8) we obtain $q_t + u(x)q_x = -u'(x)q$, so that along any characteristic curve $X(t)$ satisfying $X'(t) = u(X(t))$ we have

$$\frac{d}{dt}q(X(t), t) = -u'(X(t))\, q(X(t), t). \tag{9.13}$$

9.2 Finite Volume Methods

Natural upwind finite difference methods for the equations (9.8) and (9.12) are easy to derive and take the following form in the case $U > 0$:

$$Q_i^{n+1} = Q_i^n - \frac{\Delta t}{\Delta x}\left[u(x_i)Q_i^n - u(x_{i-1})Q_{i-1}^n\right] \tag{9.14}$$

and

$$\bar{Q}_i^{n+1} = \bar{Q}_i^n - \frac{\Delta t}{\Delta x}u(x_i)\left(\bar{Q}_i^n - \bar{Q}_{i-1}^n\right) \tag{9.15}$$

respectively. Variations of each formula are possible in which we use $u(x_{i-1/2})$ in place of $u(x_i)$. Either choice gives a first-order accurate approximation. In this section we will see how to derive these methods using the finite volume framework of Godunov's method, in a manner that allows extension to high-resolution methods.

The Riemann problem at the cell interface $x_{i-1/2}$ now takes the form of an advection equation with piecewise constant coefficients as well as piecewise constant initial data. The solution depends on what form of the advection equation we are solving. It also depends on how we discretize the velocity $u(x)$. Two natural possibilities are:

1. Associate a velocity u_i with each grid cell. We might define u_i as the pointwise value $u_i = u(x_i)$, or we could compute the cell average of $u(x)$, or we could view the pipe as having a piecewise constant cross-sectional area with value κ_i in the ith cell, and then take $u_i = U/\kappa_i$.

2. Associate a velocity $u_{i-1/2}$ with each cell interface $x_{i-1/2}$.

There are certain advantages and disadvantages of each form, and which is chosen may depend on the nature of the problem being solved. To begin, we will concentrate on the first choice. For some problems, however, it is more natural to use edge velocities $u_{i-1/2}$, and this choice is discussed in Section 9.5. This is true in particular when we extend the algorithms to more than one space dimension; see Section 18.2.

9.3 The Color Equation

In the case of the color equation (9.12), the Riemann problem models a jump in color from $\bar{Q}_{i-1}$ to $\bar{Q}_i$ at $x_{i-1/2}$. Since color is constant along particle paths, this jump in color will simply propagate into cell C_i at the velocity u_i in this cell, and so in the notation of Chapter 4 we have a wave $\mathcal{W}_{i-1/2}$ (omitting the superscript 1, since there is only one wave) with jump

$$\mathcal{W}_{i-1/2} = \bar{Q}_i - \bar{Q}_{i-1},$$

and speed

$$s_{i-1/2} = u_i.$$

Again we are assuming positive velocities u_i in each cell C_i. If $U < 0$, then we would have $s_{i-1/2} = u_{i-1}$, since the wave would enter cell $i - 1$. In spite of the fact that the color equation is not in conservation form, we can still view $\bar{Q}_i^n$ as an approximation to the cell average of $\bar{q}(x, t)$ at time t_n. In this case $\bar{Q}_i^n$ is simply the "average color" over this cell, and $\Delta x \, \bar{Q}_i^n$ is not the total mass of any conserved quantity. It is no longer true that the change in this cell average is given by a flux difference, but it *is* true that the change can be computed from the wave and speed. Note that (for $U > 0$) only the wave $\mathcal{W}_{i-1/2}$ enters cell C_i, and the value of $\bar{q}$ in this cell is changed by $\bar{Q}_i - \bar{Q}_{i-1}$ at each point the wave has reached. The wave propagates a distance $s_{i-1/2} \Delta t$ over the time step and hence has covered the fraction $s_{i-1/2} \Delta t/\Delta x$ of the cell. The cell average is thus modified by

$$\bar{Q}_i^{n+1} = \bar{Q}_i^n - \frac{\Delta t}{\Delta x} s_{i-1/2} \mathcal{W}_{i-1/2}.$$

This gives the upwind method (9.15). Also note that this has exactly the form (4.43),

$$\bar{Q}_i^{n+1} = \bar{Q}_i^n - \frac{\Delta t}{\Delta x} \left(\mathcal{A}^+ \Delta \bar{Q}_{i-1/2} + \mathcal{A}^- \Delta \bar{Q}_{i+1/2} \right),$$

if we define

$$\mathcal{A}^+ \Delta \bar{Q}_{i-1/2} = s_{i-1/2} \mathcal{W}_{i-1/2},$$
$$\mathcal{A}^- \Delta \bar{Q}_{i-1/2} = 0. \tag{9.16}$$

These formulas have been presented for the case $U > 0$, so that all velocities u_i are positive. The more general formulas that can also be used in the case $U < 0$ are

$$\mathcal{W}_{i-1/2} = \bar{Q}_i^n - \bar{Q}_{i-1}^n,$$

$$s_{i-1/2} = \begin{cases} u_i & \text{if } U > 0; \\ u_{i-1} & \text{if } U < 0, \end{cases}$$

$$= u_i^+ + u_{i-1}^-,$$ (9.17)

$$\mathcal{A}^+ \Delta \bar{Q}_{i-1/2} = s_{i-1/2}^+ \mathcal{W}_{i-1/2},$$

$$\mathcal{A}^- \Delta \bar{Q}_{i-1/2} = s_{i-1/2}^- \mathcal{W}_{i-1/2},$$

using the notation (4.40).

9.3.1 High-Resolution Corrections

The first-order upwind method for the color equation can be greatly improved by including high-resolution corrections as developed in Chapter 6. The improved method again has the form (5.7),

$$\bar{Q}_i^{n+1} = \bar{Q}_i^n - \frac{\Delta t}{\Delta x}\left(\mathcal{A}^+ \Delta Q_{i-1/2} + \mathcal{A}^- \Delta Q_{i+1/2}\right) - \frac{\Delta t}{\Delta x}\left(\tilde{F}_{i+1/2} - \tilde{F}_{i-1/2}\right),$$ (9.18)

where

$$\tilde{F}_{i-1/2} = \frac{1}{2}|s_{i-1/2}|\left(1 - \frac{\Delta t}{\Delta x}|s_{i-1/2}|\right)\tilde{\mathcal{W}}_{i-1/2}.$$ (9.19)

The modified wave $\tilde{\mathcal{W}}_{i-1/2}$ is obtained by applying a wave limiter as described in Chapter 6.

It is important to note, however, that this method is *not* formally second-order accurate when applied to a smooth solution, even when no limiter function is used. The local truncation error is found to be (see Exercise 9.1)

$$\text{local truncation error} = -\frac{1}{2}[\Delta x - u(x)\,\Delta t]u'(x)q_x(x, t) + \mathcal{O}(\Delta x^2).$$ (9.20)

The method is formally second-order accurate only in the case of constant coefficients, where $u'(x) \equiv 0$. However, this truncation error has a quite different form than the truncation error for the first-order upwind method (9.15), which is

$$\Delta x\, u(x)q_{xx}(x, t) + \mathcal{O}(\Delta x^2).$$ (9.21)

This truncation error depends on q_{xx} and leads to substantial numerical dissipation, as with any first-order upwind method. The truncation error (9.20), on the other hand, is a multiple of q_x instead. This can be viewed as corresponding to a small error in the propagation speed. The modified equation (see Section 8.6) for the high-resolution method is

$$q_t + u(x)q_x = -\frac{1}{2}[\Delta x - u(x)\,\Delta t]u'(x)q_x,$$

which we can rewrite (ignoring higher-order terms) as

$$q_t + u\left(x + \frac{1}{2}[\Delta x - u(x)\Delta t]\right)q_x = 0 \qquad (9.22)$$

This is again the color equation, but with the velocity evaluated at a point that is shifted by less than a grid cell. The solution to this equation looks essentially the same as for the correct color equation, but slightly offset. In particular there is no diffusion or smearing of the solution (at least not at the $\mathcal{O}(\Delta x)$ level), and hence we expect the solution obtained with this method to look much better than what would be obtained from the first-order upwind method.

A true second-order accurate method can be derived using the Lax–Wendroff approach (Exercise 9.2), but results are typically similar to what is produced by the method above when the solution is smooth. The high-resolution method with limiters included works very well in practice, and much better than the formally second-order method in cases where the solution is discontinuous.

9.3.2 Discontinuous Velocities

For flow in a pipe, reducing the problem to a one-dimensional advection equation is only valid if the diameter of the pipe, and hence the velocity, is very slowly varying. Otherwise the true fluid velocity may have a substantial radial component. This is particularly true at a point where the diameter is discontinuous. Here the motion will be fully multidimensional and perhaps even turbulent. However, one-dimensional approximations are frequently used even in this case in many engineering applications involving transport through complicated networks of pipes, for example.

There are also other applications where a one-dimensional advection equation with rapidly varying or even discontinuous velocities is physically relevant. One example, discussed in Section 9.4.2, arises in traffic flow along a one-lane road with a discontinuous speed limit. Another example is discussed in Section 9.4.1.

An advantage of the high-resolution finite volume methods is that they typically perform very well even when the coefficients in the equation are discontinuous. You may wish to experiment with the examples in [claw/book/chap9/color/discontu].

9.4 The Conservative Advection Equation

Now suppose we are solving the conservative equation (9.8), $q_t + (u(x)q)_x = 0$, with velocities u_i given in each grid cell. Again we can use the form (5.7) after defining the fluctuations $\mathcal{A}^+\Delta Q_{i-1/2}$ and $\mathcal{A}^-\Delta Q_{i-1/2}$ in terms of waves via the solution to the Riemann problem at $x_{i-1/2}$. To do so properly we must carefully consider the Riemann problem, which now consists of using piecewise constant data defined by Q_{i-1} and Q_i and also piecewise constant coefficients in the advection equation,

$$
\begin{aligned}
q_t + u_{i-1}q_x = 0 \qquad & \text{if } x < x_{i-1/2}, \\
q_t + u_i q_x = 0 \qquad & \text{if } x > x_{i-1/2}.
\end{aligned}
\qquad (9.23)
$$

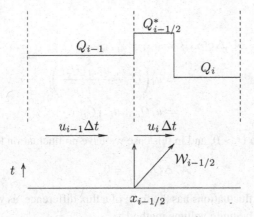

Fig. 9.1. Solution to the Riemann problem for the conservative advection equation when $u_{i-1} > u_i$.

The solution is *not* simply a single wave propagating at velocity u_i with strength $Q_i - Q_{i-1}$. The correct solution is indicated in Figure 9.1, and consists of *two* discontinuities in the solution: a stationary one at $x_{i-1/2}$ where Q_{i-1} jumps to some new value $Q^*_{i-1/2}$, and the wave

$$\mathcal{W}_{i-1/2} = Q_i - Q^*_{i-1/2} \qquad (9.24)$$

propagating at the expected speed

$$s_{i-1/2} = u_i \qquad (9.25)$$

(again we assume $U > 0$). The discontinuity at $x_{i-1/2}$ arises from the fact that the jump in u at $x_{i-1/2}$ corresponds to a jump in the cross-sectional area of the pipe. (A jump to a larger area $\kappa_i > \kappa_{i-1}$ is shown in Figure 9.1, in which case $u_i < u_{i-1}$.) Consequently, the density of tracer q, which is measured in the one-dimensional manner as mass per unit length, must increase, since there is suddenly more liquid and hence more tracer per unit length of pipe. The "color" $\bar{q}$ measured in mass per unit volume, on the other hand, does not experience any discontinuity at $x_{i-1/2}$, and so we did not have to worry about this effect in solving the color equation.

The new value $Q^*_{i-1/2}$ that arises in solving the Riemann problem can be found using the fact that the flux $f = uq$ must be continuous at $x_{i-1/2}$ in the Riemann solution. The same amount of tracer is leaving cell $\mathcal{C}_{i-1}$ as entering cell $\mathcal{C}_i$. This yields

$$u_{i-1} Q_{i-1} = u_i Q^*_{i-1/2}$$

and hence

$$Q^*_{i-1/2} = \frac{u_{i-1} Q_{i-1}}{u_i}. \qquad (9.26)$$

This value is used in the wave (9.24). The fluctuation is computed in terms of this wave and the speed (9.25) as

$$\mathcal{A}^+ \Delta Q_{i-1/2} = s_{i-1/2} \mathcal{W}_{i-1/2}$$

$$= u_i \left(Q_i - \frac{u_{i-1} Q_{i-1}}{u_i} \right)$$

$$= u_i Q_i - u_{i-1} Q_{i-1}. \tag{9.27}$$

Again this is for the case $U > 0$, and in this case we have no fluctuation to the left:

$$\mathcal{A}^- \Delta Q_{i-1/2} = 0. \tag{9.28}$$

Note that the sum of the fluctuations has the form of a flux difference, as we expect for this conservative equation. The finite volume method

$$Q_i^{n+1} = Q_i - \frac{\Delta t}{\Delta x} \left(\mathcal{A}^+ \Delta Q_{i-1/2} + \mathcal{A}^- \Delta Q_{i+1/2} \right) \tag{9.29}$$

reduces to the natural upwind method (9.14).

If $U < 0$, then we find that

$$Q_{i-1/2}^* = \frac{u_i Q_i}{U_{i-1}},$$

$$\mathcal{W}_{i-1/2} = Q_{i-1/2}^* - Q_{i-1},$$

$$s_{i-1/2} = u_{i-1}, \tag{9.30}$$

$$\mathcal{A}^- \Delta Q_{i-1/2} = s_{i-1/2} \mathcal{W}_{i-1/2} = u_i Q_i - u_{i-1} Q_{i-1},$$

$$\mathcal{A}^+ \Delta Q_{i-1/2} = 0.$$

In this case (9.29) again reduces to the natural upwind method. High-resolution correction terms can be added using (9.18) and (9.19).

Since the Riemann solution now involves two jumps, one might wonder why we don't need two waves $\mathcal{W}_{i-1/2}^1$ and $\mathcal{W}_{i-1/2}^2$ instead of just one, as introduced above. The answer is that the second wave moves at speed $s = 0$. We could include it in the Riemann solution, but it would drop out of the expression for $\mathcal{A}^+ \Delta Q_{i-1/2}$, since we sum $s^p \mathcal{W}^p$. The corresponding high-resolution corrections also drop out when $s = 0$. So nothing would be gained by including this wave except additional work in applying the updating formulas.

9.4.1 Conveyer Belts

The solution illustrated in Figure 9.1 can perhaps be better understood by considering a different context than fluid flow in a pipe. Consider a series of conveyer belts, each going for the length of one grid cell with velocity u_i, and let $q(x, t)$ measure the density of packages traveling down this series of belts. See Figure 9.2. If the velocity u_i is less than u_{i-1}, then the density will increase at the junction $x_{i-1/2}$ just as illustrated in Figure 9.1. See also Figure 9.3 for another interpretation, discussed in the next subsection.

The structure illustrated in Figure 9.1 with two waves can also be derived by solving the Riemann problem for a system of two conservation laws, obtained by introducing u as a second conserved variable; see Exercise 13.11.

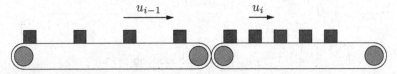

Fig. 9.2. Packages moving from a conveyer belt with speed u_{i-1} to a slower belt with speed u_i. The density of packages increases at the junction.

9.4.2 Traffic Flow

Similar advection equations arise in studying the flow of cars along a one-lane highway. To obtain a linear equation, we assume traffic is light enough that cars are always traveling at the speed limit $u(x)$, which can vary with x, the distance along the highway. If $u(x)$ is piecewise constant, then this is exactly the same problem as the conveyer-belt problem mentioned in Section 9.4.1, with cars replacing packages.

For heavier traffic the speed typically also depends on the density of cars and the problem becomes nonlinear. This more interesting case is examined in detail in Chapter 11.

Let $q(x, t)$ be the density of cars at point x at time t. Of course, at any particular point x there is either a car or no car, so the concept of density at a point appears to be meaningless. What we mean should be clear, however: over a reasonably long stretch of road from x_1 to x_2 the number of cars present at time t should be roughly

$$\text{number of cars between } x_1 \text{ and } x_2 \approx \int_{x_1}^{x_2} q(x, t)\,dx. \tag{9.31}$$

It will be convenient to measure the density in units of cars per car length, assuming all cars have the same length. Then empty highway corresponds to $q = 0$, bumper-to-bumper traffic corresponds to $q = 1$, and in general $0 \le q \le 1$. Then for (9.31) to make sense we must measure distance x in car lengths. Also we measure the speed limit $u(x)$ in car lengths per time unit. The flux of cars is given by $f(q) = u(x)q$ (in cars per time unit) and we obtain the conservation law (9.8).

One way to simulate traffic flow is to track the motion of individual vehicles, assuming a finite set of m cars on the highway. Let $X_k(t)$ be the location of the kth car at time t. Then the motion of each car can be obtained by solving the ordinary differential equation

$$X_k'(t) = u(X_k(t))$$

with initial conditions corresponding to its initial location $X_k(0)$. Note that cars move along characteristics of the advection equation. (This is not true more generally if the velocity depends on the density; see Chapter 11.)

Figure 9.3 illustrates the simulation of a line of cars with an initial density variation corresponding to a Riemann problem, traveling along a highway with a discontinuous speed limit

$$u(x) = \begin{cases} 2 & \text{if } x < 0, \\ 1 & \text{if } x > 0. \end{cases}$$

The paths of individual cars are plotted over time $0 \le t \le 20$. The *density* $q_k(t)$ observed by each driver is also plotted at both the initial and final times. For this discrete simulation

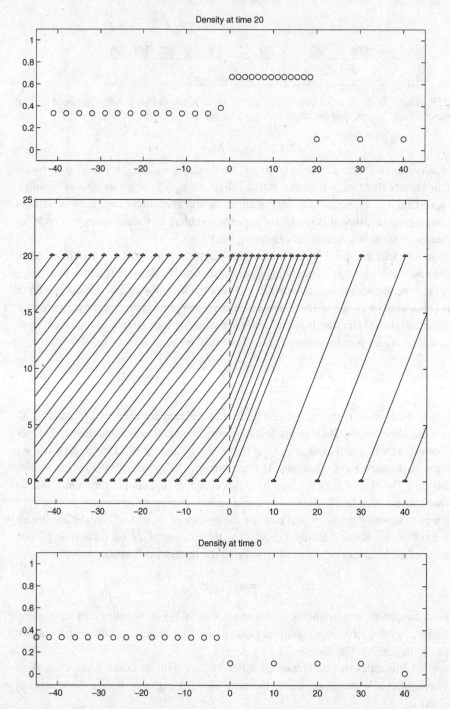

Fig. 9.3. Traffic flow model corresponding to the Riemann problem of Figure 9.1, with a change in speed limit at $x = 0$. Vehicle trajectories are shown, along with the density observed by each driver. We can also view the cars as representing packages moving from one conveyer belt to a slower one as in Figure 9.2. [`claw/book/chap9/traffic`]

a discrete density is defined for each car by

$$q_k(t) = \frac{1}{X_{k+1}(t) - X_k(t)}, \tag{9.32}$$

based on the distance to the car immediately ahead. This can be shown to be consistent with the previous notion of density; see Exercise 9.4.

The density at intermediate times can also be observed from the car paths in Figure 9.3. Taking a slice at any fixed time, the density is inversely proportional to the distance between paths.

Note that the density of cars changes discontinuously at the point $x = 0$ where the speed limit changes. In this simple model we assume cars and drivers can adjust speed instantaneously in response to the speed limit. This viewpoint may help to understand Figure 9.1. Of course if u_{i-1}/u_i is sufficiently large, then $Q^*_{i-1/2}$ in (9.26) may be greater than 1 even if Q_{i-1} is not. Although this causes no problem mathematically, with our definition of density this shouldn't happen physically. This is a case where cars are approaching the discontinuity in speed limit too rapidly to be accommodated in the slower region even at the bumper-to-bumper density of $q = 1$. In this case the linear model is definitely inadequate, and instead a nonlinear shock wave would arise in the real world. See Section 16.4.1 for a discussion of nonlinear traffic flow with a discontinuous speed limit.

For the linear traffic flow problem, the density naturally satisfies a conservative advection equation of the form (9.8). One can, however, define a new variable $\bar{q}(x, t)$ that instead satisfies the color equation (9.12), by setting

$$\bar{q}(x, t) = u(x)q(x, t).$$

This is simply the flux of cars. We have

$$\bar{q}_t = (uq)_t = uq_t = -u(uq)_x = -u\bar{q}_x,$$

and so (9.12) is satisfied. You should convince yourself that it makes sense for the flux of cars to be constant along characteristics, as must hold for the color equation.

9.5 Edge Velocities

So far we have assumed that the variable velocity $u(x)$ is specified by a constant value u_i within the ith grid cell. In some cases it is more natural to instead assume that a velocity $u_{i-1/2}$ is specified at each cell interface. This can be viewed as a transfer rate between the cells C_{i-1} and C_i.

9.5.1 The Color Equation

Solving the color equation $\bar{q}_t + u(x)\bar{q}_x = 0$ with edge velocities specified is very simple. We need only set

$$\begin{aligned} \mathcal{W}_{i-1/2} &= \bar{Q}_i - \bar{Q}_{i-1}, \\ s_{i-1/2} &= u_{i-1/2}, \end{aligned} \tag{9.33}$$

and use the usual formula

$$A^+ \Delta Q_{i-1/2} = (s_{i-1/2})^+ \mathcal{W}_{i-1/2},$$
$$A^- \Delta Q_{i-1/2} = (s_{i-1/2})^- \mathcal{W}_{i-1/2}. \tag{9.34}$$

9.5.2 The Conservative Equation

The conservative advection equation $q_t + (u(x)q)_x = 0$ can also be solved using the waves and speeds

$$\mathcal{W}_{i-1/2} = Q_i - Q_{i-1},$$
$$s_{i-1/2} = u_{i-1/2}, \tag{9.35}$$

to define second-order corrections (perhaps modified by limiters). However, we cannot use the expressions (9.34) to define the fluctuations. This would not lead to a conservative method, since (4.54) would not be satisfied. There are several different approaches that can be used to define fluctuations that do satisfy (4.54) and lead to successful methods. One approach is to notice that an upwind flux $F_{i-1/2}$ is naturally defined by

$$F_{i-1/2} = u_{i-1/2}^+ Q_{i-1} + u_{i-1/2}^- Q_i$$

at the edge between the cells. Recall that the goal in defining $A^\pm \Delta Q_{i-1/2}$ is to implement the flux-differencing algorithm

$$Q_i^{n+1} = Q_i^n - \frac{\Delta t}{\Delta x} \left(F_{i+1/2}^n - F_{i-1/2}^n \right) \tag{9.36}$$

via the formula

$$Q_i^{n+1} = Q_i^n - \frac{\Delta t}{\Delta x} \left(A^+ \Delta Q_{i-1/2} + A^- \Delta Q_{i+1/2} \right). \tag{9.37}$$

We can accomplish this by setting

$$A^+ \Delta Q_{i-1/2} = F_i - F_{i-1/2},$$
$$A^- \Delta Q_{i-1/2} = F_{i-1/2} - F_{i-1}, \tag{9.38}$$

where the cell-centered flux values F_i are chosen in an arbitrary way. When inserted into (9.37) the value F_i cancels out and (9.36) results. For simplicity one could even use $F_i = 0$ in each cell.

Aesthetically it is nicer to use some approximation to the flux in cell i, for example

$$F_i = \left(u_{i-1/2}^+ + u_{i+1/2}^- \right) Q_i, \tag{9.39}$$

so that $A^\pm \Delta Q_{i-1/2}$ have the physical interpretation of flux differences. Note that if all velocities are positive, then these formulas reduce to

$$F_{i-1/2} = u_{i-1/2} Q_{i-1}, \qquad F_i = u_{i-1/2} Q_i,$$

so that

$$\mathcal{A}^+ \Delta Q_{i-1/2} = u_{i-1/2}(Q_i - Q_{i-1}),$$
$$\mathcal{A}^- \Delta Q_{i-1/2} = (u_{i-1/2} - u_{i-3/2})Q_{i-1}. \tag{9.40}$$

The right-going fluctuation is $s_{i-1/2}\mathcal{W}_{i-1/2}$ (as for the color equation), while the left-going fluctuation accounts for the variation of u across cell $\mathcal{C}_{i-1}$. Note that even though all flow is to the right, some information must flow to the left, since the accumulation rate in cell $\mathcal{C}_{i-1}$ depends on the outflow velocity $u_{i-1/2}$ as well as the inflow velocity $u_{i-3/2}$. These fluctuations result in the first-order method

$$Q_i^{n+1} = Q_i^n - \frac{\Delta t}{\Delta x}\left(u_{i+1/2}Q_i^n - u_{i-1/2}Q_{i-1}^n\right). \tag{9.41}$$

Note also that the splitting of (9.40) gives distinct approximations to the two terms that arise when we rewrite the conservative equation in the form

$$q_t + u(x)q_x + u'(x)q = 0. \tag{9.42}$$

Another way to derive the method (9.37) with the splitting (9.40) is to start with the form (9.42) and view this as the color equation with a source term $-u'(x)q$. Then $\mathcal{W}_{i-1/2}, s_{i-1/2}$, and $\mathcal{A}^+\Delta Q_{i-1/2}$ all come from the color equation. The term $\mathcal{A}^-\Delta Q_{i-1/2}$ from (9.40) is different from the value 0 that would be used for the color equation, and can be viewed as a device for introducing the appropriate source term into cell $\mathcal{C}_{i-1}$ (rather than using a fractional-step method as discussed in Chapter 17). This approach has the advantage of maintaining conservation, while a fractional-step method might not.

If second-order correction terms are added to this algorithm using the waves and speeds (9.35) and the formula (9.18), then it can be shown that the method is formally second-order accurate if no limiters are used (Exercise 9.6).

9.6 Variable-Coefficient Acoustics Equations

Previously we have studied the equations of linear acoustics in a uniform medium where the density ρ_0 and the bulk modulus K_0 are the same at every point. We now consider acoustics in a one-dimensional heterogeneous medium in which the values $\rho(x)$ and $K(x)$ vary with x. The theory and numerical methods developed here also apply to elasticity or electromagnetic waves in heterogeneous media, and to other hyperbolic systems with variable coefficients.

An important special case is a *layered medium* in which ρ and K are piecewise constant. In applying finite volume methods we will assume the medium has this structure, at least at the level of the grid cells. We assume the ith cell contains a uniform material with density ρ_i and bulk modulus K_i. Smoothly varying material parameters can be approximated well in this manner on a sufficiently fine grid. Our discussion will focus primarily on the case of piecewise constant media, since we need to understand this case in detail in order to implement methods based on Riemann solvers. See Section 9.14 for discussion of how to choose the ρ_i and K_i if the coefficients vary on the subgrid scale.

It is possible to solve the variable-coefficient acoustics equations in conservation form by writing them in terms of momentum and strain; see [273] for example. Here we explore the nonconservative formulation that arises when pressure and velocity are used as the dependent variables. These are the quantities that must be continuous at an interface between different materials and are also physically more intuitive. This example provides a nice case study in solving a nonconservative variable-coefficient system, which can form a basis for developing algorithms for other hyperbolic systems that perhaps cannot be rewritten in conservation form.

The form (2.52) of the constant-coefficient acoustic equations generalizes naturally to

$$q_t + A(x)q_x = 0, \tag{9.43}$$

where

$$q(x,t) = \begin{bmatrix} p(x,t) \\ u(x,t) \end{bmatrix}, \qquad A(x) = \begin{bmatrix} 0 & K(x) \\ 1/\rho(x) & 0 \end{bmatrix}. \tag{9.44}$$

At each point x we can diagonalize this matrix as in Section 2.8,

$$A(x) = R(x)\Lambda(x)R^{-1}(x). \tag{9.45}$$

If we define the *sound speed*

$$c(x) = \sqrt{K(x)/\rho(x)} \tag{9.46}$$

and the *impedance*

$$Z(x) = \rho(x)c(x) = \sqrt{K(x)\rho(x)}, \tag{9.47}$$

then the eigenvector and eigenvalue matrix of (9.45) are

$$R(x) = \begin{bmatrix} -Z(x) & Z(x) \\ 1 & 1 \end{bmatrix}, \qquad \Lambda(x) = \begin{bmatrix} -c(x) & 0 \\ 0 & c(x) \end{bmatrix}. \tag{9.48}$$

9.7 Constant-Impedance Media

An interesting special case arises if $\rho(x)$ and $K(x)$ are related in such a way that $Z(x) = Z_0$ is constant, which happens if $K(x) = Z_0^2/\rho(x)$ everywhere. Then the eigenvector matrix $R(x) \equiv R$ is constant in space. In this case we can diagonalize the system (9.43) by multiplying by R^{-1} to obtain

$$(R^{-1}q)_t + [R^{-1}A(x)R](R^{-1}q)_x = 0.$$

Defining

$$w(x,t) = R^{-1}q(x,t)$$

as in Section 2.9, we can write the above system as

$$w_t + \Lambda(x)w_x = 0. \tag{9.49}$$

This diagonal system decouples into two scalar advection equations (color equations in the terminology of Section 9.1)

$$w_t^1 - c(x)w_x^1 = 0 \quad \text{and} \quad w_t^2 + c(x)w_x^2 = 0.$$

Left-going and right-going sound waves propagate independently of one another with the variable sound speed $c(x)$.

Example 9.1. Consider a piecewise constant medium with a single discontinuity at $x = 0$,

$$\rho(x) = \begin{cases} \rho_l & \text{if } x < 0, \\ \rho_r & \text{if } x > 0, \end{cases} \qquad K(x) = \begin{cases} K_l & \text{if } x < 0, \\ K_r & \text{if } x > 0. \end{cases} \tag{9.50}$$

Take $\rho_l = 1$, $\rho_r = 2$ and $K_l = 1$, $K_r = 0.5$, so that the impedance is $Z_0 = 1$ everywhere although the sound speed jumps from $c_l = 1$ to $c_r = 0.5$. Figure 9.4 shows a right-going acoustic wave having $p(x, t) = Z_0 u(x, t)$ everywhere. As it passes through the interface the pulse is compressed due to the change in velocity (just as would happen with the scalar color equation), but the entire wave is transmitted and there is no reflection at the interface. Compare with Figure 9.5, where there is also a jump in the impedance.

9.8 Variable Impedance

Note that the diagonalization (9.49) is possible only because R is constant in space. If $R(x)$ varies with x then $R^{-1}(x)q_x = (R^{-1}(x)q)_x - R_x^{-1}(x)q$, where $R_x^{-1}(x)$ is the derivative of $R^{-1}(x)$ with respect to x. In this case multiplying (9.43) by $R^{-1}(x)$ yields

$$w_t + \Lambda(x)w_x = R_x^{-1}(x)q$$
$$= B(x)w, \tag{9.51}$$

where the matrix $B(x)$ is given by $B(x) = R_x^{-1}(x)R(x)$ and is not a diagonal matrix. In this case we still have advection equations for w^1 and w^2, but they are now coupled together by the source terms. The left-going and right-going acoustic waves no longer propagate independently of one another.

Rather than attempting to work with this system, we will see that the Riemann problem can be posed and solved at a general interface between two different materials. By assuming the medium is layered (so the impedance is piecewise constant), we can reduce the general problem with variable coefficients to solving such Riemann problems.

Before considering the general Riemann problem, consider what happens when a wave hits an interface in a medium of the form (9.50) if there is a jump in impedance at the interface.

Example 9.2. Consider the medium (9.50) with $\rho_l = 1$, $\rho_r = 4$ and $K_l = 1$, $K_r = 1$. As in Example 9.1, the velocity jumps from $c_l = 1$ to $c_r = 0.5$, but now the impedance is also discontinuous with $Z_l = 1$ and $Z_r = 2$. Figure 9.5 shows what happens as a right-going wave hits the interface. Part of the wave is transmitted, but part is reflected as a left-going wave.

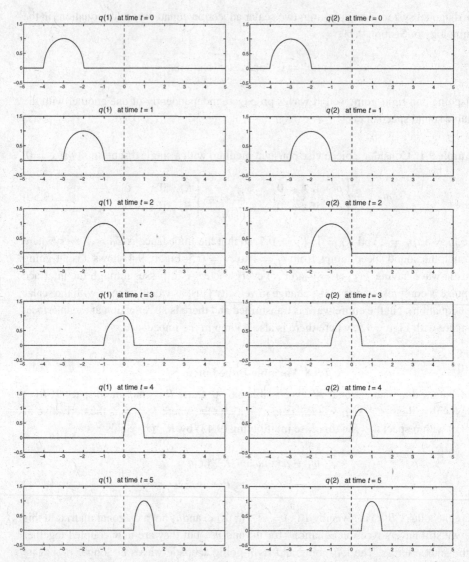

Fig. 9.4. Right-going acoustic pulse hitting a material interface (dashed line) where the sound speed changes from 1 to 0.5 but the impedance is the same. Left column: pressure. Right column: velocity. [claw/book/chap9/acoustics/interface]

If the incident pressure pulse has magnitude p_0, then the transmitted pulse has magnitude $C_T p_0$ and the reflected pulse has magnitude $C_R p_0$, where the *transmission* and *reflection coefficients* are given by

$$C_T = \frac{2Z_r}{Z_l + Z_r}, \qquad C_R = \frac{Z_r - Z_l}{Z_l + Z_r}. \tag{9.52}$$

For the example in Figure 9.5, we have $C_T = 4/3$ and $C_R = 1/3$. Partial reflection always occurs at any interface where there is an *impedance mismatch* with $Z_l \neq Z_r$. There are several ways to derive these coefficients. We will do so below by solving an appropriate Riemann problem, as motivated by the next example.

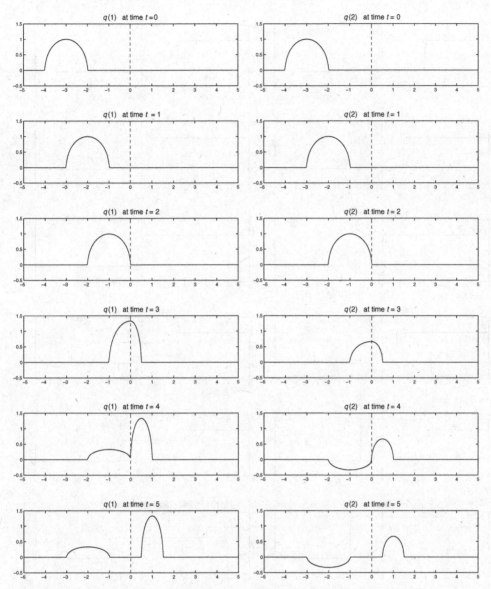

Fig. 9.5. Right-going acoustic pulse hitting a material interface (dashed line) where the sound speed changes from 1 to 0.5 and the impedance changes from 1 to 2. Part of the wave is reflected at the interface. Left column: pressure. Right column: velocity. [claw/book/chap9/acoustics/interface]

Example 9.3. To see how the transmission and reflection coefficients are related to a Riemann problem, consider the situation where the incident wave is a jump discontinuity as shown in Figure 9.6. When this discontinuity hits the interface, part of it is transmitted and part is reflected. Now suppose we ignore the first two frames of the time sequence shown in Figure 9.6 (at times $t = -2$ and $t = -1$) and consider what happens only from time $t = 0$ onwards. At time $t = 0$ we have data that is piecewise constant with a jump discontinuity at $x = 0$ in a medium that also has a jump discontinuity at $x = 0$. This is the proper generalization of the classic Riemann problem to the case of a heterogeneous medium. As time advances the discontinuity in the data resolves into two waves, one

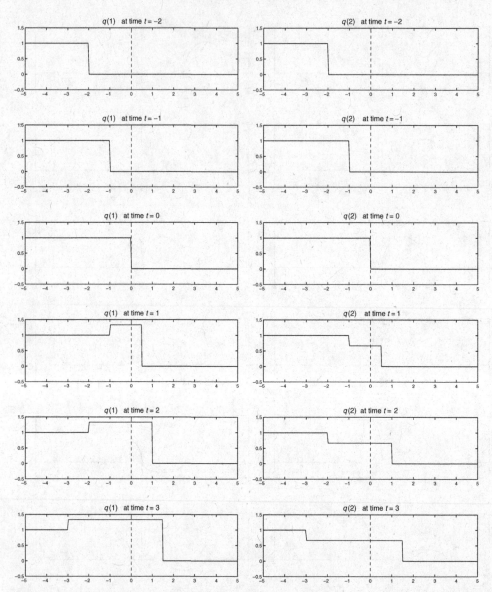

Fig. 9.6. Right-going square pulse hitting a material interface (dashed line) where the sound speed changes from 1 to 0.5 and the impedance changes from 1 to 2. The sequence from $t = 0$ onwards corresponds to the Riemann problem for variable-coefficient acoustics. Left column: pressure. Right column: velocity. [claw/book/chap9/acoustics/interface]

moving to the left at the sound speed of the medium on the left and the other moving to the right at the sound speed of that medium. Determining the magnitude of these two waves relative to the original jump will give the expressions (9.52) for the transmission and reflection coefficients. Being able to solve the general Riemann problem will also allow us to apply high-resolution finite volume methods to the general variable-coefficient acoustics problem.

9.9 Solving the Riemann Problem for Acoustics

The general Riemann problem for acoustics in a heterogeneous medium is defined by considering a piecewise constant medium of the form (9.50) together with piecewise constant initial data

$$q(x, 0) = \begin{cases} q_l & \text{if } x < 0, \\ q_r & \text{if } x > 0. \end{cases} \tag{9.53}$$

The solution to the Riemann problem consists of two acoustic waves, one moving to the left with velocity $-c_l$ and the other to the right with velocity $+c_r$. Each wave moves at the sound speed of the material in which it propagates, as indicated in Figure 9.7. At the interface $x = 0$ the pressure and velocity perturbations are initially discontinuous, but they should be continuous for $t > 0$, after the waves have departed. Hence there should be a single state q_m between the two waves as indicated in Figure 9.7 and as seen in Figure 9.6.

We know from the theory of Section 2.8 that the jump across each wave must be an eigenvector of the coefficient matrix from the appropriate material. Hence we must have

$$q_m - q_l = \alpha^1 \begin{bmatrix} -Z_l \\ 1 \end{bmatrix} \quad \text{and} \quad q_r - q_m = \alpha^2 \begin{bmatrix} Z_r \\ 1 \end{bmatrix} \tag{9.54}$$

for some scalar coefficients α^1 and α^2. Adding these two equations together yields

$$q_r - q_l = \alpha^1 \begin{bmatrix} -Z_l \\ 1 \end{bmatrix} + \alpha^2 \begin{bmatrix} Z_r \\ 1 \end{bmatrix}.$$

This leads to a linear system of equations to solve for α^1 and α^2,

$$R_{lr}\alpha = q_r - q_l,$$

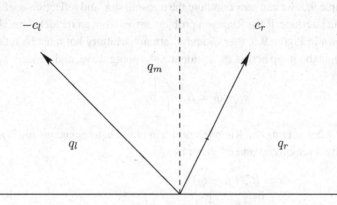

Fig. 9.7. Structure of the solution to the Riemann problem for variable-coefficient acoustics, in the x–t plane. The dashed line shows the interface between two different materials. The waves propagate at the speed of sound in each material. Between the waves is a single state q_m.

where

$$R_{lr} = \begin{bmatrix} -Z_l & Z_r \\ 1 & 1 \end{bmatrix}. \tag{9.55}$$

This is essentially the same process used to solve the Riemann problem for constant-coefficient acoustics, but with an eigenvector matrix R_{lr} that is not the matrix $R(x)$ from either side of the interface, but rather consists of the left-going eigenvector from the left medium and the right-going eigenvector from the right medium.

Since

$$R_{lr}^{-1} = \frac{1}{Z_l + Z_r} \begin{bmatrix} -1 & Z_r \\ 1 & Z_l \end{bmatrix}, \tag{9.56}$$

we find that

$$\alpha^1 = \frac{-(p_r - p_l) + Z_r(u_r - u_l)}{Z_l + Z_r},$$

$$\alpha^2 = \frac{(p_r - p_l) + Z_l(u_r - u_l)}{Z_l + Z_r}. \tag{9.57}$$

We also find that the intermediate state $q_m = q_l + \alpha^1 \mathcal{W}^1$ is given by

$$p_m = p_l - \alpha^1 Z_l = \frac{Z_r p_l + Z_l p_r}{Z_l + Z_r} - \left(\frac{Z_r Z_l}{Z_l + Z_r}\right)(u_r - u_l),$$

$$u_m = u_l + \alpha^1 = \frac{Z_l u_l + Z_r u_r}{Z_l + Z_r} - \left(\frac{1}{Z_l + Z_r}\right)(p_r - p_l). \tag{9.58}$$

If $Z_l = Z_r$, then these reduce to the formulas (3.31) and (3.32) for acoustics in a constant medium.

9.10 Transmission and Reflection Coefficients

Returning to Example 9.3, we can now compute the transmission and reflection coefficients for a wave hitting an interface. If the Riemann problem arises from an incident wave hitting an interface as shown in Figure 9.6, then q_l and q_r are not arbitrary but must be related by the fact that $q_r - q_l$ is the jump across the incident right-going wave, and hence

$$q_r - q_l = \beta \begin{bmatrix} Z_l \\ 1 \end{bmatrix} \tag{9.59}$$

for some scalar β. After solving the Riemann problem at the heterogeneous interface, the outgoing waves have strengths α^1 and α^2 given by

$$\alpha = R_{lr}^{-1}(q_r - q_l)$$

$$= \frac{\beta}{Z_l + Z_r} \begin{bmatrix} -1 & Z_r \\ 1 & Z_l \end{bmatrix} \begin{bmatrix} Z_l \\ 1 \end{bmatrix}$$

$$= \frac{\beta}{Z_l + Z_r} \begin{bmatrix} Z_r - Z_l \\ 2Z_l \end{bmatrix}. \tag{9.60}$$

We thus find that the magnitudes of the reflected and transmitted waves are

$$\alpha^1 = \left(\frac{Z_r - Z_l}{Z_l + Z_r}\right)\beta \quad \text{and} \quad \alpha^2 = \left(\frac{2Z_l}{Z_l + Z_r}\right)\beta \tag{9.61}$$

respectively. The waves are thus

$$\text{reflected wave:} \quad \alpha^1 \begin{bmatrix} -Z_l \\ 1 \end{bmatrix} = \beta \left(\frac{Z_r - Z_l}{Z_l + Z_r}\right)\begin{bmatrix} -Z_l \\ 1 \end{bmatrix},$$

$$\text{transmitted wave:} \quad \alpha^2 \begin{bmatrix} Z_r \\ 1 \end{bmatrix} = \beta \left(\frac{2Z_l}{Z_l + Z_r}\right)\begin{bmatrix} Z_r \\ 1 \end{bmatrix}.$$

Note that the pressure jump in the transmitted wave can be written as

$$\beta \left(\frac{2Z_l}{Z_l + Z_r}\right)Z_r = \left(\frac{2Z_r}{Z_l + Z_r}\right)\beta Z_l.$$

Comparing this with the pressure jump βZ_l of the incident wave (9.59) gives the transmission coefficient C_T of (9.52). The reflected wave has a pressure jump $-C_R(\beta Z_l)$, where C_R is given in (9.52). The minus sign arises from the fact that we measure the pressure jump by subtracting the value to the left of the wave from the value to the right. Since the wave is left-going, however, the gas experiences a jump in pressure of the opposite sign as the wave passes by, and it is this jump that is normally considered in defining the reflection coefficient C_R.

9.11 Godunov's Method

Having determined the solution to the Riemann problem for acoustics in a heterogeneous medium, it is easy to implement Godunov's method and other finite volume methods. Each grid cell is assigned a density ρ_i and a bulk modulus K_i, and Q_i^n represents the cell average of q over this cell at time t_n. The Riemann problem at $x_{i-1/2}$ is solved as described in Section 9.9 with $\rho_l = \rho_{i-1}$, $\rho_r = \rho_i$, etc. We obtain two waves

$$\mathcal{W}_{i-1/2}^1 = \alpha_{i-1/2}^1 r_{i-1/2}^1, \qquad \mathcal{W}_{i-1/2}^2 = \alpha_{i-1/2}^2 r_{i-1/2}^2,$$

where

$$r_{i-1/2}^1 = \begin{bmatrix} -Z_{i-1} \\ 1 \end{bmatrix}, \qquad r_{i-1/2}^2 = \begin{bmatrix} Z_i \\ 1 \end{bmatrix}.$$

These waves propagate at speeds $s_{i-1/2}^1 = -c_{i-1}$ and $s_{i-1/2}^2 = +c_i$. The coefficients $\alpha_{i-1/2}^1$ and $\alpha_{i-1/2}^2$ are given by (9.57):

$$\alpha_{i-1/2}^1 = \frac{-(p_i - p_{i-1}) + Z_i(u_i - u_{i-1})}{Z_{i-1} + Z_i},$$

$$\alpha_{i-1/2}^2 = \frac{(p_i - p_{i-1}) + Z_{i-1}(u_i - u_{i-1})}{Z_{i-1} + Z_i}. \tag{9.62}$$

Godunov's method can be implemented in the fluctuation form

$$Q_i^{n+1} = Q_i^n - \frac{\Delta t}{\Delta x}\left(\mathcal{A}^+ \Delta Q_{i-1/2} + \mathcal{A}^- \Delta Q_{i+1/2}\right)$$

if we define

$$\mathcal{A}^- \Delta Q_{i-1/2} = s_{i-1/2}^1 \mathcal{W}_{i-1/2}^1, \qquad \mathcal{A}^+ \Delta Q_{i-1/2} = s_{i-1/2}^2 \mathcal{W}_{i-1/2}^2, \qquad (9.63)$$

following the general prescription (4.42). Note that the variable-coefficient acoustic equations (9.43) are not in conservation form, and so this is not a flux-difference splitting in this case, but it still leads to the proper updating of cell averages based on waves moving into the grid cells.

For constant-coefficient acoustics, we saw in Section 4.12 that we could interpret the fluctuation $\mathcal{A}^- \Delta Q_{i-1/2}$ as $A^-(Q_i - Q_{i-1})$, the "negative part" of the matrix A multiplied by the jump in Q, with A^- defined in (4.45). For variable-coefficient acoustics it is interesting to note that we can make a similar interpretation in spite of the fact that the matrix A varies. In this case we have $\mathcal{A}^- \Delta Q_{i-1/2} = A_{i-1/2}^-(Q_i - Q_{i-1})$, where the matrix $A_{i-1/2}$ is different than the coefficient matrices

$$A_i = \begin{bmatrix} 0 & K_i \\ 1/\rho_i & 0 \end{bmatrix} = R_i \Lambda_i R_i^{-1}$$

defined in the grid cells. Define

$$R_{i-1/2} = \begin{bmatrix} -Z_{i-1} & Z_i \\ 1 & 1 \end{bmatrix},$$

analogous to R_{lr} in (9.55). Then the wave strengths $\alpha_{i-1/2}^1$ and $\alpha_{i-1/2}^2$ are determined by

$$\alpha_{i-1/2} = R_{i-1/2}^{-1}(Q_i - Q_{i-1}).$$

The fluctuations (9.63) are thus given by

$$\mathcal{A}^- \Delta Q_{i-1/2} = R_{i-1/2} \Lambda_{i-1/2}^- R_{i-1/2}^{-1}(Q_i - Q_{i-1}) = A_{i-1/2}^-(Q_i - Q_{i-1}),$$

$$\mathcal{A}^+ \Delta Q_{i-1/2} = R_{i-1/2} \Lambda_{i-1/2}^+ R_{i-1/2}^{-1}(Q_i - Q_{i-1}) = A_{i-1/2}^+(Q_i - Q_{i-1}), \qquad (9.64)$$

where we define

$$\Lambda_{i-1/2} = \begin{bmatrix} s_{i-1/2}^1 & 0 \\ 0 & s_{i-1/2}^2 \end{bmatrix} = \begin{bmatrix} -c_{i-1} & 0 \\ 0 & c_i \end{bmatrix}$$

and hence

$$A_{i-1/2} = R_{i-1/2} \Lambda_{i-1/2} R_{i-1/2}^{-1}$$

$$= \frac{1}{Z_{i-1} + Z_i} \begin{bmatrix} c_i Z_i - c_{i-1} Z_{i-1} & (c_{i-1} + c_i) Z_{i-1} Z_i \\ c_{i-1} + c_i & c_i Z_{i-1} - c_{i-1} Z_i \end{bmatrix}. \qquad (9.65)$$

If $A_{i-1} = A_i$, then $A_{i-1/2}$ also reduces to this same coefficient matrix. In general it is different from the coefficient matrix on either side of the interface, and can be viewed as a special average of the coefficient matrix from the two sides. If the coefficients are smoothly varying, then $A_{i-1/2} \approx A(x_{i-1/2})$.

Note that implementing the method does not require working with this matrix $A_{i-1/2}$. We work directly with the waves and speeds. The Riemann solver is implemented in [çlaw/book/chap9/acoustics/interface/rp1acv.f].

9.12 High-Resolution Methods

Godunov's method for variable-coefficient acoustics is easily extended to a high-resolution method via a natural extension of the methods presented in Section 6.13, together with appropriate wave limiters. The method has the general form

$$Q_i^{n+1} = Q_i^n - \frac{\Delta t}{\Delta x}(\mathcal{A}^+ \Delta Q_{i+1/2} + \mathcal{A}^- \Delta Q_{i-1/2}) - \frac{\Delta t}{\Delta x}(\tilde{F}_{i+1/2} - \tilde{F}_{i-1/2}), \quad (9.66)$$

where

$$\tilde{F}_{i-1/2} = \frac{1}{2} \sum_{p=1}^{m} |s_{i-1/2}^p| \left(1 - \frac{\Delta t}{\Delta x}|s_{i-1/2}^p|\right) \tilde{\mathcal{W}}_{i-1/2}^p. \quad (9.67)$$

Before discussing limiters, first consider the unlimited case in which $\tilde{\mathcal{W}}_{i-1/2}^p = \mathcal{W}_{i-1/2}$. Then a calculation shows that the method (9.66) can be rewritten as

$$Q_i^{n+1} = Q_i^n - \frac{\Delta t}{2\Delta x}[A_{i-1/2}(Q_i - Q_{i-1}) + A_{i+1/2}(Q_{i+1} - Q_i)]$$

$$+ \frac{\Delta t^2}{2\Delta x^2}[A_{i+1/2}^2(Q_{i+1} - Q_i) - A_{i-1/2}^2(Q_i - Q_{i-1})]. \quad (9.68)$$

This is a Lax–Wendroff-style method based on the coefficient matrices defined at the interface as in (9.65). As for the case of the variable-coefficient color equation discussed in Section 9.5, this method is not formally second-order accurate when A varies, because the final term approximates $\frac{1}{2}\Delta t^2(A^2 q_x)_x$ while the Taylor series expansion requires $\frac{1}{2}\Delta t^2 q_{tt} = \frac{1}{2}\Delta t^2 A(Aq_x)_x$. These differ by $\frac{1}{2}\Delta t^2 A_x Aq_x$. However, we have captured the $\frac{1}{2}\Delta t^2 A^2 q_{xx}$ term correctly, which is essential in eliminating the numerical diffusion of the Godunov method.

This approach yields good high-resolution results in practice in spite of the lack of formal second-order accuracy (as we've seen before, e.g., Section 8.5, Section 9.3.1). Since these high-resolution methods are of particular interest for problems where the solution or coefficients are discontinuous, formal second-order accuracy cannot be expected in any case. If the problem is in conservation form $q_t + (A(x)q)_x = 0$, then this approach in fact gives formal second-order accuracy. See Section 16.4, where this is discussed in the more general context of a conservation law with a spatially varying flux function, $q_t + f(q, x)_x = 0$.

9.13 Wave Limiters

Applying limiters to the waves $\mathcal{W}_{i-1/2}^p$ is slightly more difficult for variable-coefficient systems of equations than in the constant-coefficient case. To obtain a high-resolution method, we wish to replace the wave $\mathcal{W}^p_{i-1/2}$ by a limited version in (9.67),

$$\widetilde{\mathcal{W}}^p_{i-1/2} = \phi(\theta^p_{i-1/2})\mathcal{W}^p_{i-1/2}, \tag{9.69}$$

as described in Chapter 6. Again $\theta^p_{i-1/2}$ should be some measure of the smoothness of the pth characteristic component of the solution. When A is constant the eigenvectors r^1 and r^2 are the same everywhere and so $\mathcal{W}^p_{i-1/2} \doteq \alpha^p_{i-1/2}r^p$ can easily be compared to the corresponding wave $\mathcal{W}^p_{I-1/2} = \alpha^p_{I-1/2}r^p$ arising from the neighboring Riemann problem by simply comparing the two scalar wave strengths,

$$\theta^p_{i-1/2} = \frac{\alpha^p_{I-1/2}}{\alpha^p_{i-1/2}}. \tag{9.70}$$

(Recall that $I = i \pm 1$ as in (6.61), looking in the upwind direction as determined by the sign of $s^p_{i-1/2}$.) Note that in this constant-coefficient case we have

$$|\theta^p_{i-1/2}| = \frac{\|\mathcal{W}^p_{I-1/2}\|}{\|\mathcal{W}^p_{i-1/2}\|} \tag{9.71}$$

in any vector norm.

For a variable-coefficient problem, the two waves

$$\mathcal{W}^p_{i-1/2} = \alpha^p_{i-1/2}r^p_{i-1/2} = \left(\ell^p_{i-1/2}\,\Delta Q_{i-1/2}\right)r^p_{i-1/2} \tag{9.72}$$

and

$$\mathcal{W}^p_{I-1/2} = \alpha^p_{I-1/2}r^p_{I-1/2} = \left(\ell^p_{I-1/2}\,\Delta Q_{I-1/2}\right)r^p_{I-1/2} \tag{9.73}$$

are typically not scalar multiples of one another, since the eigenvector matrices $R(x)$ and $L(x) = R^{-1}(x)$ vary with x. In this case it may make no sense to compare $\alpha^p_{i-1/2}$ with $\alpha^p_{I-1/2}$, since these coefficients depend entirely on the normalization of the eigenvectors, which may vary with x. It would make more sense to compare the magnitudes of the full waves, using the expression (9.71) for some choice of vector norm, but some way must be found to assign a sign, which we expect to be negative near extreme points. More generally, a problem with (9.71) is that it would give $|\theta| \approx 1$ whenever the two waves are of similar magnitude, regardless of whether they are actually similar vectors in phase space or not. If the solution is truly smooth, then we expect $\mathcal{W}^p_{i-1/2} \approx \mathcal{W}^p_{I-1/2}$ and not just that these vectors have nearly the same magnitude (which might be true for two vectors pointing in very different directions in phase space at some discontinuity in the medium, or in the solution for a nonlinear problem).

We would like to define $\theta^p_{i-1/2}$ in a way that reduces to (9.70) in the constant-coefficient case but that takes into account the degree of alignment of the wave vectors as well as their magnitudes. This can be accomplished by projecting the vector $\mathcal{W}^p_{I-1/2}$ onto the vector $\mathcal{W}^p_{i-1/2}$ to obtain a vector $\theta^p_{i-1/2}\mathcal{W}^p_{i-1/2}$ that is aligned with $\mathcal{W}^p_{i-1/2}$. Using the scalar coefficient of this projection as $\theta^p_{i-1/2}$ gives

$$\theta^p_{i-1/2} = \frac{\mathcal{W}^p_{I-1/2} \cdot \mathcal{W}^p_{i-1/2}}{\mathcal{W}^p_{i-1/2} \cdot \mathcal{W}^p_{i-1/2}}, \tag{9.74}$$

where $\cdot$ is the dot product in $\mathbb{R}^m$. This works well in most cases, also for nonlinear problems, and this general procedure is implemented in the `limiter.f` routine of CLAWPACK.

Another approach has been proposed by Lax & Liu [264], [313] in defining their *positive schemes*. Recall that we have used (9.72) and (9.73) to define the limiter. Lax and Liu instead use

$$\widehat{\mathcal{W}}^p_{I-1/2} = \hat{\alpha}^p_{I-1/2} r^p_{I-1/2} = \left(\ell^p_{I-1/2} \Delta Q_{I-1/2}\right) r^p_{I-1/2} \tag{9.75}$$

in place of $\mathcal{W}^p_{I-1/2}$. The vector $\widehat{\mathcal{W}}^p_{I-1/2}$ is obtained by decomposing the jump $\Delta Q_{I-1/2}$ into eigenvectors of the matrix $R_{i-1/2}$ rather than into eigenvectors of $R_{I-1/2}$. The vector $\widehat{\mathcal{W}}^p_{I-1/2}$ is now a scalar multiple of $\mathcal{W}^p_{i-1/2}$, and so we can simply use

$$\theta^p_{i-1/2} = \frac{\hat{\alpha}^p_{I-1/2}}{\alpha^p_{i-1/2}} = \frac{\widehat{\mathcal{W}}^p_{I-1/2} \cdot \mathcal{W}^p_{i-1/2}}{\mathcal{W}^p_{i-1/2} \cdot \mathcal{W}^p_{i-1/2}}. \tag{9.76}$$

This is slightly more expensive to implement, but since we compute the left eigenvectors at each cell interface in any case in order to compute the $\mathcal{W}^p_{i-1/2}$, we can also use them to compute $\widehat{\mathcal{W}}^p_{I-1/2}$ at the same time if the computations are organized efficiently.

For most problems the simpler expression (9.74) seems to work well, but for some problems with rapid variation in the eigenvectors the Lax–Liu limiter is superior. For example, in solving the acoustics equations in a rapidly varying periodic medium an instability attributed to nonlinear resonance was observed in [139] using the standard CLAWPACK limiter. A *transmission-based* limiter, closely related to the Lax–Liu limiter, was introduced to address this problem. This is similar to (9.76) but decomposes only the wave $\mathcal{W}^p_{I-1/2}$ into eigenvectors $r^p_{i-1/2}$ rather than starting with the full jump $\Delta Q_{I-1/2}$:

$$\widehat{\mathcal{W}}^p_{I-1/2} = \hat{\alpha}^p_{I-1/2} r^p_{i-1/2} = \left(\ell^p_{i-1/2} \mathcal{W}^p_{I-1/2}\right) r^p_{i-1/2}. \tag{9.77}$$

The formula (9.76) is then used. For acoustics this has the interpretation of taking the wave from the neighboring Riemann problem that is approaching the interface $x_{i-1/2}$ and using only the portion of this wave that would be transmitted through the interface in defining the limiter. See [139] for more discussion of this and several examples where these methods are used in periodic or random media.

9.14 Homogenization of Rapidly Varying Coefficients

If the physical parameters vary substantially within a single grid cell, then it will be necessary to apply some sort of *homogenization theory* to determine appropriate averaged values to use for ρ_i and K_i. This might be the case in a seismology problem, for example, where each grid cell represents a sizable block of earth that is typically not uniform.

Since ρ is the density (mass per unit length, in one dimension), the average density of the ith cell is properly computed as the total mass in the cell divided by the length of the cell, i.e., as the *arithmetic average* of ρ within the cell,

$$\rho_i = \frac{1}{\Delta x} \int_{x_{i-1/2}}^{x_{i+1/2}} \rho(x)\,dx. \tag{9.78}$$

For the bulk modulus K it turns out that instead of an arithmetic average, it is necessary to use the *harmonic average*

$$K_i = \left(\frac{1}{\Delta x} \int_{x_{i-1/2}}^{x_{i+1/2}} \frac{1}{K(x)} \, dx \right)^{-1}. \tag{9.79}$$

Alternatively, we can view this as using the arithmetic average of the parameter $1/K$, which is called the *coefficient of compressibility* (see, e.g., [255]),

$$\frac{1}{K_i} = \frac{1}{\Delta x} \int_{x_{i-1/2}}^{x_{i+1/2}} \frac{1}{K(x)} \, dx. \tag{9.80}$$

This is most easily motivated by considering the linear elasticity interpretation of acoustics from Section 2.12, with Lamé parameters $\lambda = K$ and $\mu = 0$ and the stress–strain relation

$$\sigma = K\epsilon \tag{9.81}$$

resulting from (2.89) (dropping superscripts on σ^{11} and ϵ^{11}). Rewriting this as

$$\epsilon = \frac{1}{K}\sigma \tag{9.82}$$

shows that this has the form of Hooke's law with spring constant $1/K$. Recalling that $\epsilon = X_x - 1$, we see that for a grid cell of homogeneous material ($K_i =$ constant) with rest length $x_{i-1/2} - x_{i+1/2} = \Delta x$, applying a force σ results in a strain

$$\frac{\left[X(x_{i+1/2}) - X(x_{i-1/2}) \right] - \Delta x}{\Delta x} = \left(\frac{1}{K_i} \right)\sigma. \tag{9.83}$$

If the material in the cell is heterogeneous (with σ still constant), then instead we have

$$
\begin{aligned}
\frac{\left[X(x_{i+1/2}) - X(x_{i-1/2}) \right] - \Delta x}{\Delta x} &= \frac{1}{\Delta x} \int_{x_{i-1/2}}^{x_{i+1/2}} [X_x(x) - 1] \, dx \\
&= \frac{1}{\Delta x} \int_{x_{i-1/2}}^{x_{i+1/2}} \epsilon(x) \, dx \\
&= \left(\frac{1}{\Delta x} \int_{x_{i-1/2}}^{x_{i+1/2}} \frac{1}{K(x)} \, dx \right)\sigma.
\end{aligned}
\tag{9.84}
$$

Comparing this with (9.83) motivates the averaging (9.80).

Homogenization theory is often used to derive simpler systems of partial differential equations to model the effective behavior of a medium with rapidly varying heterogeneties. As a simple example, consider a layered medium composed of thin layers of width L_1 of material characterized by parameters (ρ_1, K_1), which alternate with thin layers of width L_2 where the parameters are (ρ_2, K_2). Numerically we could solve this problem using the high-resolution methods developed above, provided we can take Δx small enough to resolve the layers. We do this in Example 9.4 below. Analytically, however, we might like to predict the behavior of waves propagating in this layered medium by solving some simplified equation in place of the variable-coefficient acoustics equation. If we consider

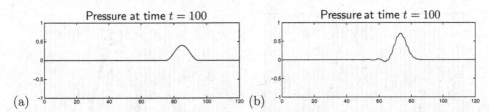

Fig. 9.8. (a) Pressure pulse in a homogeneous medium. (b) Pressure pulse in a layered medium.

waves whose wavelength is considerably longer than L_1 and L_2, then it turns out that the wave propagation can be reasonably well modeled by solving a homogenized system of equations that is simply the constant-coefficient acoustics equation with appropriate averages of ρ and $1/K$,

$$\bar{\rho} = \frac{1}{L_1 + L_2}(L_1\rho_1 + L_2\rho_2) \quad \text{and} \quad \hat{K}^{-1} = \frac{1}{L_1 + L_2}\left(\frac{L_1}{K_1} + \frac{L_2}{K_2}\right). \tag{9.85}$$

This is illustrated in Example 9.4. More accurate homogenized equations can also be derived, which for this problem requires adding higher-order dispersive terms to the constant-coefficient acoustics equation, as shown by Santosa & Symes [393].

In practical problems the material parameters often vary randomly, but with some slowly varying mean values $\rho(x)$ and $K^{-1}(x)$. This might be the case in a seismology problem, for example, where the basic type of rock or soil varies slowly (except for certain sharp interfaces) but is full of random heterogeneous structures at smaller scales. Similar problems arise in studying ultrasound waves in biological tissue, electromagnetic waves in a hazy atmosphere, and many other applications.

Example 9.4. Figure 9.8 shows the results from two acoustics problems with the same initial data and boundary conditions but different material properties. In each case the initial data is $p \equiv u \equiv 0$, and at the left boundary

$$u(0, t) = \begin{cases} 0.2\left[1 + \cos\left(\frac{\pi(t-15)}{10}\right)\right] & \text{if } |t - 15| < 10, \\ 0 & \text{otherwise.} \end{cases} \tag{9.86}$$

This in-and-out wall motion creates a pressure pulse that propagates to the right. In Figure 9.8(a) the material is uniform with $\rho \equiv 1$ and $K \equiv 1$, and so the sound speed is $c \equiv 1$. The peak pressure is generated at time $t = 15$, and at $t = 100$ is visible at $x = 85$, as should be expected.

Figure 9.8(b) shows a case where the material varies periodically with

$$\rho = K = \begin{cases} 3 & \text{for } 2i < x < 2i + 1, \\ 1 & \text{for } 2i + 1 < x < 2i + 2. \end{cases} \tag{9.87}$$

The layers have width $L_1 = L_2 = 1$ as seen in the zoomed view of Figure 9.9(a), where $\rho = K = 3$ in the dark layers and $\rho = K = 1$ in the light layers. For the choice (9.87), the sound speed is $c \equiv 1$ in all layers. However, the pressure pulse does not propagate at

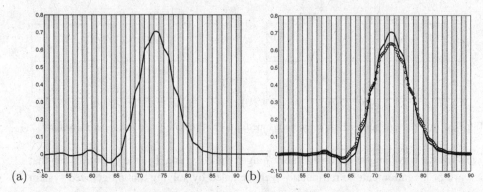

Fig. 9.9. (a) Zoom view of Figure 9.8(b) with the layers indicated. (b) The solid line is the "exact" solution. The symbols indicate the numerical solution computed with only four grid cells per layer. [claw/book/chap9/acoustics/layered]

speed 1. Instead, it propagates at the speed predicted by the homogenized acoustic equation as described above. The effective speed is

$$\bar{c} = \sqrt{\hat{K}/\bar{\rho}} \approx 0.866, \qquad (9.88)$$

based on the arithmetic average $\bar{\rho} = \frac{1}{2}(3+1) = 2$ and the harmonic average $\hat{K} = (\frac{1}{2}(\frac{1}{3} + 1))^{-1} = 1.5$. The peak pressure generated at $t = 15$ is now observed at $x = (85)(0.866) \approx 73.6$, as is clearly seen in Figure 9.9(a).

What's going on? Even though the sound speed $c \equiv 1$ is constant everywhere, the impedance $Z = \rho c = \rho$ is not, and so the eigenvectors of the coefficient matrix are different in each layer. Recall from Section 9.8 that the wave will be partially reflected at each interface in this case. The wave observed in Figure 9.8(a) is not a simple wave in the 2-characteristic family translating to the right as in the constant-coefficient case. Instead it is composed of waves moving in both directions that are constantly bouncing back and forth between the interfaces, so that the energy moves more slowly to the right than we would expect from the sound speed. It is the group velocity that we are observing rather than the phase velocity.

Notice from Figure 9.9(a) that the waveform is not completely smooth, but appears to have jump discontinuities in p_x at each interface. This is expected, since

$$u_t + \frac{1}{\rho} p_x = 0.$$

The velocity u must be continuous at all times, from which it follows that u_t and hence p_x/ρ must also be continuous. This means that p_x will be discontinuous at points where ρ is discontinuous.

Figure 9.9(a) also shows dispersive oscillations beginning to form behind the propagating pressure pulse in the periodic medium. This is *not* due to numerical dispersion, but rather is a correct feature of the exact solution, as predicted by the dispersive homogenized equation derived in [393].

In fact, the curve shown in Figure 9.8(b) and Figure 9.9(a) is essentially the exact solution. It was calculated on a fine grid with 2400 cells, and hence 20 grid cells in each layer. More importantly, the time step $\Delta t = \Delta x$ was used. Since $c \equiv 1$ everywhere, this means that each wave propagates through exactly one grid cell each time step, and there are no numerical errors introduced by averaging onto the grid. In essence we are using the method of characteristics, with the Riemann solver providing the correct characteristic decomposition at each cell interface.

We might now ask how accurate the high-resolution finite volume method would be on this problem, since for practical problems it will not always be possible to choose the time step so that $c \, \Delta t = \Delta x$ everywhere. This can be tested by choosing $\Delta t < \Delta x$, so that the method no longer reduces to the method of characteristics. Now the piecewise linear reconstruction with limiter functions is being used together with averaging onto grid cells.

Using $\Delta t = 0.8 \, \Delta x$ on the same grid with 2400 cells gives numerical results that are nearly indistinguishable from the exact solution shown in Figure 9.9(a). As a much more extreme test, the calculation was performed on a grid with only four grid cells in each layer. The results are still reasonable, as shown by the circles in Figure 9.9(b). Some accuracy is lost at this resolution, but the basic structure of the solution is clearly visible, with the wave propagating at the correct effective velocity.

Exercises

9.1. Compute the local truncation error for the method (9.18) in the case where no limiter is applied. Show that when $u(x)$ is not constant the method fails to be formally second-order accurate.

9.2. Use the Lax–Wendroff approach described in Section 6.1, expanding in Taylor series, to derive a method that is formally second-order accurate for the variable-coefficient color equation (9.12). Compare this method with (9.18) using (9.17).

9.3. Show that if we add second-order correction terms (with no limiters) to (9.41) based on (9.35), then we obtain a second-order accurate method for the conservative advection equation.

9.4. Explain why (9.32) is consistent with the notion of density as "cars per unit length."

9.5. In the discussion of variable-coefficient advection we have assumed $u(x)$ has the same sign everywhere. Consider the conservative equation $q_t + (u(x)q)_x = 0$ with $q(x, 0) \equiv 1$ and

$$u(x) = \begin{cases} -1 & \text{if } x < 0, \\ +1 & \text{if } x > 0, \end{cases} \quad \text{or} \quad u(x) = \begin{cases} +1 & \text{if } x < 0, \\ -1 & \text{if } x > 0. \end{cases}$$

What is the solution to each? (You might want to consider the conveyer-belt interpretation of Section 9.4.1.) How well will the methods presented here work for these problems? (See also Exercise 13.11.)

9.6. Show that the method developed in Section 9.5.2 with the correction terms (9.18) added is formally second-order accurate if no limiter is applied.

Other Approaches to High Resolution

In this book we concentrate on one particular approach to developing high-resolution finite volume methods, based on solving Riemann problems, using the resulting waves to define a second-order accurate method, and then applying wave limiters to achieve nonoscillatory results. A variety of closely related approaches have also been developed for achieving high-resolution results, and in this chapter a few alternatives are briefly introduced.

10.1 Centered-in-Time Fluxes

Recall the general form (4.4) of a flux-differencing method,

$$Q_i^{n+1} = Q_i^n - \frac{\Delta t}{\Delta x}\left(F_{i+1/2}^n - F_{i-1/2}^n\right), \tag{10.1}$$

where

$$F_{i-1/2}^n \approx \frac{1}{\Delta t}\int_{t_n}^{t_{n+1}} f\left(q\left(x_{i-1/2}, t\right)\right) dt. \tag{10.2}$$

The formula (10.1) can be arbitrarily accurate if we can approximate the integral appearing in (10.2) well enough. With the approach taken in Chapter 6, we approximate this integral using data at time t_n and the Taylor series, in a manner that gives second-order accuracy. There are many ways to derive methods of this type, and the approach taken in Chapter 6 is one way. Another is to first approximate

$$Q_{i-1/2}^{n+1/2} \approx q\left(x_{i-1/2}, t_{n+1/2}\right) \tag{10.3}$$

by some means and then use

$$F_{i-1/2}^n = f\left(Q_{i-1/2}^{n+1/2}\right) \tag{10.4}$$

in (10.1). Since this is centered in time, the resulting method will be second-order accurate provided that the approximation (10.3) is sufficiently good. Again Taylor series expansions can be used to approximate $q(x_{i-1/2}, t_{n+1/2})$ based on the data at time t_n. The Richtmyer two-step method (4.23) gives a second-order accurate method of this form.

To obtain high-resolution results we want to use the ideas of *upwinding* and *limiting*. An approach that is often used (e.g., [80]) is to first define two edge states $Q_{i-1/2}^{L,n+1/2}$ and

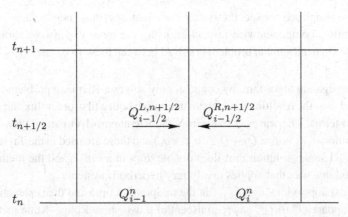

Fig. 10.1. The cell averages Q^n are used to obtain two approximations $Q_{i-1/2}^{L,n+1/2}$ and $Q_{i-1/2}^{R,n+1/2}$ to the states adjacent to $x_{i-1/2}$ at time $t_{n+1/2}$. Solving the Riemann problem between these states gives the flux (10.5).

$Q_{i-1/2}^{R,n+1/2}$ as indicated in Figure 10.1, and then solve the Riemann problem between these states to obtain $Q_{i-1/2}^{n+1/2}$ as the value along $x/t = 0$ in the solution to this Riemann problem,

$$Q_{i-1/2}^{n+1/2} = q^\vee \left(Q_{i-1/2}^{L,n+1/2}, Q_{i-1/2}^{R,n+1/2} \right). \tag{10.5}$$

In order to obtain two approximate values at time $t_{n+1/2}$ we can use the Taylor series, expanding from both x_i and x_{i+1}:

$$Q_{i-1/2}^{L,n+1/2} \approx q\left(x_{i-1/2}, t_{n+1/2} \right) = q(x_{i-1}, t_n) + \frac{\Delta x}{2} q_x(x_{i-1}, t_n) + \frac{\Delta t}{2} q_t(x_{i-1}, t_n) + \cdots$$

$$\tag{10.6}$$

and

$$Q_{i-1/2}^{R,n+1/2} \approx q\left(x_{i-1/2}, t_{n+1/2} \right) = q(x_i, t_n) - \frac{\Delta x}{2} q_x(x_i, t_n) + \frac{\Delta t}{2} q_t(x_i, t_n) + \cdots. \tag{10.7}$$

The cell averages Q_{i-1}^n and Q_i^n are used to approximate $q(x_{i-1}, t_n)$ and $q(x_i, t_n)$. The q_x-terms are approximated by reconstructing slopes in each cell, typically using limiter functions to avoid nonphysical oscillations. The time-derivative terms are replaced by spatial derivatives using the differential equation. For a general conservation law $q_t + f(q)_x = 0$ we have

$$q_t = -f'(q)q_x, \tag{10.8}$$

so that the same approximations to q_x can be used in these terms. For systems of equations, a characteristic decomposition of q is often used in the process of estimating q_x, so that limiting can be done by looking in the appropriate upwind direction for each eigencomponent of q_x.

A disadvantage of this approach is that we must first perform some sort of characteristic decomposition in order to estimate q_x in each cell and obtain the values $Q_{i-1/2}^{L,n+1/2}$ and $Q_{i-1/2}^{R,n+1/2}$. This characteristic decomposition is similar to solving a Riemann problem, though it may be simplified for specific systems of equations (or q_x may be approximated by applying limiters componentwise instead on using the more expensive characteristic decomposition). Then a Riemann problem is solved between these two states $Q_{i-1/2}^{L,n+1/2}$ and $Q_{i-1/2}^{R,n+1/2}$.

The wave-propagation algorithm, by contrast, only solves a Riemann problem between Q_{i-1} and Q_i and uses the resulting structure to determine both a first-order flux and second-order correction terms. The waves coming from the Riemann solver can be viewed as an eigendecomposition of q_x (since $Q_i - Q_{i-1} \approx \Delta x \, q_x$) and these are used in the Taylor series expansion. The Riemann solution thus does double duty in a sense, and the methods are easily formulated in a way that applies to arbitrary hyperbolic systems.

The idea of first approximating $Q_{i-1/2}^{n+1/2}$ at the midpoint in time and then evaluating f at these points to update Q^n to Q^{n+1} is reminiscent of a two-stage Runge–Kutta method. In Section 10.4 we will see a more direct application of Runge–Kutta methods that also allows the derivation of higher-order accurate methods.

10.2 Higher-Order High-Resolution Methods

For a linear problem with a smooth (many times differentiable) solution, such as the advection equation with wave-packet data as used in Figure 6.3, it is possible to obtain better accuracy by using a higher-order method. Fourth-order accurate methods are often used in practice for such problems, or even spectral methods that have a formal order of accuracy higher than any fixed power of Δx. For constant-coefficient linear problems these are often a more appropriate choice than the high-resolution methods developed in this book. However, in many contexts these high-order methods are not easily or successfully applied, such as when the solution or the coefficients of the problem are discontinuous.

One might wonder, however, why we start with a second-order method and then apply limiters to improve the behavior. Why not start with higher-order methods instead, and develop high-resolution versions of these? Ideally one would like to have a method that gives a higher order of accuracy in regions where the solution is smooth together with good resolution of discontinuities. In fact this can be done with some success, and a number of different approaches have been developed.

One possible approach is to follow the REA algorithm of Section 4.10 but to use a piecewise quadratic reconstruction of the solution in each grid cell in place of a piecewise constant or linear function. A popular method based on this approach is the *piecewise parabolic method* (PPM). This method was originally developed in [84] for gas dynamics and has been adapted to various other problems.

Another approach that is widely used is the class of *essentially nonoscillatory* (ENO) methods, which are briefly described later in this chapter. For one-dimensional linear problems, methods of this class can be developed with high order of accuracy that give very nice results on problems where the solution has both smooth regions and discontinuities.

Some care is required in assessing the accuracy of these higher-order methods in general. They are typically developed following the same general approach we used above for second-order methods: first for the scalar advection equation, generalizing to linear systems via characteristic decomposition, and then to nonlinear problems by solving Riemann problems, which gives a local characteristic decomposition. It is important to keep in mind that this decomposition is only local and that the problem cannot be decoupled into independent scalar equations except in the simplest constant-coefficient linear case. As a result there is a coupling between the characteristic components that can make it very difficult to maintain the formal high order of accuracy even on smooth solutions. Moreover, any limiting that is applied to one family near a discontinuity or steep gradient can adversely affect the accuracy of smooth waves flowing past in other families. For these reasons, such methods may not exhibit higher accuracy even in regions of the solution that appear to be smooth. See [111], [113], [120], [125], [444] for some further discussion.

For practical problems with interesting solutions involving shock waves, it appears difficult to formally achieve high-order accuracy in smooth regions with any method. This is not to say that high-resolution methods based on higher-order methods are not worth using. As stressed in Section 8.5, the order of accuracy is not the only consideration and these methods may give better resolution of the solution even if not formally higher order. But the accuracy and the efficiency of a method must be carefully assessed without assuming *a priori* that a higher-order method will be better.

10.3 Limitations of the Lax–Wendroff (Taylor Series) Approach

The Lax–Wendroff approach to obtaining second-order accuracy (see Section 6.1) is to expand in a Taylor series in time and replace the terms q_t and q_{tt} by expressions involving spatial derivatives of x. The first comes directly from the PDE, whereas the expression for q_{tt} is obtained by manipulating derivatives of the PDE. For the linear system $q_t + Aq_x = 0$ one easily obtains $q_{tt} = A^2 q_{xx}$, but for more complicated problems this may not be so easy to do; e.g., see Section 17.2.1, where an advection equation with a source term is considered. In Chapter 19 we will see that extending these methods to more than one space dimension leads to cross-derivative terms that require special treatment. In spite of these limitations, methods based on second-order Taylor series expansion are successfully used in practice and are the basis of the CLAWPACK software, for example.

There are also other approaches, however. One of the most popular is described in the next section: an approach in which the spatial and temporal discretizations are decoupled. This is particularly useful when trying to derive methods with greater than second-order accuracy, where the Lax–Wendroff approach becomes even more cumbersome. For the linear system above, we easily find that $q_{ttt} = -A^3 q_{xxx}$, but it is not so clear how this should be discretized in order to maintain the desired high-resolution properties near discontinuities, and this term may be difficult to evaluate at all for other problems.

10.4 Semidiscrete Methods plus Time Stepping

The methods discussed so far have all been fully discrete methods, discretized in both space and time. At times it is useful to consider the discretization process in two stages, first discretizing only in space, leaving the problem continuous in time. This leads to a system of ordinary differential equations in time, called the *semidiscrete equations*. We then discretize in time using any standard numerical method for systems of ordinary differential equations (ODEs). This approach of reducing a PDE to a system of ODEs, to which we then apply an ODE solver, is often called the *method of lines*.

This approach is particularly useful in developing methods with order of accuracy greater than 2, since it allows us to decouple the issues of spatial and temporal accuracy. We can define high-order approximations to the flux at a cell boundary at one instant in time using high-order interpolation in space, and then achieve high-order temporal accuracy by applying any of the wide variety of high-order ODE solvers.

10.4.1 Evolution Equations for the Cell Averages

Let $Q_i(t)$ represent a discrete approximation to the cell average of q over the ith cell at time t, i.e.,

$$Q_i(t) \approx \bar{q}_i(t) \equiv \frac{1}{\Delta x} \int_{x_{i-1/2}}^{x_{i+1/2}} q(x, t)\,dx. \tag{10.9}$$

We know from the integral form of the conservation law (2.2) that the cell average $\bar{q}_i(t)$ evolves according to

$$\bar{q}_i'(t) = -\frac{1}{\Delta x}\big[f\big(q(x_{i+1/2}, t)\big) - f\big(q(x_{i-1/2}, t)\big)\big]. \tag{10.10}$$

We now let $F_{i-1/2}(Q(t))$ represent an approximation to $f(q(x_{i-1/2}, t))$, obtained from the discrete data $Q(t) = \{Q_i(t)\}$. For example, taking the Godunov approach of Section 4.11, we might solve the Riemann problem with data $Q_{i-1}(t)$ and $Q_i(t)$ for the intermediate state $q^\vee(Q_{i-1}(t), Q_i(t))$ and then set

$$F_{i-1/2}(Q(t)) = f(q^\vee(Q_{i-1}(t), Q_i(t))). \tag{10.11}$$

Replacing the true fluxes in (10.10) by $F_{i\pm1/2}(Q(t))$ and the exact cell average $\bar{q}_i(t)$ by $Q_i(t)$, we obtain a discrete system of ordinary differential equations for the $Q_i(t)$,

$$Q_i'(t) = -\frac{1}{\Delta x}\big[F_{i+1/2}(Q(t)) - F_{i-1/2}(Q(t))\big] \equiv \mathcal{L}_i(Q(t)). \tag{10.12}$$

This is the ith equation in a coupled system of equations

$$Q'(t) = \mathcal{L}(Q(t)), \tag{10.13}$$

since each of the fluxes $F_{i\pm1/2}(t)$ depends on two or more of the $Q_i(t)$.

We can now discretize in time. For example, if we discretize (10.12) using Euler's method with time step Δt, and let Q_i^n now represent our fully discrete approximation to $Q_i(t_n)$, then we obtain

$$Q_i^{n+1} = Q_i^n + \Delta t\, \mathcal{L}_i(Q^n)$$

$$= Q_i^n - \frac{\Delta t}{\Delta x}\big[F_{i+1/2}(Q^n) - F_{i-1/2}(Q^n)\big], \tag{10.14}$$

which is in the familiar form of a conservative method (4.4). In particular, if F is given by (10.11), then (10.14) is simply Godunov's method. More generally, however, $F_{i-1/2}(Q^n)$ represents an approximation to the value of $f(q(x_{i-1/2}, t))$ at *one point in time*, whereas the numerical flux $F_{i-1/2}^n$ used before has always represented an approximation to the *time average* of $f(q(x_{i-1/2}, t))$ over the time interval $[t_n, t_{n+1}]$. In Godunov's method these are the same, since $q(x_{i-1/2}, t)$ is constant in time in the Riemann solution. With the Lax–Wendroff approach the time average is approximated more accurately using approximations to time derivatives, which have been rewritten in terms of spatial derivatives. With the semidiscrete approach we do not attempt to approximate this time average at all, but rather use only pointwise values of the flux.

To obtain higher-order accuracy, we must make two improvements: the value $F_{i-1/2}$ obtained by piecewise constant approximations in (10.11) must be improved to give better spatial accuracy at this one time, and the first-order accurate Euler method must be replaced by a higher-order time-stepping method in solving the system of ODEs (10.13). One advantage of the method-of-lines approach is that the spatial and temporal accuracy are decoupled and can be considered separately. This is particularly useful in several space dimensions.

One way to obtain greater spatial accuracy is to use a piecewise linear approximation in defining $F_{i-1/2}(Q(t))$. From the data $\{Q_i(t)\}$ we can construct a piecewise linear function $\tilde{q}(x)$ using slope limiters as discussed in Chapter 6. Then at the interface $x_{i-1/2}$ we have values on the left and right from the two linear approximations in each of the neighboring cells (see Figure 10.2). Denote these values by

$$Q_{i-1}^R = Q_{i-1} + \frac{\Delta x}{2}\sigma_{i-1} \quad \text{and} \quad Q_i^L = Q_i - \frac{\Delta x}{2}\sigma_i.$$

A second-order accurate approximation to the flux at this cell boundary at time t is then obtained by solving the Riemann problem with left and right states given by these two values, and setting

$$F_{i-1/2}(Q) = f(q^\vee(Q_{i-1}^R, Q_i^L)). \tag{10.15}$$

This type of semidiscrete MUSCL scheme is discussed in more detail by Osher [350].

If we use this flux in the fully discrete method (10.14), then the method is second-order accurate in space but only first-order accurate in time, i.e., the global error is $\mathcal{O}(\Delta x^2 + \Delta t)$, since the time discretization is still Euler's method. For time-dependent problems this improvement in spatial accuracy alone is usually not advantageous, but for steady-state

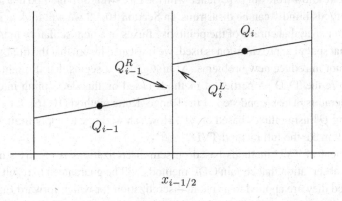

Fig. 10.2. Piecewise linear reconstruction of $\tilde{q}(x)$ used to define left and right states at $x_{i-1/2}$.

problems this type of method will converge as time evolves to a second-order accurate approximation to the steady-state solution, in spite of the fact that it is not second-order accurate in time along the way.

To obtain a method that is second-order accurate in time as well as space, we can discretize the ODEs (10.12) using a second-order accurate ODE method. One possibility is a two-stage explicit Runge–Kutta method. One standard Runge–Kutta method is the following (which is *not* recommended, for reasons explained in Section 10.4.2):

$$Q_i^* = Q_i^n - \frac{\Delta t}{2\Delta x}\left[F_{i+1/2}(Q^n) - F_{i-1/2}(Q^n)\right] = Q_i^n + \frac{1}{2}\Delta t \, \mathcal{L}_i(Q^n),$$

$$Q_i^{n+1} = Q_i^n - \frac{\Delta t}{\Delta x}\left[F_{i+1/2}(Q^*) - F_{i-1/2}(Q^*)\right] = Q_i^n + \Delta t \, \mathcal{L}_i(Q^*).$$

(10.16)

Note that this requires solving two Riemann problems at each cell boundary in each time step. We first solve Riemann problems based on the cell averages Q^n and use these to obtain Q^*, an approximation to the cell averages at time $t_n + \Delta t/2$. We next solve Riemann problems based on these cell averages in order to compute the fluxes used to advance Q^n to Q^{n+1}. This second set of fluxes can now be viewed as approximations to the time integral of the true flux over $[t_n, t_{n+1}]$, but it is based on estimating the pointwise flux at the midpoint in time rather than on a Taylor series expansion at time t_n. This is similar in spirit to the method described in Section 10.1. But now higher-order ODE methods can be used to obtain better approximations based only on pointwise values rather than attempting to compute more terms in the Taylor series. This is the basic idea behind most high-order methods for ODEs, such as Runge–Kutta methods and linear multistep methods, which are generally easier to apply than high-order Taylor series methods. (See for example [145], [177], [253] for some general discussions of ODE methods.)

10.4.2 TVD Time Stepping

Some care must be exercised in choosing a time-stepping method for the system of ODEs (10.13). In order to obtain a high-resolution method we would wish to avoid spurious oscillations, and we know from our experience with the Lax–Wendroff method that applying higher-order methods blindly can be disastrous. In Section 10.4.3 we will look at methods of achieving better spatial accuracy of the pointwise fluxes in a nonoscillatory manner, but regardless of what spatial approximation is used, we must also insure that the time-stepping algorithm does not introduce new problems. At first glance it seems that it might be very difficult to analyze the TVD properties of methods based on this decoupling into spatial and temporal operators. The second step of the Runge–Kutta method (10.16), for example, involves updating Q^n using fluxes based on Q^*. How can we hope to apply limiters to Q^* in a way that will make the full method TVD?

The development of TVD methods based on semidiscretizations is greatly simplified, however, by the observation that certain ODE methods will be guaranteed to result in TVD methods provided they are applied to a spatial discretization for which forward Euler time stepping is TVD. In other words, suppose we know that $\mathcal{L}(Q)$ is a discretization for which the forward Euler method

$$Q^{n+1} = Q^n + \Delta t\, \mathcal{L}(Q^n) \tag{10.17}$$

is a TVD method, which may be easy to check. Then if we apply a *TVD Runge–Kutta method* to this operator, the resulting method will also be TVD. The Runge–Kutta method (10.16) is not one of these special methods, but the following two-stage second-order Runge–Kutta method is, and is equally easy to apply:

$$
\begin{aligned}
Q^* &= Q^n + \Delta t\, \mathcal{L}(Q^n), \\
Q^{**} &= Q^* + \Delta t\, \mathcal{L}(Q^*), \\
Q^{n+1} &= \frac{1}{2}(Q^n + Q^{**}).
\end{aligned}
\tag{10.18}
$$

Again two applications of $\mathcal{L}$, and hence two sets of Riemann problems, are required. It is easy to verify that this method is TVD using the property (10.17), since we then have $\mathrm{TV}(Q^*) \le \mathrm{TV}(Q^n)$ and $\mathrm{TV}(Q^{**}) \le \mathrm{TV}(Q^*)$, and hence

$$\mathrm{TV}(Q^{n+1}) \le \frac{1}{2}[\mathrm{TV}(Q^n) + \mathrm{TV}(Q^{**})] \le \frac{1}{2}[\mathrm{TV}(Q^n) + \mathrm{TV}(Q^n)] = \mathrm{TV}(Q^n).$$

Shu & Osher [410] and Shu [408] present a number of methods of this type and also TVD multistep methods. See also [165], [214], [295], [409]. More recently such methods have been called *strong stability-preserving* (SSP) time discretizations, e.g., [166], [423].

10.4.3 Reconstruction by Primitive Functions

To obtain high spatial accuracy we need to define $F_{i-1/2}$ in such a way that it is a good approximation to $f(q(x_{i-1/2}, t))$. Recall that $\bar{q}(t)$ is the vector of exact *cell averages*, and from these we want to obtain a value $Q_{i-1/2}$ that approximates the *pointwise* value $q(x_{i-1/2}, t)$. One approach to this was outlined above: define Q_{i-1}^R and Q_i^L using slope-limited piecewise linears, and then set

$$Q_{i-1/2} = q^{\vee}(Q_{i-1}^R, Q_i^L).$$

To obtain higher-order accuracy we can take the same approach but define Q_{i-1}^L and Q_i^R via some higher-order polynomial approximation to q over the cells to the left and right of $x_{i-1/2}$.

This raises the following question: *Given only the cell averages $\bar{q}_i(t)$, how can we construct a polynomial approximation to q that is accurate pointwise to high order?*

A very elegant solution to this problem uses the *primitive function* for $q(x, t)$. This approach was apparently first introduced by Colella and Woodward [84] in their PPM method and has since been used in a variety of other methods, particularly the ENO methods discussed below.

At a fixed time t, the primitive function $w(x)$ is defined by

$$w(x) = \int_{x_{1/2}}^{x} q(\xi, t)\, d\xi. \tag{10.19}$$

The lower limit $x_{1/2}$ is arbitrary; any fixed point could be used. Changing the lower limit only shifts $w(x)$ by a constant, and the property of w that we will ultimately use is that

$$q(x, t) = w'(x), \qquad (10.20)$$

which is unaffected by a constant shift. Equation (10.20) allows us to obtain pointwise values of q if we have a good approximation to w.

Now the crucial observation is that knowing *cell averages* of q gives us *pointwise* values of w at the particular points $x_{i+1/2}$. Set

$$W_i = w(x_{i+1/2}) = \int_{x_{1/2}}^{x_{i+1/2}} q(\xi, t)\, d\xi. \qquad (10.21)$$

This is Δx times the average of q over a collection of j cells, and hence

$$W_i = \Delta x \sum_{j=1}^{i} \bar{q}_j(t).$$

Of course this only gives us pointwise values of w at the points $x_{i+1/2}$, but it gives us the *exact* values at these points (assuming we start with the exact cell averages $\bar{q}_i$, as we do in computing the truncation error). If w is sufficiently smooth (i.e., if q is sufficiently smooth), we can then approximate w more globally to arbitrary accuracy using polynomial interpolation. In particular, to approximate w in the ith cell C_i, we can use an interpolating polynomial of degree s passing through some $s + 1$ points $W_{i-j}, W_{i-j+1}, \ldots, W_{i-j+s}$ for some j. (The choice of j is discussed below.) If we call this polynomial $p_i(x)$, then we have

$$p_i(x) = w(x) + \mathcal{O}(\Delta x^{s+1}) \qquad (10.22)$$

for $x \in C_i$, provided $w \in C^{s+1}$ (which requires $q(\cdot, t) \in C^s$).

Using the relation (10.20), we can obtain an approximation to $q(x, t)$ by differentiating $p_i(x)$. We lose one order of accuracy by differentiating the interpolating polynomial, and so

$$p_i'(x) = q(x, t) + \mathcal{O}(\Delta x^s) \qquad \text{on } C_i.$$

We can now use this to obtain approximations to q at the left and right cell interfaces, setting

$$Q_i^L = p_i'(x_{i-1/2}),$$
$$Q_i^R = p_i'(x_{i+1/2}).$$

Performing a similar reconstruction on the cell $[x_{i-3/2}, x_{i-1/2}]$ gives $p_{i-1}(x)$, and we set

$$Q_{i-1}^R = p_{i-1}'(x_{i-1/2})$$

and then define $F_{i-1/2}$ as in (10.15). This gives spatial accuracy of order s for sufficiently smooth q.

10.4.4 ENO Methods

In the above description of the interpolation process, the value of j was left unspecified (recall that we interpolate $W_{i-j}, \ldots, W_{i-j+s}$ to approximate q on cell C_i). When $q(\cdot, t) \in C^s$ in the vicinity of x_i, interpolation based on any value of j between 1 and s will give sth-order accuracy. However, for a high-resolution method we must be able to automatically cope with the possibility that the data is not smooth. Near discontinuities we do not expect to maintain the high order of accuracy, but want to choose a stencil of interpolation points that avoids introducing oscillations. It is well known that a high-degree polynomial interpolant can be highly oscillatory even on smooth data, and certainly will be on nonsmooth data.

In the piecewise linear version described initially, this was accomplished by using a slope limiter. For example, the minmod slope compares linear interpolants based on cells to the left and right and takes the one that is less steep. This gives a global piecewise linear approximation that is nonoscillatory in the sense that its total variation is no greater than that of the discrete data.

This same idea can be extended to higher-order polynomials by choosing the value of j for each i so that the interpolant through $W_{i-j}, \ldots, W_{i-j+s}$ has the least oscillation over all possible choices $j = 1, \ldots, s$. This is the main idea in the ENO methods originally developed by Chakravarthy, Engquist, Harten, and Osher. Complete details, along with several variations and additional references, can be found in the papers [182], [183], [189], [188], [410], [411].

One variant uses the following procedure. Start with the linear function passing through W_{i-1} and W_i to define $p_i^{(1)}(x)$ (where superscripts now indicate the degree of the polynomial). Next compute the divided difference based on $\{W_{i-2}, W_{i-1}, W_i\}$ and the divided difference based on $\{W_{i-1}, W_i, W_{i+1}\}$. Either of these can be used to extend $p_i^{(1)}(x)$ to a quadratic polynomial using the Newton form of the interpolating polynomial. We define $p_i^{(2)}(x)$ by choosing the divided difference that is smaller in magnitude.

We continue recursively in this manner, adding either the next point to the left or to the right to our stencil depending on the magnitude of the divided difference, until we have a polynomial of degree s based on some $s + 1$ points.

Note that the first-order divided differences of W are simply the values $\bar{q}_i$,

$$\frac{W_i - W_{i-1}}{\Delta x} = \bar{q}_i,$$

and so divided differences of W are directly related to divided differences of the cell averages $\bar{q}_i$. In practice we need never compute the W_i. (The zero-order divided difference, W_i itself, enters $p_i(x)$ only as the constant term, which drops out when we compute $p_i'(x)$.)

More recently, *weighted ENO* (WENO) methods have been introduced, which combine the results obtained using all possible stencils rather than choosing only one. A weighted combination of the results from all stencils is used, where the weights are based on the magnitudes of the divided differences in such a way that smoother approximations receive greater weight. This is more robust than placing all the weight on a single stencil, since it responds more smoothly to changes in the data. These methods are developed in [219] and [314]. See the survey [409] for an overview and other references.

10.5 Staggered Grids and Central Schemes

For nonlinear systems of equations, solving a Riemann problem can be an expensive oper-
ation, as we will see in Chapter 13, and may not even be possible exactly. A variety of approximate Riemann solvers have been developed to simplify this process (see Section 15.3),
but algorithms based on Riemann problems are still typically expensive relative to approaches that only require evaluating the flux function.

The Lax–Friedrichs (LxF) method (Section 4.6) is one very simple method that only
uses flux evaluation and is broadly applicable. However, it is only first-order accurate and
is very dissipative. Consequently a very fine grid must be used to obtain good approximate
solutions, which also leads to an expensive algorithm.

Recently a class of algorithms known as *central schemes* have been introduced, starting
with the work of Nessyahu and Tadmor [338], which extends the LxF idea to higher-
order accuracy. The basic idea is most easily explained using a staggered grid as shown in
Figure 10.3(a), starting with an interpretation of the LxF method on this grid.

Note that the LxF method

$$Q_i^{n+1} = \frac{1}{2}\left(Q_{i-1}^n + Q_{i+1}^n\right) - \frac{\Delta t}{2\,\Delta x}\left[f\left(Q_{i+1}^n\right) - f\left(Q_{i-1}^n\right)\right] \tag{10.23}$$

computes Q_i^{n+1} based only on Q_{i-1}^n and Q_{i+1}^n, so it makes sense to use a grid on which only
odd-numbered indices appear at one time and only even-numbered indices at the next. It is
interesting to note that on this grid we can view the LxF method as a variant of Godunov's
method in which Riemann problems are solved and the resulting solution averaged over
grid cells. Figure 10.3(b) indicates how this can be done. We integrate the conservation law
over the rectangle $[x_{i-1}, x_{i+1}] \times [t_n, t_{n+1}]$ and obtain

$$
\begin{aligned}
Q_i^{n+1} &\approx \frac{1}{2\,\Delta x}\int_{x_{i-1}}^{x_{i+1}} \tilde{q}^n(x, t_{n+1})\,dx \\
&= \frac{1}{2\,\Delta x}\int_{x_{i-1}}^{x_{i+1}} \tilde{q}^n(x, t_n)\,dx \\
&\quad - \frac{1}{2\,\Delta x}\left[\int_{t_n}^{t_{n+1}} f(\tilde{q}^n(x_{i+1}, t))\,dt - \int_{t_n}^{t_{n+1}} f(\tilde{q}^n(x_{i-1}, t))\,dt\right],
\end{aligned} \tag{10.24}
$$

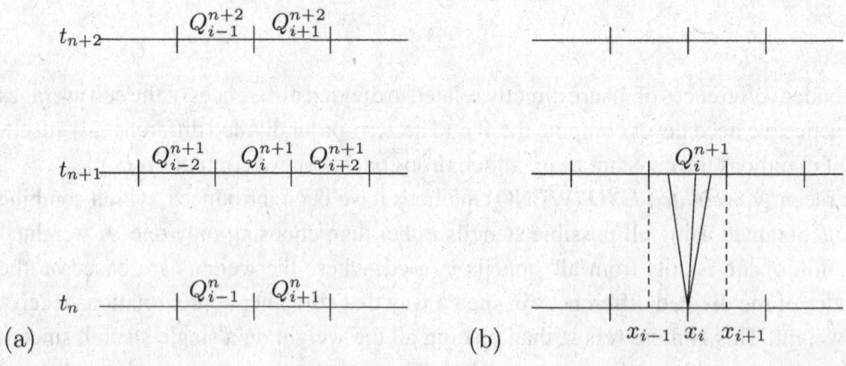

Fig. 10.3. Interpretation of the Lax–Friedrichs method on a staggered grid. (a) The grid labeling.
(b) Integrating the conservation law over the region $[x_{i-1}, x_{i+1}] \times [t_n, t_{n+1}]$, which includes the full
Riemann solution at x_i.

where $\tilde{q}^n(x, t)$ is the exact solution to the conservation law with the piecewise constant data Q^n. This is analogous to Godunov's method as described in Section 4.10 and Section 4.11, but now implemented over a cell of width $2\,\Delta x$. Because of the grid staggering we have

$$\frac{1}{\Delta x} \int_{x_{i-1}}^{x_{i+1}} \tilde{q}^n(x, t_n)\,dx = \frac{1}{2}\left(Q_{i-1}^n + Q_{i+1}^n\right). \tag{10.25}$$

As in Godunov's method, the flux integrals in (10.24) can be calculated exactly because $\tilde{q}^n(x_{i\pm1}, t)$ are constant in time. However, these are even simpler than in Godunov's method because we do not need to solve Riemann problems to find the values. The Riemann problems are now centered at x_i, $x_{i\pm2}$, ..., and the waves from these Riemann solutions do not affect the values we need at $x_{i\pm1}$ provided the Courant number is less than 1. We simply have $\tilde{q}^n(x_{i\pm1}, t) = Q_{i\pm1}^n$, and so evaluating the integrals in (10.24) exactly gives the Lax–Friedrichs method (10.23).

To obtain better accuracy we might construct approximations to $\tilde{q}^n(x, t_n)$ in each cell on the staggered grid by a piecewise linear function, or some higher-order polynomial, again using limiters or an ENO reconstruction to choose these functions. As the solution $\tilde{q}^n$ evolves the values $\tilde{q}^n(x_{i\pm1}, t)$ will no longer be constant and the flux integrals in (10.24) must typically be approximated. But the large jump discontinuities in the piecewise polynomial $\tilde{q}^n(x, t_n)$ are still at the points x_i, $x_{i\pm2}$, ... and the solution remains smooth at $x_{i\pm1}$ over time Δt, and so simple approximations to the flux integrals can be used.

The original second-order *Nessyahu–Tadmor scheme* is based on using a piecewise linear representation in each grid cell, choosing slopes σ_{i-1}^n and σ_{i+1}^n in the two cells shown in Figure 10.3(a), for example. For a scalar problem this can be done using any standard limiter, such as those described in Chapter 6. For systems of equations, limiting is typically done componentwise rather than using a characteristic decomposition, to avoid using any characteristic information. The updating formula takes the form

$$Q_i^{n+1} = \bar{Q}_i^n - \frac{\Delta t}{\Delta x}\left(F_{i+1}^{n+1/2} - F_{i-1}^{n+1/2}\right). \tag{10.26}$$

Now $\bar{Q}_i^n$ is the cell average at time t_n based on integrating the piecewise linear function over the cell x_i as in (10.25), but now taking into account the linear variation. This yields

$$\bar{Q}_i^n = \frac{1}{2}\left(Q_{i-1}^n + Q_{i+1}^n\right) + \frac{1}{8}\Delta x\left(\sigma_{i-1}^n - \sigma_{i+1}^n\right). \tag{10.27}$$

The flux $F_{i-1}^{n+1/2}$ is computed by evaluating f at some approximation to $\tilde{q}^n(x_{i-1}, t_{n+1/2})$ to give give second-order accuracy in time. Since $\tilde{q}^n$ remains smooth near x_{i-1}, we can approximate this well by

$$\tilde{q}^n(x_{i-1}, t_{n+1/2}) \approx Q_{i-1}^n - \frac{1}{2}\Delta t\,\frac{\partial f}{\partial x}\left(Q_{i-1}^n\right). \tag{10.28}$$

We use

$$F_{i+1}^{n+1/2} = f\left(Q_{i-1}^n - \tfrac{1}{2}\Delta t\,\phi_{i-1}^n\right), \tag{10.29}$$

where

$$\phi_{i-1}^n \approx \frac{\partial f}{\partial x}\left(Q_{i-1}^n\right). \tag{10.30}$$

This can be chosen either as

$$\phi_{i-1}^n = f'(Q_{i-1}^n)\sigma_{i-1}^n \tag{10.31}$$

or as a direct estimate of the slope in $f(q)$ based on nearby cell values $f(Q_j)$ and the same limiting procedure used to obtain slopes σ_{i-1}^n from the data Q_j. The latter approach avoids the need for the Jacobian matrix $f'(q)$.

This is just the basic idea of the high-resolution central schemes. Several different variants of this method have been developed, including nonstaggered versions, higher-order methods, and multidimensional generalizations. For some examples see the papers [13], [36], [206], [220], [221], [250], [294], [315], [391], and references therein.

Exercises

10.1. Show that the ENO method described in Section 10.4.4 with $s = 2$ (quadratic interpolation of W) gives a piecewise linear reconstruction of q with slopes that agree with the minmod formula (6.26).

10.2. Derive the formula (10.27).

Part two

Nonlinear Equations

11

Nonlinear Scalar Conservation Laws

We begin our study of nonlinear conservation laws by considering scalar conservation laws of the form

$$q_t + f(q)_x = 0. \tag{11.1}$$

When the flux function $f(q)$ is linear, $f(q) = \bar{u}q$, then this is simply the advection equation with a very simple solution. The equation becomes much more interesting if $f(q)$ is a nonlinear function of q. The solution no longer simply translates uniformly. Instead, it deforms as it evolves, and in particular *shock waves* can form, across which the solution is discontinuous. At such points the differential equation (11.1) cannot hold in the classical sense. We must remember that this equation is typically derived from a more fundamental integral conservation law that does still make sense even if q is discontinuous. In this chapter we develop the mathematical theory of nonlinear scalar equations, and then in the next chapter we will see how to apply high-resolution finite volume methods to these equations. In Chapter 13 we turn to the even more interesting case of nonlinear hyperbolic systems of conservation laws.

In this chapter we assume that the flux function $f(q)$ has the property that $f''(q)$ does not change sign, i.e., that $f(q)$ is a convex or concave function. This is often called a *convex flux* in either case. The nonconvex case is somewhat trickier and is discussed in Section 16.1.

For the remainder of the book we consider only conservation laws. In the linear case, hyperbolic problems that are not in conservation form can be solved using techniques similar to those developed for conservation laws, and several examples were explored in Chapter 9. In the nonlinear case it is possible to apply similar methods to a nonconservative quasilinear hyperbolic problem $q_t + A(q)q_x = 0$, but equations of this form must be approached with caution for reasons discussed in Section 16.5. The vast majority of physically relevant nonlinear hyperbolic problems arise from integral conservation laws, and it is generally best to keep the equation in conservation form.

11.1 Traffic Flow

As a specific motivating example, we consider the flow of cars on a one-lane highway, as introduced in Section 9.4.2. This is frequently used as an example of nonlinear conservation laws, and other introductory discussions can be found in [175], [457], [486], for example.

This model provides a good introduction to the effects of nonlinearity in a familiar context where the results will probably match readers' expectations. Moreover, a good understanding of this example provides at least some physical intuition for the types of solutions seen in gas dynamics, an important application area for this theory. A gas is a dilute collection of molecules that are quite far apart and hence can be compressed or expanded. The density (measured in mass or molecules per unit volume) can vary by many orders of magnitude depending on how much space is available, and so a gas is a *compressible* fluid.

Cars on the highway can also be thought of as a compressible fluid. The density (measured in cars per car length, as introduced in Section 9.4.2), can vary from 0 on an empty highway to 1 in bumper-to-bumper traffic. (We ignore the compressibility of individual cars in a major collision)

Since the flux of cars is given by uq, we obtain the conservation law

$$q_t + (uq)_x = 0. \tag{11.2}$$

In Section 9.4.2 we considered the variable-coefficient linear problem in which cars always travel at the speed limit $u(x)$. This is a reasonable assumption only for very light traffic. As traffic gets heavier, it is more reasonable to assume that the speed also depends on the density. At this point we will suppose the speed limit and road conditions are the same everywhere, so that u depends only on q and not explicitly on x. Suppose that the velocity is given by some specific known function $u = U(q)$ for $0 \le q \le 1$. Then (11.2) becomes

$$q_t + (q\, U(q))_x = 0, \tag{11.3}$$

which in general is now a *nonlinear conservation law* since the flux function

$$f(q) = qU(q) \tag{11.4}$$

will be nonlinear in q. This is often called the LWR model for traffic flow, after Lighthill & Whitham [299] and Richards [368]. (See Section 17.17 for a brief discussion of other models.)

Various forms of $U(q)$ might be hypothesized. One simple model is to assume that $U(q)$ varies linearly with q,

$$U(q) = u_{max}(1 - q) \quad \text{for } 0 \le q \le 1. \tag{11.5}$$

At zero density (empty road) the speed is u_{max}, but decreases to zero as q approaches 1. We then have the flux function

$$f(q) = u_{max}q(1 - q). \tag{11.6}$$

This is a quadratic function. Note that the flux of cars is greatest when $q = 1/2$. As q decreases, the flux decreases because there are few cars on the road, and as q increases, the flux decreases because traffic slows down.

Before attempting to solve the nonlinear conservation law (11.3), we can develop some intuition for how solutions behave by doing simulations of traffic flow in which we track the motion of individual vehicles $X_k(t)$, as described in Section 9.4.2. If we define the density

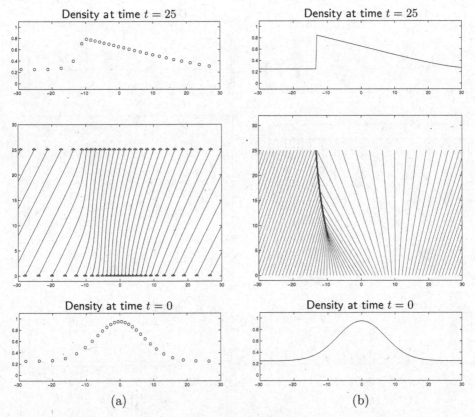

Fig. 11.1. Solution to the traffic flow model starting with a bulge in the density. (a) Trajectories of individual vehicles and the density observed by each driver. (b) The characteristic structure is shown along with the density as computed by CLAWPACK. [claw/book/chap11/congestion]

seen by the kth driver as $q_k(t)$, as in (9.32), then the velocity of the kth car at time t will be $U(q_k(t))$. This is a reasonable model of driver behavior: the driver chooses her speed based only on the distance to the car she is following. This viewpoint can be used to justify the assumption that U depends on q. The ordinary differential equations for the motion of the cars now become a coupled set of nonlinear equations:

$$X'_k(t) = U(q_k(t)) = U([X_{k+1}(t) - X_k(t)]^{-1}) \tag{11.7}$$

for $k = 1, 2, \ldots, m$, where m is the number of cars.

Figure 11.1(a) shows an example of one such simulation. The speed limit is $u_{\max} = 1$, and the cars are initially distributed so there is a Gaussian increase in the density near $x = 0$. Note the following:

- The disturbance in density does not move with the individual cars, the way it would for the linear advection equation. Individual cars move through the congested region, slowing down and later speeding up.

- The hump in density changes shape with time. By the end of the simulation a *shock wave* is visible, across which the density increases and the velocity decreases very quickly (drivers who were zipping along in light traffic must suddenly slam on their brakes).

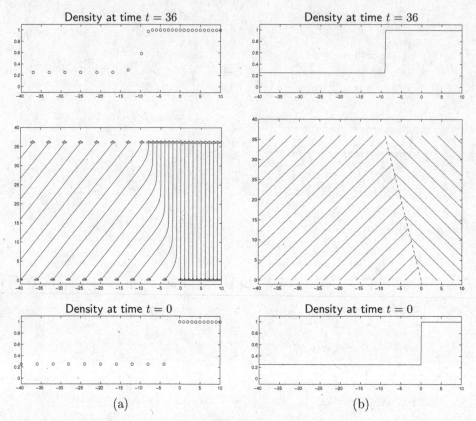

Fig. 11.2. Riemann problem for the traffic flow model at a red light. (a) Trajectories of individual vehicles and the density observed by each driver. (b) The characteristic structure is shown along with the density as computed by CLAWPACK. [claw/book/chap11/redlight]

- As cars move out of the congested region, they accelerate smoothly and the density decreases smoothly. This is called a *rarefaction wave*, since the fluid is becoming more rarefied as the density decreases.

Figures 11.2 and 11.3 show two more simulations for special cases corresponding to *Riemann problems* in which the initial data is piecewise constant. Figure 11.2 can be interpreted as cars approaching a traffic jam, or a line of cars waiting for a light to change; Figure 11.3 shows how cars accelerate once the light turns green. Note that the solution to the Riemann problem may consist of either a shock wave as in Figure 11.2 or a rarefaction wave as in Figure 11.3, depending on the data.

11.2 Quasilinear Form and Characteristics

By differentiating the flux function $f(q)$ we obtain the quasilinear form of the conservation law (11.1),

$$q_t + f'(q)q_x = 0. \tag{11.8}$$

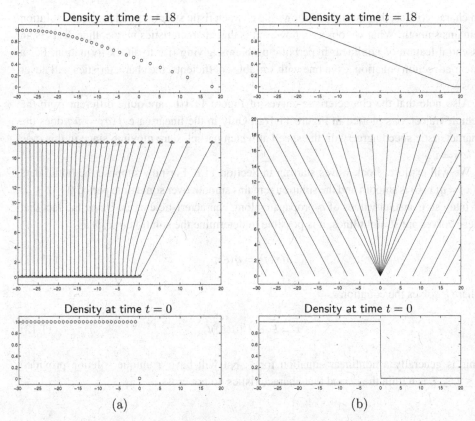

Fig. 11.3. Riemann problem for the traffic flow model at a green light. (a) Trajectories of individual vehicles and the density observed by each driver. (b) The characteristic structure is shown, along with the density as computed by CLAWPACK. [claw/book/chap11/greenlight]

It is normally preferable to work directly with the conservative form, but it is the quasilinear form that determines the characteristics of the equation. If we assume that $q(x, t)$ is smooth, then along any curve $X(t)$ satisfying the ODE

$$X'(t) = f'(q(X(t), t)), \tag{11.9}$$

we have

$$\frac{d}{dt} q(X(t), t) = X'(t)q_x + q_t$$

$$= 0 \tag{11.10}$$

by (11.8) and (11.9). Hence q is constant along the curve $X(t)$, and consequently $X'(t)$ is also constant along the curve, and so the characteristic curve must be a straight line.

We thus see that *for a scalar conservation law, q is constant on characteristics, which are straight lines, as long as the solution remains smooth.* The structure of the characteristics depends on the initial data $q(x, 0)$. Figure 11.1(b) shows the characteristics for the traffic flow problem of Figure 11.1(a). Actually, Figure 11.1(b) shows a contour plot (in x and t) of the density q as computed using CLAWPACK on a very fine grid. Since q is constant on characteristics, this essentially shows the characteristic structure, as long as the solution remains smooth. What we observe, however, is that characteristics may collide. This is an essential feature of nonlinear hyperbolic problems, giving rise to shock formation. For a linear advection equation, even one with variable coefficients, the characteristics will never cross.

Also note that the characteristic curves of Figure 11.1(b) are quite different from the vehicle trajectories plotted in Figure 11.1(a). Only in the linear case $f(q) = \bar{u}q$ does the characteristic speed agree with the speed at which particles are moving, since in this case $f'(q) = \bar{u}$.

We will explore shock waves starting in Section 11.8. For now suppose the initial data $\overset{\circ}{q}(x) = q(x, 0)$ is smooth and the solution remains smooth over sóme time period $0 \le t \le T$ of interest. Constant values of q propagate along characteristic curves. Using the fact that these curves are straight lines, it is possible to determine the solution $q(x, t)$ as

$$q(x, t) = \overset{\circ}{q}(\xi), \qquad (11.11)$$

where ξ solves the equation

$$x = \xi + f'(\overset{\circ}{q}(\xi))t. \qquad (11.12)$$

This is generally a nonlinear equation for ξ, but will have a unique solution provided $0 \le t \le T$ is within the period that characteristics do not cross.

11.3 Burgers' Equation

The traffic flow model gives a scalar conservation law with a quadratic flux function. An even simpler scalar equation of this form is the famous *Burgers equation*

$$u_t + \left(\frac{1}{2}u^2\right)_x = 0. \qquad (11.13)$$

This should more properly be called the *inviscid Burgers equation*, since Burgers [54] actually studied the viscous equation

$$u_t + \left(\frac{1}{2}u^2\right)_x = \epsilon u_{xx}. \qquad (11.14)$$

Rather than modeling a particular physical problem, this equation was introduced as the simplest model equation that captures some key features of gas dynamics: the nonlinear hyperbolic term and viscosity. In the literature of hyperbolic equations, the inviscid problem (11.13) has been widely used for developing both theory and numerical methods.

Around 1950, Hopf, and independently Cole, showed that the *exact* solution of the nonlinear equation (11.14) could be found using what is now called the *Cole–Hopf transformation*. This reduces (11.14) to a linear heat equation. See Chapter 4 of Whitham [486] for details and Exercise 11.2 for one particular solution.

Solutions to the inviscid Burgers' equation have the same basic structure as solutions to the traffic flow problem considered earlier. The quasilinear form of Burgers' equation,

$$u_t + uu_x = 0, \tag{11.15}$$

shows that in smooth portions of the solution the value of u is constant along characteristics traveling at speed u. Note that (11.15) looks like an advection equation in which the value of u is being carried at velocity u. This is the essential nonlinearity that appears in the conservation-of-momentum equation of fluid dynamics: the velocity or momentum is carried in the fluid at the fluid velocity.

Detailed discussions of Burgers' equation can be found in many sources, e.g., [281], [486]. All of the issues illustrated below for the traffic flow equation can also be illustrated with Burgers' equation, and the reader is encouraged to explore this equation in the process of working through the remainder of this chapter.

11.4 Rarefaction Waves

Suppose the initial data for the traffic flow model satisfies $q_x(x, 0) < 0$, so the density falls with increasing x. In this case the characteristic speed

$$f'(q) = U(q) + q\, U'(q) = u_{\max}(1 - 2q) \tag{11.16}$$

is increasing with x. Hence the characteristics are spreading out and will never cross. The density observed by each car will decrease with time, and the flow is being *rarefied*.

A special case is the Riemann problem shown in Figure 11.3. In this case the initial data is discontinuous, but if we think of smoothing this out very slightly as in Figure 11.4(a), then each value of q between 0 and 1 is taken on in the initial data, and each value propagates with its characteristic speed $f'(q)$, as is also illustrated in Figure 11.4(a). Figure 11.3(b) shows the characteristics in the x–t plane. This is called a *centered rarefaction wave*, or rarefaction fan, emanating from the point $(0, 0)$.

For Burgers' equation (11.13), the characteristics are all spreading out if $u_x(x, 0) > 0$, since the characteristic speed is $f'(u) = u$ for this equation.

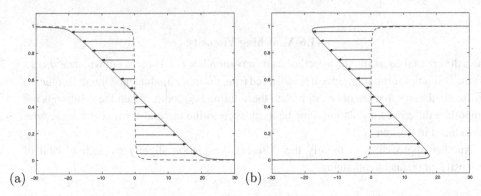

Fig. 11.4. Tracing each value of q at its characteristic speed from the initial data (shown by the dashed line) to the solution at time $t = 20$. (a) When q is decreasing, we obtain a rarefaction wave. (b) When q is increasing, we obtain an unphysical triple-valued solution.

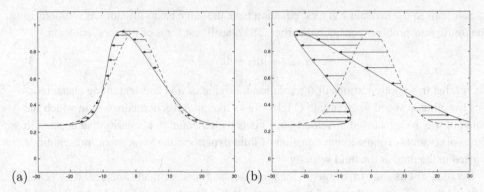

Fig. 11.5. Tracing each value of q at its characteristic speed from the initial data from Figure 11.1 (shown by the dashed line) to the solution at two different times. (a) At $t = 5$ we see a compression wave and a rarefaction wave. (b) At $t = 25$ we obtain an unphysical triple-valued solution.

11.5 Compression Waves

Figure 11.5(a) shows a picture similar to Figure 11.4(a) for initial data consisting of a hump of density as in the example of Figure 11.1. In this case the right half of the hump, where the density falls, is behaving as a rarefaction wave. The left part of the hump, where the density is rising with increasing x, gives rise to a *compression wave*. A driver passing through this region will experience increasing density with time (this can be observed in Figure 11.1).

If we try to draw a similar picture for larger times then we obtain Figure 11.5(b). As in the case of Figure 11.4(b), this does not make physical sense. At some points x the density appears to be triple-valued. At such points there are three characteristics reaching (x, t) from the initial data, and the equation (11.12) would have three solutions.

An indication of what should instead happen physically is seen in Figure 11.1(a). Recall that this was obtained by simulating the traffic flow directly rather than by solving a conservation law. The compression wave should steepen up into a discontinuous shock wave. At some time T_b the slope $q_x(x, t)$ will become infinite somewhere. This is called the *breaking time* (by analogy with waves breaking on a beach; see Section 13.1). Beyond time T_b characteristics may cross and a shock wave appears. It is possible to determine T_b from the initial data $q(x, 0)$; see Exercise 11.1.

11.6 Vanishing Viscosity

As a differential equation, the hyperbolic conservation law (11.1) breaks down once shocks form. This is not surprising, since it was derived from the more fundamental integral equation by manipulations that are only valid when the solution is smooth. When the solution is not smooth, a different formulation must be used, such as the integral form or the *weak form* introduced in Section 11.11.

Another approach is to modify the differential equation slightly by adding a bit of viscosity, or diffusion, obtaining

$$q_t + f(q)_x = \epsilon q_{xx}, \tag{11.17}$$

where $\epsilon > 0$ is a small parameter. The term "viscosity" is motivated by fluid dynamics

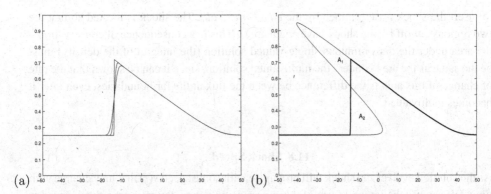

Fig. 11.6. (a) Solutions to the viscous traffic flow equation (11.17) for two values of $\epsilon > 0$, along with the limiting shock-wave solution when $\epsilon = 0$. (b) The shock wave (heavy line) can be determined by applying the equal-area rule to the triple-valued solution of Figure 11.5(b) so that the areas A_1 and A_2 are equal.

equations, as discussed below. If ϵ is extremely small, then we might expect solutions to (11.17) to be very close to solutions of (11.1), which has $\epsilon = 0$. However, the equation (11.17) is *parabolic* rather than hyperbolic, and it can be proved that this equation has a unique solution for all time $t > 0$, for any set of initial conditions, provided only that $\epsilon > 0$. Away from shock waves, q_{xx} is bounded and this new term is negligible. If a shock begins to form, however, the derivatives of q begin to blow up and the ϵq_{xx} term becomes important. Figure 11.6 shows solutions to the traffic flow model from Figure 11.1 with this new term added, for various values of ϵ. As $\epsilon \to 0$ we approach a limiting solution that has a discontinuity corresponding to the shock wave seen in Figure 11.1.

The idea of introducing the small parameter ϵ and looking at the limit $\epsilon \to 0$ is called the *vanishing-viscosity* approach to defining a sensible solution to the hyperbolic equation. To motivate this, remember that in general any mathematical equation is only a model of reality, in which certain effects may have been modeled well but others are necessarily neglected. In gas dynamics, for example, the hyperbolic equations (2.38) only hold if we ignore thermal diffusion and the effects of viscosity in the gas, which is the frictional force of molecules colliding and converting kinetic energy into internal energy. For a dilute gas this is reasonable most of the time, since these forces are normally small. Including the (parabolic) viscous terms in the equations would only complicate them without changing the solution very much. Near a shock wave, however, the viscous terms are important. In the real world shock waves are not sharp discontinuities but rather smooth transitions over very narrow regions. By using the hyperbolic equation we hope to capture the big picture without modeling exactly how the solution behaves within this thin region.

11.7 Equal-Area Rule

From Figure 11.1 it is clear that after a shock forms, the solution contains a discontinuity, but that away from the shock the solution remains smooth and still has the property that it is constant along characteristics. This solution can be constructed by taking the unphysical solution of Figure 11.4(b) and eliminating the triple-valued portion by inserting a discontinuity at some point as indicated in Figure 11.6(b). It can be shown that the correct place to insert the shock is determined by the *equal-area rule*: The shock is located so that the two regions cut off by the shock have equal areas. This is a consequence of conservation – the area under the discontinuous single-valued solution (the integral of the density) must be the same as the area "under" the multivalued solution, since it can be shown that the rate of change of this area is the difference between the flux at the far boundaries, even after it becomes multivalued.

11.8 Shock Speed

As the solution evolves the shock will propagate with some speed $s(t)$ that may change with time. We can use the integral form of the conservation law to determine the shock speed at any time in terms of the states $q_l(t)$ and $q_r(t)$ immediately to the left and right of the shock. Suppose the shock is moving as shown in Figure 11.7, where we have zoomed in on a very short time increment from t_1 to $t_1 + \Delta t$ over which the shock speed is essentially a constant value s. Then the rectangle $[x_1, x_1 + \Delta x] \times [t_1, t_1 + \Delta t]$ shown in Figure 11.7 is split by the shock into two triangles and the value of q is roughly constant in each. If we apply the integral form of the conservation law (2.6) to this region, we obtain

$$\int_{x_1}^{x_1+\Delta x} q(x, t_1 + \Delta t)\, dx - \int_{x_1}^{x_1+\Delta x} q(x, t_1)\, dx$$

$$= \int_{t_1}^{t_1+\Delta t} f(q(x_1, t))\, dt - \int_{t_1}^{t_1+\Delta t} f(q(x_1 + \Delta x, t))\, dt. \qquad (11.18)$$

Since q is essentially constant along each edge, this becomes

$$\Delta x\, q_r - \Delta x\, q_l = \Delta t f(q_l) - \Delta t f(q_r) + \mathcal{O}(\Delta t^2), \qquad (11.19)$$

where the $\mathcal{O}(\Delta t^2)$ term accounts for the variation in q. If the shock speed is s, then $\Delta x = -s\, \Delta t$ (for the case $s < 0$ shown in the figure). Using this in (11.19), dividing by $-\Delta t$, and

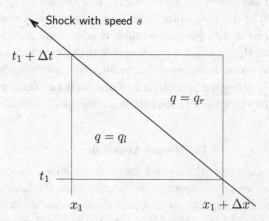

Fig. 11.7. The Rankine–Hugoniot jump conditions are determined by integrating over an infinitesimal rectangular region in the x–t plane.

taking the limit as $\Delta t \to 0$ gives

$$s(q_r - q_l) = f(q_r) - f(q_l). \tag{11.20}$$

This is called the *Rankine–Hugoniot jump condition*. This is sometimes written simply as $s[\![q]\!] = [\![f]\!]$, where $[\![\cdot]\!]$ represents the jump across the shock.

For a scalar conservation law we can divide by $q_r - q_l$ and obtain the *shock speed*:

$$s = \frac{f(q_r) - f(q_l)}{q_r - q_l}. \tag{11.21}$$

In general $q_l(t)$ and $q_r(t)$, the states just to the left and the right of the shock, vary with time and the shock speed also varies. For the special case of a Riemann problem with piecewise constant data q_l and q_r, the resulting shock moves with constant speed, given by (11.21).

Note that if $q_l \approx q_r$, then the expression (11.21) approximates $f'(q_l)$. A weak shock, which is essentially an acoustic wave, thus propagates at the characteristic velocity, as we expect from linearization.

For the traffic flow flux (11.6) we find that

$$s = u_{\max}[1 - (q_l + q_r)] = \frac{1}{2}[f'(q_l) + f'(q_r)], \tag{11.22}$$

since $f'(q) = u_{\max}(1 - 2q)$. In this case, and for any quadratic flux function, the shock speed is simply the average of the characteristic speeds on the two sides. Another quadratic example is Burgers' equation (11.13), for which we find

$$s = \frac{1}{2}(u_l + u_r). \tag{11.23}$$

11.9 The Rankine–Hugoniot Conditions for Systems

The derivation of (11.20) is valid for systems of conservation laws as well as for scalar equations. However, for a system of m equations, $q_r - q_l$ and $f(q_r) - f(q_l)$ will both be m-vectors and we will not be able to simply divide as in (11.21) to obtain the shock speed. In fact, for arbitrary states q_l and q_r there will be no scalar value s for which (11.20) is satisfied. Special relations must exist between the two states in order for them to be connected by a shock: the vector $f(q_l) - f(q_r)$ must be a scalar multiple of the vector $q_r - q_l$. We have already seen this condition in the case of a linear system, where $f(q) = Aq$. In this case the Rankine–Hugoniot condition (11.20) becomes

$$A(q_r - q_l) = s(q_r - q_l),$$

which means that $q_r - q_l$ must be an *eigenvector* of the matrix A. The propagation speed s of the discontinuity is then the corresponding eigenvalue λ. In Chapter 3 we used this condition to solve the linear Riemann problem, and in Chapter 13 we will see how this theory can be extended to nonlinear systems.

11.10 Similarity Solutions and Centered Rarefactions

For the special case of a Riemann problem with data

$$q(x,0) = \begin{cases} q_l & \text{if } x < 0, \\ q_r & \text{if } x > 0, \end{cases} \tag{11.24}$$

the solution to a conservation law is a *similarity solution*, a function of x/t alone that is self-similar at different times. The solution

$$q(x,t) = \tilde{q}(x/t) \tag{11.25}$$

is constant along any ray $x/t = \text{constant}$ through the origin, just as in the case of a linear hyperbolic system (see Chapter 3). From (11.25) we compute

$$q_t(x,t) = -\frac{x}{t^2}\tilde{q}'(x/t) \quad \text{and} \quad f(q)_x = \frac{1}{t}f'(\tilde{q}(x/t))\tilde{q}'(x/t).$$

Inserting these in the quasilinear equation $q_t + f'(q)q_x = 0$ shows that

$$f'(\tilde{q}(x/t))\tilde{q}'(x/t) = \frac{x}{t}\tilde{q}'(x/t). \tag{11.26}$$

For the scalar equation we find that either $\tilde{q}'(x/t) = 0$ ($\tilde{q}$ is constant) or that

$$f'(\tilde{q}(x/t)) = x/t. \tag{11.27}$$

This allows us to determine the solution through a centered rarefaction wave explicitly. (For a system of equations we cannot simply cancel $\tilde{q}'(x/t)$ from equation (11.26). See Section 13.8.3 for the construction of a rarefaction wave in a nonlinear system.)

Consider the traffic flow model, for example, with f given by (11.6). The solution to a Riemann problem with $q_l > q_r$ consists of a rarefaction fan. The left and right edges of this fan propagate with the characteristic speeds $f'(q_l)$ and $f'(q_r)$ respectively (see Figure 11.3), and so we have

$$\tilde{q}(x/t) = \begin{cases} q_l & \text{for } x/t \leq f'(q_l), \\ q_r & \text{for } x/t \geq f'(q_r). \end{cases} \tag{11.28}$$

In between, $\tilde{q}$ varies and so (11.27) must hold, which gives

$$u_{\max}[1 - 2\tilde{q}(x/t)] = x/t$$

and hence

$$\tilde{q}(x/t) = \frac{1}{2}\left(1 - \frac{x}{u_{\max}t}\right) \quad \text{for } f'(q_l) \leq x/t \leq f'(q_r). \tag{11.29}$$

Note that at any fixed time t the solution $q(x,t)$ is linear in x as observed in Figure 11.3(b). This is a consequence of the fact that $f(q)$ is quadratic and so $f'(q)$ is linear. A different flux function could give rarefaction waves with more interesting structure (see Exercise 11.8).

11.11 Weak Solutions

We have observed that the differential equation (11.1) is not valid in the classical sense for solutions containing shocks (though it still holds in all regions where the solution is smooth). The integral form of the conservation law (2.6) does hold, however, even when q is discontinuous, and it was this form that we used to determine the Rankine–Hugoniot condition (11.20) that must hold across shocks. More generally we can say that a function $q(x, t)$ is a solution of the conservation law if (2.2) holds for all t and any choice of x_1, x_2.

This formulation can be difficult to work with mathematically. In this section we look at a somewhat different integral formulation that is useful in proving results about solutions. In particular, in Section 12.10 we will investigate the convergence of finite volume methods as the grid is refined and will need this formulation to handle discontinuous solutions.

To motivate this *weak form*, first suppose that $q(x, t)$ is smooth. In Chapter 2 we derived the differential equation (11.1) by rewriting (2.7) as (2.9). Integrating this latter equation in time between t_1 and t_2 gives

$$\int_{t_1}^{t_2} \int_{x_1}^{x_2} [q_t + f(q)_x] \, dx \, dt = 0. \tag{11.30}$$

Rather than considering this integral for arbitrary choices of x_1, x_2, t_1, and t_2, we could instead consider

$$\int_0^\infty \int_{-\infty}^\infty [q_t + f(q)_x] \phi(x, t) \, dx \, dt \tag{11.31}$$

for a certain set of functions $\phi(x, t)$. In particular, if $\phi(x, t)$ is chosen to be

$$\phi(x, t) = \begin{cases} 1 & \text{if } (x, t) \in [x_1, x_2] \times [t_1, t_2], \\ 0 & \text{otherwise,} \end{cases} \tag{11.32}$$

then this integral reduces to the one in (11.30). We can generalize this notion by letting $\phi(x, t)$ be any function that has *compact support*, meaning it is identically zero outside of some bounded region of the x–t plane. If we now also assume that ϕ is a smooth function (unlike (11.32)), then we can integrate by parts in (11.31) to obtain

$$\int_0^\infty \int_{-\infty}^\infty [q\phi_t + f(q)\phi_x] \, dx \, dt = -\int_{-\infty}^\infty q(x, 0)\phi(x, 0) \, dx. \tag{11.33}$$

Only one boundary term along $t = 0$ appears in this process, since we assume ϕ vanishes at infinity.

A nice feature of (11.33) is that the derivatives are on ϕ, and no longer of q and $f(q)$. So (11.33) continues to make sense even if q is discontinuous. This motivates the following definition.

Definition 11.1. *The function $q(x, t)$ is a* weak solution *of the conservation law (11.1) with given initial data $q(x, 0)$ if (11.33) holds for all functions ϕ in C_0^1.*

The function space C_0^1 denotes the set of all functions that are C^1 (continuously differentiable) and have compact support. By assuming ϕ is smooth we rule out the special choice

(11.32) that gave (11.30), but we can approximate this function arbitrarily well by a slightly smoothed-out version. It can be shown that any weak solution also satisfies the integral conservation laws and vice versa.

11.12 Manipulating Conservation Laws

One danger to observe in dealing with conservation laws is that transforming the differential form into what appears to be an equivalent differential equation may not give an equivalent equation in the context of weak solutions.

Example 11.1. If we write Burgers' equation

$$u_t + \left(\frac{1}{2}u^2\right)_x = 0 \tag{11.34}$$

in the quasilinear form $u_t + uu_x = 0$ and multiply by $2u$, we obtain $2uu_t + 2u^2u_x = 0$, which can be rewritten as

$$(u^2)_t + \left(\frac{2}{3}u^3\right)_x = 0. \tag{11.35}$$

This is again a conservation law, now for u^2 rather than u itself, with flux function $f(u^2) = \frac{2}{3}(u^2)^{3/2}$. The differential equations (11.34) and (11.35) have precisely the same smooth solutions. However, they have different weak solutions, as we can see by considering the Riemann problem with $u_l > u_r$. The unique weak solution of (11.34) is a shock traveling at speed

$$s_1 = \frac{[\![\frac{1}{2}u^2]\!]}{[\![u]\!]} = \frac{1}{2}(u_l + u_r), \tag{11.36}$$

whereas the unique weak solution to (11.35) is a shock traveling at speed

$$s_2 = \frac{[\![\frac{2}{3}u^3]\!]}{[\![u^2]\!]} = \frac{2}{3}\left(\frac{u_r^3 - u_l^3}{u_r^2 - u_l^2}\right). \tag{11.37}$$

It is easy to check that

$$s_2 - s_1 = \frac{1}{6}\frac{(u_l - u_r)^2}{u_l + u_r}, \tag{11.38}$$

and so $s_2 \neq s_1$ when $u_l \neq u_r$, and the two equations have different weak solutions. The derivation of (11.35) from (11.34) requires manipulating derivatives in a manner that is valid only when u is smooth.

11.13 Nonuniqueness, Admissibility, and Entropy Conditions

The Riemann problem shown in Figure 11.3 has a solution consisting of a rarefaction wave, as determined in Section 11.10. However, this is not the only possible weak solution to the equation with this data. Another solution is

$$q(x, t) = \begin{cases} q_l & \text{if } x/t < s, \\ q_r & \text{if } x/t > s, \end{cases} \tag{11.39}$$

where the speed s is determined by the Rankine–Hugoniot condition (11.21). This function consists of the discontinuity propagating at speed s, the *shock speed*. We don't expect a shock wave in this case, since characteristics are spreading out rather than colliding, but we did not use any information about the characteristic structure in deriving the Rankine–Hugoniot condition. The function (11.39) is a weak solution of the conservation law regardless of whether $q_l < q_r$ or $q_l > q_r$. The discontinuity in the solution (11.39) is sometimes called an *expansion shock* in this case.

We see that the weak solution to a conservation law is not necessarily unique. This is presumably another failing of our mathematical formulation, since physically we expect only one thing to happen for given initial data, and hence a unique solution. Again this results from the fact that the hyperbolic equation is an imperfect model of reality. A better model might include "viscous terms" as in Section 11.6, for example, and the resulting parabolic equation would have a unique solution for any set of data. As in the case of shock waves, what we hope to capture with the hyperbolic equation is the correct limiting solution as the viscosity ϵ vanishes. As Figure 11.4(a) suggests, if the discontinuous data is smoothed only slightly (as would happen immediately if the equation were parabolic), then there is a unique solution determined by the characteristics. This solution clearly converges to the rarefaction wave as $\epsilon \to 0$. So the expansion shock solution (11.39) is an artifact of our formulation and is not physically meaningful.

The existence of these spurious solution is not merely a mathematical curiosity. Under some circumstances nonphysical solutions of this type are all too easily computed numerically, in spite of the fact that numerical methods typically contain some "numerical viscosity." See Section 12.3 for a discussion of these numerical difficulties.

In order to effectively use the hyperbolic equations we must impose some additional condition along with the differential equation in order to insure that the problem has a unique weak solution that is physically correct. Often the condition we want to impose is simply that the weak solution must be the vanishing-viscosity solution to the proper viscous equation. However, this condition is hard to work with directly in the context of the hyperbolic equation. Instead, a variety of other conditions have been developed that can be applied directly to weak solutions of hyperbolic equations to check if they are physically admissible. Such conditions are sometimes called *admissibility conditions*, or more often *entropy conditions*. This name again comes from gas dynamics, where the second law of thermodynamics demands that the entropy of a system must be nondecreasing with time (see Section 14.5). Across a physically admissible shock the entropy of the gas increases. Across an expansion shock, however, the entropy of the gas would decrease, which is not allowed. The entropy at each point can be computed as a simple function of the pressure and density, (2.35), and the behavior of this function can be used to test a weak solution for admissibility. For other conservation laws it is sometimes possible to define a function $\eta(q)$, called an *entropy function*, which has similar diagnostic properties. This approach to developing entropy conditions is discussed in Section 11.14. Expansion shocks are often called *entropy-violating shocks*, since they are weak solutions that fail to satisfy the appropriate entropy condition.

First we discuss some other admissibility conditions that relate more directly to the characteristic structure. We discuss only a few possibilities here and concentrate on the scalar case. For some systems of equations the development of appropriate admissibility conditions remains a challenging research question. In some cases it is not even well understood what the appropriate viscous regularization of the conservation law is, and it may be that different choices of the viscous term lead to different vanishing-viscosity solutions.

For scalar equations there is an obvious condition suggested by Figures 11.2(b) and 11.3(b). A shock should have characteristics going *into* the shock, as time advances. A propagating discontinuity with characteristics coming *out of* it would be unstable to perturbations. Either smearing out the initial profile a little, or adding some viscosity to the system, will cause this to be replaced by a rarefaction fan of characteristics, as in Figure 11.3(b). This gives our first version of the entropy condition:

Entropy Condition 11.1 (Lax). For a convex scalar conservation law, a discontinuity propagating with speed s given by (11.21) satisfies the Lax entropy condition if

$$f'(q_l) > s > f'(q_r). \tag{11.40}$$

Note that $f'(q)$ is the characteristic speed. For convex or concave f, the Rankine–Hugoniot speed s from (11.21) must lie between $f'(q_l)$ and $f'(q_r)$, so (11.40) reduces to simply the requirement that $f'(q_l) > f'(q_r)$. For the traffic flow flux (11.4), this then implies that we need $q_l < q_r$ in order for the solution to be an admissible shock since $f''(q) < 0$. If $q_l > q_r$ then the correct solution would be a rarefaction wave. For Burgers' equation with $f(u) = u^2/2$, on the other hand, the Lax Entropy Condition 11.1 requires $u_l > u_r$ for an admissible shock, since $f''(u)$ is everywhere positive rather than negative.

A more general form of (11.40), due to Oleinik, also applies to nonconvex scalar flux functions and is given in Section 16.1.2. The generalization to systems of equations is discussed in Section 13.7.2.

Another form of the entropy condition is based on the spreading of characteristics in a rarefaction fan. We state this for the convex case with $f''(q) > 0$ (such as Burgers' equation), since this is the form usually seen in the literature. If $q(x, t)$ is an increasing function of x in some region, then characteristics spread out in this case. The rate of spreading can be quantified, and gives the following condition, also due to Oleinik [346].

Entropy Condition 11.2 (Oleinik). $q(x, t)$ is the entropy solution to a scalar conservation law $q_t + f(q)_x = 0$ with $f''(q) > 0$ if there is a constant $E > 0$ such that for all $a > 0, t > 0$, and $x \in \mathbb{R}$,

$$\frac{q(x + a, t) - q(x, t)}{a} < \frac{E}{t}. \tag{11.41}$$

Note that for a discontinuity propagating with constant left and right states q_l and q_r, this can be satisfied only if $q_r - q_l \leq 0$, so this agrees with (11.40). The form of (11.41) also gives information on the rate of spreading of rarefaction waves as well as on the form of allowable jump discontinuities, and is easier to apply in some contexts. In particular, this formulation has advantages in studying numerical methods, and is related to one-sided Lipschitz conditions and related stability concepts [339], [340], [437].

11.14 Entropy Functions

Another approach to the entropy condition is to define an entropy function $\eta(q)$, motivated by thermodynamic considerations in gas dynamics as described in Chapter 14. This approach often applies to systems of equations, as in the case of gas dynamics, and is also used in studying numerical methods; see Section 12.11.1.

In general an entropy function should be a function that is conserved (i.e., satisfies some conservation law) whenever the function $q(x, t)$ is smooth, but which has a source or a sink at discontinuities in q. The entropy of a gas has this property: entropy is produced in an admissible shock but would be reduced across an expansion shock. The second law of thermodynamics requires that the total entropy must be nondecreasing with time. This entropy condition rules out expansion shocks.

We now restate this in mathematical form. Along with an entropy function $\eta(q)$, we need an *entropy flux* $\psi(q)$ with the property that whenever q is smooth an integral conservation law holds,

$$\int_{x_1}^{x_2} \eta(q(x, t_2)) \, dx = \int_{x_1}^{x_2} \eta(q(x, t_1)) \, dx$$
$$+ \int_{t_1}^{t_2} \psi(q(x_1, t)) \, dt - \int_{t_1}^{t_2} \psi(q(x_2, t)) \, dt. \tag{11.42}$$

The entropy function and flux must also be chosen in such a way that if q is discontinuous in $[x_1, x_2] \times [t_1, t_2]$, then the equation (11.42) does *not* hold with equality, so that the total entropy in $[x_1, x_2]$ at time t_2 is either less or greater than what would be predicted by the fluxes at x_1 and x_2. Requiring that an inequality of the form

$$\int_{x_1}^{x_2} \eta(q(x, t_2)) \, dx \leq \int_{x_1}^{x_2} \eta(q(x, t_1)) \, dx$$
$$+ \int_{t_1}^{t_2} \psi(q(x_1, t)) \, dt - \int_{t_1}^{t_2} \psi(q(x_2, t)) \, dt \tag{11.43}$$

always holds gives the entropy condition. (For the physical entropy in gas dynamics one would require an inequality of this form with $\geq$ in place of $\leq$, but in the mathematical literature $\eta(q)$ is usually chosen to be a convex function with $\eta''(q) > 0$, leading to the inequality (11.43).)

If $q(x, t)$ is smooth, then the conservation law (11.42) can be manipulated as in Chapter 2 to derive the differential form

$$\eta(q)_t + \psi(q)_x = 0. \tag{11.44}$$

Moreover, if η and ψ are smooth function of q, we can differentiate these to rewrite (11.44) as

$$\eta'(q)q_t + \psi'(q)q_x = 0. \tag{11.45}$$

On the other hand, the smooth function q satisfies

$$q_t + f'(q)q_x = 0. \qquad (11.46)$$

Multiplying (11.46) by $\eta'(q)$ and comparing with (11.45) yields

$$\psi'(q) = \eta'(q)f'(q) \qquad (11.47)$$

as a relation that should hold between the entropy function and the entropy flux. For a scalar conservation law this equation admits many solutions $\eta(q)$, $\psi(q)$. One trivial choice of η and ψ satisfying (11.47) would be $\eta(q) = q$ and $\psi(q) = f(q)$, but then η would be conserved even across discontinuities and this would not help in defining an admissibility criterion. Instead we also require that $\eta(q)$ be a *convex function* of q with $\eta''(q) > 0$ for all q. This will give an entropy function for which the inequality (11.43) should hold.

For a system of equations η and ψ are still *scalar* functions, but now $\eta'(q)$ and $\psi'(q)$ must be interpreted as the row-vector gradients of η and ψ with respect to q, e.g.,

$$\eta'(q) = \left[\frac{\partial \eta}{\partial q^1}, \frac{\partial \eta}{\partial q^2}, \ldots, \frac{\partial \eta}{\partial q^m} \right], \qquad (11.48)$$

and $f'(q)$ is the $m \times m$ Jacobian matrix. In general (11.47) is a system of m equations for the two variables η and ψ. Moreover, we also require that $\eta(q)$ be convex, which for a system requires that the Hessian matrix $\eta''(q)$ be positive definite. For $m > 2$ this may have no solutions. Many physical systems do have entropy functions, however, including of course the Euler equations of gas dynamics, where the negative of physical entropy can be used. See Exercise 13.6 for another example. For symmetric systems there is always an entropy function $\eta(q) = q^T q$, as observed by Godunov [158]. Conversely, if a system has a convex entropy, then its Hessian matrix $\eta''(q)$ symmetrizes the system [143]; see also [433].

In order to see that the physically admissible weak solution should satisfy (11.43), we go back to the more fundamental condition that the admissible $q(x, t)$ should be the vanishing-viscosity solution. Consider the related viscous equation

$$q_t^\epsilon + f(q^\epsilon)_x = \epsilon q_{xx}^\epsilon \qquad (11.49)$$

for $\epsilon > 0$. We will investigate how $\eta(q^\epsilon)$ behaves for solutions $q^\epsilon(x, t)$ to (11.49) and take the limit as $\epsilon \to 0$. Since solutions to the parabolic equation (11.49) are always smooth, we can derive the corresponding evolution equation for the entropy following the same manipulations we used for smooth solutions of the inviscid equation, multiplying (11.49) by $\eta'(q^\epsilon)$ to obtain

$$\eta(q^\epsilon)_t + \psi(q^\epsilon)_x = \epsilon \eta'(q^\epsilon)q_{xx}^\epsilon.$$

We can now rewrite the right-hand side to obtain

$$\eta(q^\epsilon)_t + \psi(q^\epsilon)_x = \epsilon \left(\eta'(q^\epsilon)q_x^\epsilon \right)_x - \epsilon \eta''(q^\epsilon) \left(q_x^\epsilon \right)^2.$$

Integrating this equation over the rectangle $[x_1, x_2] \times [t_1, t_2]$ gives

$$\int_{x_1}^{x_2} \eta(q^\epsilon(x, t_2))\, dx = \int_{x_1}^{x_2} \eta(q^\epsilon(x, t_1))\, dx$$

$$- \left(\int_{t_1}^{t_2} \psi(q^\epsilon(x_2, t))\, dt - \int_{t_1}^{t_2} \psi(q^\epsilon(x_1, t))\, dt \right)$$

$$+ \epsilon \int_{t_1}^{t_2} \left[\eta'(q^\epsilon(x_2, t)) q_x^\epsilon(x_2, t) - \eta'(q^\epsilon(x_1, t)) q_x^\epsilon(x_1, t) \right] dt$$

$$- \epsilon \int_{t_1}^{t_2} \int_{x_1}^{x_2} \eta''(q^\epsilon) \left(q_x^\epsilon \right)^2 dx\, dt. \qquad (11.50)$$

In addition to the flux differences, the total entropy is modified by two terms involving ϵ. As $\epsilon \to 0$, the first of these terms vanishes. (This is clearly true if the limiting function $q(x, t)$ is smooth at x_1 and x_2, and can be shown more generally.) The other term, however, involves integrating $(q_x^\epsilon)^2$ over the rectangle $[x_1, x_2] \times [t_1, t_2]$. If the limiting weak solution is discontinuous along a curve in this rectangle, then this term will not vanish in the limit. However, since $\epsilon > 0$, $(q_x^\epsilon)^2 > 0$, and $\eta'' > 0$ (by our convexity assumption), we can conclude that this term is nonpositive in the limit and hence the vanishing-viscosity weak solution satisfies (11.43).

Just as for the conservation law, an alternative weak form of the entropy condition can be formulated by integrating against smooth test functions ϕ, now required to be nonnegative, since the entropy condition involves an inequality. A weak solution q satisfies the *weak form of the entropy inequality* if

$$\int_0^\infty \int_{-\infty}^\infty [\phi_t \eta(q) + \phi_x \psi(q)]\, dx\, dt + \int_{-\infty}^\infty \phi(x, 0)\eta(q(x, 0))\, dx \geq 0 \qquad (11.51)$$

for all $\phi \in C_0^1(\mathbb{R} \times \mathbb{R})$ with $\phi(x, t) \geq 0$ for all x, t. In Section 12.11.1 we will see that this form of the entropy condition is convenient to work with in proving that certain numerical methods converge to entropy-satisfying weak solutions.

The entropy inequalities (11.43) and (11.51) are often written informally as

$$\eta(q)_t + \psi(q)_x \leq 0, \qquad (11.52)$$

with the understanding that where q is smooth (11.44) is in fact satisfied and near discontinuities the inequality (11.52) must be interpreted in the integral form (11.43) or the weak form (11.51).

Another form that is often convenient is obtained by applying the integral inequality (11.43) to an infinitesimal rectangle near a shock as illustrated in Figure 11.7. When the integral form of the original conservation law was applied over this rectangle, we obtained the Rankine–Hugoniot jump condition (11.20),

$$s(q_r - q_l) = f(q_r) - f(q_l).$$

Going through the same steps using the inequality (11.43) leads to the inequality

$$s(\eta(q_r) - \eta(q_l)) \geq \psi(q_r) - \psi(q_l). \qquad (11.53)$$

We thus see that a discontinuity propagating with speed s satisfies the entropy condition if and only if the inequality (11.53) is satisfied.

Example 11.2. For Burgers' equation (11.34), the discussion of Section 11.12 shows that the convex function $\eta(u) = u^2$ can be used as an entropy function, with entropy flux $\psi(u) = \frac{2}{3}u^3$. Note that these satisfy (11.47). Consider a jump from u_l to u_r propagating with speed $s = (u_l + u_r)/2$, the shock speed for Burgers' equation. The entropy condition (11.53) requires

$$\frac{1}{2}(u_l + u_r)(u_r^2 - u_l^2) \geq \frac{2}{3}(u_r^3 - u_l^3).$$

This can be rearranged to yield

$$\frac{1}{6}(u_l - u_r)^3 \geq 0,$$

and so the entropy condition is satisfied only if $u_l > u_r$, as we already found from the Lax Entropy Condition 11.1.

11.14.1 The Kružkov Entropies

In general an entropy function should be strictly convex, with a second derivative that is strictly positive at every point. This is crucial in the above analysis of (11.50), since the $(q_x^\epsilon)^2$ term that gives rise to the inequality in (11.51) is multiplied by $\eta''(q^\epsilon)$.

Rather than considering a single strictly convex function that can be used to investigate every value of q^ϵ, a different approach was adopted by Kružkov [248], who introduced the idea of entropy inequalities. He used a whole family of entropy functions and corresponding entropy fluxes,

$$\eta_k(q) = |q - k|, \qquad \psi_k(q) = \text{sgn}(q - k)[f(q) - f(k)], \qquad (11.54)$$

where k is any real number. Each function $\eta_k(q)$ is a piecewise linear function of q with a kink at $q = k$, and hence $\eta''(q) = \delta(q - k)$ is a delta function with support at $q = k$. It is a weakly convex function whose nonlinearity is concentrated at a single point, and hence it is useful only for investigating the behavior of weak solutions near the value $q = k$. However, it is sometimes easier to obtain entropy results by studying the simple piecewise linear function $\eta_k(q)$ than to work with an arbitrary entropy function. If it can be shown that if a weak solution satisfies the entropy inequality (11.51) for an arbitrary choice of $\eta_k(q)$ (i.e., that the entropy condition is satisfied for *all* the Kružkov entropies), then (11.51) also holds more generally for any strictly convex entropy $\eta(q)$.

11.15 Long-Time Behavior and N-Wave Decay

Figure 11.8 shows the solution to Burgers' equation with some smooth initial data having compact support. As time evolves, portions of the solution where $u_x < 0$ steepen into shock waves while portions where $u_x > 0$ spread out as rarefaction waves. Over time these shocks

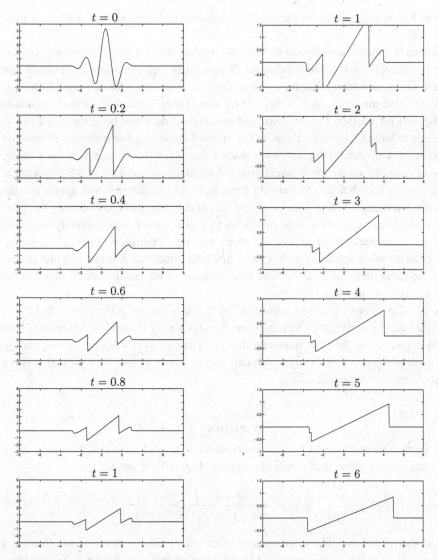

Fig. 11.8. Decay of general initial data to an N-wave with Burgers' equation. The left column shows the initial behavior from time $t = 0$ to $t = 1$ (going down). The right column shows later times $t = 1$ to $t = 6$ on a different scale. [claw/book/chap11/burgers]

and rarefactions interact. The shocks travel at different speeds, and those going in the same direction merge into stronger shocks. Meanwhile, the rarefaction waves weaken the shocks. The most striking fact is that in the long run all the structure of the initial data completely disappears and the solution behaves like an *N-wave*: one shock propagates to the left, another to the right, and in between the rarefaction wave is essentially linear. Similar long-time behavior would be seen starting from any other initial data with compact support. The position and strength of the two shocks does depend on the data, but this same N-wave shape will always arise. The same is true for other nonlinear conservation laws, provided the flux function is convex, although the shape of the rarefaction wave between the two shocks will depend on this flux. Only for a quadratic flux (such as Burgers' equation) will it be linear. For more details about the long-time behavior of solutions, see for example [98], [308], [420].

Note that this loss of structure in the solution implies that in general a nonlinear conservation law models irreversible phenomena. With a linear equation, such as the advection equation or the acoustics equations, the solution is reversible. We can start with arbitrary data, solve the equation $q_t + Aq_x = 0$ for some time T, and then solve the equation $q_t - Aq_x = 0$ for another T time units, and the original data will be recovered. (This is equivalent to letting t run backwards in the original equation.) For a nonlinear equation $q_t + f(q)_x = 0$, we can do this only over times T for which the solution remains smooth. Over such time periods the solution is constant along the characteristics, which do not cross, and the process is reversible. Once shocks form, however, characteristics disappear into the shock and information is lost. Many different sets of data can give rise to exactly the same shock wave. If the equation is now solved backwards in time (or, equivalently, we solve $q_t - f(q)_x = 0$) then compression at the shock becomes expansion. There are infinitely many different weak solutions in this case. Spreading back out into the original data is one of these, but the solution at the final time contains no information about which of the infinitely many possibilities is "correct."

The fact that information is irretrievably lost in a shock wave is directly related to the notion that the physical entropy increases across a shock wave. Entropy is a measure of the amount of disorder in the system (see Section 14.5) and a loss of structure corresponds to an increase in entropy. Since entropy can only increase once shocks have formed, it is not possible to recover the initial data.

Exercises

11.1. Show that in solving the scalar conservation law $q_t + f(q)_x = 0$ with smooth initial data $q(x, 0)$, the time at which the solution "breaks" is given by

$$T_b = \frac{-1}{\min_x [f''(q(x,0))q_x(x,0)]} \tag{11.55}$$

if this is positive. If this is negative, then characteristics never cross. *Hint:* Use $q(x, t) = q(\xi(x, t), 0)$ from (11.11), differentiate this with respect to x, and determine where q_x becomes infinite. To compute ξ_x, differentiate the equation (11.12) with respect to x.

11.2. Show that the viscous Burgers equation $u_t + uu_x = \epsilon u_{xx}$ has a *traveling-wave solution* of the form $u^\epsilon(x, t) = w^\epsilon(x - st)$, by deriving an ODE for w and verifying that this ODE has solutions of the form

$$w(\xi) = u_r + \frac{1}{2}(u_l - u_r)\left[1 - \tanh\left(\frac{(u_l - u_r)\xi}{4\epsilon}\right)\right], \tag{11.56}$$

when $u_l > u_r$, with the propagation speed s agreeing with the shock speed (11.23). Note that $w(\xi) \to u_l$ as $\xi \to -\infty$, and $w(\xi) \to u_r$ as $\xi \to +\infty$. Sketch this solution and indicate how it varies as $\epsilon \to 0$. What happens to this solution if $u_l < u_r$, and why is there no traveling-wave solution with limiting values of this form?

11.3. For a general smooth scalar flux functions $f(q)$, show by Taylor expansion of (11.21) that the shock speed is approximately the average of the characteristic speed on each side,

$$s = \frac{1}{2}[f'(q_l) + f'(q_r)] + \mathcal{O}(|q_r - q_l|^2).$$

The exercises below require determining the exact solution to a scalar conservation law with particular initial data. In each case you should sketch the solution at several instants in time as well as the characteristic structure and shock-wave locations in the x–t plane.

You may wish to solve the problem numerically by modifying the CLAWPACK codes for this chapter in order to gain intuition for how the solution behaves and to check your formulas.

11.4. Determine the exact solution to Burgers' equation $u_t + (\frac{1}{2}u^2)_x = 0$ for all $t > 0$ when each of the following sets of initial data is used:

(a)

$$\overset{\circ}{u}(x) = \begin{cases} 1 & \text{if } x < -1, \\ 0 & \text{if } -1 < x < 1, \\ -1 & \text{if } x > 1. \end{cases}$$

(b)

$$\overset{\circ}{u}(x) = \begin{cases} -1 & \text{if } x < -1, \\ 0 & \text{if } -1 < x < 1, \\ 1 & \text{if } x > 1. \end{cases}$$

11.5. Determine the exact solution to Burgers' equation for $t > 0$ with initial data

$$\overset{\circ}{u}(x) = \begin{cases} 2 & \text{if } 0 < x < 1, \\ 0 & \text{otherwise.} \end{cases}$$

Note that the rarefaction wave catches up to the shock at some time T_c. For $t > T_c$ determine the location of the shock by two different approaches:

(a) Let $x_s(t)$ represent the shock location at time t. Determine and solve an ODE for $x_s(t)$ by using the Rankine–Hugoniot jump condition (11.21), which must hold across the shock at each time.

(b) For $t > T_c$ the exact solution is triangular-shaped. Use conservation to determine $x_s(t)$ based on the area of this triangle. Sketch the corresponding "overturned" solution, and illustrate the equal-area rule (as in Figure 11.6).

11.6. Repeat Exercise 11.5 with the data

$$\overset{\circ}{u}(x) = \begin{cases} 2 & \text{if } 0 < x < 1, \\ 4 & \text{otherwise.} \end{cases}$$

Note that in this case the shock catches up with the rarefaction wave.

11.7. Determine the exact solution to Burgers' equation for $t > 0$ with the data

$$\mathring{u}(x) = \begin{cases} 12 & \text{if } 0 < x, \\ 8 & \text{if } 0 < x < 14, \\ 4 & \text{if } 14 < x < 17, \\ 2 & \text{if } 17 < x. \end{cases}$$

Note that the three shocks eventually merge into one shock.

11.8. Consider the scalar conservation law $u_t + (e^u)_x = 0$. Determine the exact solution with the following sets of initial data:

(a)

$$\mathring{u}(x) = \begin{cases} 1 & \text{if } x < 0, \\ 0 & \text{if } x > 0. \end{cases}$$

(b)

$$\mathring{u}(x) = \begin{cases} 0 & \text{if } x < 0, \\ 1 & \text{if } x > 0. \end{cases}$$

(c)

$$\mathring{u}(x) = \begin{cases} 2 & \text{if } 0 < x < 1, \\ 0 & \text{otherwise.} \end{cases}$$

Hint: Use the approach outlined in Exercise 11.5(a).

11.9. Determine an entropy function and entropy flux for the traffic flow equation with flux (11.4). Use this to show that $q_l < q_r$ is required for a shock to be admissible.

12

Finite Volume Methods for Nonlinear Scalar
Conservation Laws

We now turn to the development of finite volume methods for nonlinear conservation laws. We will build directly on what has already been developed for linear systems in Chapter 4, concentrating in this chapter on scalar equations, although much of what is developed will also apply to nonlinear systems of equations. In Chapter 15 we consider some additional issues arising for systems of equations, particularly the need for efficient *approximate* Riemann solvers in Section 15.3.

Nonlinearity introduces a number of new difficulties not seen for the linear problem. Stability and convergence theory are more difficult than in the linear case, particularly in that we are primarily interested in discontinuous solutions involving shock waves. This theory is taken up in Section 12.10. Moreover, we must ensure that we are converging to the *correct* weak solution of the conservation law, since the weak solution may not be unique. This requires that the numerical method be consistent with a suitable entropy condition; see Section 12.11.

For a nonlinear conservation law $q_t + f(q)_x = 0$ it is very important that the method be in *conservation form*, as described in Section 4.1,

$$Q_i^{n+1} = Q_i^n - \frac{\Delta t}{\Delta x}\left(F_{i+1/2}^n - F_{i-1/2}^n\right), \qquad (12.1)$$

in order to insure that weak solutions to the conservation law are properly approximated. Recall that this form is derived directly from the integral form of the conservation laws, which is the correct equation to model when the solution is discontinuous. In Section 12.9 an example is given to illustrate that methods based instead on the quasilinear form $q_t + f'(q)q_x = 0$ may be accurate for smooth solutions but may completely fail to approximate a weak solution when the solution contains shock waves.

12.1 Godunov's Method

Recall from Section 4.11 that Godunov's method is obtained by solving the Riemann problem between states Q_{i-1}^n and Q_i^n in order to determine the flux $F_{i-1/2}^n$ as

$$F_{i-1/2}^n = f\left(Q_{i-1/2}^{\downarrow}\right).$$

Fig. 12.1. Five possible configurations for the solution to a scalar Riemann problem between states Q_{i-1} and Q_i, shown in the x–t plane: (a) left-going shock, $Q^\vee_{i-1/2} = Q_i$; (b) left-going rarefaction, $Q^\vee_{i-1/2} = Q_i$; (c) transonic rarefaction, $Q^\vee_{i-1/2} = q_s$; (d) right-going rarefaction, $Q^\vee_{i-1/2} = Q_{i-1}$; (e) right-going shock, $Q^\vee_{i-1/2} = Q_{i-1}$.

The value $Q^\vee_{i-1/2} = q^\vee(Q^n_{i-1}, Q^n_i)$ is the value obtained along the ray $x \equiv x_{i-1/2}$ in this Riemann solution. This value is constant for $t > t_n$, since the Riemann solution is a similarity solution. To keep the notation less cluttered, we will often drop the superscript n on Q below.

To begin, we assume the flux function $f(q)$ is *convex* (or concave), i.e., $f''(q)$ does not change sign over the range of q of interest. Then the Riemann solution consists of a single shock or rarefaction wave. (See Section 16.1.3 for the nonconvex case.) For a scalar conservation law with a convex flux function there are five possible forms that the Riemann solution might take, as illustrated in Figure 12.1. In most cases the solution $Q^\vee_{i-1/2}$ is either Q_i (if the solution is a shock or rarefaction wave moving entirely to the left, as in Figure 12.1(a) or (b)), or Q_{i-1} (if the solution is a shock or rarefaction wave moving entirely to the right, as in Figure 12.1(d) or (e)).

The only case where $Q^\vee_{i-1/2}$ has a different value than Q_i or Q_{i-1} is if the solution consists of a rarefaction wave that spreads partly to the left and partly to the right, as shown in Figure 12.1(c). Suppose for example that $f''(q) > 0$ everywhere, in which case $f'(q)$ is increasing with q, so that a rarefaction wave arises if $Q_{i-1} < Q_i$. In this case the situation shown in Figure 12.1(c) occurs only if

$$Q_{i-1} < q_s < Q_i,$$

where q_s is the (unique) value of q for which $f'(q_s) = 0$. This is called the *stagnation point*, since the value q_s propagates with velocity 0. It is also called the *sonic point*, since in gas dynamics the eigenvalues $u \pm c$ can take the value 0 only when the fluid speed $|u|$ is equal to the sound speed c. The solution shown in Figure 12.1(c) is called a *transonic rarefaction* since in gas dynamics the fluid is accelerated from a subsonic velocity to a supersonic velocity through such a rarefaction. In a transonic rarefaction the value along $x/t = x_{i-1/2}$ is simply q_s.

For the case $f''(q) > 0$ we thus see that the Godunov flux function for a convex scalar conservation law is

$$F^n_{i-1/2} = \begin{cases} f(Q_{i-1}) & \text{if } Q_{i-1} > q_s \text{ and } s > 0, \\ f(Q_i) & \text{if } Q_i < q_s \text{ and } s < 0, \\ f(q_s) & \text{if } Q_{i-1} < q_s < Q_i. \end{cases} \tag{12.2}$$

Here $s = [f(Q_i) - f(Q_{i-1})]/(Q_i - Q_{i-1})$ is the shock speed given by (11.21).

Note in particular that if $f'(q) > 0$ for both Q_{i-1} and Q_i then $F^n_{i-1/2} = f(Q_{i-1})$ and Godunov's method reduces to the *first-order upwind* method

$$Q^{n+1}_i = Q_i - \frac{\Delta t}{\Delta x}[f(Q_i) - f(Q_{i-1})]. \tag{12.3}$$

The natural upwind method is also obtained if $f'(q) < 0$ for both values of Q, involving one-sided differences in the other direction. Only in the case where $f'(q)$ changes sign between Q_{i-1} and Q_i is the formula more complicated, as we should expect, since the "upwind direction" is ambiguous in this case and information must flow both ways.

The formula (12.2) can be written more compactly as

$$
F_{i-1/2}^n = \begin{cases} \min\limits_{Q_{i-1} \le q \le Q_i} f(q) & \text{if } Q_{i-1} \le Q_i, \\ \max\limits_{Q_i \le q \le Q_{i-1}} f(q) & \text{if } Q_i \le Q_{i-1}, \end{cases} \tag{12.4}
$$

since the stagnation point q_s is the global minimum or maximum of f in the convex case. This formula is valid also for the case $f''(q) < 0$ and even for nonconvex fluxes, in which case there may be several stagnation points at each maximum and minimum of f (see Section 16.1).

Note that there is one solution structure not illustrated in Figure 12.1, a stationary shock with speed $s = 0$. In this case the value $Q_{i-1/2}^{\vee}$ is ambiguous, since the Riemann solution is discontinuous along $x = x_{i-1/2}$. However, if $s = 0$ then $f(Q_{i-1}) = f(Q_i)$ by the Rankine–Hugoniot condition, and so $F_{i-1/2}^n$ is still well defined and the formula (12.4) is still valid.

12.2 Fluctuations, Waves, and Speeds

Godunov's method can be implemented in our standard form

$$
Q_i^{n+1} = Q_i - \frac{\Delta t}{\Delta x} \left(\mathcal{A}^+ \Delta Q_{i-1/2} + \mathcal{A}^- \Delta Q_{i+1/2} \right) \tag{12.5}
$$

if we define the fluctuations $\mathcal{A}^{\pm} \Delta Q_{i-1/2}$ by

$$
\begin{aligned}
\mathcal{A}^+ \Delta Q_{i-1/2} &= f(Q_i) - f(Q_{i-1/2}^{\vee}), \\
\mathcal{A}^- \Delta Q_{i-1/2} &= f(Q_{i-1/2}^{\vee}) - f(Q_{i-1}).
\end{aligned} \tag{12.6}
$$

In order to define high-resolution correction terms, we also wish to compute a wave $\mathcal{W}_{i-1/2}$ and speed $s_{i-1/2}$ resulting from this Riemann problem. The natural choice is

$$
\mathcal{W}_{i-1/2} = Q_i - Q_{i-1},
$$

$$
s_{i-1/2} = \begin{cases} [f(Q_i) - f(Q_{i-1})]/(Q_i - Q_{i-1}) & \text{if } Q_{i-1} \ne Q_i, \\ f'(Q_i) & \text{if } Q_{i-1} = Q_i, \end{cases} \tag{12.7}
$$

although the value of $s_{i-1/2}$ is immaterial when $Q_{i-1} = Q_i$. The speed chosen is the Rankine–Hugoniot shock speed (11.21) for this data. If the Riemann solution is a shock wave, this is clearly the right thing to do. If the solution is a rarefaction wave, then the wave is not simply a jump discontinuity propagating at a single speed. However, this is still a suitable definition of the wave and speed to use in defining correction terms that yield second-order accuracy in the smooth case. This can be verified by a truncation-error analysis of the resulting method; see Section 15.6. Note that when a smooth solution is being approximated, we expect $\mathcal{W}_{i-1/2} = \mathcal{O}(\Delta x)$, and there is very little spreading of the rarefaction wave in any case. Moreover, a wave consisting of this jump discontinuity propagating with speed $s_{i-1/2}$

does define a weak solution to the Riemann problem, although it is an expansion shock that does not satisfy the entropy condition. However, provided it is not a transonic rarefaction, the same result will be obtained in Godunov's method whether we use the entropy-satisfying rarefaction wave or an expansion shock as the solution to the Riemann problem. When the wave lies entirely within one cell, the cell average is determined uniquely by conservation and does not depend on the structure of the particular weak solution chosen. We can compute the fluctuations $\mathcal{A}^{\pm}\Delta Q_{i-1/2}$ using

$$
\begin{aligned}
\mathcal{A}^{+}\Delta Q_{i-1/2} &= s_{i-1/2}^{+}\mathcal{W}_{i-1/2}, \\
\mathcal{A}^{-}\Delta Q_{i-1/2} &= s_{i-1/2}^{-}\mathcal{W}_{i-1/2}
\end{aligned}
\tag{12.8}
$$

in place of (12.6), provided that $Q_{i-1/2}^{\vee} = Q_{i-1}$ or Q_i. Only in the case of a transonic rarefaction is it necessary to instead use the formulas (12.6) with $Q_{i-1/2}^{\vee} = q_s$.

12.3 Transonic Rarefactions and an Entropy Fix

Note that if we were to always use (12.8), even for transonic rarefactions, then we would still be applying Godunov's method using an exact solution to the Riemann problem; the Rankine–Hugoniot conditions are always satisfied for the jump and speed determined in (12.7). The only problem is that the entropy condition would not be satisfied and we would be using the wrong solution. For this reason the modification to $\mathcal{A}^{\pm}\Delta Q_{i-1/2}$ required in the transonic case is often called an *entropy fix*.

This approach is used in the Riemann solver [claw/book/chap12/efix/rp1.f]. The wave and speed are first calculated, and the fluctuations are set using (12.8). Only in the case of a transonic rarefaction is the entropy fix applied to modify the fluctuations. If $f'(Q_{i-1}) < 0 < f'(Q_i)$ then the fluctuations in (12.8) are replaced by

$$
\begin{aligned}
\mathcal{A}^{+}\Delta Q_{i-1/2} &= f(Q_i) - f(q_s), \\
\mathcal{A}^{-}\Delta Q_{i-1/2} &= f(q_s) - f(Q_{i-1}).
\end{aligned}
\tag{12.9}
$$

We will see in Section 15.3 that this approach generalizes to nonlinear systems of equations in a natural way. An *approximate Riemann solver* can often be used that gives an approximate solution involving a finite set of waves that are jump discontinuities $\mathcal{W}_{i-1/2}^p$ propagating at some speeds $s_{i-1/2}^p$. These are used to define fluctuations and also high-resolution correction terms, as indicated in Section 6.15. A check is then performed to see if any of the waves should really be transonic rarefactions, and if so, the fluctuations are modified by performing an entropy fix.

For the scalar equation this approach may seem a rather convoluted way to specify the true Godunov flux, which is quite simple to determine directly. For nonlinear systems, however, it is generally too expensive to determine the rarefaction wave structure exactly and this approach is usually necessary. This is discussed further in Section 15.3, where various entropy fixes are presented.

Dealing with transonic rarefactions properly is an important component in the development of successful methods. This is illustrated in Figure 12.2, which shows several computed solutions to Burgers' equation (11.13) at time $t = 1$ for the same set of data, a Riemann problem with $u_l = -1$ and $u_r = 2$. (Note that for Burgers' equation the sonic point is at

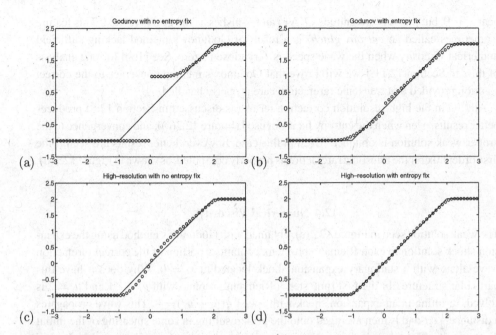

Fig. 12.2. The solid line is the entropy-satisfying solution to Burgers' equation with a transonic rarefaction wave. The circles show computed solutions. (a) Godunov's method using expansion-shock solutions to each Riemann problem. This method converges to a weak solution that does not satisfy the entropy condition. (b) Godunov's method using entropy-satisfying solutions to each Riemann problem. (c) High-resolution corrections added to the expansion-shock Godunov method. (d) High-resolution corrections added to the entropy-satisfying Godunov method. [claw/book/chap12/efix]

$u_s = 0$.) The top row shows results obtained with Godunov's method and the bottom row shows results with the high-resolution method, using the MC limiter. The plots on the left were obtained using (12.8) everywhere, with no entropy fix. The plots on the right were obtained using the entropy fix, which means that the fluctuations were redefined at a single grid interface each time step, the one for which $U_{i-1}^n < 0$ and $U_i^n > 0$. This modification at a single grid interface makes a huge difference in the quality of the results. In particular, the result obtained using Godunov's method with the expansion-shock Riemann solution looks entirely wrong. In fact it is a reasonable approximation to a weak solution to the problem, the function

$$u(x, t) = \begin{cases} -1 & \text{if } x < 0, \\ 1 & \text{if } 0 < x \leq t, \\ x/t & \text{if } t \leq x \leq 2t, \\ 2 & \text{if } x \geq 2t. \end{cases}$$

This solution consists of an entropy-violating stationary shock at $x = 0$ and also a rarefaction wave. If the grid is refined, the computed solution converges nicely to this weak solution. However, this is not the physically relevant vanishing-viscosity solution that we had hoped to compute.

When the correct rarefaction-wave solution to each Riemann problem is used (i.e., the entropy fix (12.9) is employed), Godunov's method gives a result that is much closer to the weak solution we desire. There is still a small expansion shock visible in Figure 12.2(b) near $x = 0$, but this is of magnitude $\mathcal{O}(\Delta x)$ and vanishes as the grid is refined. This feature (sometimes called an *entropy glitch*) is a result of Godunov's method lacking sufficient numerical viscosity when the wave speed is very close to zero. See [160] for one analysis of this. In Section 12.11.1 we will prove that Godunov's method converges to the correct solution provided that transonic rarefactions are properly handled.

Adding in the high-resolution correction terms (as discussed in Section 12.8) produces better results, even when the entropy fix is not used (Figure 12.2(c)), and convergence to the proper weak solution is obtained. Even in this case, however, better results are seen if the first-order flux for the transonic rarefaction is properly computed, as shown in Figure 12.2(d).

12.4 Numerical Viscosity

The weak solution seen in Figure 12.2(a), obtained with Godunov's method using the expansion shock solution to each Riemann problem, contains a portion of the correct rarefaction wave along with a stationary expansion shock located at $x = 0$. Why does it have this particular structure? In the first time step a Riemann problem with $u_l = -1$ and $u_r = 2$ is solved, resulting in an expansion shock with speed $\frac{1}{2}(u_l + u_r) = \frac{1}{2}$. This wave propagates a distance $\frac{1}{2}\Delta t$ and is then averaged onto the grid, resulting in some smearing of the initial discontinuity. The *numerical viscosity* causing this smearing acts similarly to the *physical viscosity* of the viscous Burgers equation (11.14), and tends to produce a rarefaction wave. However, unlike the viscosity of fixed magnitude ϵ appearing in (11.14), the magnitude of the numerical viscosity depends on the local Courant number $s_{i-1/2}\Delta t/\Delta x$, since it results from the averaging process (where $s_{i-1/2}$ is the Rankine–Hugoniot shock speed for the data Q_{i-1} and Q_i). In particular, if $s_{i-1/2} = 0$, then there is no smearing of the discontinuity and no numerical viscosity. For Burgers' equation this happens whenever $Q_{i-1} = -Q_i$, in which case the expansion-shock weak solution is stationary. The solution shown in Figure 12.2(a) has just such a stationary shock.

This suggests that another way to view the entropy fix needed in the transonic case is as the addition of extra numerical viscosity in the neighborhood of such a point. This can be examined further by noting that the fluctuations (12.8) result in the numerical flux function

$$F_{i-1/2} = \frac{1}{2}\left[f(Q_{i-1}) + f(Q_i) - |s_{i-1/2}|(Q_i - Q_{i-1}) \right], \tag{12.10}$$

as in the derivation of (4.61). Recall from Section 4.14 that this is the central flux (4.18) with the addition of a viscous flux term. However, when $s_{i-1/2} = 0$ this viscous term disappears. This viewpoint is discussed further in Section 15.3.5 where a variety of entropy fixes for nonlinear systems are discussed.

12.5 The Lax–Friedrichs and Local Lax–Friedrichs Methods

The Lax-Friedrichs (LxF) method was introduced in Section 4.6. The flux function (4.21) for this method,

$$F_{i-1/2} = \frac{1}{2}[f(Q_{i-1}) + f(Q_i) - a(Q_i - Q_{i-1})], \tag{12.11}$$

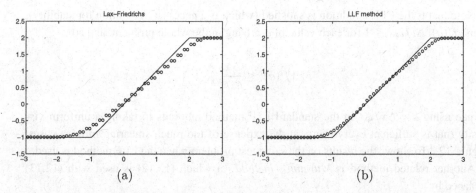

Fig. 12.3. The solid line is the entropy-satisfying solution to Burgers' equation with a transonic rarefaction wave. The symbols show computed solutions. (a) Lax–Friedrichs method. (b) Local Lax–Friedrichs (LLF) method. [claw/book/chap12/llf]

has a numerical viscosity $a = \Delta x/\Delta t$ with a fixed magnitude that does not vanish near a sonic point. As a result, this method always converges to the correct vanishing viscosity solution as the grid is refined; see Section 12.7.

If the LxF method is applied to the same transonic rarefaction problem considered in Figure 12.2, we obtain the results shown in Figure 12.3(a). Note that this method is more dissipative than Godunov's method. It also exhibits a curious stair-step pattern in which $Q_{2j} = Q_{2j+1}$ for each value of j. This results from the fact that the formula (4.20) for Q_i^{n+1} involves only Q_{i-1}^n and Q_{i+1}^n, so there is a decoupling of even and odd grid points. With the piecewise constant initial data used in this example, the even and odd points evolve in exactly the same manner, so each solution value appears twice. (See Section 10.5 for a discussion of the LxF method on a staggered grid in which only half the points appear. This viewpoint allows it to be related more directly to Godunov's method.)

An improvement to the LxF method is obtained by replacing the value $a = \Delta x/\Delta t$ in (12.11) by a locally determined value,

$$F_{i-1/2} = \frac{1}{2}[f(Q_{i-1}) + f(Q_i) - a_{i-1/2}(Q_i - Q_{i-1})], \qquad (12.12)$$

where

$$a_{i-1/2} = \max(|f'(q)|) \quad \text{over all } q \text{ between } Q_{i-1} \text{ and } Q_i. \qquad (12.13)$$

For a convex flux function this reduces to

$$a_{i-1/2} = \max(|f'(Q_{i-1})|, |f'(Q_i)|).$$

This resulting method is *Rusanov's method* [387], though recently it is often called the *local Lax–Friedrichs* (LLF) *method* because it has the same form as the LxF method but the viscosity coefficient is chosen locally at each Riemann problem. It can be shown that this is sufficient viscosity to make the method converge to the vanishing-viscosity solution; see Section 12.7.

Note that if the CFL condition is satisfied (which is a necessary condition for stability), then $|f'(q)| \, \Delta t / \Delta x \leq 1$ for each value of q arising in the whole problem, and so

$$|f'(q)| \leq \frac{\Delta x}{\Delta t}.$$

Hence using $a = \Delta x / \Delta t$ in the standard LxF method amounts to taking a uniform viscosity that is sufficient everywhere, at the expense of too much smearing in most cases. Figure 12.3(b) shows the results on the same test problem when the LLF method is used.

Another related method is *Murman's method*, in which (12.12) is used with (12.13) replaced by

$$a_{i-1/2} = \left| \frac{f(Q_i) - f(Q_{i-1})}{Q_i - Q_{i-1}} \right|. \tag{12.14}$$

Unlike the LLF scheme, solutions generated with this method may fail to satisfy the entropy condition because $a_{i-1/2}$ vanishes for the case of a stationary expansion shock. In fact, this is exactly the method (12.5) with $\mathcal{A}^{\pm} \Delta Q_{i-1/2}$ defined by (12.8), expressed in a different form (see Exercise 12.1).

Note that all these methods are easily implemented in CLAWPACK by taking

$$\mathcal{A}^{-} \Delta Q_{i-1/2} = \frac{1}{2}[f(Q_i) - f(Q_{i-1}) - a_{i-1/2}(Q_i - Q_{i-1})],$$

$$\mathcal{A}^{+} \Delta Q_{i-1/2} = \frac{1}{2}[f(Q_i) - f(Q_{i-1}) + a_{i-1/2}(Q_i - Q_{i-1})], \tag{12.15}$$

as is done in [claw/book/chap12/llf/rp1.f].

12.6 The Engquist–Osher method

We have seen that the first-order method (12.5) with the fluctuations (12.8) can be interpreted as an implementation of Godunov's method in which we always use the shock-wave solution to each Riemann problem, even when this violates the entropy condition. The *Engquist–Osher method* [124] takes the opposite approach and always assumes the solution is a "rarefaction wave", even when this wave must be triple-valued as in Figure 11.4(b). This can be accomplished by setting

$$\mathcal{A}^{+} \Delta Q_{i-1/2} = \int_{Q_{i-1}}^{Q_i} (f'(q))^{+} \, dq,$$

$$\mathcal{A}^{-} \Delta Q_{i-1/2} = \int_{Q_{i-1}}^{Q_i} (f'(q))^{-} \, dq. \tag{12.16}$$

Here the $\pm$ superscript on $f'(q)$ means the positive and negative part as in (4.40). These fluctuations result in an interface flux $F_{i-1/2}$ that can be expressed in any of the following ways:

$$F_{i-1/2} = f(Q_{i-1}) + \int_{Q_{i-1}}^{Q_i} (f'(q))^- \, dq$$

$$= f(Q_i) - \int_{Q_{i-1}}^{Q_i} (f'(q))^+ \, dq$$

$$= \frac{1}{2}[f(Q_{i-1}) + f(Q_i)] - \frac{1}{2}\int_{Q_{i-1}}^{Q_i} |f'(q)| \, dq. \qquad (12.17)$$

If $f'(q)$ does not change sign between Q_{i-1} and Q_i, then one of the fluctuations in (12.16) will be zero and these formulas reduce to the usual upwind fluxes as in (12.2). In the sonic rarefaction case both fluctuations are nonzero and we obtain the desired value $F_{i-1/2} = f(q_s)$ as in (12.2). It is only in the *transonic shock* case, when $f'(Q_{i-1}) > 0 > f'(Q_i)$, that the Engquist–Osher method gives a value different from (12.2). In this case both fluctuations are again nonzero and we obtain

$$F_{i-1/2} = f(Q_{i-1}) + f(Q_i) - f(q_s) \qquad (12.18)$$

rather than simply $f(Q_{i-1})$ or $f(Q_i)$. This is because the triple-valued solution of Figure 11.4(b) spans the interface $x_{i-1/2}$ in this case, so that the integral picks up three different values of f. This flux is still consistent with the conservation law, however, and by assuming the rarefaction structure, the entropy condition is always satisfied. This approach can be extended to systems of equations to derive approximate Riemann solvers that satisfy the entropy condition, giving the *Osher scheme* [349], [352].

12.7 E-schemes

Osher [349] introduced the notion of an *E-scheme* as one that satisfies the inequality

$$\text{sgn}(Q_i - Q_{i-1}) \left[F_{i-1/2} - f(q) \right] \leq 0 \qquad (12.19)$$

for all q between Q_{i-1} and Q_i. In particular, Godunov's method with flux $F_{i-1/2}^G$ defined by (12.4) is clearly an E-scheme. In fact it is the limiting case, in the sense that E-schemes are precisely those for which

$$F_{i-1/2} \leq F_{i-1/2}^G \quad \text{if} \, Q_{i-1} \leq Q_i,$$
$$F_{i-1/2} \geq F_{i-1/2}^G \quad \text{if} \, Q_{i-1} \geq Q_i. \qquad (12.20)$$

It can be shown that any E-scheme is TVD if the Courant number is sufficiently small (Exercise 12.3). Osher [349] proves that E-schemes are convergent to the entropy-satisfying weak solution. In addition to Godunov's method, the LxF, LLF, and Engquist–Osher methods are all E-schemes. Osher also shows that E-schemes are at most first-order accurate.

12.8 High-Resolution TVD Methods

The methods described so far are only first-order accurate and not very useful in their own right. They are, however, used as building blocks in developing certain high-resolution methods. Godunov's method, as described in Section 12.2, can be easily extended to a high-resolution method of the type developed in Chapter 6,

$$Q_i^{n+1} = Q_i - \frac{\Delta t}{\Delta x}\left(A^+ \Delta Q_{i-1/2} + A^- \Delta Q_{i+1/2}\right) - \frac{\Delta t}{\Delta x}\left(\tilde{F}_{i+1/2} - \tilde{F}_{i-1/2}\right). \qquad (12.21)$$

We set

$$\tilde{F}_{i-1/2} = \frac{1}{2}\left|s_{i-1/2}\right|\left(1 - \frac{\Delta t}{\Delta x}\left|s_{i-1/2}\right|\right)\tilde{\mathcal{W}}_{i-1/2}, \qquad (12.22)$$

just as we have done for the variable-coefficient advection equation in Section 9.3.1. Again $\tilde{\mathcal{W}}_{i-1/2}$ is a limited version of the wave (12.7) obtained by comparing $\tilde{\mathcal{W}}_{i-1/2}$ to $\tilde{\mathcal{W}}_{i-3/2}$ or $\tilde{\mathcal{W}}_{i-1/2}$, depending on the sign of $s_{i-1/2}$. Note that the method remains conservative with such a modification, which is crucial in solving nonlinear conservation laws (see Section 12.9).

It is also possible to prove that the resulting method will be TVD provided that one of the TVD limiters presented in Section 6.9 is used. This is a very important result, since it means that these methods can be applied with confidence to nonlinear problems with shock waves where we wish to avoid spurious oscillations. Moreover, this TVD property allows us to prove stability and hence convergence of the methods as the grid is refined, as shown in Section 12.12. (Unfortunately, these claims are valid only for *scalar* conservation laws. Extending these ideas to systems of equations gives methods that are often very successful in practice, but for which much less can be proved in general.)

Here we will prove that the limiter methods are TVD under restricted conditions to illustrate the main ideas. (See [160] for more details.) We assume that data is monotone (say nonincreasing) and that $f'(q)$ does not change sign over the range of the data (say $f'(Q_i^n) > 0$). A similar approach can be used near extreme points of Q^n and sonic points, but more care is required, and the formulas are more complicated. We will also impose the time-step restriction

$$\frac{\Delta t}{\Delta x}\max|f'(q)| < \frac{1}{2}, \qquad (12.23)$$

although this can also be relaxed to the usual CFL limit of 1 with some modification of the method.

The main idea is to again use the REA Algorithm 4.1 to interpret the high-resolution method. The first step is to reconstruct the piecewise linear function $\tilde{q}^n(x, t_n)$ from the cell averages Q_i^n. This is where the limiters come into play, and this reconstruction does not increase the total variation of the data. The second step is to evolve the conservation law with this data. If we were to solve the conservation law *exactly* and then average onto the grid, then the resulting method would clearly be TVD, because the exact solution operator for a scalar conservation law is TVD (and so is the averaging process). But unlike the methods we developed in Chapter 6 for the advection equation, we are not able to solve the original conservation law exactly in step 2 of Algorithm 4.1 (except in the case of zero slopes, where Godunov's method does this). In principle one could do so also for more general slopes, but the resulting correction formula would be much more complicated than (12.22). However, the formula (12.22) can be interpreted as what results from solving a slightly different conservation law exactly and then averaging onto the grid. Exact solutions of this modified conservation law also have the TVD property, and it follows that the method is TVD.

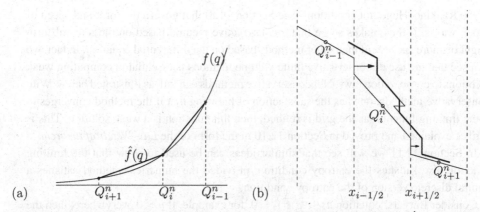

Fig. 12.4. (a) The flux function $f(q)$ is approximated by a piecewise linear function $\hat{f}(q)$. (b) The piecewise linear data $\tilde{q}^n(x, t_n)$ is evolved by solving the conservation law with flux $\hat{f}(q)$. The result is the heavy line. Note that a shock forms near the value Q_i^n where the characteristic velocity $\hat{f}'(q)$ is discontinuous.

The modified conservation law used in the time step from t_n to t_{n+1} is obtained by replacing the flux function $f(q)$ by a piecewise linear function $\hat{f}(q)$ that interpolates the values $(Q_i^n, f(Q_i^n))$, as shown in Figure 12.4(a). Then in step 2 we solve the conservation law with this flux function to evolve $\tilde{q}^n(x, t_n)$, as shown in Figure 12.4(b). This flux is still nonlinear, but the nonlinearity has been concentrated at the points Q_i^n. Shocks form immediately at the points x_i (the midpoints of the grid cells), but because of the time-step restriction (12.23), these shocks do not reach the cell boundary during the time step.

Near each interface $x_{i-1/2}$ the data lies between Q_{i-1}^n and Q_i^n, and so the flux function is linear with constant slope $s_{i-1/2} = [f(Q_i^n) - f(Q_{i-1}^n)]/(Q_i^n - Q_{i-1}^n)$, as arises from the piecewise linear interpolation of f. Hence the conservation law with flux $\hat{f}$ behaves locally like the scalar advection equation with velocity $s_{i-1/2}$. This is exactly the velocity that appears in the updating formula (12.22), and it can be verified that this method produces the correct cell averages at the end of the time step for this modified conservation law.

In this informal analysis we have assumed the data is monotone near Q_i^n. The case where Q_i^n is a local extreme point must be handled differently, since we would not be able to define a single function $\hat{f}$ in the same manner. However, in this case the slope in cell C_i is zero if a TVD limiter is used, and we can easily show that the total variation can not increase in this case.

Some more details may be found in Goodman & LeVeque [160]. Recently Morton [332] has performed a more extensive analysis of this type of method, including also similar methods with piecewise quadratic reconstructions as well as methods on nonuniform grids and multidimensional versions.

12.9 The Importance of Conservation Form

In Section 4.1 we derived the conservative form of a finite volume method based on the integral form of the conservation law. Using a method in this form guarantees that the discrete solution will be conservative in the sense that (4.8) will be satisfied. For weak solutions involving shock waves, this integral form is more fundamental than the differential equation and forms the basis for the mathematical theory of weak solutions, including the derivation of the Rankine–Hugoniot conditions (see Section 11.8) that govern the form and speed of shock waves. It thus makes sense that a conservative method based on the integral form might be more successful than other methods based on the differential equation. In fact, we will see that the use of conservative finite volume methods is essential in computing weak solutions to conservation laws. Nonconservative methods can fail, as illustrated below. With conservative methods, one has the satisfaction of knowing that if the method converges to some limiting function as the grid is refined, then this function is a weak solution. This is further explained and proved in Section 12.10 in the form of the *Lax–Wendroff theorem*.

In Section 12.11 we will see that similar ideas can be used to show that the limiting function also satisfies the entropy condition, provided the numerical method satisfies a natural discrete version of the entropy condition.

Consider Burgers' equation $u_t + \frac{1}{2}(u^2)_x = 0$, for example. If $u > 0$ everywhere, then the conservative upwind method (Godunov's method) takes the form

$$U_i^{n+1} = U_i^n - \frac{\Delta t}{\Delta x}\left(\frac{1}{2}(U_i^n)^2 - \frac{1}{2}(U_{i-1}^n)^2\right). \tag{12.24}$$

On the other hand, using the quasilinear form $u_t + uu_x = 0$, we could derive the nonconservative upwind method

$$U_i^{n+1} = U_i^n - \frac{\Delta t}{\Delta x}U_i^n(U_i^n - U_{i-1}^n). \tag{12.25}$$

On smooth solutions, both of these methods are first-order accurate, and they give comparable results. When the solution contains a shock wave, the method (12.25) fails to converge to a weak solution of the conservation law. This is illustrated in Figure 12.5. The conservative method (12.24) gives a slightly smeared approximation to the shock, but it is smeared about the correct location. We can easily see that it must be, since the method has the discrete conservation property (4.8). The nonconservative method (12.25), on the other hand, gives the results shown in Figure 12.5(b). These clearly do not satisfy (4.8), and as the grid is

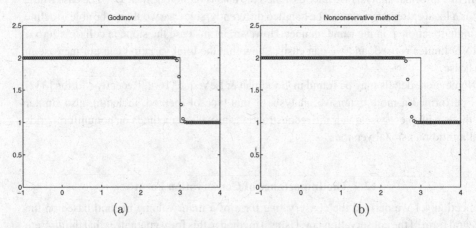

Fig. 12.5. True and computed solutions to a Riemann problem for Burgers' equation with data $u_l = 2$, $u_r = 1$, shown at time $t = 2$: (a) using the conservative method (12.24), (b) using the nonconservative method (12.25). [claw/book/chap12/nonconservative]

refined the approximation converges to a discontinuous function that is not a weak solution to the conservation law.

This is not surprising if we recall that it is possible to derive a variety of conservation laws that are equivalent for smooth solutions but have different weak solutions. For example, the equations (11.34) and (11.35) have exactly the same smooth solutions, but the Rankine–Hugoniot condition gives different shock speeds, and hence different weak solutions. Consider a finite difference method that is consistent with one of these equations, say (11.34), using the definition of consistency introduced in Section 8.2 for linear problems (using the local truncation error derived by expanding in Taylor series). Then the method is also consistent with (11.35), since the Taylor series expansion (which assumes smoothness) gives the same result in either case. So the method is consistent with both (11.34) and (11.35), and while we might then expect the method to converge to a function that is a weak solution of both, that is impossible when the two weak solutions differ. Similarly, if we use a nonconservative method based on the quasilinear form, then there is no reason to expect to obtain the correct solution, except in the case of smooth solutions.

Note that the nonconservative method (12.25) can be rewritten as

$$U_i^{n+1} = U_i^n - \frac{\Delta t}{\Delta x}\left(\frac{1}{2}(U_i^n)^2 - \frac{1}{2}(U_{i-1}^n)^2\right) + \frac{1}{2}\Delta t\,\Delta x\left(\frac{U_i^n - U_{i-1}^n}{\Delta x}\right)^2. \quad (12.26)$$

Except for the final term, this is identical to the conservative method (12.24). The final term approximates the time integral of $\frac{1}{2}\Delta x\,(u_x)^2$. For smooth solutions, where u_x is bounded, the effect of this term can be expected to vanish as $\Delta x \to 0$. For a shock wave, however, it does not. Just as in the derivation of the weak form of the entropy inequality (11.51), this term can give a finite contribution in the limit, leading to a different shock speed.

The final term in (12.26) can also be viewed as a singular source term that is being added to the conservation law, an approximation to a delta function concentrated at the shock. This leads to a change in the shock speed as discussed in Section 17.12. See [203] for further analysis of the behavior of nonconservative methods.

12.10 The Lax–Wendroff Theorem

The fact that conservative finite volume methods are based on the integral conservation law suggests that we can hope to correctly approximate discontinuous weak solutions to the conservation law by using such a method. Lax and Wendroff [265] proved that this is true, at least in the sense that *if* the approximation converges to some function $q(x, t)$ as the grid is refined, through some sequence $\Delta t^{(j)}, \Delta x^{(j)} \to 0$, then this function will in fact be a weak solution of the conservation law. The theorem does not guarantee that convergence occurs. For that we need some form of stability, and even then, if there is more than one weak solution, it might be that one sequence of approximations will converge to one weak solution, while another sequence converges to a different weak solution (and therefore a third sequence, obtained for example by merging the first two sequences, will not converge at all!).

Nonetheless, this is a very powerful and important theorem, for it says that we can have confidence in solutions we compute. In practice we typically do not consider a whole sequence of approximations. Instead we compute a single approximation on one fixed grid. If this solution looks reasonable and has well-resolved discontinuities (an indication that the method is stable and our grid is sufficiently fine), then we can believe that it is in fact a good approximation to *some* weak solution.

Before stating the theorem, we note that it is valid for systems of conservation laws $q_t + f(q)_x = 0$ as well as for scalar equations.

Theorem 12.1 (Lax and Wendroff [265]). *Consider a sequence of grids indexed by $j = 1$, $2, \ldots$, with mesh parameters $\Delta t^{(j)}, \Delta x^{(j)} \to 0$ as $j \to \infty$. Let $Q^{(j)}(x, t)$ denote the numerical approximation computed with a consistent and conservative method on the jth grid. Suppose that $Q^{(j)}$ converges to a function q as $j \to \infty$, in the sense made precise below. Then $q(x, t)$ is a weak solution of the conservation law.*

The proof of this theorem does not use smoothness of the solution, and so we do not define consistency in terms of Taylor series expansions. Instead we need the form of consistency discussed in Section 4.3.1.

In the statement of this theorem, $Q^{(j)}(x, t)$ denotes a piecewise constant function that takes the value Q_i^n on the space–time mesh cell $(x_{i-1/2}, x_{i+1/2}) \times [t_n, t_{n+1})$. It is indexed by j corresponding to the particular mesh used, with $\Delta x^{(j)}$ and $\Delta t^{(j)}$ both approaching zero as $j \to \infty$. We assume that we have convergence of the function $Q^{(j)}(x, t)$ to $q(x, t)$ in the following sense:

1. Over every bounded set $\Omega = [a, b] \times [0, T]$ in x–t space,

$$\int_0^T \int_a^b \left| Q^{(j)}(x, t) - q(x, t) \right| dx \, dt \to 0 \quad \text{as } j \to \infty. \tag{12.27}$$

This is the 1-norm over the set Ω, so we can simply write

$$\left\| Q^{(j)} - q \right\|_{1,\Omega} \to 0 \quad \text{as } j \to \infty. \tag{12.28}$$

2. We also assume that for each T there is an $R > 0$ such that

$$\text{TV}\left(Q^{(j)}(\cdot, t) \right) < R \quad \text{for all } 0 \le t \le T, \qquad j = 1, 2, \ldots, \tag{12.29}$$

where TV denotes the total variation function introduced in Section 6.7.

Lax and Wendroff assumed a slightly different form of convergence, namely that $Q^{(j)}$ converges to q almost everywhere (i.e., except on a set of measure zero) in a uniformly bounded manner. Using the fact that each $Q^{(j)}$ is a piecewise constant function, it can be shown that this requirement is essentially equivalent to (12.28) and (12.29) above. The advantage of assuming (12.28) and (12.29) is twofold: (a) it is these properties that are really needed in the proof, and (b) for certain important classes of methods (e.g., the total variation diminishing methods), it is this form of convergence that we can most directly prove.

Proof. We will show that the limit function $q(x, t)$ satisfies the weak form (11.33), i.e., for all $\phi \in C_0^1$,

$$\int_0^\infty \int_{-\infty}^{+\infty} [\phi_t q + \phi_x f(q)] \, dx \, dt = -\int_{-\infty}^\infty \phi(x, 0) q(x, 0) \, dx. \qquad (12.30)$$

Let ϕ be a C_0^1 test function. On the jth grid, define the discrete version $\Phi^{(j)}$ by $\Phi_i^{(j)n} = \phi(x_i^{(j)}, t_n^{(j)})$ where $(x_i^{(j)}, t_n^{(j)})$ is a grid point on this grid. Similarly $Q_i^{(j)n}$ denotes the numerical approximation on this grid. To simplify notation, we will drop the superscript (j) below and simply use Φ_i^n and Q_i^n, but remember that (j) is implicitly present, since in the end we must take the limit as $j \to \infty$.

Multiply the conservative numerical method

$$Q_i^{n+1} = Q_i^n - \frac{\Delta t}{\Delta x} \left(F_{i+1/2}^n - F_{i-1/2}^n \right)$$

by Φ_i^n to obtain

$$\Phi_i^n Q_i^{n+1} = \Phi_i^n Q_i^n - \frac{\Delta t}{\Delta x} \Phi_i^n \left(F_{i+1/2}^n - F_{i-1/2}^n \right). \qquad (12.31)$$

This is true for all values of i and n on each grid j. If we now sum (12.31) over all i and $n \geq 0$, we obtain

$$\sum_{n=0}^\infty \sum_{i=-\infty}^\infty \Phi_i^n \left(Q_i^{n+1} - Q_i^n \right) = -\frac{\Delta t}{\Delta x} \sum_{n=0}^\infty \sum_{i=-\infty}^\infty \Phi_i^n \left(F_{i+1/2}^n - F_{i-1/2}^n \right). \qquad (12.32)$$

We now use *summation by parts*, which just amounts to recombining the terms in each sum. A simple example is

$$\sum_{i=1}^m a_i(b_i - b_{i-1}) = (a_1 b_1 - a_1 b_0) + (a_2 b_2 - a_2 b_1) + \cdots + (a_m b_m - a_m b_{m-1})$$

$$= -a_1 b_0 + (a_1 b_1 - a_2 b_1) + (a_2 b_2 - a_3 b_2)$$
$$+ \cdots + (a_{m-1} b_{m-1} - a_m b_{m-1}) + a_m b_m$$

$$= a_m b_m - a_1 b_0 - \sum_{i=1}^{m-1} (a_{i+1} - a_i) b_i. \qquad (12.33)$$

Note that the original sum involved the product of a_i with differences of b's, whereas the final sum involves the product of b_i with differences of a's. This is completely analogous to integration by parts, where the derivative is moved from one function to the other. Just as in integration by parts, there are also boundary terms $a_m b_m - a_1 b_0$ that arise.

We will use this on both sides of (12.32) (for the n-sum on the left and for the i-sum on the right). By our assumption that ϕ has compact support, $\Phi_i^n = 0$ for $|i|$ or n sufficiently large, and hence the boundary terms at $i = \pm\infty$, $n = \infty$ all drop out. The only boundary term that remains is at $n = 0$, where $t_0 = 0$. This gives

$$-\sum_{i=-\infty}^\infty \Phi_i^0 Q_i^0 - \sum_{n=1}^\infty \sum_{i=-\infty}^\infty \left(\Phi_i^n - \Phi_i^{n-1} \right) Q_i^n = \frac{\Delta t}{\Delta x} \sum_{n=0}^\infty \sum_{i=-\infty}^\infty \left(\Phi_{i+1}^n - \Phi_i^n \right) F_{i-1/2}^n.$$

Note that each of these sums is in fact a finite sum, since ϕ has compact support. Multiplying by Δx and rearranging this equation gives

$$\Delta x \, \Delta t \left[\sum_{n=1}^{\infty} \sum_{i=-\infty}^{\infty} \left(\frac{\Phi_i^n - \Phi_i^{n-1}}{\Delta t} \right) Q_i^n \right.$$
$$\left. + \sum_{n=0}^{\infty} \sum_{i=-\infty}^{\infty} \left(\frac{\Phi_{i+1}^n - \Phi_i^n}{\Delta x} \right) F_{i-1/2}^n \right] = -\Delta x \sum_{i=-\infty}^{\infty} \Phi_i^0 Q_i^0. \quad (12.34)$$

This transformation using summation by parts is completely analogous to the derivation of (11.33) from (11.31).

Now let $j \to \infty$, so that $\Delta t^{(j)}, \Delta x^{(j)} \to 0$ in (12.34). (Recall that all of the symbols in that equation should also be indexed by (j) as we refine the grid.) It is reasonably straightforward, using the 1-norm convergence of $Q^{(j)}$ to q and the smoothness of ϕ, to show that the term on the top line of (12.34) converges to $\int_0^\infty \int_{-\infty}^\infty \phi_t(x, t) q(x, t) \, dx$ as $j \to \infty$. If we define initial data Q_i^0 by taking cell averages of the data $\overset{\circ}{q}(x)$, for example, then the right-hand side converges to $-\int_{-\infty}^\infty \phi(x, 0) q(x, 0) \, dx$ as well.

The remaining term in (12.34), involving $F_{i-1/2}^n$, is more subtle and requires the additional assumptions on F and Q that we have imposed. For a three-point method (such as Godunov's method), we have

$$F_{i-1/2}^n \equiv F_{i-1/2}^{(j)n} = \mathcal{F}\big(Q_{i-1}^{(j)n}, Q_i^{(j)n}\big),$$

and the consistency condition (4.15), with the choice $\bar{q} = Q_i^{(j)n}$, gives

$$\big|F_{i-1/2}^{(j)n} - f\big(Q_i^{(j)n}\big)\big| \le L \big| Q_i^{(j)n} - Q_{i-1}^{(j)n} \big|, \quad (12.35)$$

where L is the Lipschitz constant for the numerical flux function. Since $Q^{(j)n}$ has bounded total variation, uniformly in j, it must be that

$$\big|F_{i-1/2}^{(j)n} - f\big(Q_i^{(j)n}\big)\big| \to 0 \quad \text{as } j \to \infty$$

for almost all values of i. Using this and the fact that $Q^{(j)n}$ converges to q, it can be shown that

$$\Delta x \, \Delta t \sum_{n=1}^{\infty} \sum_{i=-\infty}^{\infty} \left(\frac{\Phi_{i+1}^n - \Phi_i^n}{\Delta x} \right) F_{i-1/2}^n \to \int_0^\infty \int_{-\infty}^\infty \phi_x(x, t) \, f(q(x, t)) \, dx \, dt$$

as $j \to \infty$, which completes the demonstration that (12.34) converges to the weak form (12.30). Since this is true for any test function $\phi \in C_0^1$, we have proved that q is in fact a weak solution. $\qquad \square$

For simplicity we assumed the numerical flux $F_{i-1/2}^n$ depends only on the two neighboring values Q_{i-1}^n and Q_i^n. However, the proof is easily extended to methods with a wider stencil provided a more general consistency condition holds, stating that the flux function is uniformly Lipschitz-continuous in all values on which it depends.

12.11 The Entropy Condition

The Lax–Wendroff theorem does not guarantee that weak solutions obtained using conservative methods satisfy the entropy condition. As we have seen in Section 12.3, some additional care is required to insure that the correct weak solution is obtained.

For some numerical methods, it is possible to show that any weak solution obtained by refining the grid will satisfy the entropy condition. Of course this supposes that we have a suitable entropy condition for the system to begin with, and the most convenient form is typically the entropy inequality introduced in Section 11.14. Recall that this requires a convex scalar entropy function $\eta(q)$ and entropy flux $\psi(q)$ for which

$$\frac{\partial}{\partial t}\eta\,(q(x,t)) + \frac{\partial}{\partial x}\psi\,(q(x,t)) \leq 0 \tag{12.36}$$

in the weak sense, i.e., for which the inequality (11.51) holds for all $\phi \in C_0^1$ with $\phi(x,t) \geq 0$ for all x, t:

$$\int_0^\infty \int_{-\infty}^\infty [\phi_t(x,t)\eta(q(x,t)) + \phi_x(x,t)\psi(q(x,t))]\,dx\,dt$$

$$+ \int_{-\infty}^\infty \phi(x,0)\eta(q(x,0))\,dx \geq 0. \tag{12.37}$$

In order to show that the weak solution $q(x,t)$ obtained as the limit of $Q^{(j)}$ satisfies this inequality, it suffices to show that a discrete entropy inequality holds, of the form

$$\eta\big(Q_i^{n+1}\big) \leq \eta\big(Q_i^n\big) - \frac{\Delta t}{\Delta x}\big(\Psi_{i+1/2}^n - \Psi_{i-1/2}^n\big). \tag{12.38}$$

Here $\Psi_{i-1/2}^n = \Psi(Q_{i-1}^n, Q_i^n)$, where $\Psi(q_l, q_r)$ is some numerical entropy flux function that must be consistent with ψ in the same manner that we require F to be consistent with f. If we can show that (12.38) holds for a suitable Ψ, then mimicking the proof of the Lax–Wendroff theorem (i.e., multiplying (12.38) by Φ_i^n, summing over i and n, and using summation by parts), we can show that the limiting weak solution $q(x,t)$ obtained as the grid is refined satisfies the entropy inequality (12.37).

12.11.1 Entropy Consistency of Godunov's Method

For Godunov's method we can show that the numerical approximation will always satisfy the entropy condition provided that the Riemann solution used to define the flux at each cell interface satisfies the entropy condition. Recall that we can interpret Godunov's method as an implementation of the REA Algorithm 4.1. The piecewise constant function $\tilde{q}^n(x, t_n)$ is constructed from the data Q^n, and the exact solution $\tilde{q}^n(x, t_{n+1})$ to the conservation law is then averaged on the grid to obtain Q^{n+1}. What we now require is that the solution $\tilde{q}^n(x, t)$ satisfy the entropy condition. If so, then integrating (12.36) over the rectangle

$(x_{i-1/2}, x_{i+1/2}) \times (t_n, t_{n+1})$ gives

$$\int_{x_{i-1/2}}^{x_{i+1/2}} \eta(\tilde{q}^n(x, t_{n+1})) \, dx \leq \int_{x_{i-1/2}}^{x_{i+1/2}} \eta(\tilde{q}^n(x, t_n)) \, dx$$

$$+ \int_{t_n}^{t_{n+1}} \psi\left(\tilde{q}^n(x_{i-1/2}, t)\right) dt - \int_{t_n}^{t_{n+1}} \psi\left(\tilde{q}^n(x_{i+1/2}, t)\right) dt.$$

This is almost what we need. Since $\tilde{q}^n$ is constant along three of the four sides of this rectangle, all integrals on the right-hand side can be evaluated. Doing so, and dividing by Δx, yields

$$\frac{1}{\Delta x} \int_{x_{i-1/2}}^{x_{i+1/2}} \eta(\tilde{q}^n(x, t_{n+1})) \, dx \leq \eta\left(Q_i^n\right) - \frac{\Delta t}{\Delta x}\left[\psi\left(Q_{i+1/2}^{\downarrow}\right) - \psi\left(Q_{i-1/2}^{\downarrow}\right)\right]. \quad (12.39)$$

Again $Q_{i-1/2}^{\downarrow}$ represents the value propagating with velocity 0 in the solution of the Riemann problem. If we define the numerical entropy flux by

$$\Psi_{i-1/2}^n = \psi\left(Q_{i-1/2}^{\downarrow}\right), \quad (12.40)$$

then Ψ is consistent with ψ, and the right-hand side of (12.39) agrees with that of (12.38).

The left-hand side of (12.39) is not equal to $\eta(Q_i^{n+1})$, because $\tilde{q}^n$ is not constant in this interval. However, since the entropy function η is convex with $\eta''(q) > 0$, we can use *Jensen's inequality*. This states that the value of η evaluated at the average value of $\tilde{q}^n$ is less than or equal to the average value of $\eta(\tilde{q}^n)$, i.e.,

$$\eta\left(\frac{1}{\Delta x} \int_{x_{i-1/2}}^{x_{i+1/2}} \tilde{q}^n(x, t_{n+1}) \, dx\right) \leq \frac{1}{\Delta x} \int_{x_{i-1/2}}^{x_{i+1/2}} \eta(\tilde{q}^n(x, t_{n+1})) \, dx. \quad (12.41)$$

The left-hand side here is simply $\eta(Q_i^{n+1})$, while the right-hand side is bounded by (12.39). Combining (12.39), (12.40), and (12.41) thus gives the desired entropy inequality (12.38).

This shows that weak solutions obtained by Godunov's method satisfy the entropy condition, provided we use entropy-satisfying Riemann solutions at each cell interface. This result is valid not only for scalar conservation laws. It holds more generally for any nonlinear system for which we have an entropy function.

For the special case of a convex scalar conservation law, this simply means that we must use a rarefaction wave when possible rather than an expansion shock in defining the state $Q_{i-1/2}^{\downarrow}$ used to compute the Godunov flux. However, as we have seen in Section 12.2, this affects the value of $Q_{i-1/2}^{\downarrow}$ only in the case of a transonic rarefaction. So we conclude that Godunov's method will always produce the vanishing-viscosity solution to a convex scalar conservation law provided that transonic rarefactions are handled properly.

12.12 Nonlinear Stability

The Lax–Wendroff theorem presented in Section 12.10 does not say anything about whether the method converges, only that if a sequence of approximations converges, then the limit is a weak solution. To guarantee convergence, we need some form of stability, just as for linear problems. Unfortunately, the Lax equivalence theorem mentioned in Section 8.3.2 no longer holds, and we cannot use the same approach (which relies heavily on linearity) to prove convergence.

The convergence proof of Section 8.3.1 can be used in the nonlinear case if the numerical method is contractive in some norm. In particular, this is true for the class of *monotone methods*. These are methods with the property that

$$\frac{\partial Q_i^{n+1}}{\partial Q_j^n} \geq 0 \qquad (12.42)$$

for all values of j. This means that if we increase the value of any Q_j^n at time t_n, then the value of Q_i^{n+1} at the next time step cannot decrease as a result. This is suggested by the fact that the true vanishing-viscosity solution of a scalar conservation law has an analogous property: If $\overset{\circ}{q}(x)$ and $\overset{\circ}{p}(x)$ are two sets of initial data and $\overset{\circ}{q}(x) \geq \overset{\circ}{p}(x)$ for all x, then $q(x, t) \geq p(x, t)$ for all x at later times as well. Unfortunately, this monotone property holds only for certain first-order accurate methods, and so this approach cannot be applied to the high-resolution methods of greatest interest. For more details on monotone methods see, for example, [96], [156], [185], [281].

In this chapter we consider a form of nonlinear stability based on total-variation bounds that allows us to prove convergence results for a wide class of practical TVD or TVB methods. So far, this approach has been completely successful only for scalar problems. For general systems of equations with arbitrary initial data no numerical method has been proved to be stable or convergent in general, although convergence results have been obtained in some special cases (see Section 15.8.2).

12.12.1 Convergence Notions

To discuss the convergence of a grid function with discrete values Q_i^n to a function $q(x, t)$, it is convenient to define a piecewise-constant function $Q^{(\Delta t)}(x, t)$ taking the values

$$Q^{(\Delta t)}(x, t) = Q_i^n \quad \text{for } (x, t) \in [x_{i-1/2}, x_{i+1/2}) \times [t_n, t_{n+1}). \qquad (12.43)$$

We index this function by Δt because it depends on the particular grid being used. It should really be indexed by Δx as well, but to simplify notation we suppose there is a fixed relation between Δx and Δt as we refine the grid and talk about convergence as $\Delta t \to 0$. In Section 12.10 a similar sequence of functions was considered and labeled $Q^{(j)}$, corresponding to a a grid with mesh spacing $\Delta t^{(j)}$ and $\Delta x^{(j)}$. The same notation could be used here, but it will be more convenient below to use Δt as the index rather than j.

One difficulty immediately presents itself when we contemplate the convergence of a numerical method for conservation laws. The global error $Q^{(\Delta t)}(x, t) - q(x, t)$ is not well defined when the weak solution q is not unique. Instead, we measure the global error in our approximation by the distance from $Q^{(\Delta t)}(x, t)$ to the set of *all* weak solutions $\mathcal{W}$,

$$\mathcal{W} = \{q : q(x, t) \text{ is a weak solution to the conservation law}\}. \qquad (12.44)$$

To measure this distance we need a norm, for example the 1-norm over some finite time interval $[0, T]$, denoted by

$$\|v\|_{1,T} = \int_0^T \|v(\cdot, t)\|_1 \, dt$$

$$= \int_0^T \int_{-\infty}^{\infty} |v(x, t)| \, dx \, dt. \tag{12.45}$$

The global error is then defined by

$$\text{dist}\left(Q^{(\Delta t)}, \mathcal{W}\right) = \inf_{w \in \mathcal{W}} \left\|Q^{(\Delta t)} - q\right\|_{1,T}. \tag{12.46}$$

The convergence result we would now like to prove takes the following form:

If $Q^{(\Delta t)}$ is generated by a numerical method in conservation form, consistent with the conservation law, and if the method is stable in some appropriate sense, then

$$\text{dist}\left(Q^{(\Delta t)}, \mathcal{W}\right) \to 0 \quad \text{as } \Delta t \to 0.$$

Note that there is no guarantee that $\|Q^{(\Delta t)} - q\|_{1,T} \to 0$ as $\Delta t \to 0$ for any fixed weak solution $q(x, t)$. The computed $Q^{(\Delta t)}$ might be close to one weak solution for one value of the time step Δt and close to a completely different weak solution for a slightly smaller value of Δt. This is of no great concern, since in practice we typically compute only on one particular grid, not a sequence of grids with $\Delta t \to 0$, and what the convergence result tells us is that by taking a fine enough grid, we can be assured of being arbitrarily close to *some* weak solution.

Of course, in situations where there is a unique physically relevant weak solution satisfying some entropy condition, we would ultimately like to prove convergence to this particular weak solution. This can be done if we also know that the method satisfies a discrete form of the entropy condition, such as (12.38). For then we know that any limiting solution obtained by refining the grid must satisfy the entropy condition (see Section 12.11). Since the entropy solution $q(x, t)$ to the conservation law is unique, this can be used to prove that in fact any sequence $Q^{(\Delta t)}$ must converge to this function q as $\Delta t \to 0$.

12.12.2 Compactness

In order to prove a convergence result of the type formulated above for nonlinear problems, we must define an appropriate notion of stability. For nonlinear problems one very useful tool for proving convergence is *compactness*, and so we will take a slight detour to define this concept and indicate its use.

There are several equivalent definitions of a compact set within some normed space. One definition, which describes the most important property of compact sets in relation to our goals of defining stability and proving convergence, is the following.

Definition 12.1. $\mathcal{K}$ *is a compact set in some normed space if any infinite sequence of elements of $\mathcal{K}$, $\{\kappa_1, \kappa_2, \kappa_3, \ldots\}$, contains a subsequence that converges to an element of $\mathcal{K}$.*

This means that from the original sequence we can, by selecting certain elements from this sequence, construct a new infinite sequence

$$\{\kappa_{i_1}, \kappa_{i_2}, \kappa_{i_3}, \ldots\} \qquad (\text{with } i_1 < i_2 < i_3 < \cdots)$$

that converges to some element $\kappa \in \mathcal{K}$,

$$\|\kappa_{i_j} - \kappa\| \to 0 \qquad \text{as } j \to \infty.$$

The fact that compactness guarantees the existence of convergent subsequences, combined with the Lax–Wendroff theorem 12.1, will give us a convergence proof of the type formulated above.

Example 12.1. In the space $\mathbb{R}$ with norm given by the absolute value, any closed interval is a compact set. So, for example, any sequence of real numbers in $[0, 1]$ contains a subsequence that converges to a number between 0 and 1. Of course, there may be several different subsequences one could extract, converging perhaps to different numbers. For example, the sequence

$$\{0, 1, 0, 1, 0, 1, \ldots\}$$

contains subsequences converging to 0 and subsequences converging to 1.

Example 12.2. In the same space as the previous example, an open interval is *not* compact. For example, the sequence

$$\{1, 10^{-1}, 10^{-2}, 10^{-3}, \ldots\}$$

of elements lying in the open interval $(0, 1)$ contains no subsequences convergent to an element of $(0, 1)$. Of course the whole sequence, and hence every subsequence, converges to 0, but this number is not in $(0,1)$.

Example 12.3. An unbounded set, e.g., $[0, \infty)$, is *not* compact, since the sequence $\{1, 2, 3, \ldots\}$ contains no convergent subsequence.

Generalizing these examples, it turns out that in any finite-dimensional normed linear space, any closed and bounded set is compact. Moreover, these are the only compact sets.

Example 12.4. In the n-dimensional space $\mathbb{R}^n$ with any vector norm $\|\cdot\|$, the closed ball

$$B_R = \{x \in \mathbb{R}^n : \|x\| \le R\}$$

is a compact set.

12.12.3 Function Spaces

Since we are interested in proving the convergence of a sequence of functions $Q^{(\Delta t)}(x, t)$, our definition of stability will require that all the functions lie within some compact set in some normed *function space*. Restricting our attention to the time interval $[0, T]$, the natural function space is the space $L_{1,T}$ consisting of all functions of x and t for which the norm (12.45) is finite,

$$L_{1,T} = \{v : \|v\|_{1,T} < \infty\}.$$

This is an infinite-dimensional space, and so it is not immediately clear what constitutes a compact set in this space. Recall that the *dimension* of a linear space is the number of elements in a basis for the space, and that a *basis* is a linearly independent set of elements with the property that any element can be expressed as a linear combination of the basis elements. Any space with n linearly independent elements has dimension at least n.

Example 12.5. The space of functions of x alone with finite 1-norm is denoted by L_1,

$$L_1 = \{v(x) : \|v\|_1 < \infty\}.$$

This space is clearly infinite-dimensional, since the functions

$$v_j(x) = \begin{cases} 1 & \text{if } j < x < j+1, \\ 0 & \text{otherwise} \end{cases} \tag{12.47}$$

for $j = 0, 1, 2, \ldots$ are linearly independent, for example.

Unfortunately, in an infinite-dimensional space, a closed and bounded set is not necessarily compact, as the next example shows.

Example 12.6. The sequence of functions $\{v_1, v_2, \ldots\}$ with v_j defined by (12.47) all lie in the closed and bounded unit ball

$$B_1 = \{v \in L_1 : \|v\|_1 \leq 1\},$$

and yet this sequence has no convergent subsequences.

The difficulty here is that the support of these functions is nonoverlapping and marches off to infinity as $j \to \infty$. We might try to avoid this by considering a set of the form

$$\{v \in L_1 : \|v\|_1 \leq R \quad \text{and} \quad \text{Supp}(v) \subset [-M, M]\}$$

for some $R, M > 0$, where $\text{Supp}(v)$ denotes the support of the function v, i.e., $\text{Supp}(v) \subset [-M, M]$ means that $v(x) \equiv 0$ for $|x| > M$. However, this set is also not compact, as shown by the sequence of functions $\{v_1, v_2, \ldots\}$ with

$$v_j(x) = \begin{cases} \sin(jx) & \text{if } |x| \leq 1, \\ 0 & \text{if } |x| > 1. \end{cases}$$

Again this sequence has no convergent subsequences, now because the functions become more and more oscillatory as $j \to \infty$.

12.12.4 Total-Variation Stability

In order to obtain a compact set in L_1, we will put a bound on the total variation of the functions, a quantity already defined in (6.19) through (6.21). The set

$$\{v \in L_1 : \mathrm{TV}(v) \leq R \quad \text{and} \quad \mathrm{Supp}(v) \subset [-M, M]\} \tag{12.48}$$

is a compact set, and any sequence of functions with uniformly bounded total variation and support must contain convergent subsequences. (Note that the 1-norm will also be uniformly bounded as a result, with $\|v\|_1 \leq MR$.)

Since our numerical approximations $Q^{(\Delta t)}$ are functions of x and t, we need to bound the total variation in both space and time. We define the total variation over $[0, T]$ by

$$\mathrm{TV}_T(q) = \limsup_{\epsilon \to 0} \frac{1}{\epsilon} \int_0^T \int_{-\infty}^{\infty} |q(x + \epsilon, t) - q(x, t)| \, dx \, dt$$

$$+ \limsup_{\epsilon \to 0} \frac{1}{\epsilon} \int_0^T \int_{-\infty}^{\infty} |q(x, t + \epsilon) - q(x, t)| \, dx \, dt. \tag{12.49}$$

It can be shown that the set

$$\mathcal{K} = \{q \in L_{1,T} : \mathrm{TV}_T(q) \leq R \text{ and } \mathrm{Supp}(q(\cdot, t)) \subset [-M, M] \; \forall t \in [0, T]\} \tag{12.50}$$

is a compact set in $L_{1,T}$.

Since our functions $Q^{(\Delta t)}(x, t)$ are always piecewise constant, the definition (12.49) of TV_T reduces to simply

$$\mathrm{TV}_T\big(Q^{(\Delta t)}\big) = \sum_{n=0}^{T/\Delta t} \sum_{j=-\infty}^{\infty} \left[\Delta t \big|Q_{i+1}^n - Q_i^n\big| + \Delta x \big|Q_i^{n+1} - Q_i^n\big| \right]. \tag{12.51}$$

Note that we can rewrite this in terms of the one-dimensional total variation and 1-norm as

$$\mathrm{TV}_T\big(Q^{(\Delta t)}\big) = \sum_{n=0}^{T/\Delta t} [\Delta t \, \mathrm{TV}(Q^n) + \|Q^{n+1} - Q^n\|_1]. \tag{12.52}$$

Definition 12.2. *We will say that a numerical method is* total-variation-stable, *or simply* TV-stable, *if all the approximations $Q^{(\Delta t)}$ for $\Delta t < \Delta t_0$ lie in some fixed set of the form (12.50) (where R and M may depend on the initial data $\mathring{q}(x)$, the time T, and the flux function $f(q)$, but not on Δt).*

Note that our requirement in (12.50) that $\mathrm{Supp}(q)$ be uniformly bounded over $[0, T]$ is always satisfied for any explicit method if the initial data $\mathring{q}$ has compact support and $\Delta t/\Delta x$ is constant as $\Delta t \to 0$. This follows from the finite speed of propagation for such a method.

The other requirement for TV-stability can be simplified considerably by noting the following theorem. This says that for the special case of functions generated by conservative numerical methods, it suffices to insure that the *one-dimensional* total variation at each time t_n is uniformly bounded (independent of n). Uniform boundedness of TV_T then follows.

Theorem 12.2. *Consider a conservative method with a Lipschitz-continuous numerical flux $F^n_{i-1/2}$, and suppose that for each initial data $\overset{\circ}{q}$ there exist some Δt_0, $R > 0$ such that*

$$\mathrm{TV}(Q^n) \le R \qquad \forall n, \Delta t \quad \text{with } \Delta t < \Delta t_0, \quad n \, \Delta t \le T. \tag{12.53}$$

Then the method is TV-stable.

To prove this theorem we use the following lemma, which is proved below.

Lemma 1. *If Q^n is generated by a conservative method with a Lipschitz-continuous numerical flux function, then the bound (12.53) implies that there exists $\alpha > 0$ such that*

$$\|Q^{n+1} - Q^n\|_1 \le \alpha \Delta t \qquad \forall n, \Delta t \quad \text{with } \Delta t < \Delta t_0, \quad n \, \Delta t \le T. \tag{12.54}$$

Proof of Theorem 12.2. Using (12.53) and (12.54) in (12.52) gives

$$\mathrm{TV}_T\big(Q^{(\Delta t)}\big) = \sum_{n=0}^{T/\Delta t} [\Delta t \, \mathrm{TV}(Q^n) + \|Q^{n+1} - Q^n\|_1]$$

$$\le \sum_{n=0}^{T/\Delta t} [\Delta t \, R + \alpha \, \Delta t]$$

$$\le \Delta t \, (R + \alpha) T / \Delta t = (R + \alpha) T$$

for all $\Delta t < \Delta t_0$, showing that $\mathrm{TV}_T(Q^{(\Delta t)})$ is uniformly bounded as $\Delta t \to 0$. This, together with the finite-speed-of-propagation argument outlined above, shows that all $Q^{(\Delta t)}$ lie in a set of the form (12.50) for all $\Delta t < \Delta t_0$ and the method is TV-stable. $\qquad\square$

Proof of Lemma 1. Recall that a method in conservation form has

$$Q^{n+1}_i - Q^n_i = \frac{\Delta t}{\Delta x}\big(F^n_{i+1/2} - F^n_{i-1/2}\big)$$

and hence

$$\|Q^{n+1} - Q^n\|_1 = \Delta t \sum_{j=-\infty}^{\infty} \big|F^n_{i+1/2} - F^n_{i-1/2}\big|. \tag{12.55}$$

The flux $F^n_{i-1/2}$ depends on a finite number of values $Q_{i-p}, \dots, Q_{i+r}$. The bound (12.53) together with the compact support of each Q^n easily gives

$$\big|Q^n_i\big| \le R/2 \qquad \forall i, n \quad \text{with } n \, \Delta t \le T. \tag{12.56}$$

This uniform bound on Q^n_i, together with the Lipschitz continuity of the flux function, allows us to derive a bound of the form

$$\big|F^n_{i+1/2} - F^n_{i-1/2}\big| \le K \max_{-p \le j \le r} \big|Q^n_{i+j} - Q^n_{i+j-1}\big|. \tag{12.57}$$

It follows that

$$\left| F^n_{i+1/2} - F^n_{i-1/2} \right| \le K \sum_{j=-p}^{r} \left| Q^n_{i+j} - Q^n_{i+j-1} \right|,$$

and so (12.55) gives

$$\| Q^{n+1} - Q^n \|_1 \le \Delta t\, K \sum_{j=-p}^{r} \sum_{i=-\infty}^{\infty} \left| Q^n_{i+j} - Q^n_{i+j-1} \right|$$

after interchanging sums. But now the latter sum is simply $TV(Q^n)$ for any value of j, and so

$$\| Q^{n+1} - Q^n \|_1 \le \Delta t\, K \sum_{j=-p}^{r} TV(Q^n)$$

$$\le \Delta t\, K(p+r+1)R,$$

yielding the bound (12.54). $\qquad\qquad\square$

We are now set to prove our convergence theorem, which requires TV-stability along with consistency.

Theorem 12.3. *Suppose $Q^{(\Delta t)}$ is generated by a numerical method in conservation form with a Lipschitz continuous numerical flux, consistent with some scalar conservation law. If the method is TV-stable, i.e., if $TV(Q^n)$ is uniformly bounded for all n, Δt with $\Delta t < \Delta t_0$, $n\,\Delta t \le T$, then the method is convergent, i.e., $\mathrm{dist}(Q^{(\Delta t)}, \mathcal{W}) \to 0$ as $\Delta t \to 0$.*

Proof. To prove this theorem we suppose that the conclusion is false, and obtain a contradiction. If $\mathrm{dist}(Q^{(\Delta t)}, \mathcal{W})$ does not converge to zero, then there must be some $\epsilon > 0$ and some sequence of approximations $\{Q^{(\Delta t_1)}, Q^{(\Delta t_2)}, \dots\}$ such that $\Delta t_j \to 0$ as $j \to \infty$ while

$$\mathrm{dist}\left(Q^{(\Delta t_j)}, \mathcal{W} \right) > \epsilon \quad \text{for all } j. \qquad (12.58)$$

Since $Q^{(\Delta t_j)} \in \mathcal{K}$ (the compact set of (12.50)) for all j, this sequence must have a convergent subsequence, converging to some function $v \in \mathcal{K}$. Hence far enough out in this subsequence, $Q^{(\Delta t_j)}$ must satisfy

$$\left\| Q^{(\Delta t_j)} - v \right\|_{1,T} < \epsilon \quad \text{for all } j \text{ sufficiently large} \qquad (12.59)$$

for the ϵ defined above. Moreover, since the $Q^{(\Delta t)}$ are generated by a conservative and consistent method, it follows from the Lax–Wendroff theorem (Theorem 12.1) that the limit v must be a weak solution of the conservation law, i.e., $v \in \mathcal{W}$. But then (12.59) contradicts (12.58), and hence a sequence satisfying (12.58) cannot exist, and we conclude that $\mathrm{dist}(Q^{(\Delta t)}, \mathcal{W}) \to 0$ as $\Delta t \to 0$. $\qquad\square$

There are other ways to prove convergence of approximate solutions to nonlinear scalar problems that do not require TV-stability. This is particularly useful in more than one dimension, where the total variation is harder to bound; see Section 20.10.2. For nonlinear systems of equations it is impossible to bound the total variation in most cases, and convergence results are available only for special systems; see Section 15.8.2.

Exercises

12.1. (a) Show that the method (12.5) with $\mathcal{A}^\pm \Delta Q_{i-1/2}$ defined by (12.8) corresponds to the flux function (12.12) with $a_{i-1/2}$ given by Murman's formula (12.14).

 (b) Show that this has entropy-violating solutions by computing the flux $F_{i-1/2}$ everywhere for Burgers' equation with Riemann data $q_l = -1$ and $q_r = 1$.

12.2. Show that the LLF method (12.12) is an E-scheme.

12.3. Show that any E-scheme is TVD if the Courant number is sufficiently small. Hint: Use Theorem 6.1 with

$$C_{i-1} = \frac{\Delta t}{\Delta x} \left(\frac{f(Q_{i-1}) - F_{i-1/2}}{Q_i - Q_{i-1}} \right) \tag{12.60}$$

and a suitable choice for D_i.

12.4. Show that Jensen's inequality (12.41) need not hold if $\eta(q)$ is not convex.

13

Nonlinear Systems of Conservation Laws

In Chapter 3 we developed the theory of *linear systems* of hyperbolic equations, in which case the general Riemann problem can be solved by decomposing the jump in states into eigenvectors of the coefficient matrix. Each eigenvector corresponds to a wave traveling at one of the characteristic speeds of the system, which are given by the corresponding eigenvalues of the coefficient matrix.

In Chapter 11 we explored *nonlinear scalar* problems, and saw that when the wave speed depends on the solution, then waves do not propagate unchanged, but in general will deform as compression waves or expansion waves, and that shock waves can form from smooth initial data. The solution to the Riemann problem (in the simplest case where the flux function is convex) then consists of a single shock wave or centered rarefaction wave.

In this chapter we will see that these two theories can be melded together into an elegant general theory for *nonlinear systems* of equations. As in the linear case, solving the Riemann problem for a system of m equations will typically require splitting the jump in states into m separate waves. Each of these waves, however, can now be a shock wave or a centered rarefaction wave.

We will develop this general theory using the one-dimensional shallow water equations as a concrete example. The same theory will later be illustrated for several other systems of equations, including the Euler equations of gas dynamics in Chapter 14. The shallow water equations are a nice example to consider first, for several reasons. It is a system of only two equations, and hence the simplest step up from the scalar case. The nonlinear structure of the equations is fairly simple, making it possible to solve the Riemann problem explicitly, and yet the structure is typical of what is seen in other important examples such as the Euler equations. Finally, it is possible to develop intuition quite easily for how solutions to the shallow water equations should behave, and even perform simple experiments illustrating some of the results we will see.

In [281] the isothermal equations are used as the primary example. This is also a simple system of two equations with a nonlinear structure very similar to that of the shallow water equations. This system is discussed briefly in Section 14.6. More details and development of the theory of nonlinear systems based on the isothermal equations may be found in [281].

13.1 The Shallow Water Equations

To derive the one-dimensional shallow water equations, we consider fluid in a channel of unit width and assume that the vertical velocity of the fluid is negligible and the horizontal velocity $u(x, t)$ is roughly constant throughout any cross section of the channel. This is true if we consider small-amplitude waves in a fluid that is shallow relative to the wavelength.

We now assume the fluid is incompressible, so the density $\bar{\rho}$ is constant. Instead we allow the depth of the fluid to vary, and it is this depth, or *height* $h(x, t)$, that we wish to determine. The total mass in $[x_1, x_2]$ at time t is

$$\int_{x_1}^{x_2} \bar{\rho} h(x, t) \, dx.$$

The density of momentum at each point is $\bar{\rho} u(x, t)$, and integrating this vertically gives the mass flux to be $\bar{\rho} u(x, t) h(x, t)$. The constant $\bar{\rho}$ drops out of the conservation-of-mass equation, which then takes the familiar form (compare (2.32))

$$h_t + (uh)_x = 0. \tag{13.1}$$

The quantity hu is often called the *discharge* in shallow water theory, since it measures the flow rate of water past a point.

The conservation-of-momentum equation also takes the same form as in gas dynamics (see (2.34)),

$$(\bar{\rho} hu)_t + (\bar{\rho} hu^2 + p)_x = 0, \tag{13.2}$$

but now p is determined from a hydrostatic law, stating that the pressure at distance $h - y$ below the surface is $\bar{\rho} g(h - y)$, where g is the gravitational constant. This pressure arises simply from the weight of the fluid above. Integrating this vertically from $y = 0$ to $y = h(x, t)$ gives the total pressure felt at (x, t), the proper pressure term in the momentum flux:

$$p = \frac{1}{2} \bar{\rho} g h^2. \tag{13.3}$$

Using this in (13.2) and canceling $\bar{\rho}$ gives

$$(hu)_t + \left(hu^2 + \frac{1}{2} gh^2 \right)_x = 0. \tag{13.4}$$

We can combine equations (13.1), (13.4) into the system of *one-dimensional shallow water equations*

$$\begin{bmatrix} h \\ hu \end{bmatrix}_t + \begin{bmatrix} uh \\ hu^2 + \frac{1}{2} gh^2 \end{bmatrix}_x = 0. \tag{13.5}$$

Note that this is equivalent to the isentropic equations of gas dynamics (discussed in Section 2.6) with the value $\gamma = 2$, since setting $P(\rho) = \frac{1}{2} g \rho^2$ in (2.38) gives the same system.

If we assume that h and u are smooth, then equation (13.4) can be simplified by expanding the derivatives and using (13.1) to replace the h_t term. Then several terms drop out, and (13.4) is reduced to

$$u_t + \left(\frac{1}{2}u^2 + gh\right)_x = 0. \tag{13.6}$$

Finally, the explicit dependence on g can be eliminated by introducing the variable $\varphi = gh$ into (13.1) and (13.6). The system then becomes

$$\begin{bmatrix} u \\ \varphi \end{bmatrix}_t + \begin{bmatrix} u^2/2 + \varphi \\ u\varphi \end{bmatrix}_x = 0. \tag{13.7}$$

This set of equations is equivalent to the previous set (13.5) for smooth solutions, but it is important to note that the manipulations performed above depend on smoothness. For problems with shock waves, the two sets of conservation laws are not equivalent, and we know from Section 12.9 that it is crucial that we use the correct set in calculating shock waves. The form (13.5), which is derived directly from the original integral equations, is the correct set to use.

Since we will be interested in studying shock waves, we use the form (13.5) and take

$$q(x, t) = \begin{bmatrix} h \\ hu \end{bmatrix} = \begin{bmatrix} q^1 \\ q^2 \end{bmatrix}, \qquad f(q) = \begin{bmatrix} hu \\ hu^2 + \frac{1}{2}gh^2 \end{bmatrix} = \begin{bmatrix} q^2 \\ (q^2)^2/q^1 + \frac{1}{2}g(q^1)^2 \end{bmatrix}.$$

For smooth solution, these equations can equivalently be written in the quasilinear form

$$q_t + f'(q)q_x = 0,$$

where the Jacobian matrix $f'(q)$ is

$$f'(q) = \begin{bmatrix} 0 & 1 \\ -(q^2/q^1)^2 + gq^1 & 2q^2/q^1 \end{bmatrix} = \begin{bmatrix} 0 & 1 \\ -u^2 + gh & 2u \end{bmatrix}. \tag{13.8}$$

The eigenvalues of $f'(q)$ are

$$\lambda^1 = u - \sqrt{gh}, \qquad \lambda^2 = u + \sqrt{gh}, \tag{13.9}$$

with the corresponding eigenvectors

$$r^1 = \begin{bmatrix} 1 \\ u - \sqrt{gh} \end{bmatrix}, \qquad r^2 = \begin{bmatrix} 1 \\ u + \sqrt{gh} \end{bmatrix}. \tag{13.10}$$

Note that the eigenvalues and eigenvectors are functions of q for this nonlinear system.

If we wish to study waves with very small amplitude, then we can linearize these equations to obtain a linear system. Suppose the fluid is essentially at a constant depth $h_0 > 0$ and moving at a constant velocity u_0 (which may be zero), and let q now represent the perturbations from this constant state, so

$$q = \begin{bmatrix} h - h_0 \\ hu - h_0 u_0 \end{bmatrix} \quad \text{and} \quad q_0 = \begin{bmatrix} h_0 \\ h_0 u_0 \end{bmatrix}.$$

Then expanding the flux function and dropping terms of $\mathcal{O}(\|q\|^2)$ gives the linear system $q_t + A q_x = 0$ where $A = f'(q_0)$. Hence small-amplitude waves move at the characteristic velocities $\lambda_0^1 = u_0 - c_0$ and $\lambda_0^2 = u_0 + c_0$, where $c_0 = \sqrt{g h_0}$. These waves propagate at speed $\pm c_0$ relative to the fluid, exactly analogous to acoustic waves in a moving fluid as in Section 2.8. These shallow water waves should not be confused with sound waves, however. Sound does propagate in water, due to its slight compressibility, but in the shallow water equations we are ignoring this compressibility and hence ignoring sound waves. The waves we are modeling are often called *gravity waves*, since they are driven by the hydrostatic pressure resulting from gravity. They typically propagate at a speed $\sqrt{gh}$ that is much less than the speed of sound in water.

Note that λ^1 and λ^2 can be of either sign, depending on the magnitude of u relative to c. In shallow water theory the ratio

$$Fr = |u|/c \qquad (13.11)$$

is called the *Froude number*, and is analogous to the Mach number of gas dynamics.

The wave speed $\sqrt{g h_0}$ depends on the depth of the fluid; waves in deeper water move faster. Note that within a wave the depth of the fluid varies (it is deeper at a crest than in a trough), and so we should expect the crest of a wave to propagate slightly faster than a trough. If the amplitude of the wave is very small compared to h_0, then we can safely ignore this slight variation in speed, which is what we do in linearizing the equations. Then all parts of the wave travel at the same speed based on the background depth h_0, and the wave propagates with its shape unchanged. For waves with larger amplitude, however, the deformation of the wave due to differing wave speeds may be quite noticeable. In this case the linearized equations will not be an adequate model and the full nonlinear equations must be solved.

The nonlinear distortion of a wave leads to a steepening of the wave in the region where the fast-moving crest is catching up with the slower trough ahead of it (a compression wave), and a flattening of the wave (an expansion or rarefaction) in the region where the crest is pulling away from the following trough. This is similar to what is illustrated in Figure 11.1 for the nonlinear equations of traffic flow.

This behavior is familiar from watching waves break on the beach. Far from shore the waves we normally observe have a wavelength that is very small compared to the water depth, and hence they are governed by surface-wave theory rather than shallow water theory. Near the beach, however, the water depth is small enough that nonlinear shallow water theory applies. In this shallow water, the difference in h between crests and troughs is significant and the waves steepen. In fact the crest is often observed to move beyond the position of the preceding trough, somewhat like what is shown in Figure 11.4(b). At this point the assumptions of shallow water theory no longer hold, and a more complicated set of equations would have to be used to model *breakers*. Beyond the breaking time the depth h is triple-valued, a situation that obviously can't occur with other systems of conservation laws (such as traffic flow or gas dynamics) where the corresponding variable is a density that must be single-valued.

This extreme behavior of breaking waves results from the additional complication of a sloping beach. This leads to a continuous decrease in the fluid depth seen by the wave and a severe accentuation of the nonlinear effects. (The sloping beach, or more generally any variation in the bottom topography, also leads to additional source terms in the shallow water equations.) Shallow water waves in a domain with a flat bottom will typically not exhibit this type of breakers. Instead the gradual steepening of the wave due to nonlinearity would be counterbalanced by other effects such as surface tension (and also the vertical velocity, which is ignored in the one-dimensional model). Modeling these other effects would lead to higher-order derivatives in the equations (with small coefficients) and consequently the equations would have smooth solutions for all time, analogous to what is seen in Figure 11.6 for the viscous scalar equation. When these coefficients are small, the wave can become nearly discontinuous, and the shock-wave solution to the hyperbolic system gives a good approximation to such solutions. In shallow water flow, a shock wave is often called a *hydraulic jump*.

Example 13.1. Figure 13.1 shows the evolution of a hump of water (with initial velocity 0). As with the acoustics equations (see Figure 3.1), the hump gives rise to two waves, one moving in each direction. If the height of the hump were very small compared to the

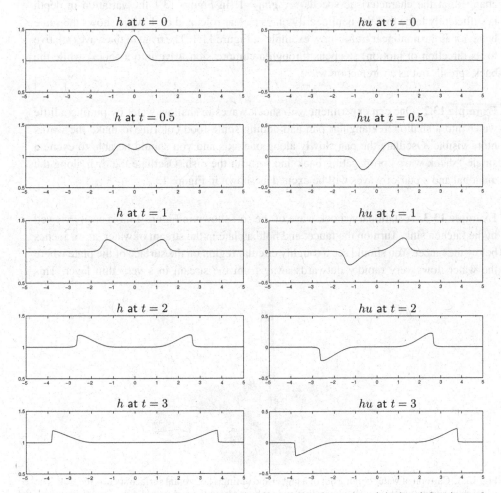

Fig. 13.1. Evolution of an initial depth perturbation, concentrated near the origin, into left-going and right-going waves. The shallow water equations are solved with $g = 1$. The left column shows the depth $q^1 = h$, the right column shows the momentum $q^2 = hu$. [claw/book/chap13/swhump]

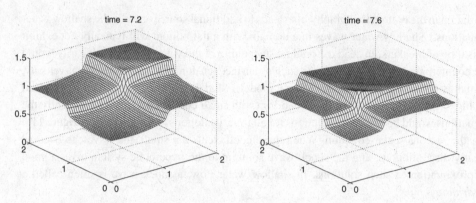

Fig. 13.2. Shallow water sloshing in a rectangular pan that is oscillated along the diagonal.
[claw/book/chap13/slosh]

background depth $h_0 = 1$, then these would propagate with their shape essentially unchanged, at the characteristic speeds $\pm\sqrt{gh_0} = 1$. In Figure 13.1 the variation in depth
is sufficiently large that the nonlinearity plays a clear role, and each wave shows the same
behavior as the nonlinear traffic-flow example of Figure 11.1. The front of the wave (relative
to its direction of motion) steepens through a compression wave into a shock, while the
back spreads out as a rarefaction wave.

Example 13.2. One can experiment with shock waves in shallow water by putting a little
water into a shallow rectangular pan and adding some food coloring to make the waves
more visible. Oscillate the pan slowly along one axis and you should be able to excite a
single "shock wave" propagating back and forth in the dish. Oscillate the dish along the
diagonal and a pair of waves will be excited as shown in Figure 13.2.

Example 13.3. A stationary shock wave of the sort shown in Figure 13.3 is easily viewed
in the kitchen sink. Turn on the faucet and hold a plate in the stream of water, a few inches
below the faucet. You should see a roughly circular region on the surface of the plate where
the water flows very rapidly outwards away from the stream in a very thin layer. This

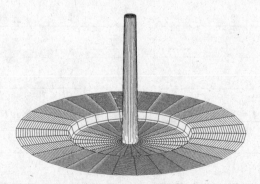

Fig. 13.3. Cartoon of water coming out of a faucet and hitting a horizontal surface such as the sink. The
water expands outwards in a thin layer that suddenly thickens through a hydraulic jump (stationary
shock wave).

region is bounded by a hydraulic jump, where the depth of the water suddenly increases and its speed abruptly decreases. By adjusting the flow rate or angle of the plate you should be able to make the location of this shock wave move around. When the conditions are fixed, the shock is stationary and has zero propagation speed. This can be modeled by the two-dimensional shallow water equations, or in the radially symmetric case by the one-dimensional equations with additional source terms incorporated to model the geometric effects as described in Section 18.9.

13.2 Dam-Break and Riemann Problems

Consider the shallow water equations (13.5) with the piecewise-constant initial data

$$h(x, 0) = \begin{cases} h_l & \text{if } x < 0, \\ h_r & \text{if } x > 0, \end{cases} \qquad u(x, 0) = 0, \tag{13.12}$$

where $h_l > h_r \geq 0$. This is a special case of the Riemann problem in which $u_l = u_r = 0$, and is called the *dam-break problem* because it models what happens if a dam separating two levels of water bursts at time $t = 0$. This is the shallow water equivalent of the shock-tube problem of gas dynamics (Section 14.13). We assume $h_r > 0$.

Example 13.4. Figure 13.4 shows the evolution of the depth and fluid velocity for the dam-break problem with data $h_l = 3$ and $h_r = 1$. Figure 13.5 shows the structure of this solution in the x–t plane. Water flows from left to right in a wedge that expands from the dam location $x = 0$. At the right edge of this wedge, moving water with some intermediate depth h_m and velocity $u_m > 0$ slams into the stationary water with $h = h_r$, accelerating it instantaneously

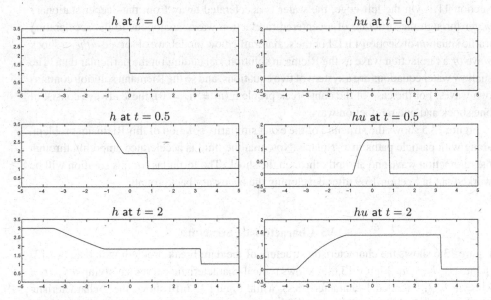

Fig. 13.4. Solution of the dam-break Riemann problem for the shallow water equations with $u_l = u_r = 0$. On the left is the depth h and on the right is the momentum hu. [claw/book/chap13/dambreak]

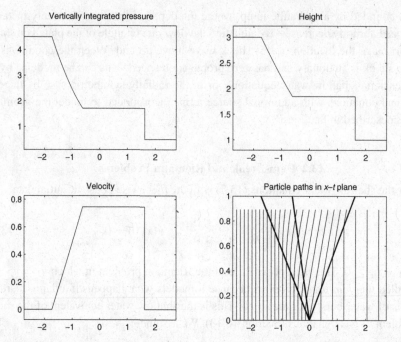

Fig. 13.5. Structure of the similarity solution of the dam-break Riemann problem for the shallow water equations with $u_l = u_r = 0$. The depth h, velocity u, and vertically integrated pressure are displayed as functions of x/t. The structure in the x–t plane is also shown with particle paths indicated for a set of particles with the spacing between particles inversely proportional to the depth. [claw/book/chap13/rpsoln]

through a shock wave. This is roughly analogous to the traffic-jam shock wave studied in Section 11.1. On the left edge, the water is accelerated away from the deeper stationary water through the structure of a centered rarefaction wave, analogous to the accelerating traffic situation of Section 11.1. For the scalar traffic flow model, we observed *either* a shock wave *or* a rarefaction wave as the Riemann solution, depending on the particular data. The shallow water equations are a system of two equations, and so the Riemann solution contains two waves. For the case of the dam-break problem ($u_l = u_r = 0$), these always consist of one shock and one rarefaction wave.

Figure 13.5 shows the structure of the exact similarity solution of this Riemann problem, along with particle paths in x–t plane. Note that the fluid is accelerated smoothly through the rarefaction wave and abruptly through the shock. The formulas for this solution will be worked out in Section 13.9 after developing the necessary background.

13.3 Characteristic Structure

Figure 13.6 shows the characteristic structure of the dam-break problem with data (13.12) in the case $h_l > h_r$. Figure 13.6(a) shows typical characteristic curves satisfying $dX/dt = \lambda^1 = u - \sqrt{gh}$ (called 1-characteristics), while Figure 13.6(b) shows the 2-characteristic curves satisfying $dX/dt = \lambda^2 = u + \sqrt{gh}$. Note that each characteristic direction is constant (the curves are straight lines) in each wedge where q is constant.

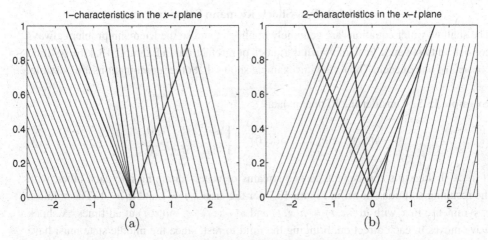

Fig. 13.6. Solution of the dam-break Riemann problem for the shallow water equations, shown in the x–t plane. The dark lines show the shock wave and the edges of the rarefaction wave seen in Figure 13.4. The lighter lines show 1-characteristics and 2-characteristics.

In Figure 13.6(a) we see that the 1-characteristics behave near the 1-rarefaction wave just as we would expect from the nonlinear scalar case. They spread out through the rarefaction wave, and the edges of this wave move with the characteristic velocity in each constant region bounding the rarefaction. Also note that the 1-characteristics *cross* the 2-waves in the sense that they are approaching the 2-wave on one side (for smaller time t) and then moving away from the 2-wave on the other side, for larger t.

On the other hand, 2-characteristics shown in Figure 13.6(b) *impinge* on the 2-shock, again as we would expect from the scalar theory. These characteristics *cross* the 1-rarefaction with a smooth change in velocity.

This is the standard situation for many nonlinear systems of equations. For a system of m equations, there will be m characteristic families and m waves in the solution to the Riemann problem. If the pth wave is a shock, then characteristics of families 1 through $p - 1$ will cross the shock from right to left, characteristics of family $p + 1$ through m will cross from left to right, and characteristics of family p will impinge on the shock from both sides. This classical situation is observed in many physical problems, including the Euler equations of gas dynamics, and is the case that is best understood mathematically. Such shocks are often called *classical Lax shocks*, because much of this theory was developed by Peter Lax.

In order for the classical situation to occur, certain conditions must be satisfied by the flux function $f(q)$. We assume the system is *strictly hyperbolic*, so the eigenvalues are always distinct. The conditions of *genuine nonlinearity* must also be satisfied, analogous to the convexity condition for scalar equations. This is discussed further in Section 13.8.4. Otherwise the Riemann solution can be more complicated. As in the nonconvex scalar case discussed in Section 16.1, there can be compound waves in some family consisting of more than a single shock or rarefaction. For systems it could also happen that the number of characteristics impinging on a shock is different from what has just been described, and the shock is *overcompressive* or *undercompressive*. See Section 16.2 for further discussion.

13.4 A Two-Shock Riemann Solution

The shallow water equations are genuinely nonlinear, and so the Riemann problem always consists of two waves, each of which is a shock or rarefaction. In Example 13.4 the solution consists of one of each. The following example shows that other combinations are possible.

Example 13.5. Consider the Riemann data

$$h(x, 0) \equiv h_0, \qquad u(x, 0) = \begin{cases} u_l & \text{if } x < 0, \\ -u_l & \text{if } x > 0. \end{cases} \tag{13.13}$$

If $u_l > 0$, then this corresponds to two streams of water slamming into each other, with the resulting solution shown in Figure 13.7 for the case $h_0 = 1$ and $u_l = 1$. The solution is symmetric in x with $h(-x, t) = h(x, t)$ and $u(-x, t) = -u(x, t)$ at all times. A shock wave moves in each direction, bringing the fluid to rest, since the middle state must have $u_m = 0$ by symmetry. The solution to this problem is computed in Section 13.7.1.

The characteristic structure of this solution is shown in Figure 13.8. Note again that 1-characteristics impinge on the 1-shock while crossing the 2-shock, whereas 2-characteristics impinge on the 2-shock.

Note that if we look at only half of the domain, say $x < 0$, then we obtain the solution to the problem of shallow water flowing into a wall located at $x = 0$ with velocity u_l. A shock wave moves out from the wall, behind which the fluid is at rest. This is now exactly analogous to traffic approaching a red light, as shown in Figure 11.2.

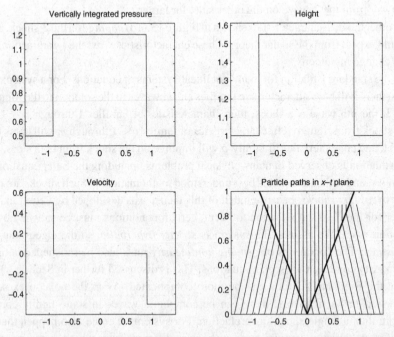

Fig. 13.7. Structure of the similarity solution of the two-shock Riemann problem for the shallow water equations with $u_l = -u_r$. The depth h, velocity u, and vertically integrated pressure are displayed as functions of x/t. The structure in the x–t plane is also shown with particle paths indicated for a set of particles with the spacing between particles inversely proportional to the depth. [claw/book/chap13/twoshock]

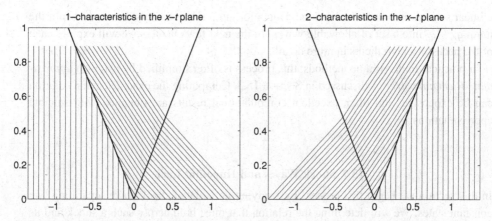

Fig. 13.8. Solution of the two-shock Riemann problem for the shallow water equations, shown in the x–t plane. The dark lines show the shocks. The lighter lines show 1-characteristics and 2-characteristics.

13.5 Weak Waves and the Linearized Problem

To understand the characteristic structure of the shallow water equations, it is useful to consider what happens in the solution to the Riemann problems discussed above in the case where the initial jump is so small that the linearized equation gives a good model. Consider the data (13.12), for example, with $h_l = h_0 + \epsilon$ and $h_r = h_0 - \epsilon$ for some $\epsilon \ll h_0$. Then if we solve the Riemann problem for the linearized equation with $u_0 = 0$ in (13.8), we find that the solution consists of two *acoustic* waves with speeds $\pm\sqrt{gh_0}$, separated by a state (h_m, u_m) with

$$h_m = h_0, \qquad u_m = \epsilon\sqrt{gh_0}.$$

The solution consists of two discontinuities. If we solved the nonlinear equations with this same data, the solution would look quite similar, but the left-going wave would be a weak rarefaction wave, spreading very slightly with time, and with the 1-characteristics spreading slightly apart rather than being parallel as in the linear problem. The right-going wave would be a weak shock wave, with slightly converging characteristics.

13.6 Strategy for Solving the Riemann Problem

Above we have presented a few specific examples of Riemann solutions for the shallow water equations. In order to apply Riemann-solver-based finite volume methods, we must be able to solve the general Riemann problem with arbitrary left and right states q_l and q_r. To compute the exact solution, we must do the following:

1. Determine whether each of the two waves is a shock or a rarefaction wave (perhaps using an appropriate entropy condition).
2. Determine the intermediate state q_m between the two waves.
3. Determine the structure of the solution through any rarefaction waves.

The first and third of these are similar to what must be done for a nonlinear scalar equation, while the second step corresponds to the procedure for solving the Riemann problem for a linear system, in which the Rankine–Hugoniot jump relations must be used to split the jump $q_r - q_l$ into a set of allowable waves. In the next few chapters we will explore each of the necessary ingredients in more detail.

In practical finite volume methods, this process is often simplified by using an *approximate Riemann solver* as discussed in Section 15.3. Computing the exact Riemann solution can be expensive, and often excellent computational results are obtained with suitable approximations.

13.7 Shock Waves and Hugoniot Loci

In this section we begin this process by studying an isolated shock wave separating two constant states. We will determine the relation that must hold across such a shock and its speed of propagation. For an arbitrary fixed state we will determine the set of all other states that could be connected to this one by a shock of a given family. This is a necessary ingredient in order to solve a general Riemann problem.

We will continue to use the shallow water equations as the primary example system. Consider, for example, a shallow water 2-shock such as the right-going shock of Example 13.4 or Example 13.5. This shock connects some state q_m to the right state q_r from the Riemann data. We will view q_r as being fixed and determine all possible states q that can be connected to q_r by a 2-shock. We will find that there is a one-parameter family of such states, which trace out a curve in state space as shown in Figure 13.9(b). Here the state space (phase plane) is the h–hu plane. This set of states is called a *Hugoniot locus*.

Which one of these possible states corresponds to q_m in the solution to the Riemann problem depends not only on q_r but also on q_l. The state q_m must lie on the curve shown in Figure 13.9(b), but it must also lie on an analogous curve of all states that can be connected to q_l by a 1-wave, as determined below. This is completely analogous to the manner in which the linear Riemann problem was solved in Chapter 3.

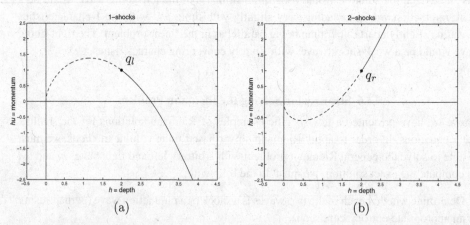

(a)   (b)

Fig. 13.9. (a) Hugoniot locus of points q in shallow water state space that can be connected to a given state q_l by a 1-shock satisfying the Rankine–Hugoniot conditions. Only some of these states (on the solid portion of the curves) satisfy the entropy condition; see Section 13.7.2. (b) Hugoniot locus of points in shallow water state space that can be connected to a given state q_r by a 2-shock satisfying the Rankine–Hugoniot conditions (13.15).

We now consider the problem of determining all states q that can be connected to some fixed state q_* (representing either q_l or q_r) by a shock. Recall from Section 11.8 that across any shock the Rankine–Hugoniot condition (11.20) must be satisfied, so

$$s(q_* - q) = f(q_*) - f(q). \tag{13.14}$$

For the shallow water equations, this gives a system of two equations that must simultaneously be satisfied:

$$s(h_* - h) = h_* u_* - hu,$$
$$s(h_* u_* - hu) = h_* u_*^2 - hu^2 + \frac{1}{2} g(h_*^2 - h^2). \tag{13.15}$$

Recall that (h_*, u_*) is fixed and we wish to find all states (h, u) and corresponding speeds s satisfying these relations. We thus have two equations with three unknowns, so we expect to find a one-parameter family of solutions. In fact there are two distinct families of solutions, corresponding to 1-shocks and 2-shocks. For the time being we use the term "shock" to refer to a discontinuous weak solution satisfying the Rankine–Hugoniot condition. Later we will consider the additional admissibility condition that is required to ensure that such a solution is truly a physical shock wave.

There are many different ways to parameterize these families. Fairly simple formulas result from using h as the parameter. For each value of h we will determine the corresponding u and s, and plotting hu against h will give the curves shown in Figure 13.9.

We first determine u by eliminating s from the system (13.15). The first equation gives

$$s = \frac{h_* u_* - hu}{h_* - h}, \tag{13.16}$$

and substituting this into the second equation gives an equation relating u to h. This is a quadratic equation in u that, after simplifying somewhat, becomes

$$u^2 - 2u_* u + \left[u_*^2 - \frac{g}{2} \left(\frac{h_*}{h} - \frac{h}{h_*} \right)(h_* - h) \right] = 0,$$

with roots

$$u(h) = u_* \pm \sqrt{\frac{g}{2} \left(\frac{h_*}{h} - \frac{h}{h_*} \right)(h_* - h)}. \tag{13.17}$$

Note that when $h = h_*$ this reduces to $u = u_*$, as we expect, since the curves we seek must pass through the point (h_*, u_*).

For each $h \neq h_*$ there are two different values of u, corresponding to the two families of shocks. In the case of a very weak shock ($q \approx q_*$) we expect the linearized theory to hold, and so we expect one of these curves to be tangent to the eigenvector $r^1(q_*)$ at q_* and the other to be tangent to $r^2(q_*)$. This allows us to distinguish which curve corresponds to the 1-shocks and which to 2-shocks. To see this more clearly, we multiply (13.17) by h and reparameterize by a value α, with

$$h = h_* + \alpha,$$

so that $h = h_*$ at $\alpha = 0$, to obtain

$$hu = h_* u_* + \alpha \left[u_* \pm \sqrt{gh_* \left(1 + \frac{\alpha}{h_*}\right)\left(1 + \frac{\alpha}{2h_*}\right)} \right]. \qquad (13.18)$$

Hence we have

$$q = q_* + \alpha \left[\frac{1}{u_* \pm \sqrt{gh_*} + \mathcal{O}(\alpha)} \right] \quad \text{as } \alpha \to 0.$$

For α very small (as q approaches q_*), we can ignore the $\mathcal{O}(\alpha)$ term and we see that these curves approach the point q_* tangent to the vectors

$$\left[\frac{1}{u_* \pm \sqrt{gh_*}} \right],$$

which are simply the eigenvectors of the Jacobian matrix (13.8) at q_*. From this we see that choosing the $-$ sign in (13.18) gives the locus of 1-shocks, while the $+$ sign gives the locus of 2-shocks. (Note: The same is not true in (13.17), where choosing a single sign gives part of one locus and part of the other as h varies.)

13.7.1 The All-Shock Riemann Solution

Now consider a general Riemann problem with data q_l and q_r, and suppose we know that the solution consists of two shocks. We can then solve the Riemann problem by finding the state q_m that can be connected to q_l by a 1-shock and also to q_r by a 2-shock.

We found in the previous section that through the point q_r there is a curve of points q that can be connected to q_r by a 2-shock. For the shallow water equations, these points satisfy (13.18) with the plus sign and with $q_* = q_r$. Since q_m must lie on this curve, we have

$$h_m u_m = h_r u_r + (h_m - h_r)\left[u_r + \sqrt{gh_r\left(1 + \frac{h_m - h_r}{h_r}\right)\left(1 + \frac{h_m - h_r}{2h_r}\right)} \right],$$

which can be simplified to give

$$u_m = u_r + (h_m - h_r)\sqrt{\frac{g}{2}\left(\frac{1}{h_m} + \frac{1}{h_r}\right)}. \qquad (13.19)$$

Similarly, there is a curve through q_l of states that can be connected to q_l by a 1-shock, obtained by setting $q_* = q_l$ and taking the minus sign in (13.18). Since q_m must lie on this curve, we find that

$$u_m = u_l - (h_m - h_l)\sqrt{\frac{g}{2}\left(\frac{1}{h_m} + \frac{1}{h_l}\right)}. \qquad (13.20)$$

We thus have a system of two equations (13.19) and (13.20) for the two unknowns h_m and u_m. Solving this system gives the desired intermediate state in the Riemann solution. We

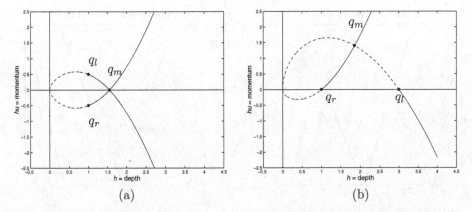

Fig. 13.10. All-shock solutions to the shallow water Riemann problem can be constructed by finding the intersection of the appropriate Hugoniot loci. (a) For Example 13.6. (b) An entropy-violating Riemann solution for Example 13.7.

can easily eliminate u_m from this system by noting that this appears only on the left of each equation, and the left-hand sides are equal, so equating the right-hand sides gives a single equation involving only the one unknown h_m. This can be solved by an iterative method for nonlinear equations, such as Newton's method.

This is analogous to solving the Riemann problem for a linear hyperbolic system, as discussed in Chapter 3, but in that case a linear system of equations results, which is more easily solved for the intermediate state.

Example 13.6. Consider the shallow water Riemann problem with $h_l = h_r = 1$, $u_l = 0.5$, and $u_r = -0.5$, as in Example 13.5. Figure 13.10(a) shows the states q_l, q_r, and q_m in the phase plane, together with the Hugoniot loci of 1-shocks through q_l and 2-shocks through q_r. In this case we could use our knowledge that $u_m = 0$ to simplify the above system further, using either (13.19) or (13.20) with the left-hand side replaced by 0. We find that the solution is $h_m = 1.5514$.

Example 13.7. What happens if we apply this same procedure to a Riemann problem where the physical solution should *not* consist of two shocks? For example, consider the dam-break Riemann problem of Example 13.4, where the solution should consist of a 1-rarefaction and a 2-shock. We can still solve the problem in terms of two "shock waves" that satisfy the Rankine–Hugoniot jump conditions, as illustrated in Figure 13.10(b). This gives a weak solution of the conservation laws, but one that does not satisfy the proper *entropy condition* for this system, as discussed in the next section. The procedure for finding the physically correct solution with a rarefaction wave is given in Section 13.9.

13.7.2 The Entropy Condition

Figure 13.6 shows the characteristic structure for the physically correct solution to the dam-break Riemann problem of Example 13.4. Figure 13.11 shows the structure for the weak solution found in Example 13.7, which consists of two discontinuities. The 1-characteristics

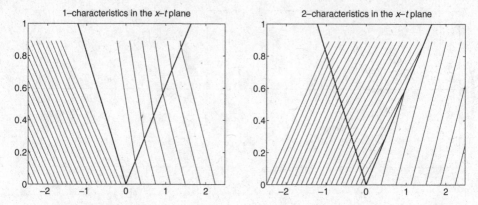

Fig. 13.11. Entropy-violating solution of the dam-break Riemann problem for the shallow water equations, shown in the x–t plane. The dark lines show the shocks. The lighter lines show 1-characteristics and 2-characteristics.

are not impinging on the 1-shock as they should, an indication that this structure is not stable to small perturbations and that this shock should be replaced by a rarefaction wave.

This suggests the following criterion for judging whether a given weak solution is in fact the physically correct solution, a generalization of the Lax Entropy Condition 11.1 to systems of equations.

Entropy Condition 13.1 (Lax). A discontinuity separating states q_l and q_r, propagating at speed s, satisfies the *Lax entropy condition* if there is an index p such that

$$\lambda^p(q_l) > s > \lambda^p(q_r), \tag{13.21}$$

so that p-characteristics are impinging on the discontinuity, while the other characteristics are crossing the discontinuity,

$$\begin{aligned} \lambda^j(q_l) < s \quad &\text{and} \quad \lambda^j(q_r) < s \qquad \text{for } j < p, \\ \lambda^j(q_l) > s \quad &\text{and} \quad \lambda^j(q_r) > s \qquad \text{for } j > p. \end{aligned} \tag{13.22}$$

In this definition we assume the eigenvalues are ordered so that $\lambda^1 < \lambda^2 < \cdots < \lambda^m$ in each state.

This condition can be shown to be correct for strictly hyperbolic conservation laws in which each field is genuinely nonlinear (as defined in Section 13.8.4). For the nonstrictly-hyperbolic case shocks may instead be overcompressive or undercompressive with a different number of characteristics impinging, as described in Section 16.2.

For the shallow water equations there is a simple criterion that can be applied to determine which parts of each Hugoniot locus give physically correct shock waves satisfying the Lax entropy condition. Across a 1-shock connecting q_l to a state q_m, we require that the characteristic velocity $\lambda^1 = u - \sqrt{gh}$ must decrease. In conjunction with the Rankine–Hugoniot condition, it can be shown that this implies that h must increase, so we require

$h_m > h_l$. Similarly, a 2-shock connecting q_m to q_r satisfies the Lax entropy condition if $h_m > h_r$. Note from Figure 13.5 and Figure 13.7 that this also means that fluid particles experience an *increase* in depth as they pass through a shock. This is similar to the physical entropy condition for gas dynamics, that gas particles must experience an increase in physical entropy as they pass through a shock wave.

Figure 13.10 shows the portions of each Hugoniot locus along which the entropy condition is satisfied as solid lines. These are simply the portions along which h is increasing. The portions indicated by dashed lines are states that can be connected by a discontinuity that satisfies the Rankine–Hugoniot condition, but not the entropy condition.

We see from Figure 13.10(b) that the solution to the dam-break Riemann problem consisting of two shocks fails to satisfy the entropy condition. Instead we must find a solution to the Riemann problem that consists of a 1-rarefaction and a 2-shock. In the next section we investigate rarefaction waves and will see that the Hugoniot locus through q_l must be replaced by a different curve, the integral curve of r^1. The intersection of this curve with the 1-shock Hugoniot locus will give the correct intermediate state q_m, as described in Section 13.9.

The shallow water equations also possess a convex entropy function $\eta(q)$; see Exercise 13.6. From this it follows that Godunov's method will converge to the physically correct solution, if the correct entropy solution to each Riemann problem is used; see Section 12.11.1.

13.8 Simple Waves and Rarefactions

Solutions to a hyperbolic system of m equations are generally quite complicated, since at any point there are typically m waves passing by, moving at different speeds, and what we observe is some superposition of these waves. In the nonlinear case the waves are constantly interacting with one another as well as deforming separately, leading to problems that generally cannot be solved analytically. It is therefore essential to look for special situations in which a single wave from one of the characteristic families can be studied in isolation. A shock wave consisting of piecewise constant data (satisfying the Rankine–Hugoniot conditions across the discontinuity) is one important example that was investigated in the previous section.

In this section we will investigate solutions that are smoothly varying (rather than discontinuous) but which also have the property that they are associated with only one characteristic family of the system. Such waves are called *simple waves*. These have already been introduced for linear systems in Section 3.4. In the linear case simple waves have fixed shape and propagate with fixed speed according to a scalar advection equation. In the nonlinear case they will deform due to the nonlinearity, but their evolution can be modeled by *scalar nonlinear equations*.

In particular, the *centered rarefaction waves* that arise in the solution to Riemann problems for nonlinear systems are simple waves, but these are just one special case. They are special in that they also have the property that they are similarity solutions of the equations and are constant along every ray $x/t = $ constant. They arise naturally from Riemann problems because of the special data used, which varies only at a single point $x = 0$, and hence all variation in the solution flows out from the point $x = t = 0$. Recall from Section 11.10 that if $q(x, t) = \tilde{q}(x/t)$ then the function $\tilde{q}(\xi)$ must satisfy (11.26),

$$f'(\tilde{q}(x/t))\,\tilde{q}'(x/t) = \left(\frac{x}{t}\right)\tilde{q}'(x/t). \qquad (13.23)$$

For a scalar equation we could cancel $\tilde{q}'(x/t)$ from this equation. For a system of equations $\tilde{q}'$ is a vector and (13.23) requires that it be an eigenvector of the Jacobian matrix $f'(\tilde{q}(x/t))$ for each value of x/t. We will see how to determine this function for centered rarefaction waves and employ it in solving the Riemann problem, but first we study some more basic ideas.

Again we will concentrate on the shallow water equations to illustrate this theory. For this system of two equations we can easily draw diagrams in the two-dimensional state space that help to elucidate the theory.

13.8.1 Integral Curves

Let $\tilde{q}(\xi)$ be a smooth curve through state space parameterized by a scalar parameter ξ. We say that this curve is an *integral curve of the vector field* r^p if at each point $\tilde{q}(\xi)$ the tangent vector to the curve, $\tilde{q}'(\xi)$, is an eigenvector of $f'(\tilde{q}(\xi))$ corresponding to the eigenvalue $\lambda^p(\tilde{q}(\xi))$. If we have chosen some particular set of eigenvectors that we call $r^p(q)$, e.g., (13.10) for the shallow water equations, then $\tilde{q}'(\xi)$ must be some scalar multiple of the particular eigenvector $r^p(\tilde{q}(\xi))$,

$$\tilde{q}'(\xi) = \alpha(\xi)r^p(\tilde{q}(\xi)). \qquad (13.24)$$

The value of $\alpha(\xi)$ depends on the particular parameterization of the curve and on the normalization of r^p, but the crucial idea is that the tangent to the curve is always in the direction of the appropriate eigenvector r^p evaluated at the point on the curve.

Example 13.8. Figure 13.12 shows integral curves of r^1 and r^2 for the shallow water equations of Section 13.1, for which the eigenvectors are given by (13.10). As an example of how these curves can be determined, consider r^1 and set $\alpha(\xi) \equiv 1$, which selects one particular parameterization for which the formulas are relatively simple. Then (13.24) reduces to

$$\tilde{q}'(\xi) = r^1(\tilde{q}(\xi)) = \begin{bmatrix} 1 \\ \tilde{q}^2/\tilde{q}^1 - \sqrt{g\tilde{q}^1} \end{bmatrix} \qquad (13.25)$$

by using (13.10). This gives two ordinary differential equations for the two components of $\tilde{q}(\xi)$:

$$(\tilde{q}^1)' = 1 \qquad (13.26)$$

and

$$(\tilde{q}^2)' = \tilde{q}^2/\tilde{q}^1 - \sqrt{g\tilde{q}^1}. \qquad (13.27)$$

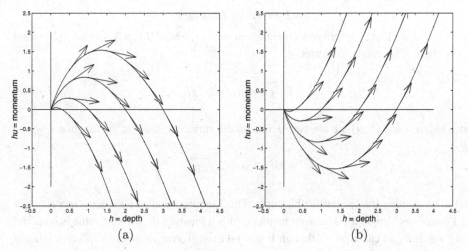

Fig. 13.12. (a) Integral curves of the eigenvector field r^1 for the shallow water equations. The eigenvector $r^1(q)$ evaluated at any point on a curve is tangent to the curve at that point. (b) Integral curves for r^2.

If we set

$$\tilde{q}^1(\xi) = \xi, \tag{13.28}$$

then (13.26) is satisfied. Note that since the first component of q is h, this means we are parameterizing the integral curve by depth. With this choice of $\tilde{q}^1$, the second equation (13.27) becomes

$$(\tilde{q}^2)' = \tilde{q}^2/\xi - \sqrt{g\xi}. \tag{13.29}$$

If we fix one point (h_*, u_*) on the integral curve and require that $\tilde{q}^2(h_*) = h_* u_*$, then solving the differential equation (13.29) with this initial value yields the solution

$$\tilde{q}^2(\xi) = \xi u_* + 2\xi(\sqrt{gh_*} - \sqrt{g\xi}). \tag{13.30}$$

Plotting $(\tilde{q}^1(\xi), \tilde{q}^2(\xi))$ from (13.28) and (13.30) gives the curves shown in Figure 13.12(a). Since ξ is just the depth h, we can also state more simply that the integral curves of r^1 have the functional form

$$hu = hu_* + 2h(\sqrt{gh_*} - \sqrt{gh}). \tag{13.31}$$

In terms of the velocity instead of the momentum, we can rewrite this as

$$u = u_* + 2(\sqrt{gh_*} - \sqrt{gh}). \tag{13.32}$$

Similarly, the integral curve of r^2 passing through the point (h_*, u_*) can be shown to have the form

$$u = u_* - 2(\sqrt{gh_*} - \sqrt{gh}). \tag{13.33}$$

13.8.2 Riemann Invariants

The expression (13.32) describes an integral curve of r^1, where (h_*, u_*) is an arbitrary point on the curve. This can be rewritten as

$$u + 2\sqrt{gh} = u_* + 2\sqrt{gh_*}.$$

Since (h_*, u_*) and (h, u) are any two points on the curve, we see that the function

$$w^1(q) = u + 2\sqrt{gh} \tag{13.34}$$

has the same value at all points on this curve. This function is called a *Riemann invariant* for the 1-family, or simply a 1-Riemann invariant. It is a function of q whose value is invariant along any integral curve of r^1, though it will take a different value on a different integral curve.

Similarly, from (13.33) we see that

$$w^2(q) = u - 2\sqrt{gh} \tag{13.35}$$

is a 2-Riemann invariant, a function whose value is constant along any integral curve of r^2.

If $\tilde{q}(\xi)$ represents a parameterization of any integral curve of r^p, then since $w^p(q)$ is constant along $\tilde{q}(\xi)$ as ξ varies, we must have $\frac{d}{d\xi} w^p(\tilde{q}(\xi)) = 0$. Expanding this out gives

$$\nabla w^p(\tilde{q}(\xi)) \cdot \tilde{q}'(\xi) = 0,$$

where ∇w^p is the gradient of w^p with respect to q. By (13.24) this gives

$$\nabla w^p(\tilde{q}(\xi)) \cdot r^p(\tilde{q}(\xi)) = 0. \tag{13.36}$$

This relation must hold at any point on every integral curve, and hence in general $\nabla w^p \cdot r^p = 0$ everywhere. This gives another way to characterize a p-Riemann invariant – it is a function whose gradient is orthogonal to the eigenvector r^p at each point q.

We can also view the integral curves of r^p as being level sets of the function $w^p(q)$, so that the curves in Figure 13.12(a), for example, give a contour plot of $w^1(q)$. The gradient of $w^1(q)$ is orthogonal to the contour lines, as expressed by (13.36).

Note that all the integral curves shown in Figure 13.12 appear to meet at the origin $h = hu = 0$. This may seem odd in that they are level sets of a function w^p that takes different values on each curve. But note that each w^p involves the velocity $u = (hu)/h$, which has different limiting values depending on how the point $h = hu = 0$ is approached. The integral curves would look different if plotted in the h–u plane; see Exercise 13.2.

For a system of $m > 2$ equations, the integral curves of r^p will still be curves through the m-dimensional state space, and can be determined by solving the system (13.24) with $\alpha(\xi) \equiv 1$, for example. This is now a system of m ODEs. In general there will now be $m - 1$ distinct functions $w^p(\xi)$ that are p-Riemann invariants for each family p.

13.8.3 Simple Waves

A *simple wave* is a special solution to the conservation law in which

$$q(x, t) = \tilde{q}(\xi(x, t)), \tag{13.37}$$

where $\tilde{q}(\xi)$ traces out an integral curve of some family of eigenvectors r^p and $\xi(x, t)$ is a smooth mapping from (x, t) to the parameter ξ. This means that all states $q(x, t)$ appearing in the simple wave lie on the same integral curve. Note that any p-Riemann invariant is constant throughout the simple wave.

Of course not every function of the form (13.37) will satisfy the conservation law. The function $\xi(x, t)$ must be chosen appropriately. We compute

$$q_t = \tilde{q}'(\xi(x, t)) \xi_t \quad \text{and} \quad q_x = \tilde{q}'(\xi(x, t)) \xi_x,$$

so to satisfy $q_t + f(q)_x = 0$ we must have

$$\xi_t \, \tilde{q}'(\xi) + \xi_x \, f'(\tilde{q}(\xi)) \tilde{q}'(\xi) = 0.$$

Since $\tilde{q}'(\xi)$ is always an eigenvector of $f'(\tilde{q}(\xi))$, this yields

$$[\xi_t + \xi_x \lambda^p(\tilde{q}(\xi))] \, \tilde{q}'(\xi) = 0,$$

and hence the function $\xi(x, t)$ must satisfy

$$\xi_t + \lambda^p(\tilde{q}(\xi)) \xi_x = 0. \tag{13.38}$$

Note that this is a scalar quasilinear hyperbolic equation for ξ.

In particular, if we choose initial data $q(x, 0)$ that is restricted entirely to this integral curve, so that

$$q(x, 0) = \tilde{q}(\overset{\circ}{\xi}(x))$$

for some smooth choice of $\overset{\circ}{\xi}(x)$, then (13.37) will be a solution to the conservation law for $t > 0$ provided that $\xi(x, t)$ solves (13.38) with initial data $\xi(x, 0) = \overset{\circ}{\xi}(x)$, at least for as long as the function $\xi(x, t)$ remains smooth. In a simple wave the nonlinear system of equations reduces to the scalar nonlinear equation (13.38) for $\xi(x, t)$.

Since (13.38) is nonlinear, the smooth solution may eventually break down at some time T_b. At this time a shock forms in $q(x, t)$ and the solution is in general no longer a simple wave for later times. States q that do not lie on the same integral curve will typically appear near the shock.

In the special case where $\lambda^p(\tilde{q}(\xi(x, 0)))$ is monotonically increasing in x, the characteristics will always be spreading out and a smooth solution will exist for all time. This is a pure rarefaction wave. If the characteristic speed $\lambda^p(\tilde{q}(\xi(x, 0)))$ is decreasing in x over some region, then a compression wave arises that will eventually break.

Note that $\xi(x, t)$ is constant along characteristic curves of the equation (13.38), curves $X(t)$ that satisfy $X'(t) = \lambda^p(\tilde{q}(\xi(X(t), t)))$. Since ξ is constant on this curve, so is $X'(t)$, and hence the characteristics are straight lines. Along these characteristics the value of

$q(x, t)$ is also constant, since $q(x, t)$ is determined by (13.37) and ξ is constant. Hence a simple wave behaves exactly like the solution to a scalar conservation law, as described in Chapter 11, up to the time it breaks.

For the special case of a linear hyperbolic system, simple waves satisfy scalar advection equations (λ^p is constant, independent of $\tilde{q}(\xi)$), and the theory just developed agrees with what was presented in Section 3.4.

13.8.4 Genuine Nonlinearity and Linear Degeneracy

For the shallow water equations, the characteristic speed $\lambda^p(\tilde{q}(\xi))$ varies monotonically as we move along an integral curve. We will verify this in Example 13.9, but it can be seen in Figure 13.13, where contours of λ^p are plotted along with a typical integral curve. This monotonicity is analogous to the situation for a scalar conservation law with a *convex* flux function $f(q)$, in which case the single characteristic speed $\lambda^1(q) = f'(q)$ is monotonic in q. As discussed in Section 16.1, solutions to the scalar conservation law can be considerably more complex if f is nonconvex. The same is true for systems of equations, but for many physical systems we have a property analogous to convexity. If $\lambda^p(\tilde{q}(\xi))$ varies monotonically with ξ along every integral curve, then we say that the pth field is *genuinely nonlinear*. Note that the variation of λ^p along the curve can be computed as

$$\frac{d}{d\xi}\lambda^p(\tilde{q}(\xi)) = \nabla\lambda^p(\tilde{q}(\xi)) \cdot \tilde{q}'(\xi). \tag{13.39}$$

Here $\nabla\lambda^p$ is the gradient vector obtained by differentiating the scalar $\lambda^p(q)$ with respect to each component of the vector q. The quantity in (13.39) must be nonzero everywhere if the field is to be genuinely nonlinear. If the value of (13.39) is positive, then the characteristic

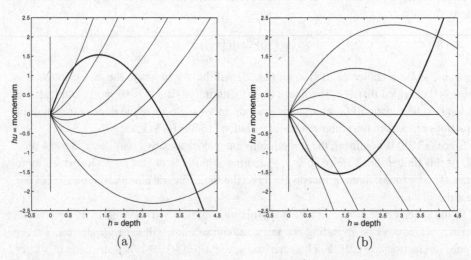

| | (a) | | (b) |

Fig. 13.13. (a) A typical integral curve of the eigenvector field r^1 for the shallow water equations is shown as the heavy line. The other lines are contour lines of $\lambda^1(q)$, curves along which λ^1 is constant. The curves shown are for values $\lambda^1 = 1, 0, -1, -1.5, -2, -2.5$ from top to bottom. Note that λ^1 varies monotonically along the integral curve. (b) An integral curve for r^2 and contours of $\lambda^2(q)$ for values $\lambda^2 = 2.5, 2, 1.5, 1, 0, -1$ from top to bottom.

speed is increasing with ξ and if it is negative the speed is decreasing. If this derivative were zero at some point, then the characteristic speed would be locally constant and characteristics would be essentially parallel as in a linear problem. The property of genuine nonlinearity insures that this linear situation never occurs and characteristics are always compressing or expanding as q varies.

Since $\tilde{q}'(\xi)$ is in the direction $r^p(\tilde{q}(\xi))$ by (13.24), we see from (13.39) that the pth field is genuinely nonlinear if

$$\nabla \lambda^p(q) \cdot r^p(q) \neq 0 \qquad (13.40)$$

for all q. Note that for a scalar problem $\lambda^1(q) = f'(q)$ and we can take $r^1(q) \equiv 1$, so that (13.40) reduces to the convexity requirement $f''(q) \neq 0$.

At the other extreme, in some physical systems there are characteristic fields for which the relation

$$\nabla \lambda^p(q) \cdot r^p(q) \equiv 0 \qquad (13.41)$$

holds for all q. This means that λ^p is identically constant along each integral curve. A trivial example occurs in a constant-coefficient linear hyperbolic system, in which case λ^p is constant everywhere and $\nabla \lambda^p(q) \equiv 0$. But in nonlinear systems it may also happen that some field satisfies the relation (13.41) even though λ^p takes different values along different integral curves. This happens if the integral curves of r^p are identical to the contour lines of λ^p. A field satisfying (13.41) is said to be *linearly degenerate*. Through a simple wave in such a field the characteristics are parallel, as in a linear system, rather than compressing or expanding. See Section 13.12 and Section 14.9 for examples of linearly degenerate fields.

13.8.5 Centered Rarefaction Waves

A *centered rarefaction wave* is a special case of a simple wave in a genuinely nonlinear field, in which $\xi(x, t) = x/t$, so that the solution is constant on rays through the origin. A centered rarefaction wave has the form

$$q(x, t) = \begin{cases} q_l & \text{if } x/t \leq \xi_1, \\ \tilde{q}(x/t) & \text{if } \xi_1 \leq x/t \leq \xi_2, \\ q_r & \text{if } x/t \geq \xi_2, \end{cases} \qquad (13.42)$$

where q_l and q_r are two points on a single integral curve with $\lambda^p(q_l) < \lambda^p(q_r)$. This condition is required so that characteristics spread out as time advances and the rarefaction wave makes physical sense. (The picture should look like Figure 11.4(a), not Figure 11.4(b).)

For a centered rarefaction wave a particular parameterization of the integral curve is forced upon us by the fact that we set $\xi = x/t$. Rewriting this as $x = \xi t$, we see that the value $\tilde{q}(\xi)$ observed along the ray $x/t = \xi$ is propagating at speed ξ, which suggests that ξ at each point on the integral curve must be equal to the characteristic speed $\lambda^p(\tilde{q}(\xi))$ at this point. This is confirmed by noting that (13.38) in this case becomes

$$-\frac{x}{t^2} + \lambda^p(\tilde{q}(x/t)) \left(\frac{1}{t} \right) = 0$$

and hence

$$\frac{x}{t} = \lambda^p(\tilde{q}(x/t)). \tag{13.43}$$

In particular, the left edge of the rarefaction fan should be the ray $x/t = \lambda^p(q_l)$ so that $\xi_1 = \lambda^p(q_l)$ in (13.42), while the right edge should be the ray $x/t = \lambda^p(q_r)$ so that $\xi_2 = \lambda^p(q_r)$. We thus have

$$\begin{aligned} \xi_1 &= \lambda^p(q_l), & \tilde{q}(\xi_1) &= q_l, \\ \xi_2 &= \lambda^p(q_r), & \tilde{q}(\xi_2) &= q_r. \end{aligned} \tag{13.44}$$

To determine how $\tilde{q}(\xi)$ varies for $\xi_1 < \xi < \xi_2$ through the rarefaction wave (13.42), rewrite (13.43) as

$$\xi = \lambda^p(\tilde{q}(\xi)) \tag{13.45}$$

and differentiate this with respect to ξ to obtain

$$1 = \nabla\lambda^p(\tilde{q}(\xi)) \cdot \tilde{q}'(\xi). \tag{13.46}$$

Using (13.24) in (13.46) gives

$$1 = \alpha(\xi)\nabla\lambda^p(\tilde{q}(\xi)) \cdot r^p(\tilde{q}(\xi)),$$

and hence

$$\alpha(\xi) = \frac{1}{\nabla\lambda^p(\tilde{q}(\xi)) \cdot r^p(\tilde{q}(\xi))}. \tag{13.47}$$

Using this in (13.24) gives a system of ODEs for $\tilde{q}(\xi)$:

$$\tilde{q}'(\xi) = \frac{r^p(\tilde{q}(\xi))}{\nabla\lambda^p(\tilde{q}(\xi)) \cdot r^p(\tilde{q}(\xi))}. \tag{13.48}$$

This system must be solved over the interval $\xi_1 \le \xi \le \xi_2$ using either of the conditions in (13.44) as an initial condition. Note that the denominator is nonzero provided that λ^p is monotonically varying. A rarefaction wave would not make sense past a point where the denominator vanishes. In particular, if the pth field is genuinely nonlinear, then the denominator is always nonzero by (13.40).

Example 13.9. For the shallow water equations we have

$$\lambda^1 = u - \sqrt{gh} = q^2/q^1 - \sqrt{gq^1},$$

$$\nabla\lambda^1 = \begin{bmatrix} -q^2/(q^1)^2 - \frac{1}{2}\sqrt{g/q^1} \\ 1/q^1 \end{bmatrix},$$

$$r^1 = \begin{bmatrix} 1 \\ q^2/q^1 - \sqrt{gq^1} \end{bmatrix}, \tag{13.49}$$

and hence

$$\nabla \lambda^1 \cdot r^1 = -\frac{3}{2}\sqrt{g/q^1}, \tag{13.50}$$

so that the equations (13.48) become

$$\tilde{q}' = -\frac{2}{3}\sqrt{\tilde{q}^1/g}\left[\begin{array}{c} 1 \\ \tilde{q}^2/\tilde{q}^1 - \sqrt{g\tilde{q}^1} \end{array}\right]. \tag{13.51}$$

The first equation of this system is

$$\tilde{h}'(\xi) = -\frac{2}{3}\sqrt{\tilde{h}(\xi)/g}.$$

The general solution is

$$\tilde{h} = \frac{1}{9g}(A - \xi)^2, \tag{13.52}$$

for some constant A. This constant must be chosen so that (13.44) is satisfied, i.e., so that $\tilde{h} = h_l$ at $\xi = u_l - \sqrt{gh_l}$ and also $\tilde{h} = h_r$ at $\xi = u_r - \sqrt{gh_r}$. Provided that q_l and q_r both lie on an integral curve of r^1, as they must if they can be joined by a centered rarefaction wave, we can satisfy both of these conditions by taking

$$A = u_l + 2\sqrt{gh_l} = u_r + 2\sqrt{gh_r}. \tag{13.53}$$

Recall that $u + 2\sqrt{gh}$ is a 1-Riemann invariant, which has the same value at all points on the integral curve. We see that $\tilde{h}$ varies quadratically with $\xi = x/t$ through a rarefaction wave (13.42).

Once we know h as a function of ξ, we can use the formula (13.33), which holds through any simple wave, to determine how u varies through the rarefaction wave. (i.e., we use the fact that the Riemann invariant is constant). Note that since we know the relation between h and u from having previously found the Riemann invariants, we do not need to solve both the ODEs in the system (13.51). We have chosen the simpler one to solve. This trick is often useful for other systems as well.

Note that (13.50) is nonzero for all physically meaningful states $q^1 = h > 0$, showing that this field is genuinely nonlinear. The expressions for the 2-characteristic field are very similar (with only a few minus signs changed), and this field is also genuinely nonlinear; see Exercise 13.3.

13.8.6 The All-Rarefaction Riemann Solution

Now suppose we wish to solve a Riemann problem for which we know the solution consists of two rarefaction waves, as in the following example.

Example 13.10. Again consider the Riemann problem for the shallow water equations with data (13.13), but now take $u_l < 0$. This corresponds to two streams of water that are moving apart from one another. Again the solution will be symmetric but will consist of

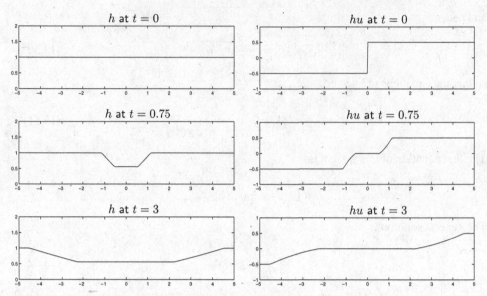

Fig. 13.14. Solution of the Riemann problem for the shallow water equations with $u_l = -u_r < 0$. [claw/book/chap13/tworaref]

two rarefaction waves as shown in Figure 13.14. (Looking at only half the domain gives the solution to the boundary-value problem with water flowing away from a wall.)

To solve this Riemann problem, we can proceed in a manner similar to what we did in Section 13.7.1 for the all-shock solution. There is an integral curve of r^1 through q_l consisting of all states that can be connected to q_l by a 1-rarefaction, and an integral curve of r^2 through q_r consisting of all states that can be connected to q_r by a 2-rarefaction. These are illustrated in Figure 13.15(a) for the Riemann data

$$u_l = -0.5, \quad u_r = 0.5, \quad \text{and} \quad h_l = h_r = 1. \tag{13.54}$$

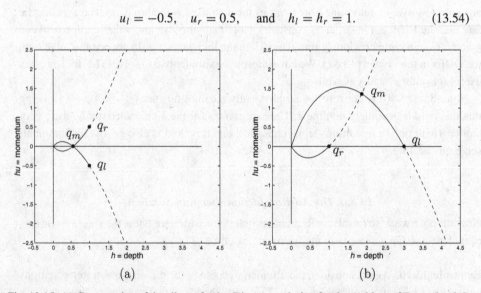

Fig. 13.15. (a) Construction of the all-rarefaction Riemann solution for the problem of Example 13.5. (b) The physically incorrect all-rarefaction Riemann solution for the dam-break problem of Example 13.4.

The intermediate state q_m in the Riemann solution must lie on both of these curves, and hence is at the intersection as shown in Figure 13.15(a). For this particular example q_m lies on the h-axis due to symmetry. In general we can find the intersection by using the fact that q_m must lie on the curve described by (13.32) with $q_* = q_l$ and on the curve described by (13.33) with $q_* = q_r$, so

$$u_m = u_l + 2(\sqrt{gh_l} - \sqrt{gh_m}),$$
$$u_m = u_r - 2(\sqrt{gh_r} - \sqrt{gh_m}). \tag{13.55}$$

This is a system of two nonlinear equations for h_m and u_m. Equating the right-hand sides gives a single equation for h_m, which can be explicitly solved to obtain

$$h_m = \frac{1}{16g}[u_l - u_r + 2(\sqrt{gh_l} + \sqrt{gh_r})]^2. \tag{13.56}$$

This is valid provided that the expression being squared is nonnegative. When it reaches zero, the outflow is sufficiently great that the depth h_m goes to zero. (See Exercise 13.2.)

For the symmetric data (13.54) used in Figures 13.14 and 13.15(a), the expression (13.56) gives $h_m = (4\sqrt{g}-1)^2/16g = 9/16$, since we use $g = 1$. Then either equation from (13.55) gives $u_m = 0$.

The integral curves in Figure 13.15(a) are shown partly as dashed lines. For a given state q_l only some points on the integral curve of r^1 can be connected to q_l by a rarefaction wave that makes physical sense, since we are assuming q_l is the state on the *left* of the rarefaction wave. We must have $\lambda^1(q_l) < \lambda^1(q)$ for all states q in the rarefaction wave, and hence q must lie on the portion of the integral curve shown as a solid line (see Figure 13.13(a)). Similarly, if q_r is the state to the right of a 2-rarefaction, then states q in the rarefaction must satisfy $\lambda^2(q) < \lambda^2(q_r)$ and must lie on the solid portion of the integral curve for r^2 sketched through q_r in Figure 13.15(a). For the data shown in this figure, there is a state q_m that can be connected to both q_l and q_r by physically correct rarefaction waves, and the Riemann solution consists of two rarefactions as illustrated in Figure 13.14.

For other data this might not be the case. Figure 13.15(b) shows the data for the dam-break Riemann problem of Example 13.4, with $h_l = 3$, $h_r = 1$, and $u_l = u_r = 0$. We can still use (13.56) to compute an intermediate state q_m lying at the intersection of the integral curves, as illustrated in Figure 13.15(b), but the resulting 2-wave does not make physical sense as a rarefaction wave, since $\lambda^2(q_m) > \lambda^2(q_r)$. This wave would overturn, as illustrated in Figure 13.16.

Compare Figure 13.15(b) with Figure 13.10(b), where we found an all-shock solution to this same Riemann problem. In that case the 2-shock was acceptable, but the 1-shock failed to satisfy the entropy condition. The correct solution consists of a 1-rarefaction and a 2-shock as shown in Figure 13.5 and determined in the next section.

13.9 Solving the Dam-Break Problem

We now illustrate how to construct the general solution to a nonlinear Riemann problem, using the theory of shock waves and rarefaction waves developed above.

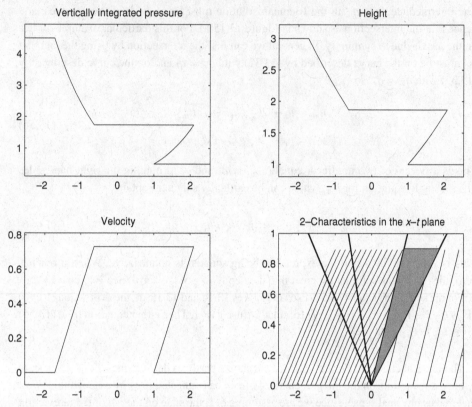

Fig. 13.16. Unphysical solution to the dam-break Riemann problem for the shallow water equations corresponding to the solution found in Figure 13.15(b). The 2-characteristics are shown in the lower right plot. Note that each point in the shaded region lies on three distinct 2-characteristics.

The dam-break Riemann problem for the shallow water equations (introduced in Example 13.4) has a solution that consists of a 1-rarefaction and a 2-shock, as illustrated in Figure 13.5. In Figure 13.10 we saw how to construct a weak solution to this problem that consists of two shock waves, one of which does not satisfy the Lax entropy condition. In Figure 13.15 we found an all-rarefaction solution to this problem that is not physically realizable. To find the correct solution we must determine an intermediate state q_m that is connected to q_l by a 1-rarefaction wave and simultaneously is connected to q_r by a 2-shock wave. The state q_m must lie on an integral curve of r^1 passing through q_l, so by (13.32) we must have

$$u_m = u_l + 2(\sqrt{gh_l} - \sqrt{gh_m}). \tag{13.57}$$

It must also lie on the Hugoniot locus of 2-shocks passing through q_r, so by (13.19) it must satisfy

$$u_m = u_r + (h_m - h_r)\sqrt{\frac{g}{2}\left(\frac{1}{h_m} + \frac{1}{h_r}\right)}. \tag{13.58}$$

We can easily eliminate u_m from these two equations and obtain a single nonlinear equation

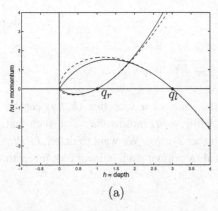
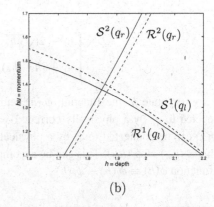

(a) (b)

Fig. 13.17. (a) The Hugoniot loci from Figure 13.10 together with the integral curves. (b) Close-up of the region where the curves intersect. $\mathcal{S}^1$: Entropy-violating 1-shocks; $\mathcal{R}^1$: 1-rarefactions; $\mathcal{S}^2$: 2-shocks; $\mathcal{R}^2$: unphysical 2-rarefactions.

to solve for h_m. The structure of the rarefaction wave connecting q_l to q_m can then be determined using the theory of Section 13.8.5.

Note that the intermediate state q_m resulting from this procedure will be slightly different from that obtained in either Figure 13.10 or Figure 13.15, since the Hugoniot loci are different from the integral curves. This is illustrated in Figure 13.17, where the curves from Figures 13.10 and 13.15 are plotted together. Figure 13.17(b) shows a close-up near the points of intersection of these curves. The correct solution to the dam-break Riemann problem has the intermediate state at the point where the two solid lines cross.

Note that both sets of curves are tangent to the eigenvector $r^1(q_l)$ at q_l and to $r^2(q_r)$ at q_r. Moreover, it can be shown that the Hugoniot locus and the integral curve through a given point have the same curvature at that point, and so the curves are really quite similar near that point. (See, e.g., Lax [263].) How rapidly the curves diverge from one another typically depends on how nonlinear the system is. For a linear system, of course, the integral curves and Hugoniot loci are identical, each being straight lines in the directions of the constant eigenvectors. Even for nonlinear systems the Hugoniot loci *may* be identical to the integral curves, though this is not the usual situation. See Exercise 13.12 for one example, and Temple [447] for some general discussion of such systems, which are often called *Temple-class systems*.

13.10 The General Riemann Solver for Shallow Water Equations

For the dam-break problem we know that the 1-wave is a rarefaction while the 2-wave is a shock, leading to the system of equations (13.57) and (13.58) to solve for h_m and u_m. For general values of q_l and q_r we might have any combination of shocks and rarefactions in the two families, depending on the specific data. To find the state q_m in general we can define two functions ϕ_l and ϕ_r by

$$\phi_l(h) = \begin{cases} u_l + 2(\sqrt{gh_l} - \sqrt{gh}) & \text{if } h < h_l, \\ u_l - (h - h_l)\sqrt{\frac{g}{2}\left(\frac{1}{h} + \frac{1}{h_l}\right)} & \text{if } h > h_l, \end{cases}$$

and

$$\phi_r(h) = \begin{cases} u_r - 2(\sqrt{gh_r} - \sqrt{gh}) & \text{if } h < h_r, \\ u_r + (h - h_r)\sqrt{\frac{g}{2}\left(\frac{1}{h} + \frac{1}{h_r}\right)} & \text{if } h > h_r. \end{cases}$$

For a given state h, the function $\phi_l(h)$ returns the value of u such that (h, hu) can be connected to q_l by a physically correct 1-wave, while $\phi_r(h)$ returns the value such that (h, hu) can be connected to q_r by a physically-correct 2-wave. We want to determine h_m so that $\phi_l(h_m) = \phi_r(h_m)$. This can be accomplished by applying a nonlinear root finder to the function $\phi(h) \equiv \phi_l(h) - \phi_r(h)$.

13.11 Shock Collision Problems

In a scalar equation, such a Burgers equation, when two shock waves collide, they simply merge into a single shock wave with a larger jump. For a system of equations, the result of a shock collision is not so simple, even if the two shocks are in the same characteristic family. The result will include a stronger shock in this same family, but the collision will typically also introduce waves in the other families. For example, consider initial data for the shallow water equations consisting of the three states shown in Figure 13.18(a). These three states all lie on the Hugoniot locus $\mathcal{S}^2(q_2)$, so there is a 2-shock connecting q_1 to q_2 and a slower 2-shock connecting q_2 to q_3. (Note that the shock speed $s = [\![hu]\!]/[\![h]\!]$ is given by the slope of the line joining the two points.) If we solve the shallow water equations with data

$$q(x, 0) = \begin{cases} q_1 & \text{if } x < x_1, \\ q_2 & \text{if } x_1 \le x \le x_2, \\ q_3 & \text{if } x > x_2 \end{cases} \tag{13.59}$$

for some initial shock locations $x_1 < x_2$, then the solution consists of the these two shocks, which eventually collide at some point x_c. At the time t_c when they collide, the state q_2

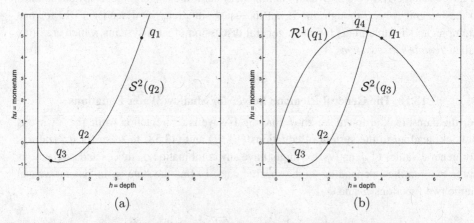

(a)   (b)

Fig. 13.18. States arising in the collision of two 2-shocks for the shallow water equations. (a) Initial states q_1 and q_3 are each connected to q_2 by a 2-shock. (b) After collision, solving the Riemann problem between q_3 and q_1 gives a new state q_4 and a reflected 1-wave.

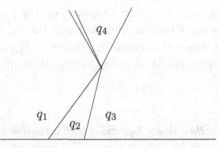

Fig. 13.19. Collision of two 2-shocks giving rise to a 1-rarefaction and a 2-shock, as seen in the x–t plane.

disappears and the solution has the form

$$q(x, t_c) = \begin{cases} q_1 & \text{if } x < x_c, \\ q_3 & \text{if } x > x_c. \end{cases} \tag{13.60}$$

To determine the solution beyond this time, note that this has the form of Riemann-problem data with left state q_1 and right state q_3. The Riemann solution is not a single 2-shock, because q_1 will not lie on the Hugoniot locus $\mathcal{S}^2(q_3)$. Instead, a 1-wave must be introduced to connect q_1 to a new state q_4 that lies on this Hugoniot locus, as illustrated in Figure 13.18(b). We see that the 1-wave must be a rarefaction wave, since $h_4 < h_1$ and h_4 is determined by the intersection of the integral curve $\mathcal{R}^1(q_1)$ with the Hugoniot locus $\mathcal{S}^2(q_3)$. (Note that q_2 does lie on the Hugoniot locus $\mathcal{S}^2(q_3)$, since q_3 lies on the Hugoniot locus $\mathcal{S}^2(q_2)$.)

A view of this collision in the x–t plane is seen in Figure 13.19. To view a numerical solution of this collision, see [claw/book/chap13/collide].

13.12 Linear Degeneracy and Contact Discontinuities

The shallow water equations are a system of two equations for which both characteristic fields are genuinely nonlinear, as defined in Section 13.8.4. A smooth simple wave in one of these fields will always distort via compression or expansion as characteristics converge or diverge. For the Riemann problem, each wave will be either a single shock or a rarefaction wave. Genuine nonlinearity of the pth field requires that the eigenvalue λ^p be monotonically varying as we move along an integral curve of r^p, and hence that $\nabla \lambda^p(q) \cdot r^p(q)$ be nonzero everywhere.

We now consider the opposite extreme, a field in which $\nabla \lambda^p(q) \cdot r^p(q)$ is identically zero for all q, so that λ^p is constant along each integral curve of r^p (but may take different values on different integral curves). Such a field is called *linearly degenerate*, since simple waves in which the variation of q is only in this field behave like solutions to linear hyperbolic equations. Since λ^p is constant throughout the wave, it simply translates with this constant speed without distorting. If the initial data is a jump discontinuity, with q_l and q_r both lying on a single integral curve of this field, then the solution will consist of this discontinuity propagating at the constant speed λ^p associated with this integral curve. Hence the Hugoniot locus for this field agrees with the integral curve. Such a discontinuity is not a shock wave, however, since the characteristic speed $\lambda^p(q_l) = \lambda^p(q_r)$ on each side agrees with the propagation speed of the wave. Characteristics are parallel to the wave in the x–t plane rather than impinging on it. Waves of this form are generally called *contact discontinuities*, for reasons that will become apparent after considering the simple example in the next section.

13.12.1 Shallow Water Equations with a Passive Tracer

Again consider the shallow water equations, but now suppose we introduce some dye into the water in order to track its motion. Let $\phi(x, t)$ represent the concentration of this passive tracer, measured in units of mass or molarity per unit volume, so that it is a *color* variable as described in Chapter 9. Then values of ϕ move with the fluid velocity u and are constant along particle paths, and ϕ satisfies the color equation (9.12),

$$\phi_t + u\phi_x = 0. \tag{13.61}$$

Since ϕ measures a passive tracer that is assumed to have no influence on the fluid dynamics, it simply satisfies this linear advection equation with the variable coefficient $u(x, t)$, which can be obtained by first solving the shallow water equations. However, to illustrate linearly degenerate fields we can couple this equation into the shallow water equations and obtain an augmented system of three conservation laws.

We first rewrite (13.61) as a conservation law by instead considering the conserved quantity $h\phi W$, where h is the depth of the water and W is the width of the channel modeled in our one-dimensional equations. This is needed solely for dimensional reasons, and we can take $W = 1$ in the appropriate length units and consider $h\phi$ as the conserved quantity, measuring mass or molarity per unit length. This is conserved in one dimension with the flux $uh\phi$, so $h\phi$ satisfies the conservation law

$$(h\phi)_t + (uh\phi)_x = 0. \tag{13.62}$$

Note that differentiating this out gives

$$h_t\phi + h\phi_t + (hu)_x\phi + (hu)\phi_x = 0,$$

and using $h_t + (hu)_x = 0$ allows us to relate this to the color equation (13.61). Numerically one can work directly with (13.61) rather than (13.62) since the wave-propagation algorithms do not require that all equations be in conservation form, and there are often advantages to doing so (see Section 16.5).

For our present purposes, however, we wish to investigate the mathematical structure of the resulting system of conservation laws. Augmenting the shallow water equations with (13.62) gives the system $q_t + f(q)_x = 0$ with

$$q = \begin{bmatrix} h \\ hu \\ h\phi \end{bmatrix} = \begin{bmatrix} q^1 \\ q^2 \\ q^3 \end{bmatrix}, \qquad f(q) = \begin{bmatrix} hu \\ hu^2 + \frac{1}{2}gh^2 \\ uh\phi \end{bmatrix} = \begin{bmatrix} q^2 \\ (q^2)^2/q^1 + \frac{1}{2}g(q^1)^2 \\ q^2q^3/q^1 \end{bmatrix}. \tag{13.63}$$

The Jacobian is now

$$
f'(q) = \begin{bmatrix} 0 & 1 & 0 \\ -(q^2)^2/(q^1)^2 + gq^1 & 2q^2/q^1 & 0 \\ -q^2 q^3/(q^1)^2 & q^3/q^1 & q^2/q^1 \end{bmatrix} = \begin{bmatrix} 0 & 1 & 0 \\ -u^2 + gh & 2u & 0 \\ -u\phi & \phi & u \end{bmatrix}. \quad (13.64)
$$

This is a block lower-triangular matrix, and the eigenvalues are given by those of the two blocks. The upper 2×2 block is simply the Jacobian matrix (13.8) for the shallow water equations, and has eigenvalues $u \pm \sqrt{gh}$. The lower 1×1 block yields the additional eigenvalue u. The eigenvectors are also easy to compute from those of (13.8), and we find that

$$
\lambda^1 = u - \sqrt{gh}, \qquad \lambda^2 = u, \qquad \lambda^3 = u + \sqrt{gh},
$$

$$
r^1 = \begin{bmatrix} 1 \\ u - \sqrt{gh} \\ \phi \end{bmatrix}, \qquad r^2 = \begin{bmatrix} 0 \\ 0 \\ 1 \end{bmatrix}, \qquad r^3 = \begin{bmatrix} 1 \\ u + \sqrt{gh} \\ \phi \end{bmatrix}. \qquad (13.65)
$$

The fact that the scalar ϕ is essentially decoupled from the shallow water equations is clearly apparent. Fields 1 and 3 correspond to the nonlinear waves in the shallow water equations and involve ϕ only because the conserved quantity $h\phi$ has a jump discontinuity where there is a jump in h. The tracer concentration ϕ is continuous across these waves. Field 2 carries a jump in ϕ alone, since the first two components of r^2 are 0. The speed of the 2-wave, $\lambda^2 = u$, depends on the shallow water behavior, just as we expect.

If we considered very small variations in h and u, we could linearize this system and obtain a form of the acoustics equations coupled with the advection equation for ϕ. This linear system has already been considered in Section 3.10 and shows the same basic structure.

For the nonlinear system, fields 1 and 3 are still genuinely nonlinear, as in the standard shallow water equations, but field 2 is linearly degenerate, since it corresponds to the linear advection equation. This is easily verified by computing

$$
\nabla \lambda^2 = \begin{bmatrix} -u/h \\ 1/h \\ 0 \end{bmatrix}
$$

and observing that $\nabla \lambda^2 \cdot r^2 \equiv 0$. (Recall that $\nabla \lambda^2$ means the gradient of $u = (hu)/h = q^2/q^1$ with respect to q.)

Any variation in ϕ will simply be advected with velocity u. In general the shape of ϕ may distort, since $u(x, t)$ may vary in the solution to the shallow water equations. However, if we consider a simple wave in Field 2, then variations in q can occur only along an integral curve of r^2. This vector always points in the ϕ-direction in the three-dimensional state space, and integral curves are straight lines in this direction with no variation in h or hu. So in particular u is constant along any integral curve of r^2, and simple waves consist of arbitrary variations in ϕ being carried along in water of constant depth moving at constant speed u. These are, of course, special solutions of the augmented shallow water equations.

13.12.2 The Riemann Problem and Contact Discontinuities

Now consider the Riemann problem for the augmented shallow water system (13.63), with piecewise constant data having an arbitrary jump between states q_l and q_r (which allows arbitrary jumps in h, u, and ϕ). It is clear how to solve this Riemann problem. Since ϕ does not affect h or u, the procedure of Section 13.9 can be used to determine the 1-wave and 3-wave. Each is a shock or rarefaction wave, as in the standard shallow water equations, and there is no variation in ϕ across either of these waves. The 1-wave is moving into the fluid on the left, in which $\phi \equiv \phi_l$ while the 3-wave is moving into the fluid on the right, in which $\phi \equiv \phi_r$. Between these two waves the velocity u_m is constant (obtained as in Section 13.9), and a 2-wave is now introduced with velocity $\lambda^2 = u_m$. Across this wave, $h = h_m$ and $u = u_m$ are constant while ϕ jumps from ϕ_l to ϕ_r. This wave has a simple physical interpretation: it marks the boundary between water that was initially to the left of the interface and water that was initially to the right. This is clear because ϕ satisfies the color equation and hence is constant on particle paths.

This 2-wave is called a *contact discontinuity* because it marks the point at which the two distinct fluids (e.g., with different colors) are in contact with one another. Figure 13.20

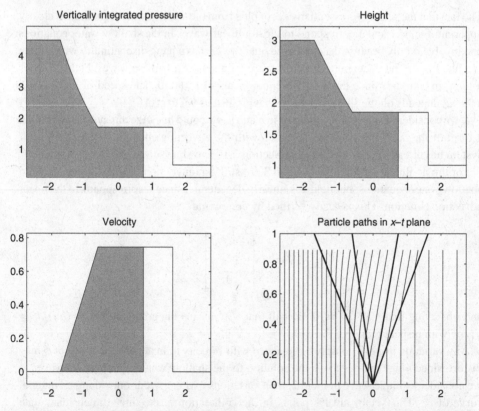

Fig. 13.20. Structure of the similarity solution of the dam-break Riemann problem for the augmented shallow water equations with $u_l = u_r = 0$. The depth h, velocity u, and vertically integrated pressure are displayed as functions of x/t. The value of ϕ is indicated by coloring the water darker where $\phi = \phi_l$. The structure in the x–t plane is also shown, and particle paths are indicated for a set of particles with the spacing between particles inversely proportional to the depth. The contact discontinuity is now indicated along with the rarefaction and shock.

shows one example for the dam-break problem of Example 13.4. This looks exactly like Figure 13.5 except that we have now also indicated the variation in ϕ by coloring the water that was initially behind the dam (i.e., in the region $x < 0$) darker. The contact discontinuity is indicated in the plot of particle paths, and it is clear that this wave separates water initially behind the dam from water initially downstream.

Since the characteristic velocity $\lambda^2 = u$ agrees with the particle velocity, we can also interpret this plot of particle paths as a plot of the 2-characteristics for the augmented system. These characteristics cross the 1-wave and 3-wave and are parallel to the 2-wave, as expected because this field is linearly degenerate.

The Euler equations of gas dynamics, discussed in the next chapter, also have a linearly degenerate field corresponding to contact discontinuities. For general initial data the Riemann solution may contain a jump in density across the surface where the two initial gases are in contact.

Exercises

13.1.   (a)   Consider an integral curve of r^1 for the shallow water equations, as illustrated in Figure 13.13(a), for example. Show that the slope tangent to this curve in the q^1–q^2 plane at any point is equal to λ^1 at that point. ($q^1 = h$ and $q^2 = hu$.)

       (b)   Consider the Hugoniot locus for 1-shocks for the shallow water equations, as illustrated in Figure 13.9(a), for example. Show that if q_l and q_r are two points lying on this curve (and hence connected by a 1-shock) then the slope of the secant line connecting these points is equal to the shock speed s.

13.2.   The graphs of Figure 13.15 show the h–hu plane. The curves look somewhat different if we instead plot them in the h–u plane.

       (a)   Redraw Figure 13.15(a) in the h–u plane.

       (b)   Draw a similar figure for $h_l = h_r = 1$ and $-u_l = u_r = 1.9$.

       (c)   Draw a similar figure for $h_l = h_r = 1$ and $-u_l = u_r = 2.1$. In this case the Riemann solution contains a region in which $h = 0$: dry land between the two outgoing rarefaction waves.

13.3.   Repeat the computations of Example 13.9 to determine the form of 2-rarefactions in the shallow water equations and show that this field is genuinely nonlinear.

13.4.   For the shallow water equations, show that when a 1-shock collides with a 2-shock the result is a new pair of shocks. Exhibit the typical solution in the phase plane and the x–t plane.

13.5.   In the shallow water equations, is it possible for two 2-rarefactions to collide with each other?

13.6.   For the shallow water equations, the total energy can be used as an entropy function in the mathematical sense. This function and the associated entropy flux are given by

$$\eta(q) = \frac{1}{2}hu^2 + \frac{1}{2}gh^2,$$
$$\psi(q) = \frac{1}{2}hu^3 + gh^2u. \tag{13.66}$$

Verify this by showing the following.

(a) $\eta(q)$ is convex: Show that the Hessian matrix $\eta''(q)$ is positive definite.

(b) $\eta(q)_t + \psi(q)_x = 0$ for smooth solutions: Verify that (11.47) holds, $\psi'(q) = \eta'(q)f'(q)$.

13.7.   Consider the *p*-system (described in Section 2.13),

$$v_t - u_x = 0,$$
$$u_t + p(v)_x = 0,$$

where $p(v)$ is a given function of v.

(a)   Compute the eigenvalues of the Jacobian matrix, and show that the system is hyperbolic provided $p'(v) < 0$.

(b)   Use the Rankine–Hugoniot condition to show that a shock connecting $q = (v, u)$ to some fixed state $q^* = (v^*, u^*)$ must satisfy

$$u = u_* \pm \sqrt{-\left(\frac{p(v) - p(v_*)}{v - v_*}\right)}(v - v_*).   \qquad (13.67)$$

(c)   What is the propagation speed for such a shock? How does this relate to the eigenvalues of the Jacobian matrix computed in part (a)?

(d)   Plot the Hugoniot loci for the point $q_* = (1, 1)$ over the range $-3 \le v \le 5$ for each of the following choices of $p(v)$. (Note that these are not physically reasonable models for pressure as a function of specific volume!)
   (i) $p(v) = -e^v$,
   (ii) $p(v) = -(2v + 0.1e^v)$,
   (iii) $p(v) = -2v$.

(e)   Determine the two-shock solution to the Riemann problem for the *p*-system with $p(v) = -e^v$ and data

$$q_l = (1, 1), \qquad q_r = (3, 4).$$

Do this in two ways:
   (i)   Plot the relevant Hugoniot loci, and estimate where they intersect.
   (ii)   Set up and solve the proper scalar nonlinear equation for v_m. You might use the Matlab command `fzero` or write your own Newton solver.

(f)   Does the Riemann solution found in the previous part satisfy the Lax entropy condition? Sketch the structure of the solution in the x–t plane, showing also some sample 1-characteristics and 2-characteristics.

(g)   For the given left state $q_l = (1, 1)$, in what region of the phase plane must the right state q_r lie in order for the two-shock Riemann solution to satisfy the Lax entropy condition?

13.8.   Consider the *p*-system of Exercise 13.7, and take $p(v) = -e^v$.

(a)   Follow the procedure of Section 13.8.1 to show that along any integral curve of r^1 the relation

$$u = u_* - 2\left(e^{v_*/2} - e^{v/2}\right)$$

must hold, where (v_*, u_*) is a particular point on the integral curve. Conclude that

$$w^1(q) = u - 2e^{v/2}$$

is a 1-Riemann invariant for this system.

(b) Follow the procedure of Section 13.8.5 to show that through a centered rarefaction wave

$$\tilde{u}(\xi) = A - 2\xi,$$

where

$$A = u_l - 2e^{v_l/2} = u_r - 2e^{v_r/2},$$

and determine the form of $\tilde{v}(\xi)$.

(c) Show that this field is genuinely nonlinear for all q.

(d) Determine the 2-Riemann invariants and the form of a 2-rarefaction.

(e) Suppose arbitrary states q_l and q_r are specified and we wish to construct a Riemann solution consisting of two "rarefaction waves" (which might not be physically realizable). Determine the point $q_m = (v_m, u_m)$ where the two relevant integral curves intersect.

(f) What conditions must be satisfied on q_l and q_r for this to be the physically correct solution to the Riemann problem?

13.9. For the general p-system of Exercise 13.7, determine the condition on the function $p(v)$ that must be satisfied in order for both fields to be genuinely nonlinear for all q.

13.10. Consider the equations (2.97) modeling a one-dimensional slice of a nonlinear elastic solid. Suppose the stress–strain relation $\sigma(\epsilon)$ has the shape indicated in Figure 2.3(a). Is the system genuinely nonlinear in this case?

13.11. The variable-coefficient scalar advection equation $q_t + (u(x)q)_x = 0$ studied in Section 9.4 can be viewed as a hyperbolic system of two equations,

$$\begin{aligned} q_t + (uq)_x &= 0, \\ u_t &= 0, \end{aligned} \qquad (13.68)$$

where we now view $u(x, t) \equiv u(x)$ as a second component of the system.

(a) Determine the eigenvalues and eigenvectors of the Jacobian matrix for this system.

(b) Show that both fields are linearly degenerate, and that in each field the integral curves and Hugoniot loci coincide. Plot the integral curves of each field in the q–u plane.

(c) Indicate the structure of a general Riemann solution in the q–u plane for the case $u_l, u_r > 0$. Relate this to Figure 9.1.

(d) Note that this system fails to be strictly hyperbolic along $u = 0$. Can the Riemann problem be solved if $u_l < 0$ and $u_r > 0$? If $u_l > 0$ and $u_r < 0$? (See Section 16.4.2 for more discussion of such problems.)

13.12. Consider the system

$$v_t + [vg(v, \phi)]_x = 0,$$
$$\phi_t + [\phi g(v, \phi)]_x = 0,$$

(13.69)

where $g(v, \phi)$ is a given function. Systems of this form arise in two-phase flow. As a simple example, take $g(v, \phi) = \phi^2$ and assume $\phi > 0$.

(a) Determine the eigenvalues and eigenvectors for this system and show that the first field is linearly degenerate while the second field is genuinely nonlinear.

(b) Show that the Hugoniot locus of any point q_* consists of a pair of straight lines, and that each line is also the integral curve of the corresponding eigenvector.

(c) Obtain the general solution to the Riemann problem consisting of one shock and one contact discontinuity. Show that this solution satisfies the Lax Entropy Condition 11.1 if and only if $\phi_l \geq \phi_r$.

Gas Dynamics and the Euler Equations

A brief introduction to gas dynamics was given in Section 2.6, where the equations for conservation of mass and momentum were stated. In this chapter we will consider the energy equation and the equation of state in more detail, along with a few other quantities of physical (and mathematical) significance, such as the entropy. We will also look at some special cases, including isentropic and isothermal flows, where systems of two equations are obtained. These provide simplified examples that can be used to illustrate the nonlinear theory.

The derivations here will be very brief, with an emphasis on the main ideas without a detailed description of the physics. More thorough introductions may be found in several books on hyperbolic equations, such as [92], [156], [420], [486], or on gas dynamics, such as [58], [70], [297], [405], [474].

Recall that ρ is the density, u the velocity, E the total energy, and p the pressure of the gas. In Section 2.6 the continuity equation

$$\rho_t + (\rho u)_x = 0 \tag{14.1}$$

was derived from the more fundamental integral form, obtained by integrating the density over a test section $[x_1, x_2]$ and using the mass flux ρu at each end. More generally, for any quantity z that is advected with the flow there will be a contribution to the flux for z of the form zu. Thus, the momentum equation has a contribution of the form $(\rho u)u = \rho u^2$, and the energy equation has a flux contribution Eu.

14.1 Pressure

The velocity $u(x, t)$ used in gas dynamics is a macroscopic quantity that represents an average over a huge number of gas molecules in the neighborhood of the point x. The advective momentum flux ρu^2 mentioned above is a macroscopic flux, the same as what would be obtained if all nearby gas molecules were moving at this same speed. In reality they are not, however, and this microscopic variation leads to an additional microscopic contribution to the momentum flux, which is given by the *pressure*. To understand this, consider a gas that is "at rest", with macroscopic velocity $u = 0$. The individual molecules are still moving, however, at least if the temperature is above absolute zero. One way to calculate the pressure at a point x_1 is to think of inserting an imaginary wall in our one-dimensional tube of gas at this point and calculate the force (per unit area) exerted on each side of this wall by the gas. These forces will normally be of equal magnitude and opposite in sign, and arise from the molecules on each side colliding with the wall and bouncing off. Of course, if the wall is not really there, then the molecules don't bounce off it. Instead, molecules that approach x_1 from $x < x_1$ with positive momentum move into the test section $[x_1, x_2]$ and by doing so will increase the momentum in this section, and hence make a positive contribution to the flux of momentum past x_1. Likewise molecules that approach x_1 from $x > x_1$ must have negative momentum, so as they pass the point they are *removing negative momentum* from the test section, which also makes a positive contribution to the momentum flux past this point. Note that the two contributions to momentum flux due to molecules moving rightwards and leftwards do not cancel out, but rather add together to give a net positive flux of momentum. The momentum in the section $[x_1, x_2]$ is thus always increasing due to the flux at the left boundary, resulting simply from the random movement of molecules near x_1. This seems rather paradoxical, but note that if the pressure is constant over the test section then there will be an equal flux of momentum at x_2, out of the section, so that the total momentum in the section remains constant (and equal to zero, since $u = 0$). Note that if the pressure differs between x_1 and x_2, then there will be a net nonzero flux of momentum into this section, and hence an apparent macroscopic acceleration of the gas. (Individual molecules are not actually accelerated, however – it's just that the distribution of velocities observed in the test section is changing.)

In general the pressure gives the microscopic momentum flux that must be added to the advective flux ρu^2 to obtain the total momentum flux,

$$\text{momentum flux} = \rho u^2 + p, \tag{14.2}$$

leading to the integral conservation law

$$\frac{d}{dt} \int_{x_1}^{x_2} \rho(x, t) u(x, t)\, dx = -[\rho u^2 + p]_{x_1}^{x_2}. \tag{14.3}$$

The differential form of the momentum equation is

$$(\rho u)_t + (\rho u^2 + p)_x = 0. \tag{14.4}$$

There may also be external forces acting on the gas, such as gravity, that do cause the acceleration of individual molecules. In this case the external force must be integrated over $[x_1, x_2]$ and this contribution added to (14.3). This leads to the addition of a source term on the right-hand side of (14.4); see Section 2.5 and Chapter 17.

14.2 Energy

The total energy E is often decomposed as

$$E = \rho e + \frac{1}{2}\rho u^2. \tag{14.5}$$

The term $\frac{1}{2}\rho u^2$ is the kinetic energy, while ρe is the internal energy. The variable e, internal energy per unit mass, is called the *specific internal energy*. (In general, "specific" means "per unit mass"). Internal energy includes translational, rotational, and vibrational energy and possibly other forms of energy in more complicated situations. In the Euler equations we assume that the gas is in local chemical and thermodynamic equilibrium and that the internal energy is a known function of pressure and density:

$$e = e(p, \rho). \tag{14.6}$$

This is the *equation of state* for the gas, which depends on the particular gas under study.

The total energy advects with the flow, leading to the macroscopic energy flux term Eu. In addition, the microscopic momentum flux measured by p leads to a flux in kinetic energy that is given by pu. In the absence of outside forces, the conservation law for total energy thus takes the differential form

$$E_t + [(E + p)u]_x = 0. \tag{14.7}$$

If outside forces act on the gas, then a source term must be included, since the total energy will be modified by work done on the gas.

14.3 The Euler Equations

Putting these equations together gives the system of *Euler equations*

$$\begin{bmatrix} \rho \\ \rho u \\ E \end{bmatrix}_t + \begin{bmatrix} \rho u \\ \rho u^2 + p \\ (E + p)u \end{bmatrix}_x = 0. \tag{14.8}$$

These are a simplification of the more realistic *Navier–Stokes* equations, which also include effects of fluid viscosity and heat conduction. The terms dropped involve second-order derivatives that would make the system parabolic rather than hyperbolic, and cause them to have smooth solutions for all time. However, when the viscosity and heat conductivity are very small, the vanishing-viscosity hyperbolic equations are a good approximation. The resulting discontinuous shock waves are good approximations to what is observed in reality – very thin regions over which the solution is rapidly varying. (In some cases viscous effects may be nonnegligible. For example, viscous boundary layers may have a substantial effect on the overall solution; see Section 21.8.4.)

To obtain a closed system of equations, we still need to specify the equation of state relating the internal energy to pressure and density.

14.4 Polytropic Ideal Gas

For an ideal gas, internal energy is a function of temperature alone, $e = e(T)$. The *temperature T* is related to p and ρ by the *ideal gas law*,

$$p = R\rho T \tag{14.9}$$

where R is a constant obtained by dividing the universal gas constant $\mathcal{R}$ by the molecular weight of the gas. To good approximation, the internal energy is simply proportional to the temperature,

$$e = c_v T, \tag{14.10}$$

where c_v is a constant known as the *specific heat at constant volume*. Such gases are called *polytropic*. If energy is added to a fixed volume of a polytropic gas, then the change in energy and change in temperature are related via

$$de = c_v dT. \tag{14.11}$$

On the other hand, if the gas is allowed to expand at constant pressure, not all of the energy goes into increasing the internal energy. The work done in expanding the volume $1/\rho$ by $d(1/\rho)$ is $p\,d(1/\rho)$, and we obtain another relation

$$de + p\,d(1/\rho) = c_p\,dT \tag{14.12}$$

or

$$d(e + p/\rho) = c_p\,dT, \tag{14.13}$$

where c_p is the *specific heat at constant pressure*. The quantity

$$h = e + p/\rho \tag{14.14}$$

is called the *(specific) enthalpy* of the gas. For a polytropic gas, c_p is constant, so that (14.13) yields

$$h = c_p T, \tag{14.15}$$

and the enthalpy is simply proportional to the temperature. Note that by the ideal gas law,

$$c_p - c_v = R. \tag{14.16}$$

The equation of state for a polytropic gas turns out to depend only on the *ratio of specific heats*, usually denoted by

$$\gamma = c_p/c_v. \tag{14.17}$$

This parameter is also often called the *adiabatic exponent*.

Internal energy in a molecule is typically split up between various degrees of freedom (translational, rotational, vibrational, etc.). How many degrees of freedom exist depends on the nature of the gas. The general *principle of equipartition of energy* says that the average energy in each of these is the same. Each degree of freedom contributes an average energy of $\frac{1}{2}kT$ per molecule, where k is *Boltzmann's constant*. This gives a total contribution of $\frac{\alpha}{2}kT$ per molecule if there are α degrees of freedom. Multiplying this by n, the number of molecules per unit mass (which depends on the gas), gives

$$e = \frac{\alpha}{2}nkT. \tag{14.18}$$

The product nk is precisely the gas constant R, so comparing this with (14.10) gives

$$c_v = \frac{\alpha}{2} R. \tag{14.19}$$

From (14.16) we obtain

$$c_p = \left(1 + \frac{\alpha}{2}\right) R, \tag{14.20}$$

and so

$$\gamma = c_p/c_v = \frac{\alpha + 2}{\alpha}. \tag{14.21}$$

Note that $T = p/R\rho$, so that

$$e = c_v T = \left(\frac{c_v}{R}\right) \frac{p}{\rho} = \frac{p}{(\gamma - 1)\rho} \tag{14.22}$$

by (14.16) and (14.17). Using this in (14.5) gives the common form of the *equation of state for an ideal polytropic gas*:

$$E = \frac{p}{\gamma - 1} + \frac{1}{2}\rho u^2. \tag{14.23}$$

An ideal gas with this equation of state is also sometimes called a *gamma-law gas*.

For a monatomic gas the only degrees of freedom are the three translational degrees, so $\alpha = 3$ and $\gamma = 5/3$. For a diatomic gas there are also two rotational degrees of freedom and $\alpha = 5$, so that $\gamma = 7/5 = 1.4$. Under ordinary circumstances air is composed primarily of N_2 and O_2, and so $\gamma \approx 1.4$.

14.5 Entropy

The fundamental thermodynamic quantity is the entropy. Roughly speaking, this measures the disorder in the system, and indicates the degree to which the internal energy is available for doing useful work. The greater the entropy, the less available the energy.

The specific entropy s (entropy per unit mass) is given by

$$s = c_v \log (p/\rho^\gamma) + \text{constant}. \tag{14.24}$$

This can be solved for p to give

$$p = \kappa e^{s/c_v} \rho^\gamma, \tag{14.25}$$

where κ is a constant.

We can manipulate the Euler equations to derive the relation

$$s_t + u s_x = 0, \tag{14.26}$$

which says that entropy is constant along particle paths in regions of smooth flow. In fact, (14.26) can be derived from fundamental principles, and this equation, together with the conservation of mass and momentum equations, gives an alternative formulation of the Euler equations for smooth flows (though not in conservation form):

$$\rho_t + (\rho u)_x = 0,$$
$$(\rho u)_t + (\rho u^2 + p)_x = 0, \tag{14.27}$$
$$s_t + u s_x = 0.$$

The equation of state in these variables gives p as a function of ρ and s, e.g. (14.25) for a polytropic gas.

From our standpoint the most important property of entropy is that in smooth flow it remains constant on each particle path, whereas if a particle crosses a shock, then the entropy may jump, but only to a *higher* value. This results from the fact that the viscous processes (molecular collisions) in the thin physical shock profile cause the entropy to increase. This gives the physical *entropy condition* for shocks. (The term "fluid particle" is used to mean an infinitesimal volume of fluid that nonetheless contains a huge number of molecules.)

Note that along a particle path in smooth flow, since s is constant, we find by (14.25) that

$$p = \hat{k}\rho^\gamma, \tag{14.28}$$

where $\hat{k} = \kappa e^{s/c_v}$ is a constant that depends only on the initial entropy of the particle. This explicit relation between density and pressure along particle paths is sometimes useful. Of course, if the initial entropy varies in space, then $\hat{k}$ will be different along different particle paths.

Note that it appears we can combine the first and third equations of (14.27) to obtain a conservation law for the entropy per unit volume, $S = \rho s$,

$$S_t + (uS)_x = 0. \tag{14.29}$$

This equation does hold for smooth solutions, but it does not follow from an integral conservation law that holds more generally, and in fact entropy is *not* conserved across shocks. Hence the apparent system of conservation laws obtained by replacing the third equation of (14.27) by (14.29) is not equivalent to the conservative Euler equations (14.8) for weak solutions, and would result in physically incorrect shock speeds.

14.5.1 Isentropic Flow

If we consider very small smooth perturbations around some background state (as in acoustics), then no shocks will form over reasonable time periods, and so we can use these nonconservative equations (14.27) and obtain the same results as with the conservative Euler equations. Moreover, since s simply advects with the flow, if s is initially uniform throughout the gas, then s will remain constant and we do not need to bother solving the third equation in (14.27). This justifies the use of the *isentropic* equations for small disturbances, which were presented in Section 2.3. Taking $s = $ constant in (14.25) gives the equation of state (14.28), as was introduced earlier in (2.35). The isentropic equations again are

$$\begin{bmatrix} \rho \\ \rho u \end{bmatrix}_t + \begin{bmatrix} \rho u \\ \rho u^2 + \hat{k}\rho^\gamma \end{bmatrix}_x = 0. \tag{14.30}$$

Recall that the sound speed c is given by

$$c = \sqrt{\gamma p/\rho}. \tag{14.31}$$

This was derived from the linearized equations using the ideal gas equation of state presented above. With a more general equation of state we can still study acoustics by linearizing for small perturbations. Assuming the entropy is constant then results in the more general expression for the sound speed,

$$c = \sqrt{\left.\frac{\partial p}{\partial \rho}\right|_{s=\text{const}}}. \tag{14.32}$$

The sound speed is computed in general from the equation of state $p = p(\rho, s)$ by taking the partial derivative with respect to ρ. This corresponds to the fact that within an acoustic wave the density and pressure vary, but the entropy does not, and the "stiffness" of the gas (i.e., its response to compression in the form of increased pressure) dictates the velocity with which an acoustic wave propagates.

Note that the isentropic equations are still nonlinear, and so in general we expect shocks to form if we take arbitrary data. In particular, if we look at a Riemann problem with discontinuous data, then we may have shocks in the solution immediately. What is the physical meaning of these shocks, in view of the fact that across a real gas-dynamic shock we know the entropy cannot remain constant?

In reducing the Euler equations to the isentropic equations, we have dropped the conservation-of-energy equation. If we study only flows for which the entropy is truly constant (no shocks), then this equation will be automatically satisfied. However, if we use the isentropic equations for a problem with shocks, then conservation of energy will not hold across the shock. Mathematically such weak solutions of the isentropic equations make perfectly good sense, but they no longer model reality properly, since they do not model conservation of energy. (However, if the shocks are weak enough, then very little entropy is produced physically and the "isentropic" shock may be a good approximation to reality.)

Another way to view this is by the following thought experiment. We could, in principle, create a physical shock wave across which there is no increase in entropy if we could find a way to reduce the entropy of each gas particle just after it passes through the shock. In principle we could accomplish this by doing some work on each fluid particle to eliminate the "disorder" created by the trauma of passing through the shock. Doing so would require an input of energy right at the shock. Hence across this hypothetical isentropic shock there must be a jump in the energy of the gas, reflecting the outside work that has been done on it.

Moreover, across an isentropic shock we see that the energy must jump to a *higher* value. This can be used as an admissibility criterion for shocks in the weak formulation of the isentropic equations. A shock is the correct vanishing-viscosity solution to the isentropic equations only if the energy *increases* across the shock. The *energy* can thus be used as an "entropy function" for the isentropic system of equations, in the sense introduced in Chapter 11 (though that terminology is particularly confusing here).

14.6 Isothermal Flow

Taking $\gamma = 1$ in the isentropic equations (14.30) gives a particularly simple set of equations. By (14.21), the case $\gamma = 1$ is not physically realizable but can be viewed as a limiting case as $\alpha \to \infty$, i.e., for very complex molecules with many degrees of freedom. Note that such gases also have large heat capacities c_v and c_p, meaning that it requires a substantial input of energy to change the temperature very much. In the limit $\gamma \to 1$ the gas becomes *isothermal*, with constant temperature.

One can also obtain the isothermal flow equations by considering an ordinary gas in a tube that is immersed in a bath at a constant temperature $\bar{T}$. If we assume that this bath maintains a constant temperature within the gas, then we again obtain isothermal flow within the gas.

In isothermal flow, the ideal gas law (14.9) reduces to

$$p = a^2 \rho, \tag{14.33}$$

where $a^2 \equiv R\bar{T}$ is a constant and a is the sound speed (which is constant in isothermal flow). Note that maintaining this constant temperature requires heat flux through the wall of the tube (to take away heat generated at a shock or supply heat to a rarefaction), and so energy is no longer conserved in the tube. But mass and momentum are still conserved, and these equations, together with the equation of state (14.33), lead to the *isothermal equations*,

$$\begin{bmatrix} \rho \\ \rho u \end{bmatrix}_t + \begin{bmatrix} \rho u \\ \rho u^2 + a^2 \rho \end{bmatrix}_x = 0. \tag{14.34}$$

Isothermal flow is also an appropriate model for some astrophysical problems, particularly when modeling shock waves traveling through low-density interstellar space. In many cases the temperature increase caused by a shock wave leads to a radiative loss of energy via electromagnetic waves (at the speed of light) and very little of this energy is reabsorbed by the gas nearby.

In practice the temperature of a gas will never stay exactly constant, but it may *relax* towards a constant temperature very quickly as energy flows in or out of the gas via radiation or other mechanisms. A better physical model can be obtained by considering the full Euler equations with a source term that models the flow of heat into or out of the gas. A discussion of this *relaxation system* can be found in Section 17.17.3.

The isothermal equations are a system of two conservation laws for which the Hugoniot loci and integral curves are easy to compute, similarly to what was done in Chapter 13 for the shallow water equations. These are worked out for isothermal flow in [281].

14.7 The Euler Equations in Primitive Variables

For smooth solutions it is possible to rewrite the Euler equations in various nonconservative forms that are sometimes easier to work with or more revealing than the conservative form

(14.8). One example is the system (14.27), which shows that entropy is constant along streamlines for smooth solutions.

Another form that is more comprehensible physically is obtained by working in the *primitive variables* ρ, u, and p instead of the conserved variables, since the density, velocity and pressure are more intuitively meaningful. (Indeed, when we plot solutions to the the the Euler equations it is generally these variables that are plotted, even if the calculation was done in terms of the conserved variables.)

From the mass and momentum conservation equations it is easy to derive the equations

$$\rho_t + u\rho_x + \rho u_x = 0 \qquad (14.35)$$

for the density and

$$u_t + uu_x + (1/\rho)p_x = 0 \qquad (14.36)$$

for the velocity. With more manipulations one can also derive the equation

$$p_t + \gamma p u_x + u p_x = 0 \qquad (14.37)$$

for a polytropic gas. These three equations yield the quasilinear hyperbolic system

$$\begin{bmatrix} \rho \\ u \\ p \end{bmatrix}_t + \begin{bmatrix} u & \rho & 0 \\ 0 & u & 1/\rho \\ 0 & \gamma p & u \end{bmatrix} \begin{bmatrix} \rho \\ u \\ p \end{bmatrix}_x = 0. \qquad (14.38)$$

This matrix has a considerably simpler form than the Jacobian matrix $f'(q)$ obtained from the conservative equations (14.8), which is given in (14.43) below. The eigenvalues and eigenvectors of the coefficient matrix in (14.38) are easily computed to be

$$\lambda^1 = u - c, \qquad \lambda^2 = u, \qquad \lambda^3 = u + c,$$

$$r^1 = \begin{bmatrix} -\rho/c \\ 1 \\ -\rho c \end{bmatrix}, \quad r^2 = \begin{bmatrix} 1 \\ 0 \\ 0 \end{bmatrix}, \quad r^3 = \begin{bmatrix} \rho/c \\ 1 \\ \rho c \end{bmatrix}, \qquad (14.39)$$

where c is the sound speed of the polytropic gas,

$$c = \sqrt{\frac{\gamma p}{\rho}}. \qquad (14.40)$$

We see a familiar pattern in these eigenvalues: information can advect at the fluid velocity or move as acoustic waves at speeds $\pm c$ relative to the gas. The ratio

$$M = |u|/c \qquad (14.41)$$

is called the *Mach number*. The flow is *transonic* at any point where M passes through 1.

Note that linearizing these equations about a state (ρ_0, u_0, p_0) gives a result that is easily related to the acoustic equations derived in Section 2.7, and the eigenvectors (14.39) have been normalized in a manner analogous to that chosen in (2.58). (Other normalizations can be used instead. In particular, multiplying r^1 and r^3 by c and r^2 by ρ would give a form where the physical units agree with those of the vector (ρ, u, p).)

From the above expressions we see that the Euler equations are hyperbolic provided ρ and p are positive. Moreover we can compute the gradients of the eigenvalues and find that the first and third characteristic field are genuinely nonlinear while the second field is linearly degenerate:

$$\nabla\lambda^1 = \begin{bmatrix} -\partial c/\partial\rho \\ 1 \\ -\partial c/\partial p \end{bmatrix} = \begin{bmatrix} c/2\rho \\ 1 \\ -c/2p \end{bmatrix} \implies \nabla\lambda^1 \cdot r^1 = \tfrac{1}{2}(\gamma+1),$$

$$\nabla\lambda^2 = \begin{bmatrix} 0 \\ 1 \\ 0 \end{bmatrix} \implies \nabla\lambda^2 \cdot r^2 = 0, \tag{14.42}$$

$$\nabla\lambda^3 = \begin{bmatrix} \partial c/\partial\rho \\ 1 \\ \partial c/\partial p \end{bmatrix} = \begin{bmatrix} -c/2\rho \\ 1 \\ c/2p \end{bmatrix} \implies \nabla\lambda^3 \cdot r^3 = \tfrac{1}{2}(\gamma+1).$$

Simple waves in the second characteristic field consist of variations in density that are advecting with constant speed u, since u and p must be constant in such a wave. Such waves are often called *entropy waves* because the entropy satisfies the advection equation (14.26) and varies along with the density if p is constant. For the Riemann problem, the second field corresponds to *contact discontinuities* as described in Section 13.12, across which the two initial gases are in contact.

Simple waves in the first or third family will deform, since λ^1 and λ^3 vary along the integral curves of these families, sharpening into shocks or spreading out as rarefactions.

14.8 The Riemann Problem for the Euler Equations

To discuss the Riemann problem, we must return to the conservative form of the equations, which are valid across shock waves. If we compute the Jacobian matrix $f'(q)$ from (14.8), with the polytropic equation of state (14.23), we obtain

$$f'(q) = \begin{bmatrix} 0 & 1 & 0 \\ \tfrac{1}{2}(\gamma-3)u^2 & (3-\gamma)u & \gamma-1 \\ \tfrac{1}{2}(\gamma-1)u^3 - uH & H-(\gamma-1)u^2 & \gamma u \end{bmatrix}, \tag{14.43}$$

where

$$H = \frac{E+p}{\rho} = h + \frac{1}{2}u^2 \tag{14.44}$$

is the *total specific enthalpy*. The eigenvalues are again

$$\lambda^1 = u-c, \quad \lambda^2 = u, \quad \lambda^3 = u+c, \tag{14.45}$$

as for the coefficient matrix resulting from the primitive equations. They agree because the two forms are equivalent and should yield the same characteristic speeds. The eigenvectors will appear different, of course, in these different variables. We have

$$r^1 = \begin{bmatrix} 1 \\ u - c \\ H - uc \end{bmatrix}, \quad r^2 = \begin{bmatrix} 1 \\ u \\ \frac{1}{2}u^2 \end{bmatrix}, \quad r^3 = \begin{bmatrix} 1 \\ u + c \\ H + uc \end{bmatrix}. \tag{14.46}$$

Note that in these variables

$$\nabla\lambda^2(q) = \begin{bmatrix} -u/\rho \\ 1/\rho \\ 0 \end{bmatrix}, \tag{14.47}$$

and so we again find that $\nabla\lambda^2 \cdot r^2 \equiv 0$ and the second field is linearly degenerate.

14.9 Contact Discontinuities

We can have neither rarefaction waves nor shocks in the 2-characteristic field. Instead we have *contact discontinuities*, which are linear discontinuities that propagate with speed equal to the characteristic speed λ^2 on each side.

Note that because $\lambda^2 = u$ is constant on the integral curves of r^2, and because r^2 depends only on u, the vector r^2 is itself constant on these curves, and hence the integral curves are straight lines in phase space. Moreover, these integral curves also form the Hugoniot loci for contact discontinuities. Along these curves u and p are constant; only ρ varies.

It may seem strange that this discontinuity can sustain a jump in density – it seems that the denser gas should try to expand into the thinner gas. But that's because our intuition tends to equate higher density with higher pressure. It is only a *pressure* difference that can provide the force for expansion, and here the pressures are equal.

We can achieve two different densities at the same pressure by taking gases at two different temperatures. In fact, from (14.9) it is clear that there must be a jump in temperature if there is a jump in density but not in pressure. There must also be a jump in entropy by (14.24). This explains why contact discontinuities do not appear in solutions to the isothermal or isentropic equations considered previously. In the reduction of the Euler equations to one of these systems of only two equations, it is this linearly degenerate characteristic field that disappears.

Also, in the shallow water equations, where the pressure is related to the depth h by (13.3), it is not possible to have a jump in depth without also having a jump in pressure. Hence we see contact discontinuities only if we introduce another passive tracer advected with the fluid, as we did in Section 13.12.

Note that in all these hyperbolic systems we are ignoring diffusive effects, such as molecular diffusion of a tracer or diffusion of heat in a gas. These effects would smear out contact discontinuities in reality. We are assuming the diffusion coefficients are sufficiently small that these effects are negligible over the time scales of interest.

14.10 Riemann Invariants

Recall that for each family, the Riemann invariants are functions of q that are constant along any integral curve of this family and hence are constant through any simple wave in this family. Knowing these functions is very helpful in constructing solutions to the Riemann problem.

Since u and p are both constant across a contact discontinuity, these functions of q are both Riemann invariants of the 2-family.

The entropy s satisfies the advection equation (14.26) and hence is constant along particle paths. It follows that s is constant through any rarefaction wave or other simple wave in the 1-family or 3-family, and hence entropy is a Riemann invariant for these families. There's also a second set of Riemann invariants for each of these families. All of the Riemann invariants for a polytropic ideal gas are summarized below:

$$
\begin{aligned}
&\text{1-Riemann invariants:}\quad s,\quad u+\frac{2c}{\gamma-1}, \\[2mm]
&\text{2-Riemann invariants:}\quad u,\quad p, \\[2mm]
&\text{3-Riemann invariants:}\quad s,\quad u-\frac{2c}{\gamma-1}.
\end{aligned}
\tag{14.48}
$$

14.11 Solution to the Riemann Problem

The solution to a Riemann problem typically has a contact discontinuity and two nonlinear waves, each of which may be either a shock or a rarefaction wave, depending on q_l and q_r. The structure of a typical Riemann solution is shown in Figure 14.1 (see also the examples in Section 14.13). The first and third characteristic fields for the Euler equations are genuinely nonlinear and have behavior similar to the two characteristic fields in the isothermal or isentropic equations, and also similar to what we have seen for the two fields in the shallow water equations in Chapter 13. The contact discontinuity is also sometimes called the *entropy wave*, since it carries a jump in entropy. The first and third wave families are called *acoustic waves*, since in the small-disturbance limit these reduce to acoustics equations.

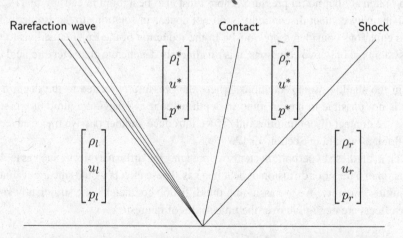

Fig. 14.1. Typical solution to the Riemann problem for the Euler equations.

Because u and p are constant across the contact discontinuity, it is often easier to work in the primitive variables (ρ, u, p) rather than $(\rho, \rho u, E)$, although of course the jump conditions must be determined using the conserved variables. The resulting Hugoniot locus and integral curves can be transformed into (ρ, u, p) space.

If the Riemann data is (ρ_l, u_l, p_l) and (ρ_r, u_r, p_r), then the two new constant states that appear in the Riemann solution will be denoted by $q_l^* = (\rho_l^*, u^*, p^*)$ and $q_r^* = (\rho_r^*, u^*, p^*)$. (See Figure 14.1.) Note that across the 2-wave we know there is a jump only in density.

Solution of the Riemann problem proceeds in principle just as in the previous chapters. Given the states q_l and q_r in the phase space, we need to determine the two intermediate states in such a way that q_l and q_l^* are connected by a 1-wave, q_l^* and q_r^* are connected by a 2-wave, and finally q_r^* and q_r are connected by a 3-wave. We need to consider three families of curves in the three-dimensional state space and find the appropriate intersections.

This seems difficult, but we can take advantage of the fact that we know the 2-wave will be a contact discontinuity across which u and p are constant to make the problem much simpler. Instead of considering the full three-dimensional (ρ, u, p) phase space, consider the p–u plane, and project the integral curves and Hugoniot loci for the 1-waves and 3-waves onto this plane. In particular, project the locus of all states that can be connected to q_l by a 1-wave (entropy satisfying shocks or rarefactions) onto this plane, and also the locus of all states that can be connected to q_r by a 3-wave. This gives Figure 14.2.

We see in this example that we can go from q_l (or actually, the projection of q_l) to q^* by a 1-rarefaction and from q^* to q_r by a 3-shock. The problem with this construction, of course, is that these curves are really curves in three-space, and the mere fact that their projections intersect does not mean the original curves intersect. However, the curve $R_1(q_l)$ must go through some state $q_l^* = (\rho_l^*, u^*, p^*)$ for some ρ_l^* (so that its projection onto the u–p plane is (u^*, p^*)). Similarly, the curve $S_3(q_r)$ must pass through some state $q_r^* = (\rho_r^*, u^*, p^*)$. But these two states differ only in ρ, and hence can be connected by a 2-wave (contact discontinuity). We have thus achieved our objective. Note that this technique depends on the fact that *any* jump in ρ is allowed across the contact discontinuity.

Based on the given state q_l, we can find a function $u = \phi_l(p)$ that describes how u varies as we adjust p in a state (p, u) that can be connected to q_l by a 1-shock or 1-rarefaction. For

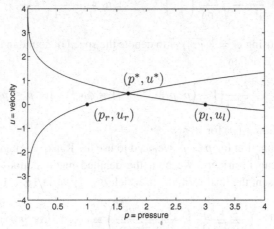

Fig. 14.2. Projection of shock and rarefaction curves onto the two-dimensional p–u plane, and determination of (p^*, u^*), for Sod's Riemann problem discussed in Section 14.13.

$p < p_l$ this function is defined by the integral curve of r^1 (projected to the p–u plane), since such states can be connected to q_l by a 1-rarefaction. For $p > p_l$ this function is defined by the Hugoniot locus, since these states can be connected by a shock. Similarly, for a given state q_r we can find a function $u = \phi_r(p)$ that describes how u varies as we adjust p in a state (p, u) that can be connected to q_r by a 3-shock or 3-rarefaction.

To solve the Riemann problem we need only solve for a pressure p^* for which $\phi_l(p^*) = \phi_r(p^*)$, which is a scalar nonlinear equation for p^*. In general this must be solved by an iterative method. Once p^* is known, u^*, ρ_l^*, and ρ_r^* are easily determined. Godunov first proposed a numerical method based on the solution of Riemann problems and presented one such iterative method in his paper [157] (also described in §12.15 of [369]). Chorin [67] describes an improvement of this method, and many other variants have been developed more recently.

We now briefly summarize how the functions $\phi_l(p)$ and $\phi_r(p)$ are determined. First consider the case of 1-rarefactions through q_l. Since the entropy s is a 1-Riemann invariant, we know that p/ρ^γ is constant through any rarefaction wave. This allows us to determine how ρ varies with p through a 1-rarefaction:

$$\rho = (p/p_l)^{1/\gamma} \rho_l. \tag{14.49}$$

We also know that the other 1-Riemann invariant from (14.48) is constant through any rarefaction wave, so

$$u + \frac{2}{\gamma - 1}\sqrt{\frac{\gamma p}{\rho}} = u_l + \frac{2}{\gamma - 1}\sqrt{\frac{\gamma p_l}{\rho_l}}. \tag{14.50}$$

We can use (14.49) to eliminate ρ on the left hand side and obtain

$$u + \frac{2}{\gamma - 1}\sqrt{\frac{\gamma p}{\rho_l}\left(\frac{p_l}{p}\right)^{1/\gamma}} = u_l + \frac{2}{\gamma - 1}\sqrt{\frac{\gamma p_l}{\rho_l}}$$

or

$$u + \frac{2}{\gamma - 1}\sqrt{\frac{\gamma p_l}{\rho_l}\left(\frac{p}{p_l}\right)^{(\gamma-1)/\gamma}} = u_l + \frac{2}{\gamma - 1}\sqrt{\frac{\gamma p_l}{\rho_l}}.$$

Rearranging this and using $c_l = \sqrt{\gamma p_l/\rho_l}$ to denote the speed of sound in the left state, we can solve this for u and obtain

$$u = u_l + \frac{2c_l}{\gamma - 1}\left[1 - (p/p_l)^{(\gamma-1)/(2\gamma)}\right] \equiv \phi_l(p) \quad \text{for } p \leq p_l. \tag{14.51}$$

This defines the function $\phi_l(p)$ for $p \leq p_l$.

To determine this function for $p > p_l$ we need to use the Rankine–Hugoniot conditions instead of the Riemann invariants. We omit the detailed manipulations (see [420], for example) and just present the final formula, in which $\beta = (\gamma + 1)/(\gamma - 1)$:

$$u = u_l + \frac{2c_l}{\sqrt{2\gamma(\gamma - 1)}}\left(\frac{1 - p/p_l}{\sqrt{1 + \beta p/p_l}}\right) \equiv \phi_l(p) \quad \text{for } p \geq p_l. \tag{14.52}$$

For each point (p, u) on the Hugoniot locus, there is a unique density ρ associated with it, given by

$$\rho = \left(\frac{1 + \beta p/p_l}{p/p_l + \beta}\right) \rho_l. \tag{14.53}$$

Similarly, the function $\phi_r(p)$ can be determined for a given state q_r by the same procedure. We obtain

$$u = u_r - \left(\frac{2c_r}{\gamma - 1}\right) \left(1 - (p/p_r)^{(\gamma-1)/(2\gamma)}\right) \equiv \phi_r(p) \quad \text{for } p \leq p_r, \tag{14.54}$$

and

$$u = u_r - \frac{2c_r}{\sqrt{2\gamma(\gamma - 1)}} \left(\frac{1 - p/p_r}{\sqrt{1 + \beta p/p_r}}\right) \equiv \phi_r(p) \quad \text{for } p \geq p_r. \tag{14.55}$$

The corresponding density for points on this 3-locus is given by

$$\rho = \left(\frac{1 + \beta p/p_r}{p/p_r + \beta}\right) \rho_r. \tag{14.56}$$

It is these functions $\phi_l(p)$ and $\phi_r(p)$ that are plotted in Figure 14.2 (for the particular case of the Sod's Riemann problem described in the Section 14.13). An iterative procedure can be used to determine the intersection (p^*, u^*).

Note that the formulas (14.55) and (14.56) are useful in setting initial data for numerical tests if we wish to specify data that corresponds to a single shock wave (a 3-shock in this case). We can choose the right state and the pressure p_l as we please, and then determine u_l and ρ_l using these formulas.

14.12 The Structure of Rarefaction Waves

In Section 14.11 we saw how to determine the Riemann solution for the Euler equations (with the polytropic equation of state) in the sense of finding the intermediate states q_l^* and q_r^*. To fully specify the Riemann solution we must also determine the structure of any rarefaction waves, i.e., determine ρ, u, and p as functions of $\xi = x/t$.

Suppose there is a 1-rarefaction connecting q_l to a state q_l^*. Then at each point in the rarefaction, $\xi = \lambda^1(q) = u - c$, and hence the sound speed c is given by

$$c = u - \xi. \tag{14.57}$$

We can use this expression to eliminate $c = \sqrt{\gamma p/\rho}$ from the equality (14.50) to obtain

$$u + \frac{2}{\gamma - 1} (u - \xi) = u_l + \frac{2}{\gamma - 1} c_l. \tag{14.58}$$

We can solve this for u as a function of ξ:

$$u(\xi) = \frac{(\gamma - 1)u_l + 2(c_l + \xi)}{\gamma + 1}. \tag{14.59}$$

From (14.57) and (14.59) we now know how c varies with ξ:

$$c(\xi) = u(\xi) - \xi. \tag{14.60}$$

Next we can use the fact that p/ρ^γ is constant to obtain the relation between c and ρ in the rarefaction:

$$\begin{aligned}
c^2 &= \gamma p/\rho \\
&= \gamma(p/\rho^\gamma)\rho^{\gamma-1} \\
&= \gamma(p_l/\rho_l^\gamma)\rho^{\gamma-1}.
\end{aligned} \tag{14.61}$$

Using (14.60), this allows us to determine how ρ varies with ξ:

$$\rho(\xi) = \left(\frac{\rho_l^\gamma [u(\xi) - \xi]^2}{\gamma p_l} \right)^{1/(\gamma-1)}. \tag{14.62}$$

Finally, again using the constancy of p/ρ^γ, we obtain

$$p(\xi) = \left(p_l/\rho_l^\gamma \right) [\rho(\xi)]^\gamma. \tag{14.63}$$

The same procedure can be used for 3-rarefactions, obtaining

$$\begin{aligned}
u(\xi) &= \frac{(\gamma - 1)u_r - 2(c_r - \xi)}{\gamma + 1}, \\
\rho(\xi) &= \left(\frac{\rho_r^\gamma [\xi - u(\xi)]^2}{\gamma p_r} \right)^{1/(\gamma-1)}, \\
p(\xi) &= \left(p_r/\rho_r^\gamma \right) [\rho(\xi)]^\gamma.
\end{aligned} \tag{14.64}$$

Having developed the formulas for the exact Riemann solution, we should note that in practical computations all the details of this structure are generally not needed. In practice methods based on Riemann solutions often use approximate Riemann solvers, as discussed in Section 15.3.

14.13 Shock Tubes and Riemann Problems

An experimental shock tube is filled with two different gases (or the same gas at different pressures and perhaps densities), separated by a membrane at $x = 0$. Initially the gas is at rest, so $u = 0$ everywhere. At time $t = 0$ the membrane is ruptured. The problem is a special case of the Riemann problem (special in that $u_l = u_r = 0$), and in this case it can be shown that the solution consists of a shock moving into the gas at lower pressure and a rarefaction wave that expands into the gas at higher pressure. The interface between the two gases moves at speed u^*, and so this interface is exactly the contact discontinuity. This is very similar to the dam-break Riemann problem for the shallow water equations described in Section 13.2.

Figure 14.3 shows particle paths in the x–t plane for one example with $\rho_l = p_l = 3$ and $\rho_r = p_r = 1$. This particular problem is called the *Sod problem* because Sod [421] used it

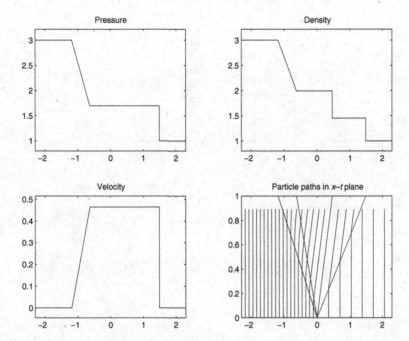

Fig. 14.3. Solution to the Sod shock-tube problem for the Euler equations. The pressure, density, and velocity are displayed as functions of x/t.

as a test in an influential early comparison of different numerical methods. At any fixed t the points are spaced proportional to $1/\rho$ (the specific volume), so wider spacing indicates lower density. Note that there is a jump in density across the contact discontinuity while the velocity is the same on both sides, and equal to u^*, so particles never cross this contact surface. Note also the decrease in density as particles go through the rarefaction wave and the compression across the shock.

By (14.9), the temperature is proportional to p/ρ. In the Sod problem the temperature is initially uniform, but in the solution there must be a jump in temperature across the contact discontinuity where ρ jumps but p is constant. Although the two gases begin at the same temperature, the gas to the left is cooled by expansion through the rarefaction wave while the gas to the right heats up as it crosses the shock.

When the initial velocity is zero everywhere, the Riemann solution typically consists of one shock and one rarefaction wave, along with a contact discontinuity. This is clear from the structure in the p–u plane shown in Figure 14.2. If the initial velocities are nonzero, then it is possible to obtain solutions that consist of two shocks or two rarefaction waves rather than one of each, depending on the data. This is analogous to what was seen in Chapter 13 for the shallow water equations.

Figure 14.4 shows one other example of a Riemann problem for adiabatic gas dynamics, in which the solution contains two shock waves. In this case the Riemann data is

$$\rho_l = 1, \quad u_l = 3, \quad p_l = 1$$

and

$$\rho_r = 2, \quad u_r = 1, \quad p_r = 1.$$

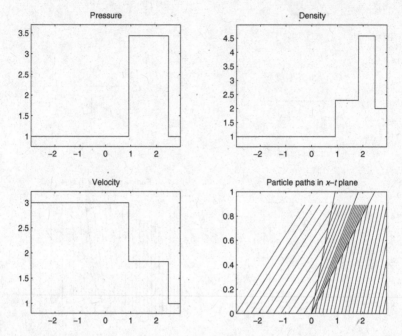

Fig. 14.4. Solution to a Riemann problem for the Euler equations where the gas to the left has a higher velocity and the Riemann solution consists of two shock waves. The pressure, density, and velocity are displayed as functions of x/t.

In this case the colliding gases lead to an increase in the density and pressure in the interme-diate states relative to the initial states. Note that again the contact discontinuity separates particle paths arising from the two initial states.

14.14 Multifluid Problems

In our discussion above, we have assumed that the Riemann problem consists of the same gas on either side of the initial discontinuity, with only a jump in the state of the gas. Many practical problems involve the interaction of more than one gas. As a starting point for developing algorithms for this multifluid or multicomponent situation, we might study a Riemann problem in which two different gases are in contact at $x = 0$ initially. In the simplest case these might be two different ideal gamma-law gases with different values of γ, e.g., a shock-tube problem with the membrane initially separating air ($\gamma_l \approx 1.4$) from a monatomic gas such as argon ($\gamma_r = 5/3$).

In this case the Riemann solution still has the same structure discussed in this chapter and seen in Figure 14.1, with the two gases always in contact at the contact discontinuity. But now the shock or rarefaction wave that is moving into each gas must satisfy the appropriate jump conditions or integral relations for that particular gas. The Riemann problem can still be solved by the procedure outlined in Section 14.11, but the function $\phi_l(p)$ must be defined using γ_l while $\phi_r(p)$ is defined using γ_r. Setting $\phi_l(p) = \phi_r(p)$ and solving for p gives the intermediate pressure p^*. Even in this more general case, the pressure and velocity must be continuous across the contact discontinuity, and only the density has a jump.

To apply a numerical method in the multifluid case we must keep track of the constituent gases so that we know what value of γ to use in each grid cell. This is further complicated by the fact that numerically the sharp interface between the two gases will be smeared due to numerical diffusion, and some cells will contain a mixture of gases. Approaches to handling this with shock-capturing finite volume methods are discussed in Section 16.6. An alternative is to use a front-tracking or moving-mesh method to insure that the fluid interface is also a grid interface, e.g., [131].

14.15 Other Equations of State and Incompressible Flow

We have considered the Riemann problem only for the simplest ideal-gas equation of state (14.23). This is valid for many problems, including aerodynamic problems at subsonic or modest supersonic speeds. In some situations it may be necessary to consider *real-gas* effects, and a more complicated equation of state must be used, either a more complicated analytic expression or perhaps an "equation of state" that is only specified by tabulated values obtained from experiments with the gas or material of interest. The use of more general equations of state can complicate the Riemann solver. See [81], [91], [144], [149], [150], [208], [326], [328], [330], [396] for some discussions of more general Riemann problems.

As an example, consider what happens to a gas if we compress it to the point where the average intermolecular distance begins to approach the size of the molecules. Then the ideal-gas equation of state will no longer be valid, since it is based on a model in which each gas molecule is viewed as a single point and the gas can be compressed indefinitely. Instead we must take into account the fact that the molecules themselves take up some space. This leads to the *covolume equation of state*

$$p = \frac{RT}{v - b}. \tag{14.65}$$

This is written in terms of the specific volume $v = 1/\rho$. For $b = 0$ this agrees with the ideal-gas equation of state (14.9), and $b > 0$ now represents the volume taken up by the molecules themselves.

Additional corrections must be made to this equation of state as v approaches b. When the molecules are very close together the intermolecular attractive forces (van der Waals forces) must be considered, which tend to reduce the pressure. This leads to the *van der Waals equation of state* for dense polytropic gases,

$$p = \frac{RT}{v - b} - \frac{a}{v^2}, \tag{14.66}$$

along with the energy equation

$$e = e(T, v) = c_v T - a/v. \tag{14.67}$$

See, for example, [191], [192], [413]. In extreme cases, the gas undergoes a phase change and becomes a liquid. Phase-change problems can lead to loss of hyperbolicity as described in Section 16.3.2.

The Euler equations of compressible flow can also be used to simulate fluid dynamics in liquids, although in most applications liquids are essentially incompressible and acoustic waves have little effect on the fluid motion. An exception is in the study of violent phenomena such as underwater explosions or cavitation, or certain other problems involving liquid–gas interfaces. For such problems the compressible equations can be used with a suitable equation of state for the liquid. See [65], [88], [138], [212], [370], [449] for some examples.

In some applications acoustic waves in liquids are of interest in their own right, e.g., underwater acoustics, or ultrasound transmission in biological tissue. In these cases the linear acoustic equations can often be used by assuming the liquid is stationary on the time scale of interest in the acoustics problem. Explicit hyperbolic solvers are appropriate in this case.

For most problems involving the dynamics of liquids, the *incompressible Navier–Stokes equations* are used instead. Viscous terms are included, since viscosity generally cannot be ignored in liquids. In these equations the pressure is typically determined by the constraint that the divergence of the velocity field must vanish everywhere ($\nabla \cdot \vec{u} = 0$), since the fluid is incompressible. This is a global constraint that must typically be imposed numerically by solving an elliptic boundary-value problem over the spatial domain each time step. Note that this couples all points in the domain and allows information to travel at infinite speed, as must occur in an incompressible fluid. Applying a force to the fluid at one point will generally cause motion everywhere instantaneously. In reality, information cannot travel at infinite speed, and this motion is in fact accomplished by acoustic waves rapidly bouncing back and forth through the domain at a much higher velocity than the observed fluid motion. See Exercise 3.7 for a similar effect in low Mach number compressible flow, and Example 22.3 for a related example in solid mechanics.

In a liquid, these acoustic waves could be explicitly modeled using the compressible Navier–Stokes equations with appropriate equations of state. However, this is generally very inefficient, since we would need to take extremely small time steps in order to properly resolve these waves. The incompressible equations correctly model the fluid dynamics of interest by capturing the effect of the acoustic waves without explicitly modeling their propagation. High-resolution hyperbolic solvers are often used as a component in these methods for the convective terms, e.g., [9], [23], [53], [56], [266], [329]. These algorithms must then be coupled with implicit solvers for the viscous terms and elliptic solvers in lieu of an equation of state for the pressure. The study of incompressible Navier–Stokes and numerical methods for these equations is a very extensive topic that will not be pursued further here.

The incompressible equations are also useful for gas dynamics at very low Mach numbers, where again the acoustic waves have little effect on the fluid dynamics and the gas density is often nearly constant. Problems in atmospheric flow and low-speed aerodynamics often have this character, for example. Challenging numerical problems arise in simulating low Mach number flow in situations where compressible effects are present but very weak. Ideally one would like to use robust methods that are efficient in the zero Mach number limit (incompressible flow) but behave like high-resolution compressible-flow algorithms as the Mach number increases. See [10], [173], [399], [404], [242], [386], [119] for some discussions of this problem and possible approaches.

15

Finite Volume Methods for Nonlinear Systems

15.1 Godunov's Method

Godunov's method has already been introduced in the context of linear systems in Chapter 4 and for scalar nonlinear problems in Chapter 12. The method is easily generalized to nonlinear systems if we can solve the nonlinear Riemann problem at each cell interface, and this gives the natural generalization of the first-order upwind method to general systems of conservation laws.

Recall that Q_i^n represents an approximation to the cell average of $q(x, t_n)$ over cell C_i,

$$Q_i^n \approx \frac{1}{\Delta x} \int_{x_{i-1/2}}^{x_{i+1/2}} q(x, t_n)\, dx,$$

and the idea is to use the piecewise constant function defined by these cell values as initial data $\tilde{q}^n(x, t_n)$ for the conservation law. Solving over time Δt with this data gives a function $\tilde{q}^n(x, t_{n+1})$, which is then averaged over each cell to obtain

$$Q_i^{n+1} = \frac{1}{\Delta x} \int_{x_{i-1/2}}^{x_{i+1/2}} \tilde{q}^n(x, t_{n+1})\, dx. \tag{15.1}$$

If the time step Δt is sufficiently small, then the exact solution $\tilde{q}^n(x, t)$ can be determined by piecing together the solutions to the Riemann problem arising from each cell interface, as indicated in Figure 15.1(a).

Recall from Section 4.11 that we do not need to perform the integration in (15.1) explicitly, which might be difficult, since $\tilde{q}^n(x, t_{n+1})$ may be very complicated as a function of x. Instead, we can use the fact that $\tilde{q}^n(x_{i-1/2}, t)$ is constant in time along each cell interface, so that the integral (4.5) can be computed exactly. Hence the cell average is updated by the formula

$$Q_i^{n+1} = Q_i^n - \frac{\Delta t}{\Delta x}\left(F_{i+1/2}^n - F_{i-1/2}^n\right), \tag{15.2}$$

with

$$F_{i-1/2}^n = \mathcal{F}\left(Q_{i-1}^n, Q_i^n\right) = f\left(q^{\vee}(Q_{i-1}^n, Q_i^n)\right). \tag{15.3}$$

As usual, $q^{\vee}(q_l, q_r)$ denotes the solution to the Riemann problem between states q_l and q_r, evaluated along $x/t = 0$.

311

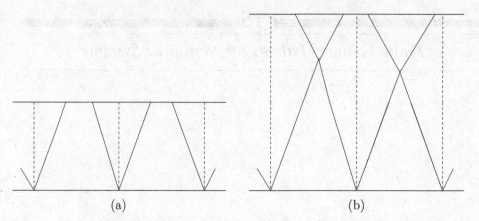

Fig. 15.1. Solving the Riemann problems at each interface for Godunov's method. (a) With Courant number less than 1/2 there is no interaction of waves. (b) With Courant number less than 1 the interacting waves do not reach the cell interfaces, so the fluxes are still constant in time.

In Figure 15.1(a) the time step is taken to be small enough that there is no interaction of waves from neighboring Riemann problems. This would be necessary if we wanted to construct the solution at $\tilde{q}^n(x, t_{n+1})$ in order to explicitly calculate the cell averages (15.1). However, in order to use the flux formula (15.3) it is only necessary that the edge value $\tilde{q}^n(x_{i-1/2}, t)$ remain constant in time over the entire time step, which allows a time step roughly twice as large, as indicated in Figure 15.1(b). If s_{max} represents the largest wave speed that is encountered, then on a uniform grid with the cell interfaces distance Δx apart, we must require

$$\frac{s_{max}\Delta t}{\Delta x} \leq 1 \tag{15.4}$$

in order to insure that the formula (15.3) is valid. Note that this is precisely the CFL condition required for stability of this three-point method, as discussed in Section 4.4. In general $s_{max}\Delta t/\Delta x$ is called the *Courant number*. Figure 15.1(a) shows a case for Courant number less than 1/2; Figure 15.1(b), for Courant number close to 1. Note that for a linear system of equations, $s_{max} = \max_p |\lambda^p|$, and this agrees with the previous definition of the Courant number in Chapter 4.

To implement Godunov's method we do not generally need to determine the full structure of the Riemann solution at each interface, only the value $Q^\vee_{i-1/2} = q^\vee(Q^n_{i-1}, Q^n_i)$ at the cell interface. Normally we only need to determine the intermediate states where the relevant Hugoniot loci and/or integral curves intersect, and $Q^\vee_{i-1/2}$ will equal one of these states. In particular we usually do not need to determine the structure of the solution within rarefaction waves at all. The only exception to this is if one of the rarefactions is transonic, so that the value of $Q^\vee_{i-1/2}$ falls within the rarefaction fan rather than being one of the intermediate states. Even in this case we only need to evaluate one value from the fan, the value corresponding to $\xi = x/t = 0$ in the theory of Section 13.8.5, since this is the value that propagates with speed 0 and gives $Q^\vee_{i-1/2}$.

Godunov's method can again be implemented in the wave propagation form

$$Q_i^{n+1} = Q_i^n - \frac{\Delta t}{\Delta x}\left(\mathcal{A}^+\Delta Q_{i-1/2} + \mathcal{A}^-\Delta Q_{i+1/2}\right) \qquad (15.5)$$

if we define the fluctuations by

$$\mathcal{A}^-\Delta Q_{i-1/2} = f\left(Q_{i-1/2}^\vee\right) - f(Q_{i-1}),$$
$$\mathcal{A}^+\Delta Q_{i-1/2} = f(Q_i) - f\left(Q_{i-1/2}^\vee\right). \qquad (15.6)$$

The fact that so little information from the Riemann solution is used in Godunov's method suggests that one may be able to approximate the Riemann solution and still obtain reasonable solutions. This is often done in practice, and some approaches are discussed in Section 15.3. In Section 15.4 we will see how to extend Godunov's method to high-resolution methods.

15.2 Convergence of Godunov's Method

The Lax–Wendroff theorem of Section 12.10 applies to conservative methods for nonlinear systems of conservation laws as well as to scalar equations. Hence if a sequence of numerical approximations converges in the appropriate sense to a function $q(x, t)$ as the grid is refined, then the limit function $q(x, t)$ must be a weak solution of the conservation law. This is a powerful and useful result, since it gives us confidence that if we compute a reasonable-looking solution on a fine grid, then it is probably close to some weak solution. In particular, we expect that shocks will satisfy the right jump conditions and be propagating at the correct speeds. This would probably not be true if we used a nonconservative method – a reasonable-looking solution might be completely wrong, as we have observed in Section 12.9.

Of course we also need some additional entropy conditions on the numerical method to conclude that the discontinuities seen are physically correct. As we have seen in Section 12.2, it is quite possible that a conservative method will converge to entropy-violating weak solutions if we don't pay attention to this point. However, if we have an entropy function for the system, as described in Section 11.14, and if the Riemann solutions we use in Godunov's method all satisfy the entropy condition (11.51), then the limiting solution produced by Godunov's method will also satisfy the entropy condition, as discussed in Section 12.11. In particular, for the Euler equations of gas dynamics the physical entropy provides an entropy function. Hence any limiting weak solution obtained via Godunov's method will be physically correct.

The limitation of the Lax–Wendroff theorem is that it doesn't guarantee that convergence will occur; it only states that *if* a sequence converges, then the limit is a weak solution. Showing that convergence occurs requires some form of *stability*. We have seen in Section 12.12 that TV-stability is one appropriate form for nonlinear problems. For scalar equations Godunov's method is total variation diminishing (TVD), and we could also develop high-resolution methods that can be shown to be TVD, and hence stable and convergent.

Unfortunately, for nonlinear systems of equations this notion cannot easily be extended. In general the true solution is not TVD in any reasonable sense, and so we cannot expect the numerical solution to be. This is explored in more detail in Section 15.8. Even for classical problems such as the shallow water equations or Euler equations, there is no proof that Godunov's method converges in general. In spite of the lack of rigorous results, this method and high-resolution variants are generally successful in practice and extensively used.

15.3 Approximate Riemann Solvers

To apply Godunov's method on a system of equations we need only determine $q^\vee(q_l, q_r)$, the state along $x/t = 0$ based on the Riemann data q_l and q_r. We do not need the entire structure of the Riemann problem. However, to compute $q^\vee$ we must typically determine something about the full wave structure and wave speeds in order to determine where $q^\vee$ lies in state space. Typically $q^\vee$ is one of the intermediate states in the Riemann solution obtained in the process of connecting q_l to q_r by a sequence of shocks or rarefactions, and hence is one of the intersections of Hugoniot loci and/or integral curves. In the special case of a transonic rarefaction, the value of $q^\vee$ will lie along the integral curve somewhere between these intersections, and additional work will be required to find it.

The process of solving the Riemann problem is thus often quite expensive, even though in the end we use very little information from this solution in defining the flux. We will see later that in order to extend the high-resolution methods of Chapter 6 to nonlinear systems of conservation laws, we must use more information, since all of the waves and wave speeds are used to define second-order corrections. Even so, it is often true that it is not necessary to compute the exact solution to the Riemann problem in order to obtain good results.

A wide variety of *approximate Riemann solvers* have been proposed that can be applied much more cheaply than the exact Riemann solver and yet give results that in many cases are equally good when used in the Godunov or high-resolution methods. In this section we will consider a few possibilities. For other surveys of approximate Riemann solvers, see for example [156], [245], [450].

Note that speeding up the Riemann solver can have a major impact on the efficiency of Godunov-type methods, since we must solve a Riemann problem at every cell interface in each time step. This will be of particular concern in more than one space dimension, where the amount of work grows rapidly. On a modest 100×100 grid in two dimensions, for example, one must solve roughly 20,000 Riemann problems in every time step to implement the simplest two-dimensional generalization of Godunov's method. Methods based on solving Riemann problems are notoriously expensive relative to other methods. The expense may pay off for problems with discontinuous solutions, if it allows good results to be obtained with far fewer grid points than other methods would require, but it is crucial that the Riemann solutions be computed or approximated as efficiently as possible.

For given data Q_{i-1} and Q_i, an approximate Riemann solution might define a function $\hat{Q}_{i-1/2}(x/t)$ that approximates the true similarity solution to the Riemann problem with data Q_{i-1} and Q_i. This function will typically consist of some set of M_w waves $\mathcal{W}_{i-1/2}^p$ propagating at some speeds $s_{i-1/2}^p$, with

$$Q_i - Q_{i-1} = \sum_{p=1}^{M_w} \mathcal{W}_{i-1/2}^p. \tag{15.7}$$

These waves and speeds will also be needed in defining high-resolution methods based on the approximate Riemann solver in Section 15.4.

To generalize Godunov's method using this function, we might take one of two different approaches:

1. Define the numerical flux by

$$F_{i-1/2} = f(\hat{Q}^{\vee}_{i-1/2}),$$

where

$$\hat{Q}^{\vee}_{i-1/2} = \hat{Q}_{i-1/2}(0) = Q_{i-1} + \sum_{p:s^p_{i-1/2}<0} \mathcal{W}^p_{i-1/2} \qquad (15.8)$$

is the value along the cell interface. Then we proceed as in Section 15.1 and set

$$\begin{aligned} \mathcal{A}^- \Delta Q_{i-1/2} &= f(\hat{Q}^{\vee}_{i-1/2}) - f(Q_{i-1}), \\ \mathcal{A}^+ \Delta Q_{i-1/2} &= f(Q_i) - f(\hat{Q}^{\vee}_{i-1/2}), \end{aligned} \qquad (15.9)$$

in order to use the updating formula (15.5). Note that this amounts to evaluating the true flux function at an approximation to $Q^{\vee}_{i-1/2}$.

2. Use the waves and speeds from the approximate Riemann solution to define

$$\begin{aligned} \mathcal{A}^- \Delta Q_{i-1/2} &= \sum_{p=1}^{M_w} (s^p_{i-1/2})^- \mathcal{W}^p_{i-1/2}, \\ \mathcal{A}^+ \Delta Q_{i-1/2} &= \sum_{p=1}^{M_w} (s^p_{i-1/2})^+ \mathcal{W}^p_{i-1/2}, \end{aligned} \qquad (15.10)$$

and again use the updating formula (15.5). Note that this amounts to implementing the REA Algorithm 4.1 with the approximate Riemann solution in place of the true Riemann solutions, averaging these solutions over the grid cells to obtain Q^{n+1}.

If the all-shock Riemann solution is used (e.g., Section 13.7.1), then these two approaches yield the same result. This follows from the fact that the Rankine–Hugoniot condition is then satisfied across each wave $\mathcal{W}^p_{i-1/2}$. In general this will not be true if an approximate Riemann solution is used. In fact, the second approach may not even be conservative unless special care is taken in defining the approximate solution. (The first approach is always conservative, since it is based on an interface flux.)

15.3.1 Linearized Riemann Solvers

One very natural approach to defining an approximate Riemann solution is to replace the nonlinear problem $q_t + f(q)_x = 0$ by some linearized problem defined locally at each cell interface,

$$\hat{q}_t + \hat{A}_{i-1/2} \hat{q}_x = 0. \qquad (15.11)$$

The matrix $\hat{A}_{i-1/2}$ is chosen to be some approximation to $f'(q)$ valid in a neighborhood of the data Q_{i-1} and Q_i. The matrix $\hat{A}_{i-1/2}$ should satisfy the following conditions:

$$\hat{A}_{i-1/2} \text{ is diagonalizable with real eigenvalues,} \qquad (15.12)$$

so that (15.11) is hyperbolic, and

$$\hat{A}_{i-1/2} \to f'(\bar{q}) \qquad \text{as } Q_{i-1}, \ Q_i \to \bar{q}, \tag{15.13}$$

so that the method is consistent with the original conservation law. The approximate Riemann solution then consists of m waves proportional to the eigenvectors $\hat{r}^p_{i-1/2}$ of $\hat{A}_{i-1/2}$, propagating with speeds $s^p_{i-1/2} = \hat{\lambda}^p_{i-1/2}$ given by the eigenvalues. Since this is a linear problem, the Riemann problem can generally be solved more easily than the original nonlinear problem, and often there are simple closed-form expressions for the eigenvectors and hence for the solution, which is obtained by solving the linear system

$$Q_i - Q_{i-1} = \sum_{p=1}^{m} \alpha^p_{i-1/2} \hat{r}^p_{i-1/2} \tag{15.14}$$

for the coefficients $\alpha^p_{i-1/2}$ and then setting $\mathcal{W}^p_{i-1/2} = \alpha^p_{i-1/2} \hat{r}^p_{i-1/2}$.
   We might take, for example,

$$\hat{A}_{i-1/2} = f'(\hat{Q}_{i-1/2}), \tag{15.15}$$

where $\hat{Q}_{i-1/2}$ is some average of Q_{i-1} and Q_i. In particular, the Roe linearization described in the next section has this form for the Euler or shallow water equations, with a very special average. This special averaging leads to some additional nice properties, but for problems where a Roe linearization is not available it is often possible to simply use $\hat{Q}_{i-1/2} = (Q_{i-1} + Q_i)/2$. Note that for any choice of $\hat{A}_{i-1/2}$ satisfying (15.12) and (15.13), we can obtain a consistent and conservative method if we use the formulas (15.8) and (15.9). The formulas (15.10) will not give a conservative method unless $\hat{A}_{i-1/2}$ satisfies an additional condition described in the next section, since (15.10) leads to a conservative method only if the condition

$$f(Q_i) - f(Q_{i-1}) = \sum_{p=1}^{M_w} s^p_{i-1/2} \mathcal{W}^p_{i-1/2} \tag{15.16}$$

is satisfied. This may not hold in general. (See Section 15.5 for an alternative approach to obtaining conservative fluctuations by directly splitting the flux difference.)
   Another obvious linearization is to take

$$\hat{A}_{i-1/2} = \frac{1}{2}\left[f'(Q_{i-1}) + f'(Q_i)\right], \tag{15.17}$$

or some other average of the Jacobian matrix between the two states. But note that in general this matrix could fail to satisfy condition (15.12) even if $f'(Q_{i-1})$ and $f'(Q_i)$ have real eigenvalues.
   Using a linearized problem can be easily motivated and justified at most cell interfaces. The solution to a conservation law typically consists of at most a few isolated shock waves or contact discontinuities separated by regions where the solution is smooth. In these regions, the variation in Q from one grid cell to the next has $\|Q_i - Q_{i-1}\| = \mathcal{O}(\Delta x)$ and the Jacobian matrix is nearly constant, $f'(Q_{i-1}) \approx f'(Q_i)$. Zooming in on the region of state space near

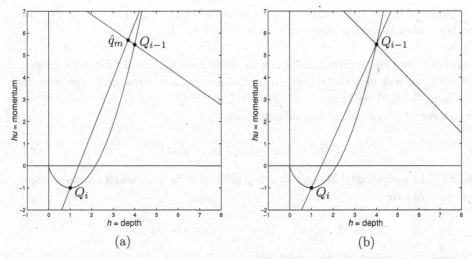

Fig. 15.2. States Q_{i-1} and Q_i are connected by a 2-shock in the shallow water equations, and lie on the same curved Hugoniot locus. The straight lines are in the directions of the eigenvectors of an averaged Jacobian matrix of the form (15.15). (a) Using the average $\hat{Q}_{i-1/2} = (Q_{i-1} + Q_i)/2$. (b) Using the Roe average.

these points, we would find that the Hugoniot loci and integral curves needed to find the exact Riemann solution are nearly straight lines pointing in the directions of the eigenvectors of these matrices. Defining $\hat{A}_{i-1/2}$ as any reasonable approximation to these matrices will yield essentially the same eigenvectors and a Riemann solution that agrees very well with the true solution. This is made more precise in Section 15.6, where we consider the truncation error of methods based on approximate Riemann solutions.

It is only near shocks that we expect Q_{i-1} and Q_i to be far apart in state space, in which case it is harder to justify the use of a Riemann solver that is linearized about one particular point such as $(Q_{i-1} + Q_i)/2$. The true nonlinear structure in state space will look very different from the eigenstructure of any one Jacobian matrix. For example, Figure 15.2(a) shows two states Q_{i-1} and Q_i that should be connected by a single 2-shock in the shallow water equations, since Q_i lies on the 2-Hugoniot locus of Q_{i-1}. If instead the linearized Riemann problem is solved using (15.15) with $\hat{Q}_{i-1/2} = (Q_{i-1} + Q_i)/2$, the state $\hat{q}_m$ indicated in the figure is obtained, with a spurious 1-wave.

15.3.2 Roe Linearization

It is important to notice that even near a shock wave the Riemann problems arising at cell interfaces will typically have a large jump in at most one wave family, say $\mathcal{W}^p$, with $\|\mathcal{W}^j\| = \mathcal{O}(\Delta x)$ for all other waves $j \neq p$. Most of the time a shock in one family is propagating through smooth flow in the other families. It is only at isolated instants in time when two shock waves collide that we expect to observe Riemann problems whose solutions contain more than one strong wave.

For this reason, the situation illustrated in Figure 15.2 is the most important to consider, along with the case where all waves are weak. This suggests the following property that we would like a linearized matrix $\hat{A}_{i-1/2}$ to possess:

If Q_{i-1} and Q_i are connected by a single wave $\mathcal{W}^p = Q_i - Q_{i-1}$ in the true Riemann solution, then $\mathcal{W}^p$ should also be an eigenvector of $\hat{A}_{i-1/2}$:

If this holds, then the "approximate" Riemann solution will also consist of this single wave and will agree with the exact solution. This condition is easy to rewrite in a more useful form using the Rankine–Hugoniot condition (11.21). If Q_{i-1} and Q_i are connected by a single wave (shock or contact discontinuity), then

$$f(Q_i) - f(Q_{i-1}) = s(Q_i - Q_{i-1}),$$

where s is the wave speed. If this is also to be a solution to the linearized Riemann problem, then we must have

$$\hat{A}_{i-1/2}(Q_i - Q_{i-1}) = s(Q_i - Q_{i-1}).$$

Combining these, we obtain the condition

$$\hat{A}_{i-1/2}(Q_i - Q_{i-1}) = f(Q_i) - f(Q_{i-1}). \tag{15.18}$$

In fact this is a useful condition to impose on $\hat{A}_{i-1/2}$ in general, for any Q_{i-1} and Q_i. It guarantees that (15.10) yields a conservative method, and in fact agrees with what is obtained by (15.9). This can be confirmed by recalling that (15.10) will be conservative provided that (6.57) is satisfied,

$$\mathcal{A}^- \Delta Q_{i-1/2} + \mathcal{A}^+ \Delta Q_{i-1/2} = f(Q_i) - f(Q_{i-1}), \tag{15.19}$$

which is satisfied for the approximate Riemann solver if and only if the condition (15.18) holds.

Another nice feature of (15.18) is that it states that the matrix $\hat{A}$, which approximates the Jacobian matrix $\partial f / \partial q$, should at least have the correct behavior in the one direction where we know the change in f that results from a change in q.

The problem now is to obtain an approximate Jacobian that will satisfy (15.18) along with (15.12) and (15.13). One way to obtain a matrix satisfying (15.18) is by integrating the Jacobian matrix over a suitable path in state space between Q_{i-1} and Q_i. Consider the straight-line path parameterized by

$$q(\xi) = Q_{i-1} + (Q_i - Q_{i-1})\xi \tag{15.20}$$

for $0 \le \xi \le 1$. Then $f(Q_i) - f(Q_{i-1})$ can be written as the line integral

$$f(Q_i) - f(Q_{i-1}) = \int_0^1 \frac{df(q(\xi))}{d\xi} \, d\xi$$

$$= \int_0^1 \frac{df(q(\xi))}{dq} q'(\xi) \, d\xi$$

$$= \left[\int_0^1 f'(q(\xi)) \, d\xi \right] (Q_i - Q_{i-1}), \tag{15.21}$$

since $q'(\xi) = Q_i - Q_{i-1}$ is constant and can be pulled out of the integral. This shows that we can define $\hat{A}_{i-1/2}$ as the average

$$\hat{A}_{i-1/2} = \int_0^1 f'(q(\xi))\,d\xi. \tag{15.22}$$

This average always satisfies (15.18) and (15.13), but in general there is no guarantee that (15.12) will be satisfied, even if the original problem is hyperbolic at each point q. An additional problem with attempting to use (15.22) is that it is generally not possible to evaluate this integral in closed form for most nonlinear problems of interest. So it cannot be used as the basis for a practical algorithm that is more efficient than using the true Riemann solver.

Roe [375] made a significant breakthrough by discovering a way to surmount this difficulty for the Euler equations (for a polytropic ideal gas) by a more clever choice of integration path. Moreover, the resulting $\hat{A}_{i-1/2}$ is of the form (15.15) and hence satisfies (15.12). His approach can also be applied to other interesting systems, and will be demonstrated for the shallow water equations in the next section. See Section 15.3.4 for the Euler equations.

Roe introduced a *parameter vector* $z(q)$, a change of variables that leads to integrals that are easy to evaluate. We assume this mapping is invertible so that we also know $q(z)$. Using this mapping, we can also view f as a function of z, and will write $f(z)$ as shorthand for $f(q(z))$.

Rather than integrating on the path (15.20), we will integrate along the path

$$z(\xi) = Z_{i-1} + (Z_i - Z_{i-1})\xi, \tag{15.23}$$

where $Z_j = z(Q_j)$ for $j = i - 1, i$. Then $z'(\xi) = Z_i - Z_{i-1}$ is independent of ξ, and so

$$f(Q_i) - f(Q_{i-1}) = \int_0^1 \frac{df(z(\xi))}{d\xi}\,d\xi$$

$$= \int_0^1 \frac{df(z(\xi))}{dz}\,z'(\xi)\,d\xi$$

$$= \left[\int_0^1 \frac{df(z(\xi))}{dz}\,d\xi\right](Z_i - Z_{i-1}). \tag{15.24}$$

We hope that this integral will be easy to evaluate. But even if it is, this does not yet give us what we need, since the right-hand side involves $Z_i - Z_{i-1}$ rather than $Q_i - Q_{i-1}$. However, we can relate these using another path integral,

$$Q_i - Q_{i-1} = \int_0^1 \frac{dq(z(\xi))}{d\xi}\,d\xi$$

$$= \int_0^1 \frac{dq(z(\xi))}{dz}\,z'(\xi)\,d\xi$$

$$= \left[\int_0^1 \frac{dq(z(\xi))}{dz}\,d\xi\right](Z_i - Z_{i-1}). \tag{15.25}$$

The goal is now to find a parameter vector $z(q)$ for which *both* the integral in (15.24) and the integral in (15.25) are easy to evaluate. Then we will have

$$f(Q_i) - f(Q_{i-1}) = \hat{C}_{i-1/2}(Z_i - Z_{i-1}),$$
$$Q_i - Q_{i-1} = \hat{B}_{i-1/2}(Z_i - Z_{i-1}), \tag{15.26}$$

where $\hat{C}_{i-1/2}$ and $\hat{B}_{i-1/2}$ are these integrals. From these we can obtain the desired relation (15.18) by using

$$\hat{A}_{i-1/2} = \hat{C}_{i-1/2}\hat{B}_{i-1/2}^{-1}. \tag{15.27}$$

Harten and Lax (see [187]) showed that an integration procedure of this form can always be used to define a matrix $\hat{A}$ satisfying (15.12) provided that the system has a convex entropy function $\eta(q)$ as described in Section 11.14. The choice $z(q) = \eta'(q)$ then works, where $\eta'(q)$ is the gradient of η with respect to q. It is shown in [187] that the resulting matrix $\hat{A}_{\text{HLL}}$ is then similar to a symmetric matrix and hence has real eigenvalues.

To make the integrals easy to evaluate, however, we generally wish to choose z in such a way that both $\partial q / \partial z$ and $\partial f / \partial z$ have components that are polynomials in the components of z. Then they will be polynomials in ξ along the path (15.23) and hence easy to integrate.

15.3.3 Roe Solver for the Shallow Water Equations

As an example, we derive the Roe matrix for the shallow water equations. In [281] the isothermal equations, which have similar structure, are used as an example.

For the shallow water equations (see Chapter 13) we have

$$q = \begin{bmatrix} h \\ hu \end{bmatrix} = \begin{bmatrix} q^1 \\ q^2 \end{bmatrix}, \qquad f(q) = \begin{bmatrix} hu \\ hu^2 + \frac{1}{2}gh^2 \end{bmatrix} = \begin{bmatrix} q^2 \\ (q^2)^2/q^1 + \frac{1}{2}g(q^1)^2 \end{bmatrix}$$

and

$$f'(q) = \begin{bmatrix} 0 & 1 \\ -(q^2/q^1)^2 + gq^1 & 2q^2/q^1 \end{bmatrix} = \begin{bmatrix} 0 & 1 \\ -u^2 + gh & 2u \end{bmatrix}.$$

As a parameter vector we choose

$$z = h^{-1/2}q, \qquad \text{so that} \qquad \begin{bmatrix} z^1 \\ z^2 \end{bmatrix} = \begin{bmatrix} \sqrt{h} \\ \sqrt{h}u \end{bmatrix}. \tag{15.28}$$

This is analogous to the parameter vector introduced by Roe for the Euler equations (see Section 15.3.4), in which case $z = \rho^{-1/2}q$. Note that the matrix $f'(q)$ involves the quotient q^2/q^1, and hence integrating along the path (15.20) would require integrating rational functions of ξ. The beauty of this choice of variables z is that the matrices we must integrate in (15.24) and (15.25) involve only polynomials in ξ. We find that

$$q(z) = \begin{bmatrix} (z^1)^2 \\ z^1 z^2 \end{bmatrix} \implies \frac{\partial q}{\partial z} = \begin{bmatrix} 2z^1 & 0 \\ z^2 & z^1 \end{bmatrix} \tag{15.29}$$

and

$$f(z) = \begin{bmatrix} z^1 z^2 \\ (z^2)^2 + \frac{1}{2}g(z^1)^4 \end{bmatrix} \implies \frac{\partial f}{\partial z} = \begin{bmatrix} z^2 & z^1 \\ 2g(z^1)^3 & 2z^2 \end{bmatrix}. \tag{15.30}$$

We now set

$$z^p = Z_{i-1}^p + \left(Z_i^p - Z_{i-1}^p \right) \xi \quad \text{for } p = 1, 2$$

and integrate each element of these matrices from $\xi = 0$ to $\xi = 1$. All elements are linear in ξ except the (2,1) element of $\partial f/\partial z$, which is cubic.

Integrating the linear terms $z^p(\xi)$ yields

$$\int_0^1 z^p(\xi)\, d\xi = \frac{1}{2}\left(Z_{i-1}^p + Z_i^p \right) \equiv \bar{Z}^p,$$

simply the average between the endpoints. For the cubic term we obtain

$$\int_0^1 (z^1(\xi))^3\, d\xi = \frac{1}{4}\left(\frac{(Z_i^1)^4 - (Z_{i-1}^1)^4}{Z_i^1 - Z_{i-1}^1} \right)$$

$$= \frac{1}{2}(Z_{i-1}^1 + Z_i^1) \cdot \frac{1}{2}\left[(Z_{i-1}^1)^2 + (Z_i^1)^2 \right]$$

$$= \bar{Z}^1 \bar{h}, \tag{15.31}$$

where

$$\bar{h} = \frac{1}{2}(h_{i-1} + h_i). \tag{15.32}$$

Hence we obtain

$$\hat{B}_{i-1/2} = \begin{bmatrix} 2\bar{Z}^1 & 0 \\ \bar{Z}^2 & \bar{Z}^1 \end{bmatrix}, \quad \hat{C}_{i-1/2} = \begin{bmatrix} \bar{Z}^2 & \bar{Z}^1 \\ 2g\bar{Z}^1 \bar{h} & 2\bar{Z}^2 \end{bmatrix} \tag{15.33}$$

and so

$$\hat{A}_{i-1/2} = \hat{C}_{i-1/2} \hat{B}_{i-1/2}^{-1} = \begin{bmatrix} 0 & 1 \\ -(\bar{Z}^2/\bar{Z}^1)^2 + g\bar{h} & 2\bar{Z}^2/\bar{Z}^1 \end{bmatrix} = \begin{bmatrix} 0 & 1 \\ -\hat{u}^2 + g\bar{h} & 2\hat{u} \end{bmatrix}. \tag{15.34}$$

Here $\bar{h}$ is the arithmetic average of h_{i-1} and h_i given in (15.32), but $\hat{u}$ is a different sort of average of the velocities, the *Roe average*:

$$\hat{u} = \frac{\bar{Z}^2}{\bar{Z}^1} = \frac{\sqrt{h_{i-1}} u_{i-1} + \sqrt{h_i} u_i}{\sqrt{h_{i-1}} + \sqrt{h_i}}. \tag{15.35}$$

Note that the matrix $\hat{A}_{i-1/2}$ in (15.34) is simply the Jacobian matrix $f'(\hat{q})$ evaluated at the special state $\hat{q} = (\bar{h}, \bar{h}\hat{u})$. In particular, if $Q_{i-1} = Q_i$ then $\hat{A}_{i-1/2}$ reduces to $f'(Q_i)$.

The eigenvalues and eigenvectors of $\hat{A}_{i-1/2}$ are known from (13.9) and (13.10):

$$\hat{\lambda}^1 = \hat{u} - \hat{c}, \qquad \hat{\lambda}^2 = \hat{u} + \hat{c}, \tag{15.36}$$

and

$$\hat{r}^1 = \begin{bmatrix} 1 \\ \hat{u} - \hat{c} \end{bmatrix}, \qquad \hat{r}^2 = \begin{bmatrix} 1 \\ \hat{u} + \hat{c} \end{bmatrix}, \tag{15.37}$$

where $\hat{c} = \sqrt{g\overline{h}}$. To use the approximate Riemann solver we decompose $Q_i - Q_{i-1}$ as in (15.14),

$$Q_i - Q_{i-1} = \alpha^1_{i-1/2}\hat{r}^1 + \alpha^2_{i-1/2}\hat{r}^2 \equiv \mathcal{W}^1_{i-1/2} + \mathcal{W}^2_{i-1/2}. \tag{15.38}$$

The coefficients $\alpha^p_{i-1/2}$ are computed by solving this linear system, which can be done explicitly by inverting the matrix $\hat{R}$ of right eigenvectors to obtain

$$\hat{L} = \hat{R}^{-1} = \frac{1}{2\hat{c}} \begin{bmatrix} \hat{u} + \hat{c} & -1 \\ -(\hat{u} - \hat{c}) & 1 \end{bmatrix}.$$

Multiplying this by the vector $\delta \equiv Q_i - Q_{i-1}$ gives the vector of α-coefficients, and hence

$$\alpha^1_{i-1/2} = \frac{(\hat{u} + \hat{c})\delta^1 - \delta^2}{2\hat{c}},$$
$$\alpha^2_{i-1/2} = \frac{-(\hat{u} - \hat{c})\delta^1 + \delta^2}{2\hat{c}}, \tag{15.39}$$

The fluctuations (15.10) are then used in Godunov's method, with the speeds s given by the eigenvalues λ of (15.36).

Alternatively, we could compute the numerical flux $F_{i-1/2}$ by

$$F_{i-1/2} = f(Q_{i-1}) + \hat{A}^-_{i-1/2}(Q_i - Q_{i-1})$$

or by

$$F_{i-1/2} = f(Q_i) - \hat{A}^+_{i-1/2}(Q_i - Q_{i-1}).$$

Averaging these two expressions gives a third version, which is symmetric in Q_{i-1} and Q_i,

$$F_{i-1/2} = \frac{1}{2}[f(Q_{i-1}) + f(Q_i)] - \frac{1}{2}|\hat{A}_{i-1/2}|(Q_i - Q_{i-1}). \tag{15.40}$$

This form is often called *Roe's method* (see Section 4.14) and has the form of the unstable centered flux plus a viscous correction term.

15.3.4 Roe Solver for the Euler Equations

For the Euler equations with the equation of state (14.23), Roe [375] proposed the parameter vector $z = \rho^{-1/2}q$, leading to the averages

$$\hat{u} = \frac{\sqrt{\rho_{i-1}}\,u_{i-1} + \sqrt{\rho_i}\,u_i}{\sqrt{\rho_{i-1}} + \sqrt{\rho_i}} \tag{15.41}$$

for the velocity,

$$\hat{H} = \frac{\sqrt{\rho_{i-1}}\, H_{i-1} + \sqrt{\rho_i}\, H_i}{\sqrt{\rho_{i-1}} + \sqrt{\rho_i}} = \frac{(E_{i-1} + p_{i-1})/\sqrt{\rho_{i-1}} + (E_i + p_i)/\sqrt{\rho_i}}{\sqrt{\rho_{i-1}} + \sqrt{\rho_i}} \quad (15.42)$$

for the total specific enthalpy, and

$$\hat{c} = \sqrt{(\gamma - 1)\left(\hat{H} - \frac{1}{2}\hat{u}^2\right)} \quad (15.43)$$

for the sound speed. The eigenvalues and eigenvectors of the Roe matrix are then obtained by evaluating (14.45) and (14.46) at this averaged state. The coefficients $\alpha^p_{i-1/2}$ in the wave decomposition

$$\delta \equiv Q_i - Q_{i-1} = \alpha^1 \hat{r}^1 + \alpha^2 \hat{r}^2 + \alpha^3 \hat{r}^3$$

can be obtained by inverting the matrix of right eigenvectors, which leads to the following formulas:

$$\alpha^2 = (\gamma - 1)\frac{(\hat{H} - \hat{u}^2)\delta^1 + \hat{u}\delta^2 - \delta^3}{\hat{c}^2},$$

$$\alpha^3 = \frac{\delta^2 + (\hat{c} - \hat{u})\delta^1 - \hat{c}\alpha^2}{2\hat{c}}, \quad (15.44)$$

$$\alpha^1 = \delta^1 - \alpha^2 - \alpha^3.$$

For other equations of state and more complicated gas dynamics problems it may also be possible to derive Roe solvers; see for example [91], [149], [172], [208], [451].

15.3.5 Sonic Entropy Fixes

One disadvantage of using a linearized Riemann solver is that the resulting approximate Riemann solution consists only of discontinuities, with no rarefaction waves. This can lead to a violation of the entropy condition, as has been observed previously for scalar conservation laws in Section 12.3.

In fact, it is worth noting that in the scalar case the Roe condition (15.18) can be satisfied by choosing the scalar $\hat{A}_{i-1/2}$ as

$$\hat{A}_{i-1/2} = \frac{f(Q_i) - f(Q_{i-1})}{Q_i - Q_{i-1}}, \quad (15.45)$$

which is simply the shock speed resulting from the scalar Rankine–Hugoniot condition. Hence using the Roe linearization in the scalar case and solving the resulting advection equation with velocity (15.45) is equivalent to always using the shock-wave solution to the scalar problem, as discussed in Section 12.2. In this scalar case (15.40) reduces to the flux for Murman's method, (12.12) with $a_{i-1/2}$ given by (12.14).

Recall that in the scalar case, the use of an entropy-violating Riemann solution leads to difficulties only in the case of a transonic rarefaction wave, in which $f'(q_l) < 0 < f'(q_r)$.

This is also typically true when we use Roe's approximate Riemann solution for a system of conservation laws. It is only for sonic rarefactions, those for which $\lambda^p < 0$ to the left of the wave while $\lambda^p > 0$ to the right of the wave, that entropy violation is a problem. In the case of a sonic rarefaction wave, it is necessary to modify the approximate Riemann solver in order to obtain entropy-satisfying solutions.

For the shallow water equations, a system of two equations, there is a single intermediate state $\hat{Q}_m$ in the approximate Riemann solution between Q_{i-1} and Q_i. We can compute the characteristic speeds in each state as

$$
\begin{aligned}
\lambda_{i-1}^1 &= u_{i-1} - \sqrt{gh_{i-1}}, & \lambda_m^1 &= u_m - \sqrt{g\hat{h}_m}, \\
\lambda_m^2 &= \hat{u}_m + \sqrt{g\hat{h}_m}, & \lambda_i^2 &= u_i + \sqrt{gh_i}.
\end{aligned}
\tag{15.46}
$$

If $\lambda_{i-1}^1 < 0 < \lambda_m^1$, then we should suspect that the 1-wave is actually a transonic rarefaction and make some adjustment to the flux, i.e., to $\mathcal{A}^-\Delta Q_{i-1/2}$ and $\mathcal{A}^+\Delta Q_{i-1/2}$, in this case. Similarly, if $\lambda_m^2 < 0 < \lambda_i^2$, then we should fix the flux to incorporate a 2-rarefaction. Note that at most one of these situations can hold, since $\lambda_m^1 < \lambda_m^2$.

For sufficiently simple systems, such as the shallow water equations, it may be easy to evaluate the true intermediate state $Q_{i-1/2}^\vee$ that lies along the interface in the Riemann solution at $x_{i-1/2}$, once we suspect it lies on a transonic rarefaction wave in a particular family. If we suspect that there should be a transonic 1-rarefaction, for example, then we can simply evaluate the state at $\xi = x/t = 0$ in the 1-rarefaction wave connected to Q_{i-1}. This is easily done using the formulas in Section 13.8.5. Evaluating (13.52) at $\xi = 0$ and then using (13.33) gives

$$
\begin{aligned}
h_{i-1/2}^\vee &= (u_{i-1} + 2\sqrt{gh_{i-1}})^2 / 9g, \\
u_{i-1/2}^\vee &= u_{i-1} - 2\left(\sqrt{gh_{i-1}} - \sqrt{gh_{i-1/2}^\vee}\right).
\end{aligned}
\tag{15.47}
$$

We can now evaluate the flux at this point and set $F_{i-1/2} = f(Q_{i-1/2}^\vee)$. Finally, the formulas (15.9) can be used to define $\mathcal{A}^\pm\Delta Q_{i-1/2}$.

Using the structure of the exact rarefaction wave is not always possible or desirable for more general systems of equations where we hope to avoid determining the exact Riemann solution. A variety of different approaches have been developed as approximate entropy fixes that work well in practice. Several of these are described below. See also [115], [349], [351], [355], [378], [382], [432], or [434] for some other discussions.

The Harten–Hyman Entropy Fix

A more general procedure was taken by Harten and Hyman [184] and modified slightly in [281]. This approach is used in many of the standard CLAWPACK solvers.

Suppose there appears to be a transonic rarefaction in the k-wave, i.e., $\lambda_l^k < 0 < \lambda_r^k$, where $\lambda_{l,r}^k$ represents the kth eigenvalue of the matrix $f'(q)$ computed in the states $q_{l,r}^k$ just to the left and right of the k-wave in the approximate Riemann solution, i.e.,

$$q_l^k = Q_{i-1} + \sum_{p=1}^{k-1} \mathcal{W}^p, \qquad q_r^k = q_l^k + \mathcal{W}^k. \tag{15.48}$$

(We suppress the subscripts $i - 1/2$ here and below for clarity, since we need to add subscripts l and r.) Then we replace the single wave $\mathcal{W}^k$ propagating at speed $\hat{\lambda}^k$ by a pair of waves $\mathcal{W}_l^k = \beta \mathcal{W}^k$ and $\mathcal{W}_r^k = (1 - \beta)\mathcal{W}^k$ propagating at speeds λ_l^k and λ_r^k. To maintain conservation we require that

$$\lambda_l^k \mathcal{W}_l^k + \lambda_r^k \mathcal{W}_r^k = \hat{\lambda}^k \mathcal{W}^k$$

and hence

$$\beta = \frac{\lambda_r^k - \hat{\lambda}^k}{\lambda_r^k - \lambda_l^k}. \tag{15.49}$$

In practice it is simpler to leave the wave $\mathcal{W}^k$ alone (and continue to use this single wave in the high-resolution correction terms; see Section 15.4) and instead modify the values $(\hat{\lambda}^k)^{\pm}$ used in defining $\mathcal{A}^{\pm}\Delta Q_{i-1/2}$ via (15.10). The formula (15.10) can still be used (with $\hat{s}^k = \hat{\lambda}^k$) if, instead of the positive and negative parts of $\hat{\lambda}^k$, we use the values

$$\begin{aligned} (\hat{\lambda}^k)^- &\equiv \beta \lambda_l^k, \\ (\hat{\lambda}^k)^+ &\equiv (1 - \beta)\lambda_r^k \end{aligned} \tag{15.50}$$

in the kth field. These still sum to $\hat{\lambda}^k$ but are both nonzero in the transonic case. We continue to use the standard definitions (4.40) or (4.63) of $(\lambda^k)^{\pm}$ in any field where λ_l^k and λ_r^k have the same sign.

In the scalar case this entropy fix can be interpreted as using a piecewise linear approximation to the flux function $f(q)$ in the neighborhood of the sonic point. This approximation lies below the true flux function in the convex case, a fact that is used in [281] to show that Roe's method with this entropy fix is an E-scheme (see Section 12.7) and hence converges to the entropy-satisfying weak solution. See also [115] for some related results.

Numerical Viscosity

From (15.40), the flux for Roe's method is

$$\begin{aligned} F_{i-1/2} &= \frac{1}{2}[f(Q_{i-1}) + f(Q_i)] - \frac{1}{2}|\hat{A}_{i-1/2}|(Q_i - Q_{i-1}) \\ &= \frac{1}{2}[f(Q_{i-1}) + f(Q_i)] - \frac{1}{2}\sum_p |\hat{\lambda}_{i-1/2}^p|\mathcal{W}_{i-1/2}^p. \end{aligned} \tag{15.51}$$

One way to view the need for an entropy fix is to recognize that the viscous term in this flux is too small in the case of a transonic rarefaction. With sufficient viscosity we should not observe entropy-violating shocks. Note that in the transonic case, where the characteristic speeds span 0, we might expect the eigenvalue $\hat{\lambda}_{i-1/2}^p$ to be close to zero. The corresponding term of the sum in (15.51) will then be close to zero, corresponding to no viscosity in this field. In fact a transonic entropy-violating shock typically has speed zero, as can be observed in Figure 12.2(a). The asymmetric portion of the rarefaction fan is moving with nonzero average speed and has positive viscosity due to the averaging process. It is only the symmetric part with speed 0 that remains as a stationary jump.

When we implement the method using $\mathcal{A}^{\pm}\Delta Q$ from (15.10), we find that the numerical flux is actually given by

$$F_{i-1/2} = \frac{1}{2}[f(Q_{i-1}) + f(Q_i)] - \frac{1}{2}\sum_p [(\hat{\lambda}^p_{i-1/2})^+ - (\hat{\lambda}^p_{i-1/2})^-]\mathcal{W}^p_{i-1/2}. \qquad (15.52)$$

With the usual definition (4.40) of $\lambda^{\pm}$, this agrees with (15.51). However, if we apply the entropy fix defined in the last section and redefine these values as in (15.50), then we have effectively increased the numerical viscosity specifically in the kth field when this field contains a transonic rarefaction.

Harten's Entropy Fix

Harten [179] proposed an entropy fix based on increasing the viscosity by modifying the absolute-value function in (15.51), never allowing any eigenvalue to be too close to zero. In this simple approach one replaces each value $|\hat{\lambda}^p_{i-1/2}|$ in (15.51) by a value $\phi_\delta(\hat{\lambda}^p_{i-1/2})$, where $\phi_\delta(\lambda)$ is a smoothed version of the absolute-value function that is always positive, staying above some value $\delta/2$:

$$\phi_\delta(\lambda) = \begin{cases} |\lambda| & \text{if } |\lambda| \geq \delta, \\ (\lambda^2 + \delta^2)/(2\delta) & \text{if } |\lambda| < \delta. \end{cases} \qquad (15.53)$$

A disadvantage of this approach is that the parameter δ must typically be tuned to the problem.

To implement this in the context of fluctuations $\mathcal{A}^{\pm}\Delta Q_{i-1/2}$, we can translate this modification of the absolute value into modifications of $(\hat{\lambda}^p_{i-1/2})^+$ and $(\hat{\lambda}^p_{i-1/2})^-$. Again we can continue to use the form (15.10) if we redefine

$$(\lambda)^- \equiv \frac{1}{2}[\lambda - \phi_\delta(\lambda)],$$

$$(\lambda)^+ \equiv \frac{1}{2}[\lambda + \phi_\delta(\lambda)]. \qquad (15.54)$$

Note that this agrees with the usual definition (4.63) of $\lambda^{\pm}$ if $\phi_\delta(\lambda) = |\lambda|$.

The LLF Entropy Fix

Another approach to introducing more numerical viscosity is to use the approximate Riemann solver in conjunction with an extension of the local Lax–Friedrichs (LLF) method to systems of equations. The formula (12.12) generalizes naturally to systems of equations as

$$F_{i-1/2} = \frac{1}{2}[f(Q_{i-1}) + f(Q_i)] - \frac{1}{2}\sum_p a^p_{i-1/2}\mathcal{W}^p_{i-1/2}, \qquad (15.55)$$

where

$$a_{i-1/2}^p = \max\left(|\lambda_{i-1}^p|, |\lambda_i^p|\right). \tag{15.56}$$

Here λ_{i-1}^p and λ_i^p are eigenvalues of the Jacobians $f'(Q_{i-1})$ and $f'(Q_i)$ respectively, while $\mathcal{W}_{i-1/2}^p$ is the wave resulting from the Roe solver. We can implement this using

$$\mathcal{A}^-\Delta Q_{i-1/2} = \frac{1}{2}\sum_p \left(\hat{\lambda}_{i-1/2}^p - a_{i-1/2}^p\right)\mathcal{W}_{i-1/2}^p,$$
$$\mathcal{A}^+\Delta Q_{i-1/2} = \frac{1}{2}\sum_p \left(\hat{\lambda}_{i-1/2}^p + a_{i-1/2}^p\right)\mathcal{W}_{i-1/2}^p. \tag{15.57}$$

Again this can be viewed as a redefinition of $\lambda^\pm$ similar to (15.54). A disadvantage of this approach is that it generally adds numerical viscosity to all fields, whether or not there is a transonic rarefaction. However, wherever the solution is smooth we have $\hat{\lambda}_{i-1/2}^p \approx \lambda_{i-1}^p \approx \lambda_i^p$ and so (15.57) essentially reduces to the standard definition of $\mathcal{A}^\pm\Delta Q_{i-1/2}$.

15.3.6 Failure of Linearized Solvers

In some situations linearized Riemann solvers such as those based on the Roe average can fail completely, giving a nonphysical solution such as negative depth in the shallow water equation or negative pressures or density in the Euler equations. This can happen in particular for data that is near the "vacuum state" or in situations where there is a strong expansion.

Figure 15.3 shows an example for the shallow water equation in the case of a Riemann problem yielding two symmetric rarefaction waves, as studied in Section 13.8.6. In Figure 15.3(a) the data is $h_l = h_r = 1$ and $u_r = u_l = 0.8$ (with $g = 1$). The straight lines

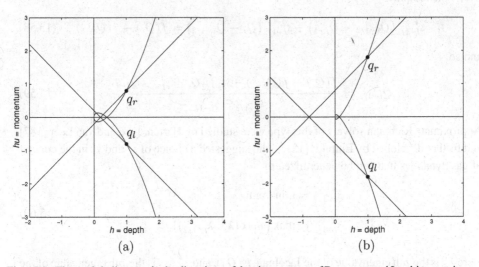

Fig. 15.3. The straight lines are in the directions of the eigenvectors of Roe-averaged Jacobian matrices $\hat{A}$ in the example of Section 15.3.6. (a) For $u_r = -u_l = 0.8$. (b) For $u_r = -u_l = 1.8$, in which case the lines intersect at an unphysical point with $\hat{h}_m < 0$.

show the eigendirections of the Roe matrix $\hat{A}$, which intersect at a point with $u_m = 0$ (as in the true solution) but with h_m about half the correct value. The solution is still positive, however, and reasonable computational results can be obtained for this data: as the solution smoothes out in later time steps the Roe average gives increasingly better estimates of the true solution, and convergence is obtained.

In Figure 15.3(b) the outflow velocity is increased to $u_r = -u_l = 1.8$. In this case the true Riemann solution still has a positive value of h_m, but in the approximate Riemann solution the curves intersect at a negative value of $\hat{h}_m$. The code will typically then fail when the sound speed is computed as $\sqrt{gh}$.

Other averages give similar results. For the Euler equations it has been shown by Einfeldt, Munz, Roe, & Sjogreen [122] that for certain Riemann problems there is no linearization that will preserve positivity, and other approaches to approximating the Riemann solution must be used. They call a method *positively conservative* for the Euler equations if the density and internal energy always remain positive for any physical data. They show that Godunov's method with the exact Riemann solver is positively conservative, and also show this for methods based on the HLLE approximate Riemann solver described in the next section.

15.3.7 The HLL and HLLE Solvers

A simple approximate Riemann solver can be based on estimating the smallest and largest wave speeds arising in the Riemann solution and then taking $\hat{Q}(x/t)$ to consist of only two waves propagating at these speeds $s_{i-1/2}^1$ and $s_{i-1/2}^2$. There will then be a single new state $\hat{Q}_{i-1/2}$ in between, and as waves we use

$$\mathcal{W}_{i-1/2}^1 = \hat{Q}_{i-1/2} - Q_{i-1} \quad \text{and} \quad \mathcal{W}_{i-1/2}^2 = Q_i - \hat{Q}_{i-1/2}.$$

We can determine the state $\hat{Q}_{i-1/2}$ by requiring that the approximate solution be conservative, which requires

$$s_{i-1/2}^1(\hat{Q}_{i-1/2} - Q_{i-1}) + s_{i-1/2}^2(Q_i - \hat{Q}_{i-1/2}) = f(Q_i) - f(Q_{i-1}) \quad (15.58)$$

and so

$$\hat{Q}_{i-1/2} = \frac{f(Q_i) - f(Q_{i-1}) - s_{i-1/2}^2 Q_i + s_{i-1/2}^1 Q_{i-1}}{s_{i-1/2}^1 - s_{i-1/2}^2}. \quad (15.59)$$

Approximate Riemann solvers of this type were studied by Harten, Lax, and van Leer [187] and further developed by Einfeldt [121], who suggested a choice of s^1 and s^2 in the context of gas dynamics that can be generalized to

$$\begin{aligned}
s_{i-1/2}^1 &= \min_p \left(\min \left(\lambda_{i-1}^p, \hat{\lambda}_{i-1/2}^p \right) \right), \\
s_{i-1/2}^2 &= \max_p \left(\max \left(\lambda_i^p, \hat{\lambda}_{i-1/2}^p \right) \right).
\end{aligned} \quad (15.60)$$

Here λ_j^p is the pth eigenvalue of the Jacobian $f'(Q_j)$, and $\hat{\lambda}_{i-1/2}^p$ is the pth eigenvalue of the Roe average (for problems where this average is easily defined). In the original HLL method of [187], the values $s_{i-1/2}^1$ and $s_{i-1/2}^2$ are chosen as some lower and upper bounds on all the characteristic speeds that might arise in the true Riemann solution. The choice (15.60) might not satisfy this, but in practice gives sharper results for shock waves since the shock speed is smaller than the characteristic speed behind the shock and in this case (15.60) reduces to the Roe approximation of the shock speed. In particular, the HLLE method shares the nice property of the Roe solver that for data connected by a single shock wave, the approximate solution agrees with the true solution. In the case where the slowest or fastest wave is a rarefaction wave, the formula (15.60) will use the corresponding characteristic speed, which is faster than the Roe average speed in this case. In general it is not necessary to use an "entropy fix" when using this solver. It is also shown in [122] that this method is positively conservative, as discussed in the previous section, and hence may be advantageous for problems where low densities are expected.

A disadvantage of this solver is that the full Riemann solution structure is modeled by only two waves based on approximate speeds of the fastest and slowest waves in the system. For a system of more than two equations this may lead to a loss of resolution for waves traveling at intermediate speeds. For the Euler equations, for example, this approximation is based only on the two acoustic waves while the contact discontinuity is ignored. The resulting numerical solutions show relatively poor resolution of the contact discontinuity as a result.

A modified HLLE method (denoted by HLLEM) is proposed in [121] that attempts to capture a contact discontinuity more accurately by introducing a piecewise linear function as the approximate solution, where the constant intermediate state (15.59) is replaced by a linear function with the same total integral for conservation. This function is based on information about the contact discontinuity. The HLLEC method described in [450] is another approach that introduces a third wave into the approximation.

The introduction of a third wave and hence two intermediate states is also discussed by Harten, Lax, and van Leer [187]. They suggest choosing the speed of this third wave as

$$V = \frac{[\eta'(Q_i) - \eta'(Q_{i-1})] \cdot [f(Q_i) - f(Q_{i-1})]}{[\eta'(Q_i) - \eta'(Q_{i-1})] \cdot [Q_i - Q_{i-1}]} \qquad (15.61)$$

for problems with a convex entropy function $\eta(q)$. It can then be shown that V lies between the smallest and largest eigenvalues of the matrix $\hat{A}_{\text{HLL}}$ discussed at the end of Section 15.3.2. Moreover, if Q_{i-1} and Q_i are connected by a single wave, i.e., if $f(Q_i) - f(Q_{i-1}) = s(Q_i - Q_{i-1})$ for some scalar s, then $V = s$, and so this solver, like the Roe solver, will reproduce the exact Riemann solution in this case. Linde [300] has recently developed this approach further.

15.4 High-Resolution Methods for Nonlinear Systems

Godunov's method (or one of the variants based on approximate Riemann solvers) can be extended to high-resolution methods for nonlinear systems using essentially the same approach as was introduced in Section 6.13 for linear systems. The formulas have already been introduced in Section 6.15. The method takes the form

$$Q_i^{n+1} = Q_i^n - \frac{\Delta t}{\Delta x}\left(\mathcal{A}^- \Delta Q_{i+1/2} + \mathcal{A}^+ \Delta Q_{i-1/2}\right) - \frac{\Delta t}{\Delta x}\left(\tilde{F}_{i+1/2} - \tilde{F}_{i-1/2}\right), \quad (15.62)$$

where $\mathcal{A}^{\pm}\Delta Q_{i-1/2}$ are the fluctuations corresponding to Godunov's method or one of its variants. The flux $\tilde{F}_{i-1/2}$ is the high-resolution correction given by

$$\tilde{F}_{i-1/2} = \frac{1}{2}\sum_{p=1}^{M_w}\left|s_{i-1/2}^p\right|\left(1 - \frac{\Delta t}{\Delta x}\left|s_{i-1/2}^p\right|\right)\widetilde{\mathcal{W}}_{i-1/2}^p. \tag{15.63}$$

Recall that $\mathcal{W}_{i-1/2}^p$ is the pth wave arising in the solution to the Riemann problem at $x_{i-1/2}$ and $\widetilde{\mathcal{W}}_{i-1/2}^p$ is a limited version of this wave. In the constant-coefficient linear case this limited wave is computed by comparing the magnitude of $\mathcal{W}_{i-1/2}^p$ to $\mathcal{W}_{I-1/2}^p$, the corresponding wave from the neighboring Riemann problem in the upwind direction ($I = i \pm 1$ depending on the sign of the wave speed s^p as in (6.61)).

If a linearized Riemann solver such as the Roe solver is used, then $M_w = m$, and we will generally assume this case below. However, high-resolution corrections of this form can also be applied to other Riemann solvers, for example the HLLE method for which $M_w = 2$. See Example 15.1 for a demonstration of the improvement this makes over the Godunov method with this solver.

Several difficulties arise with nonlinear systems that are not seen with a linear system. In the linear case each wave $\mathcal{W}^p$ is a sharp discontinuity traveling at a single speed $s^p = \lambda^p$. For a nonlinear system, shock waves and contact discontinuities have this form, but rarefaction waves do not. We can still define a wave strength $\mathcal{W}^p$ as the total jump across the wave,

$$\mathcal{W}^p = q_r^p - q_l^p, \tag{15.64}$$

where q_l^p and q_r^p are the states just to the left and right of the wave. However, there is not a single wave speed s^p to use in the formula (15.63). Instead, the characteristic speed λ^p varies continuously through the rarefaction wave. In practice an average speed, e.g.,

$$s^p = \frac{1}{2}\left(\lambda_l^p + \lambda_r^p\right),$$

can generally be used successfully. Recall that the form of the correction terms in (15.62) guarantees that the method will be conservative regardless of the manner in which s^p is chosen, so this is not an issue.

Another way to deal with rarefaction waves is to simply use a discontinuous approximation instead, such as an entropy-violating shock or, more commonly, the approximate solution obtained with a linearized Riemann solver (e.g., the Roe solver) as described in Section 15.3.2. In particular, if a linearized solver is being used to determine the fluctuations $\mathcal{A}^{\pm}\Delta Q$, these same waves $\mathcal{W}^p$ and the corresponding eigenvalues $s^p = \hat{\lambda}^p$ can be used directly in (15.63). If an entropy fix is applied to modify $\mathcal{A}^{\pm}\Delta Q$, the original waves can generally still be used for the high-resolution correction terms with good results. Even if the exact Riemann solver is used for the first-order fluctuations, as may be necessary for some difficult problems where the linearized solver does not suffice, it may still be possible to use a linearized solver to obtain the waves and speeds needed for the high-resolution corrections.

Another issue that arises for nonlinear systems is that the waves $\mathcal{W}_{i-1/2}^p$ and $\mathcal{W}_{I-1/2}^p$ are generally not collinear vectors in state space, and so applying a limiter based on comparing the magnitude of these vectors is not as simple as for a constant-coefficient linear system (where the eigenvectors r^p of $A = f'(q)$ are constant). This difficulty has already been discussed in the context of variable-coefficient linear systems in Section 9.13. Similar approaches can be taken for nonlinear systems. The default approach in CLAWPACK is to project the wave $\mathcal{W}_{I-1/2}^p$ from the neighboring Riemann problem onto $\mathcal{W}_{i-1/2}^p$ in order to obtain a vector that can be directly compared to $\mathcal{W}_{i-1/2}^p$ as described in Section 9.13.

Example 15.1. As an example we solve one standard test problem using several different methods. The Euler equations are solved with initial data $\overset{\circ}{\rho}(x) \equiv 1$, $\overset{\circ}{u}(x) \equiv 0$, and pressure

$$\overset{\circ}{p}(x) = \begin{cases} 1000 & \text{if } 0 \le x \le 0.1, \\ 0.01 & \text{if } 0.1 \le x \le 0.9, \\ 100 & \text{if } 0.9 \le x \le 1.0. \end{cases} \tag{15.65}$$

This problem was first used as a test problem by Woodward & Colella [487] and is often referred to as the *Woodward–Colella blast-wave problem*. The two discontinuities in the initial data each have the form of a shock-tube problem and yield strong shock waves and contact discontinuities going inwards and rarefaction waves going outwards. The boundaries $x = 0$ and $x = 1$ are both reflecting walls and the reflected rarefaction waves interact with the other waves.

Figure 15.4 illustrates the structure of the solution in the x–t plane, showing contour lines of both density and pressure. The two shock waves collide at about time $t = 0.27$ and generate an additional contact discontinuity. The right-going shock then collides with the contact discontinuity arising from the Riemann problem at $x = 0.9$, deflecting it to the right. Solutions are often compared at time $t = 0.038$, when the solution consists of contact discontinuities near $x = 0.6$, $x = 0.76$, and $x = 0.8$ and shock waves near $x = 0.65$ and $x = 0.87$. This is a challenging test problem because of the strength of the shocks involved and the interaction of the different waves.

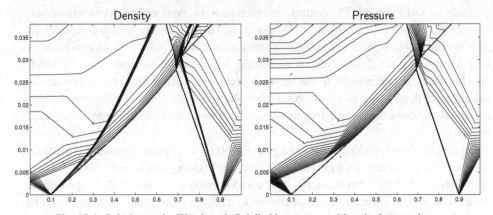

Fig. 15.4. Solution to the Woodward–Colella blast-wave problem in the x–t plane.

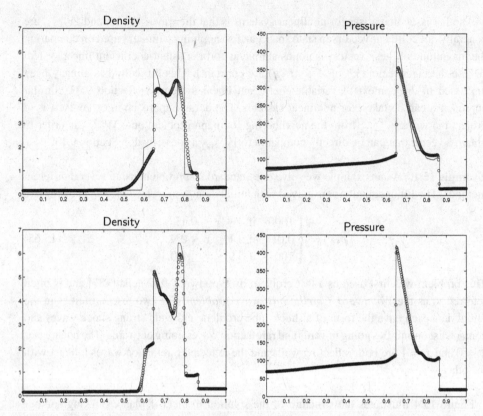

Fig. 15.5. Solution to the Woodward–Colella blast-wave problem at time $t = 0.038$ computed with the Roe solver. Top: Godunov method. Bottom: High-resolution method. [`claw/book/chap15/wcblast`]

Figure 15.5 shows results obtained on a grid with 500 cells using the Roe solver and either the first-order Godunov method (top) or the high-resolution method with the MC limiter (bottom). The solid line shows results obtained with the same method on a grid with 4000 cells. Note that with the high-resolution method the shocks are captured very sharply and are in the correct locations. The contact discontinuities are considerably more smeared out, however (even in the computation on the finer grid). This is typically seen in computations with the Euler equations. The nonlinearity that causes a shock wave to form also tends to keep it sharp numerically. A contact discontinuity is a linearly degenerate wave for which the characteristics are parallel to the wave on each side. This wave simply continues to smear further in each time step with no nonlinear sharpening effect. Notice that the pressure is continuous across the contact discontinuities and is well captured in spite of the errors in the density.

Figure 15.6 shows results obtained on a grid with 500 cells using the simpler HLLE solver and either the first-order Godunov method (top) or the high-resolution method with the MC limiter (bottom). Recall that this solver only uses two waves with speeds that approximate the acoustic speeds and hence does not attempt to model the contact discontinuity at all. In spite of this the solution has the correct structure, although with considerably more smearing of the contact discontinuities and less accuracy overall than the Roe solver provides.

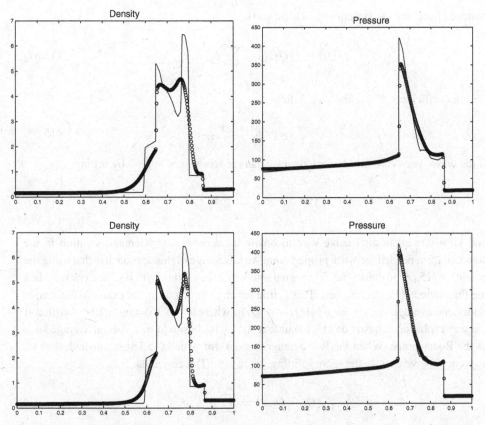

Fig. 15.6. Solution to the Woodward–Colella blast-wave problem at time $t = 0.038$ computed with the HLLE solver. Top: Godunov method. Bottom: High-resolution method. [claw/book/chap15/wcblast]

15.5 An Alternative Wave-Propagation Implementation of Approximate Riemann Solvers

The high-resolution wave-propagation method (15.62) is based on the assumption that $Q_i - Q_{i-1}$ has been split into waves as in (15.7) and the fluctuations $\mathcal{A}^{\pm} \Delta Q_{i-1/2}$ defined using either (15.9) or (15.10). An alternative approach is to first split the jump in f into "waves"

$$f(Q_i) - f(Q_{i-1}) = \sum_{p=1}^{M_w} \mathcal{Z}_{i-1/2}^p \tag{15.66}$$

moving at speed $s_{i-1/2}^p$, and then define the fluctuations and correction terms directly from the $\mathcal{Z}^p$. This viewpoint is useful in applying some approximate Riemann solvers, and will be used in showing the second-order accuracy of wave-propagation methods in the next section. It also appears to be quite useful in the context of spatially-varying flux functions (see Section 16.4), as explored in [18]. See also [288] for a more general approach where the jumps in both Q and $f(Q)$ are simultaneously split into waves.

If a linearized Riemann solver is used, then the vector $f(Q_i) - f(Q_{i-1})$ can be decomposed as a linear combination of the eigenvectors $\hat{r}_{i-1/2}^p$ of the linearized matrix $\hat{A}_{i-1/2}$.

Instead of solving the system (15.14), we solve

$$f(Q_i) - f(Q_{i-1}) = \sum_{p=1}^{m} \beta_{i-1/2}^p \hat{r}_{i-1/2}^p \tag{15.67}$$

for the coefficients $\beta_{i-1/2}^p$ and then define

$$\mathcal{Z}_{i-1/2}^p = \beta_{i-1/2}^p \hat{r}_{i-1/2}^p. \tag{15.68}$$

If the wave speeds are all nonzero, then we can recover waves $\mathcal{W}_{i-1/2}^p$ by setting

$$\mathcal{W}_{i-1/2}^p = \frac{1}{s_{i-1/2}^p} \mathcal{Z}_{i-1/2}^p, \tag{15.69}$$

and view this as an alternative way to obtain an approximate Riemann solution in the standard form needed for wave propagation. An advantage of this approach is that using the condition (15.66) to define the $\mathcal{Z}_{i-1/2}^p$ guarantees that the method will be conservative when the fluctuations (15.10) are used. This is true for any linearization, for example the simple arithmetic average $\hat{A}_{i-1/2} = f'(\frac{1}{2}(Q_{i-1} + Q_i))$, whereas (15.16) may not be satisfied if the wave splitting is based on (15.14) unless $\hat{A}_{i-1/2}$ is chosen to be a special average such as the Roe average. When the Roe average is used, for which (15.18) is satisfied, the two approaches give exactly the same splitting, since (15.67) then yields

$$\hat{A}_{i-1/2}(Q_i - Q_{i-1}) = \sum_{p=1}^{m} \beta_{i-1/2}^p \hat{r}_{i-1/2}^p$$

and applying $\hat{A}_{i-1/2}$ to (15.14) shows that $\beta_{i-1/2}^p = s_{i-1/2}^p \alpha_{i-1/2}^p$.

The wave-propagation methods can be written directly in terms of the waves $\mathcal{Z}_{i-1/2}^p$ in a manner that avoids needing to form the $\mathcal{W}_{i-1/2}^p$ at all, which is more satisfying in cases where a wave speed is near zero and (15.69) might break down. The fluctuations can be rewritten as

$$\mathcal{A}^- \Delta Q_{i-1/2} = \sum_{p:s_{i-1/2}^p < 0} \mathcal{Z}_{i-1/2}^p,$$

$$\mathcal{A}^+ \Delta Q_{i-1/2} = \sum_{p:s_{i-1/2}^p > 0} \mathcal{Z}_{i-1/2}^p. \tag{15.70}$$

The second-order correction terms (15.63) can also be rewritten in terms of the $\mathcal{Z}_{i-1/2}^p$ by combining one factor of $s_{i-1/2}^p$ with $\mathcal{W}_{i-1/2}^p$, at least in the case where no limiter is used so that $\tilde{\mathcal{W}}_{i-1/2}^p = \mathcal{W}_{i-1/2}^p$ in (15.63). We obtain

$$\tilde{F}_{i-1/2} = \frac{1}{2} \sum_{p=1}^{M_w} \text{sgn}(s_{i-1/2}^p) \left(1 - \frac{\Delta t}{\Delta x} |s_{i-1/2}^p|\right) \mathcal{Z}_{i-1/2}^p. \tag{15.71}$$

In Section 15.6 we will show that the resulting method is second-order accurate with a reasonable consistency condition on the Riemann solver.

To obtain high-resolution results, limiters can now be applied to the vectors $\mathcal{Z}_{i-1/2}^p$ rather than to the $\mathcal{W}_{i-1/2}^p$, using any standard limiting techniques. This appears to work as well in practice as the standard approach and allows a broader range of linearizations to be used. If the Roe linearization is used, then the two approaches give identical methods in the unlimited case, though they will be slightly different if the limiters are applied to the $\mathcal{Z}_{i-1/2}^p$ rather than to the $\mathcal{W}_{i-1/2}^p$.

15.6 Second-Order Accuracy

For smooth solutions we would like to confirm second-order accuracy of the method (15.62), at least if the limiters are suppressed and $\widetilde{\mathcal{W}}_{i-1/2}^p$ is replaced by $\mathcal{W}_{i-1/2}^p$ in (15.63), or the unlimited waves $\mathcal{Z}_{i-1/2}^p$ are used in the formulation of Section 15.5. Having built this method up from Godunov's method (based on Riemann solutions) and correction terms in each characteristic field (based on scalar theory), it is not obvious that second-order accuracy will be obtained for nonlinear systems, especially when approximate Riemann solvers are used. To confirm that it is, one must compute the local truncation error or, equivalently, compare the numerical updating formula with the Taylor series expansion. For the conservation law $q_t + f(q)_x = 0$, we have

$$q_t = -f(q)_x,$$
$$q_{tt} = -(f'(q)q_t)_x = \left[f'(q)f(q)_x \right]_x, \tag{15.72}$$

and so

$$q(x_i, t_{n+1}) = q(x_i, t_n) - \Delta t f(q)_x + \frac{1}{2}\Delta t^2 \left[f'(q)f(q)_x \right]_x + \mathcal{O}(\Delta t^3), \tag{15.73}$$

where all terms on the right are evaluated at (x_i, t_n).

To obtain an expression that matches this to the desired order from the numerical method, we will need to make an assumption on the accuracy of the Riemann solver. For arbitrary data Q_{i-1} and Q_i we assume that the method uses a flux-difference splitting of the form

$$f(Q_i) - f(Q_{i-1}) = \sum_{p=1}^{m} \mathcal{Z}_{i-1/2}^p \tag{15.74}$$

where the vectors $\mathcal{Z}_{i-1/2}^p$ are the eigenvectors of some matrix $\hat{A}_{i-1/2} = \hat{A}(Q_{i-1}, Q_i)$ corresponding to eigenvalues $s_{i-1/2}^p$. Either the $\mathcal{Z}_{i-1/2}^p$ are computed directly as described in Section 15.5, or else we define

$$\mathcal{Z}_{i-1/2}^p = s_{i-1/2}^p \mathcal{W}_{i-1/2}^p \tag{15.75}$$

in terms of the waves $\mathcal{W}_{i-1/2}^p$ computed from the decomposition (15.14).

To obtain second-order accuracy, we must make a mild assumption on the consistency of the matrix-valued function $\hat{A}(q_l, q_r)$ with the Jacobian $f'(q)$. If $q(x)$ is a smooth function of x, then we require that

$$\hat{A}(q(x), q(x + \Delta x)) = f'(q(x + \Delta x/2)) + E(x, \Delta x), \tag{15.76}$$

where the error $E(x, \Delta x)$ satisfies

$$E(x, \Delta x) = \mathcal{O}(\Delta x) \quad \text{as } \Delta x \to 0 \tag{15.77}$$

and

$$\frac{E(x + \Delta x, \Delta x) - E(x, \Delta x)}{\Delta x} = \mathcal{O}(\Delta x) \quad \text{as } \Delta x \to 0. \tag{15.78}$$

In particular, if

$$\hat{A}(q(x), q(x + \Delta x)) = f'(q(x + \Delta x/2)) + \mathcal{O}(\Delta x^2), \tag{15.79}$$

then both of these conditions will be satisfied. Hence we can choose

$$\hat{A}(Q_{i-1}, Q_i) = f'(\hat{Q}_{i-1/2}) \tag{15.80}$$

with $\hat{Q}_{i-1/2} = \frac{1}{2}(Q_{i-1} + Q_i)$ or with the Roe average and obtain a second-order method. The form of the conditions (15.77) and (15.78) allows more flexibility, however. The matrix $\hat{A}$ need only be a first-order accurate approximation to f' at the midpoint provided that the error is smoothly varying. This allows, for example, taking $\hat{Q}_{i-1/2} = Q_{i-1}$ or Q_i in (15.80), provided the same choice is made at all grid points.

To verify the second-order accuracy of a method satisfying this consistency condition, we write out the updating formula (15.62) for Q_i^{n+1} using the fluctuations (15.70) and the corrections (15.71). This gives

$$Q_i^{n+1} = Q_i^n - \frac{\Delta t}{\Delta x} \left[\sum_{p:s_{i-1/2}^p > 0} \mathcal{Z}_{i-1/2}^p + \sum_{p:s_{i-1/2}^p < 0} \mathcal{Z}_{i+1/2}^p \right]$$

$$- \frac{\Delta t}{2\Delta x} \left[\sum_{p=1}^m \operatorname{sgn}(s_{i+1/2}^p) \left(1 - \frac{\Delta t}{\Delta x}|s_{i+1/2}^p|\right) \mathcal{Z}_{i+1/2}^p \right.$$

$$\left. - \sum_{p=1}^m \operatorname{sgn}(s_{i-1/2}^p) \left(1 - \frac{\Delta t}{\Delta x}|s_{i-1/2}^p|\right) \mathcal{Z}_{i-1/2}^p \right]$$

$$= Q_i^n - \frac{\Delta t}{2\Delta x} \left[\sum_{p=1}^m \mathcal{Z}_{i-1/2}^p + \sum_{p=1}^m \mathcal{Z}_{i+1/2}^p \right]$$

$$+ \frac{\Delta t^2}{2\Delta x^2} \left[\sum_{p=1}^m s_{i+1/2}^p \mathcal{Z}_{i+1/2}^p - \sum_{p=1}^m s_{i-1/2}^p \mathcal{Z}_{i-1/2}^p \right]$$

$$= Q_i^n - \frac{\Delta t}{2\Delta x} \left[\sum_{p=1}^m \mathcal{Z}_{i-1/2}^p + \sum_{p=1}^m \mathcal{Z}_{i+1/2}^p \right]$$

$$+ \frac{\Delta t^2}{2\Delta x^2} \left[\hat{A}_{i+1/2} \sum_{p=1}^m \mathcal{Z}_{i+1/2}^p - \hat{A}_{i-1/2} \sum_{p=1}^m \mathcal{Z}_{i-1/2}^p \right]. \tag{15.81}$$

To obtain the last line we have used the fact that each $\mathcal{Z}^p$ is an eigenvector of the corresponding $\hat{A}$ with eigenvalue s^p. We can now use the assumption (15.66) to rewrite this as

$$Q_i^{n+1} = Q_i^n - \frac{\Delta t}{2 \Delta x}[f(Q_{i+1}) - f(Q_{i-1})]$$

$$-\frac{\Delta t^2}{2 \Delta x^2} \left\{ \hat{A}_{i+1/2}[f(Q_{i+1}) - f(Q_i)] - \hat{A}_{i-1/2}[f(Q_i) - f(Q_{i-1})] \right\}. \quad (15.82)$$

This agrees with the Taylor series expansion (15.73) to sufficient accuracy that a standard computation of the truncation error now shows that the method is second-order accurate, provided that $\hat{A}$ is a consistent approximation to $f'(q)$ as described above. This follows because the conditions (15.77) and (15.78) guarantee that

$$\hat{A}(q(x), q(x + \Delta x)) \left(\frac{f(q(x + \Delta x)) - f(q(x))}{\Delta x} \right)$$

$$= f'(q(x + \Delta x/2)) f(q(x + \Delta x/2))_x + E_2(x, \Delta x) \quad (15.83)$$

with $E_2(x, \Delta x)$ satisfying the same conditions as $E(x, \Delta x)$. This in turn is sufficient to show that the final term in (15.82) agrees with the $\mathcal{O}(\Delta t^2)$ term in (15.73) to $\mathcal{O}(\Delta t^2 \Delta x)$, as required for second-order accuracy. Note that in place of the assumptions (15.77) and (15.78) on $E(x, \Delta x)$, it would be sufficient to simply assume that (15.83) holds with $E_2(x, \Delta x)$ satisfying these conditions. This is looser in the sense that only the product of $\hat{A}$ with one particular vector is required to be well behaved, not the entire matrix. This proof carries over to spatially-varying flux functions as well, as presented in [18].

15.6.1 Two-Step Lax–Wendroff Methods

It is worth noting that there are other ways to achieve second-order accuracy that do not require approximating the Jacobian matrix or its eigenstructure, in spite of the fact that the Taylor series expansion (15.73) appears to require this for the second-order terms. The need for the Jacobian can be avoided by taking a two-step approach. One example is the *Richtmyer method* of Section 4.7. When (4.23) is inserted in (4.22) and a Taylor series expansion performed, the required Jacobian terms appear, but these are not explicitly computed in the implementation where only the flux $f(q)$ is evaluated.

Another popular variant is *MacCormack's method*, originally introduced in [318]:

$$Q_i^* = Q_i^n - \frac{\Delta t}{\Delta x}[f(Q_{i+1}^n) - f(Q_i^n)],$$

$$Q_i^{**} = Q_i^* - \frac{\Delta t}{\Delta x}[f(Q_i^*) - f(Q_{i-1}^*)], \quad (15.84)$$

$$Q_i^{n+1} = \frac{1}{2}(Q_i^n + Q_i^{**}).$$

Note that one-sided differencing is used twice, first to one side and then to the other. The order in which the two directions are used can also be switched, or one can alternate between the two orderings in successive time steps, yielding a more symmetric method. Again, Taylor series expansion shows that the method is second-order accurate, while the explicit use of Jacobian matrices or Riemann solvers is avoided.

The problem with both the Richtmyer and MacCormack methods is that they typically produce spurious oscillations unless artificial viscosity is explicitly added, as is done in most practical calculations with these methods. But adding artificial viscosity often results in the addition of too much diffusion over most of the domain. The advantage of the high-resolution methods based on Riemann solvers is that we can tune this viscosity much more carefully to the behavior of the solution. By applying limiter functions to each characteristic field separately, we are in essence applying the optimal amount of artificial viscosity at each cell interface, and only to the fields where it is needed. This often results in much better solutions, though at some expense relative to the simpler Richtmyer or MacCormack methods.

15.7 Flux-Vector Splitting

Our focus has been on flux-difference splitting methods for conservation laws, where the flux difference $f(Q_i) - f(Q_{i-1})$ is split into fluctuations $\mathcal{A}^-\Delta Q_{i-1/2}$ (which modifies Q_{i-1}) and $\mathcal{A}^+\Delta Q_{i-1/2}$ (which modifies Q_i). This splitting is typically determined by solving a Riemann problem between the states Q_{i-1} and Q_i. There is another related approach, already introduced in Section 4.13, where instead each flux vector $f(Q_i)$ is split into a left-going part $f_i^{(-)}$ and a right-going part $f_i^{(+)}$, so we have

$$f(Q_i) = f_i^{(-)} + f_i^{(+)} . \tag{15.85}$$

We can then define the interface flux

$$F_{i-1/2} = f_{i-1}^{(+)} + f_i^{(-)} \tag{15.86}$$

based on the portion of each cell-centered flux approaching the interface. A method of this form is called a *flux-vector splitting* method, since it is the flux vector $f(Q_i)$ that is split instead of the flux difference.

As noted in Section 4.13, for constant-coefficient linear systems these two approaches are identical, but for nonlinear problems they typically differ. From a flux-vector splitting method it is possible to define fluctuations $\mathcal{A}^\pm\Delta Q_{i-1/2}$ as described in Section 4.13, so that these methods can also be expressed in the form (15.5) and implemented in CLAWPACK if desired.

15.7.1 The Steger–Warming flux

Steger and Warming [424] introduced the idea of flux-vector splitting for the Euler equations and used a special property of this system of equations, that it is *homogeneous of degree 1* (at least for certain equations of state such as that of an ideal polytropic gas). This means that $f(\alpha q) = \alpha f(q)$ for any scalar α. From this it follows, by Euler's identity for homogeneous functions, that

$$f(q) = f'(q)q \tag{15.87}$$

for any state q, which is not true for nonlinear functions that do not have this homogeneity. Hence if A_i is the Jacobian matrix $f'(Q_i)$ (or some approximation to it), then a natural flux-vector splitting is given by

$$f_i^{(-)} = A_i^- Q_i = \sum_{p=1}^{m} \left(\lambda_i^p\right)^- \omega_i^p r_i^p,$$

$$f_i^{(+)} = A_i^+ Q_i = \sum_{p=1}^{m} \left(\lambda_i^p\right)^+ \omega_i^p r_i^p,$$

(15.88)

where the notation (4.45) is used and

$$Q_i = \sum_{p=1}^{m} \omega_i^p r_i^p \qquad (15.89)$$

is the eigendecomposition of Q_i. By (15.86) we thus have

$$F_{i-1/2} = A_{i-1}^+ Q_{i-1} + A_i^- Q_i . \qquad (15.90)$$

This is a natural generalization of (4.56) to nonlinear systems that are homogeneous of degree 1. For systems that don't have this property, we can still define a flux-vector splitting using the eigenvectors r_i^p of A_i. Instead of decomposing Q_i into the eigenvectors and then using (15.88), we can directly decompose $f(Q_i)$ into these eigenvectors,

$$f(Q_i) = \sum_{p=1}^{m} \phi_i^p r_i^p \qquad (15.91)$$

and then define the flux splitting by

$$f_i^{(-)} = \sum_{p=1}^{m} \phi_i^{p(-)} r_i^p, \qquad f_i^{(+)} = \sum_{p=1}^{m} \phi_i^{p(+)} r_i^p, \qquad (15.92)$$

where

$$\phi_i^{p(-)} = \begin{cases} \phi_i^p & \text{if } \lambda_i^p < 0, \\ 0 & \text{if } \lambda_i^p \geq 0, \end{cases}$$

$$\phi_i^{p(+)} = \begin{cases} 0 & \text{if } \lambda_i^p < 0, \\ \phi_i^p & \text{if } \lambda_i^p \geq 0. \end{cases}$$

(15.93)

If the system is homogeneous of degree 1, then $\phi_i^p = \lambda_i^p r_i^p$ and (15.92) reduces to the previous expression (15.88).

The above splitting for the Euler equations is called *Steger–Warming flux-vector splitting* in the aerodynamics community. An equivalent method, known as the *beam scheme*, was introduced earlier in astrophysics [392] from a different viewpoint: each state Q_i is decomposed into distinct beams of particles traveling at the different wave speeds.

For transonic flow problems in aerodynamics, the flux-vector splitting given above suffers from the fact that the splitting does not behave smoothly as the Mach number passes through

1 (where the characteristic speed $u - c$ or $u + c$ changes sign). This can cause convergence problems when solving for a steady state. A smoother flux-vector splitting was introduced by van Leer [469]. Many variants and improvements have since been introduced, such as the AUSM method of Liou and coworkers [301], [302], [479], the Marquina flux [112], [287], [323], and kinetic flux-vector splittings [39], [63], [107], [320], [488], [489].

Use of the flux function (15.90) would only give a first-order accurate method. To obtain better accuracy one might use this flux function in an ENO-based semi discrete method and then apply Runge–Kutta time stepping (see Section 10.4) to achieve higher-order accuracy.

The flux-vector splitting can also be used in conjunction with the high-resolution method developed in Section 15.4 (and CLAWPACK) by using $F_{i-1/2}$ to define fluctuations as in (4.58) and then also defining waves $\mathcal{W}^p_{i-1/2}$ and speeds $s^p_{i-1/2}$ for use in the second-order correction terms of (15.5). From (15.89) we have

$$Q_i - Q_{i-1} = \sum_{p=1}^m \left(\omega^p_i r^p_i - \omega^p_{i-1} r^p_{i-1} \right),$$

which suggests defining the pth wave $\mathcal{W}^p_{i-1/2}$ as

$$\mathcal{W}^p_{i-1/2} = \omega^p_i r^p_i - \omega^p_{i-1} r^p_{i-1}.$$

The corresponding speed $s^p_{i-1/2}$ might then be defined as

$$s^p_{i-1/2} = \frac{1}{2} \left(\lambda^p_{i-1} + \lambda^p_i \right).$$

Alternatively, the formulation of Section 15.5 can be used with this same wave speed and the waves

$$\mathcal{Z}^p_{i-1/2} = \phi^p_i r^p_i - \phi^p_{i-1} r^p_{i-1},$$

in which case the fluctuations can be defined using (15.70).

15.8 Total Variation for Systems of Equations

As noted in Section 15.2, there is no proof that even the first-order Godunov method converges on general systems of nonlinear conservation laws. This is because in general there is no analogue of the TVD property for scalar problems that allows us to prove compactness and hence stability. In fact, there is not even a proof of existence of the "true solution" for general nonlinear systems of conservation laws, unless the initial data is severely restricted, even for problems where physically we know a solution exists. In a sense, this results from our inability to prove convergence of numerical methods, since one standard way of proving existence theorems is to construct a sequence of approximations (i.e., define some algorithm that could also be used numerically) and then prove that this sequence converges to a solution of the equation. Recall, for example, that this was the context in which Courant, Friedrichs, and Lewy first developed the CFL condition [93].

In this section we will briefly explore some issues related to variation and oscillations in nonlinear systems. We start by looking at some of the difficulties inherent in trying to obtain total variation bounds. For a system of m equations, we might try to define the total variation by

$$\text{TV}(q) = \sup \sum_{j=1}^{N} \|q(\xi_j) - q(\xi_{j-1})\|,$$ (15.94)

where the supremum is taken over all subdivisions of the real line $-\infty = \xi_0 < \xi_1 < \cdots < \xi_N = \infty$, generalizing (6.19) by replacing the absolute value by some vector norm in $\mathbb{R}^m$. We will be particularly concerned with piecewise constant grid functions, in which case (15.94) reduces to

$$\text{TV}(Q) = \sum_{i=-\infty}^{\infty} \|Q_i - Q_{i-1}\|.$$ (15.95)

With this definition of TV, and similar replacement of absolute value by the vector norm elsewhere in the notions of convergence used in the Lax–Wendroff theorem, this theorem continues to hold. We might also hope that numerical methods will produce solutions that have bounded total variation in this sense, in which case we could prove stability and convergence.

In general, however, we cannot hope to develop methods that are TVD with this definition of TV, because the true solution is itself not TVD. In fact, the total variation can increase by an arbitrarily large amount over an arbitrarily short time if we choose suitable data, and so we cannot even hope to obtain a bound of the form $\text{TV}(Q^{n+1}) \le (1 + \alpha \, \Delta t) \text{TV}(Q^n)$. To see this, consider the simple example in Section 13.4, the Riemann problem for the shallow water equations in which $h_l = h_r$ and $u_l = -u_r > 0$ (two equal streams of water smashing into one another). The initial data has no variation in h, and the variation in hu is $2h_l u_l$. For any time $t > 0$ the variation in hu is still $2h_l u_l$, but the depth near $x = 0$ increases to $h_m > h_l$, and so h has total variation $2(h_m - h_l) > 0$. By choosing u_l large enough we can make this increase in variation arbitrarily large, because h_m increases with u_l.

For certain systems of equations it is possible to prove stability by measuring the total variation in terms of *wave strengths* instead of using standard vector norms in $\mathbb{R}^m$. A simple example is a constant-coefficient linear system, as considered in the next section.

15.8.1 Total Variation Estimates for Linear Systems

Consider a linear system $q_t + A q_x = 0$. Note that linear systems can exhibit exactly the same growth in TV as seen in the shallow water example above. Consider the Riemann problem for the the acoustics equations with $p_l = p_r$ and $u_l = -u_r > 0$, for example, which behaves just as described above. In spite of this, for a constant-coefficient linear system we can easily prove convergence of standard methods by diagonalizing the system, decoupling it into independent scalar advection equations that have the TVD property. For example, Godunov's method for a linear system can be written as

$$Q_i^{n+1} = Q_i^n - \frac{\Delta t}{\Delta x}\left[A^+\left(Q_i^n - Q_{i-1}^n\right) + A^-\left(Q_{i+1}^n - Q_i^n\right)\right].$$

Multiplying by R^{-1} (the matrix of left eigenvectors) and defining $W_i^n = R^{-1}Q_i^n$, we obtain

$$W_i^{n+1} = W_i^n - \frac{\Delta t}{\Delta x}\left[\Lambda^+\left(W_i^n - W_{i-1}^n\right) + \Lambda^-\left(W_{i+1}^n - W_i^n\right)\right],$$

where $\Lambda^\pm = R^{-1}A^\pm R$. This is an uncoupled set of m first-order upwind algorithms for the characteristic variables, each of which is TVD and convergent. It follows that $Q_i^n = RW_i^n$ is also convergent.

This suggests that we define the total variation of Q by using a vector norm such as

$$\|Q\|_W \equiv \|R^{-1}Q\|_1, \tag{15.96}$$

where $\|\cdot\|_1$ is the standard 1-norm in $\mathbb{R}^m$. Since R^{-1} is nonsingular, this defines a vector norm. Then we can define the corresponding total variation by

$$\begin{aligned}
\mathrm{TV}_W(Q) &= \sum_{i=-\infty}^{\infty} \|Q_i - Q_{i-1}\|_W \\
&= \sum_{i=-\infty}^{\infty} \|R^{-1}(Q_i - Q_{i-1})\|_1 \\
&= \sum_{i=-\infty}^{\infty} \|W_i - W_{i-1}\|_1 \\
&= \sum_{i=-\infty}^{\infty} \sum_{p=1}^{m} |W_i^p - W_{i-1}^p| \\
&= \sum_{p=1}^{m} \mathrm{TV}(W^p),
\end{aligned} \tag{15.97}$$

where $\mathrm{TV}(W^p)$ is the scalar total variation of the pth characteristic component. Since the scalar upwind method is TVD, we have

$$\mathrm{TV}((W^p)^{n+1}) \le \mathrm{TV}((W^p)^n),$$

and hence, with the definition (15.97) of total variation, we can show that

$$\mathrm{TV}_W(Q^{n+1}) \le \mathrm{TV}_W(Q^n). \tag{15.98}$$

For the exact solution to a linear system, this same approach shows that $\mathrm{TV}_W(q(\cdot, t))$ remains constant in time, since each scalar advection equation for the characteristic variable $w^p(x, t)$ maintains constant variation.

From this result we can also show that other forms of the total variation, such as (15.94), remain uniformly bounded even if they are not diminishing. We have

$$
\begin{aligned}
\mathrm{TV}_1(Q^n) &= \sum_i \| Q_i - Q_{i-1} \|_1 \\
&= \sum_i \| R(W_i^n - W_{i-1}^n) \|_1 \\
&\leq \| R \|_1 \sum_i \| W_i^n - W_{i-1}^n \|_1 \\
&= \| R \|_1 \, \mathrm{TV}_W(Q^n) \\
&\leq \| R \|_1 \, \mathrm{TV}_W(Q^0) \\
&= \| R \|_1 \sum_i \| R^{-1}(Q_i^0 - Q_{i-1}^0) \|_1 \\
&\leq \| R \|_1 \, \| R^{-1} \|_1 \, \mathrm{TV}_1(Q^0).
\end{aligned}
\tag{15.99}
$$

Hence TV_1 grows by at most a factor of $\| R \|_1 \| R^{-1} \|_1$, the condition number of the eigenvector matrix, over any arbitrary time period. It is natural for this condition number to appear in the bound, since it measures how nearly linearly dependent the eigenvectors of A are. Recalling the construction of the Riemann solution from Chapter 3, we know that eigenvectors that are nearly linearly dependent can give rise to large variation in the Riemann solutions. (See also the example in Section 16.3.1.)

Note that if we write the Riemann solution as a sum of waves,

$$
Q_i - Q_{i-1} = \sum_p \mathcal{W}_{i-1/2}^p = \sum_p \alpha_{i-1/2}^p r^p,
$$

as introduced in Section 3.8, then we can also express $\mathrm{TV}_W(Q)$ as

$$
\mathrm{TV}_W(Q) = \sum_i \sum_p |\alpha_{i-1/2}^p|.
\tag{15.100}
$$

Recall that in defining the basis of eigenvectors r^p (and hence the matrix R) we could choose any convenient normalization. For our present purposes it is most convenient to assume that r^p is chosen to have $\| r^p \|_1 = 1$. Then

$$
|\alpha_{i-1/2}^p| = \| \alpha_{i-1/2}^p r^p \|_1 = \| \mathcal{W}_{i-1/2}^p \|_1,
$$

and we simply have

$$
\mathrm{TV}_W(Q) = \sum_i \sum_p \| \mathcal{W}_{i-1/2}^p \|_1.
\tag{15.101}
$$

Thus another interpretation of TV_W is that it is the sum of the wave strengths over all waves arising in all Riemann problems at cell interfaces. This is illustrated in Figure 15.7. For a linear system all waves simply pass through one another (linear superposition) with no change in their strength as time advances. This is not true for nonlinear systems.

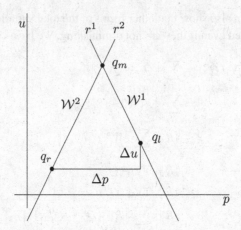

Fig. 15.7. Illustration of $TV_W(Q)$ for Riemann-problem data in the acoustics equations. $\|q_r - q_l\|_1 = |\Delta p| + |\Delta u|$, whereas $\|q_r - q_l\|_W = \|\mathcal{W}^1\|_1 + \|\mathcal{W}^2\|_1$.

15.8.2 Wave-Strength Estimates for Nonlinear Systems

For a linear system of equations we have just seen that we can prove the solution is total variation bounded by measuring the variation of a piecewise constant solution in terms of the metric

$$d_W(q_l, q_r) \equiv \sum_p \|\mathcal{W}^p\|_1, \qquad (15.102)$$

where the $\mathcal{W}^p$ represent waves arising in the Riemann solution between states q_l and q_r. Then

$$TV_W(Q) = \sum_i d_W(Q_{i-1}, Q_i). \qquad (15.103)$$

We can make the same definition of total variation in the case of a nonlinear hyperbolic system. For a linear system this metric can be rewritten in terms of a norm $\|\cdot\|_W$, since the value of $d_W(q_l, q_r)$ depends only on the value $q_r - q_l$. This is what was used in the previous section. For a nonlinear problem this is no longer true. The wave strengths in the Riemann solution between two states q_l' and q_r' might be quite different from those in the Riemann solution between q_l and q_r even if $q_r' - q_l' = q_r - q_l$.

If we attempt to use TV_W as defined in (15.103) to study the stability of Godunov's method on nonlinear problems we run into two difficulties:

1. In general the total variation TV_W may not be diminishing even in the true solution. Consider, for example, data that consists of two approaching shocks that collide and produce two outgoing shocks. For a nonlinear problem it might happen that the outgoing shocks are *stronger* than the incoming shocks (with larger $\|\mathcal{W}^p\|_1$).
2. The averaging process in Godunov's method introduces new states into the approximate solution that may not appear in the true solution. Because the structure of the Hugoniot loci and integral curves can vary rapidly in state space, this may introduce additional unphysical variation when Riemann problems are solved between these new states.

For certain special systems of equations these difficulties can be overcome and stability proved. For example, for certain systems of two equations the integral curves and Hugoniot loci are identical. In this case it can be shown that that these curves are in fact straight lines in state space. Such systems have been studied by Temple [448], [447], who obtained total variation bounds for the true solution and used these to prove existence of solutions. It is also possible to obtain TV estimates when Godunov's method is applied to such a system, and hence convergence can be proved [291]. Similar results have also been obtained for more general nonlinear systems with straight-line fields [46]. For general nonlinear systems, however, such results are not currently available.

15.8.3 Glimm's Random-Choice Method

For systems such as the Euler equations it is not possible to prove convergence of Godunov's method. In fact, it is not even possible to prove that weak solutions exist for all time if we allow arbitrary initial data. However, if the initial data is constrained to have sufficiently small total variation, then there is a famous proof of existence due to Glimm [152]. This proof is based on a constructive method that can also be implemented numerically, and is often called the *random-choice method*. Several variants of this method are described in various sources e.g., [43], [78], [98], [307], [316], [420], [450], [499]. Here we only summarize the major features:

- A finite volume grid with piecewise-constant data is introduced as in Godunov's method, and Riemann problems are solved at each cell interface to define a function $\tilde{q}^n(x, t)$ as in Section 4.10. However, instead of averaging the resulting solutions over grid cells, the value of $\tilde{q}^n(x, t_{n+1})$ at a random point x in each cell is chosen. This avoids difficulty 2 mentioned in the previous subsection.

- A functional similar to (15.103) is used to measure the solution, but an additional quadratic term is introduced that measures the potential for future interaction between each pair of waves that are approaching one another. This is necessary so as to take into account the potential increase in variation that can arise when two waves interact. The quadratic functional is the crux of Glimm's proof, since he was able to show that a suitable choice results in a functional that is nonincreasing in time (for data with sufficiently small variation).

Computationally, the random-choice method has the advantage that shocks remain sharp, since the solution is sampled rather than averaged. Of course, the method cannot be exactly conservative, for this same reason, but Glimm showed convergence to a weak solution with probability 1. Another disadvantage computationally is that smooth flow is typically not very smoothly represented. There also appear to be problems extending the method to more than one space dimension [78], and for these reasons the method is not widely used for practical problems, though it may be very useful for some special ones.

Since Glimm's paper, other approaches have been introduced to prove existence and in some cases uniqueness results for certain systems of conservation laws (under suitable restrictions on the initial data), using techniques such as compensated compactness [109], [403], [446], front tracking [44], [97], [371], and semigroup methods [35],[45], [46], [48], [49], [205], [312]. For an overview of many recent theoretical results, see Dafermos [98].

15.8.4 Oscillation in Shock Computations

Consider a simple Riemann problem in which the data lies on a Hugoniot curve, so that the solution consists of a single shock wave. In this case the exact solution is monotonically varying and has constant total variation. In this case, at least, we might hope that a sensible numerical method based on scalar TVD theory would behave well and not introduce spurious oscillations. In practice high-resolution methods of the type we have developed do work very well on problems of this sort, most of the time. However, even the simplest first-order Godunov method can produce small spurious oscillations, which can be noticeable in some computations. Here we briefly consider two common situations where this can arise.

Start-up Errors

Figure 15.8 shows the numerical solution to the Riemann problem for the Euler equations with the data

$$\begin{aligned} \rho_l &= 5.6698, & u_l &= 9.0299, & p_l &= 100 & \text{for } x < 0, \\ \rho_r &= 1.0, & u_r &= 0.0, & p_r &= 1.0 & \text{for } x > 0. \end{aligned} \tag{15.104}$$

This data has been chosen to satisfy the Rankine–Hugoniot jump relations (with $\gamma = 1.4$), so that the solution is a single 3-shock propagating with speed $s = 10.9636$. Figure 15.8 shows the computed density at time $t = 1$ on a grid with 400 cells, using the minmod limiter. We see in Figure 15.8(a) that the solution is resolved quite sharply, with only three points in the shock. The shock is in the correct location and the solution is roughly constant away from it. However, some small oscillations are visible. These are seen much more clearly in Figure 15.8(b), which shows the same solution on a different scale. Most noticeable are two dips in the density. These are waves that arose from the initial discontinuity at $x = 0$. The leftmost dip is an acoustic wave moving at speed $u_l - c_l \approx 4.06$, and the other dip is an entropy wave moving at the fluid velocity $u_l \approx 9.03$. This sort of *start-up error* is frequently seen when an exact discontinuity is used as initial data.

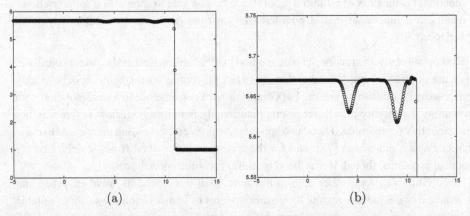

(a)   (b)

Fig. 15.8. Start-up error with Godunov's method on a 3-shock in the Euler equations. (a) Density at $t = 1$. (b) Magnification of the same results. [claw/book/chap15/startup]

How does this oscillation arise? The initial data is

$$Q_i^0 = \begin{cases} q_l & \text{if } i < I, \\ q_r & \text{if } i \geq I \end{cases}$$

for some I. In the first time step, the Riemann problem at $x_{I-1/2}$ gives rise to a single wave that is computed exactly (even if the Roe solver is used instead of the exact solver computationally). However, this wave travels a distance less than Δx in this time step, and the averaging process of Godunov's method produces a new state Q_I^1 in one grid cell that is a convex combination of q_l and q_r,

$$Q_I^1 = \frac{s \, \Delta t}{\Delta x} q_l + \left(1 - \frac{s \, \Delta t}{\Delta x}\right) q_r.$$

This state lies on the straight line connecting q_l and q_r in phase space. In the next time step, there will be two Riemann problems at $x_{I-1/2}$ and at $x_{I+1/2}$ that have nontrivial solutions. If the Hugoniot locus joining q_l to q_r happens to be a straight line (as in Temple-class systems, for example; see Section 15.8.2), then each these two Riemann problems will result in a single shock wave, since Q_I^1 will lie on the same Hugoniot locus, and the two shocks together make up the original shock. As in a scalar problem, the numerical solution will be smeared over more grid cells as time evolves, but we will still be approximating the single shock well, and no oscillations will appear.

However, for most nonlinear systems the Hugoniot curve is not a straight line, and so the state Q_I^1 does not lie on the same Hugoniot curve as $Q_{I-1}^1 = q_l$ or $Q_{I+1}^1 = q_r$. Solving these Riemann problems then results in the generation of waves in *all* families, and not just a 3-wave as expected from the initial data. It is these other waves that lead to the oscillations observed in Figure 15.8. See [14] for more analysis and some plots showing how these oscillations evolve in state space.

Slow-Moving Shocks

The start-up error discussed above tends to be damped by the numerical viscosity in Godunov's method, and so is often not very visible. In situations where the numerical viscosity is small, however, the oscillations can become significant.

In particular, oscillations are frequently seen if the shock is moving very slowly, in the sense that the shock speed is very small relative to the fastest characteristic speeds in the problem. Then it takes several time steps to cross a single grid cell even when the Courant number is close to one. Recall from (15.51) that the numerical viscosity of Roe's method vanishes as the wave speed goes to zero, and the same is true for Godunov's method based on the exact Riemann solver in the case where the solution is a single shock with speed close to zero. This can be an advantage for scalar problems: nearly stationary shocks are captured very sharply. But for nonlinear systems this lack of viscosity can lead to increased oscillations.

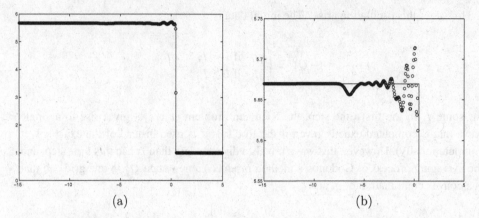

Fig. 15.9. Oscillations arising in a slow shock computed with Godunov's method. (a) Density at $t = 1$. (b) Magnification of the same results. [`claw/book/chap15/slowshock`]

As an example, suppose we take the same data (15.104) as in the previous example, but shift the velocity of u_l and u_r by an amount close to the previous shock speed s:

$$
\begin{aligned}
\rho_l &= 5.6698, & u_l &= -1.4701, & p_l &= 100 & \text{for } x < 0, \\
\rho_r &= 1.0, & u_r &= -10.5, & p_r &= 1.0 & \text{for } x > 0.
\end{aligned}
\tag{15.105}
$$

This is simply a shift in the reference frame, and so the Riemann solution is exactly the same as before, but with all velocities shifted by the same amount -10.5. So this again gives a single shock wave, now propagating with velocity $s = 0.4636$. Computational results are shown in Figure 15.9, illustrating the oscillations that appear in this case, again with the first-order Godunov method. Similar oscillations arise in high-resolution methods, even when limiters are used. In this case the shock continues to shed oscillations as it moves along.

For some discussions of oscillations due to slow-moving shocks, and ways to improve the situation by introducing additional dissipation, see [14], [112], [224], [232], [343], [373]. The lack of numerical dissipation in Godunov-type methods can also lead to some other numerical problems, particularly for multidimensional computations in regions where strong shocks are nearly aligned with the grid. This can lead to a *cross-flow instability* or *odd–even decoupling*, and to the appearance of unphysical extrusions from the shock that are often called *carbuncles*. Again the addition of more numerical dissipation may be necessary to improve the results. For some discussions of such numerical problems and their relation to physical instabilities see, for example, [195], [287], [353], [364], [374], [487].

Exercises

15.1.   Consider the p-system

$$
\begin{aligned}
v_t - u_x &= 0, \\
u_t + p(v)_x &= 0.
\end{aligned}
\tag{15.106}
$$

(a) Show that for this system the integral in (15.21) can be evaluated in order to obtain the following Roe linearization:

$$\hat{A}_{i-1/2} = \begin{bmatrix} 0 & -1 \\ \dfrac{p_i - p_{i-1}}{V_i - V_{i-1}} & 0 \end{bmatrix}.$$

(b) In particular, determine the Roe solver for $p(v) = a^2/v$, modeling isothermal flow in Lagrangian coordinates, where a is the constant sound speed.

(c) Implement and test this isothermal Riemann solver in CLAWPACK.

(d) Does this solver require an entropy fix?

15.2. Suppose an HLL approximate Riemann solver of the form discussed in Section 15.3.7 is used, but with $s_{i-1/2}^1 = -\Delta x/\Delta t$ and $s_{i-1/2}^2 = \Delta x/\Delta t$. These are the largest speeds that can be used with this grid spacing and still respect the CFL condition, so these should be upper bounds on the physical speeds. Show that if this approximate Riemann solver is used in the first-order Godunov method, then the result is the Lax–Friedrichs method (4.20).

16
Some Nonclassical Hyperbolic Problems

In the previous chapters on nonlinear problems we have concentrated on *classical* systems of conservation laws for which the wave structure is relatively simple. In particular, we have assumed that the system is strictly hyperbolic (so that there are m distinct integral curves through each point of phase space), and that each characteristic field is either linearly degenerate or genuinely nonlinear (so that the eigenvalue is constant or varies monotonically along each integral curve). Many important systems of equations satisfy these conditions, including the shallow water equations and the Euler equations for an ideal gas, as well as linear systems such as acoustics. However, there are other important applications where one or both of these conditions fail to hold, including some problems arising in nonlinear elasticity, porous-media flow, phase transition, and magnetohydrodynamics (MHD). In this chapter we explore a few of the issues that can arise with more general systems. This is only an introduction to some of the difficulties, aimed primarily at explaining why the above assumptions lead to simplifications.

We start by considering scalar conservation laws with nonconvex flux functions (which fail to be genuinely nonlinear because $f''(q)$ vanishes at one or more points). This gives a good indication of the complications that arise also in systems of more equations that fail to be genuinely nonlinear. Then in Section 16.2 we will investigate the complications that can arise if a system is not strictly hyperbolic, i.e., if some of the wave speeds coincide at one or more points in phase space. In Section 16.3 we go even further and see what can happen if the system fails to be (strongly) hyperbolic at some points, either because the matrix is not diagonalizable or because the eigenvalues are not real.

In Section 16.4 we consider nonlinear conservation laws with spatially varying flux functions $f(q, x)$, analogous to the variable-coefficient linear systems considered in Chapter 9. Finally, Section 16.5 contains some discussion of nonconservative nonlinear hyperbolic problems.

16.1 Nonconvex Flux Functions

For the scalar conservation laws studied thus far, the flux function $f(q)$ was assumed to be a convex or concave function of q, meaning that $f''(q)$ has the same sign everywhere. For the traffic flux (11.6) it is constant and negative everywhere when $u_{max} > 0$. Burgers' equation is convex, since $f''(u) \equiv 1$. As shorthand we typically refer to all such genuinely nonlinear problems as *convex*.

350

Convexity is important because it means that the characteristic speed $f'(q)$ is varying monotonically as q varies. In solving the Riemann problem with $q_l > q_r$ (Figure 11.3), we obtain a smooth rarefaction wave with cars spreading out, since $f'(q)$ is nondecreasing as x increases. On the other hand if $q_l < q_r$ as in Figure 11.2, then the Riemann solution consists of a shock joining these two values. If the function $f(q)$ is not convex, then the Riemann solution may be more complicated and can involve both shock and rarefaction waves.

16.1.1 Two-Phase Flow and the Buckley–Leverett Equation

As an example where nonconvex flux functions arise, we study a simple model for two-phase fluid flow in a porous medium. One application is to oil-reservoir simulation. When an underground source of oil is tapped, a certain amount of oil flows out on its own due to high pressure in the reservoir. After the flow stops, there is typically a large amount of oil still in the ground. One standard method of subsequent *secondary recovery* is to pump water into the oil field through some wells, forcing oil out through others. In this case the two phases are oil and water, and the flow takes place in a porous medium of rock or sand.

We can consider a one-dimensional model problem in which oil in a tube of porous material is displaced by water pumped in through one end. Let $q(x, t)$ represent the fraction of fluid that is water (the *water saturation*), so that $0 \leq q \leq 1$, and $1 - q(x, t)$ is the fraction that is oil. If we take initial data

$$q(x, 0) = \begin{cases} 1 & \text{if } x < 0, \\ 0 & \text{if } x > 0, \end{cases} \tag{16.1}$$

so that water is to the left and oil to the right, and take a positive flow rate (pumping water in from the left), then we expect the water to displace the oil. Our first guess might be that the sharp interface between pure water ($q = 1$) and pure oil ($q = 0$) will be maintained and simply advect with constant velocity. Instead, however, one observes a sharp interface between pure oil and some mixed state $q = q^* < 1$, followed by a region in which both oil and water are present ($q^* < q < 1$). Mathematically we will see that this corresponds to a shock wave followed by a rarefaction wave.

Since both fluids are essentially incompressible, we expect the total flux of fluid to be the same past any point in the tube. In regions of pure oil or pure water the velocity must thus be the same, and we will take this value to be 1. However, in regions where both fluids are present, they may be moving at different average velocities. It is this behavior that leads to the interesting mathematical structure. Physically this arises from the fact that a porous medium consists of a solid material with many pores or minute cracks through which the fluids slowly seep. Particles of the oil are initially bound to the solid substrate and must be displaced by the water molecules. The fact that oil and water do not mix (the fluids are *immiscible*) means that surface tension effects between the two fluids are crucial on the small length scales of the pores, and can inhibit the movement of water into the smallest pores filled with oil. Regions where $0 < q < 1$ correspond to regions where some of the oil is still bound to the substrate and is stationary, so the average velocity of oil molecules may be considerably less than the average velocity of water molecules.

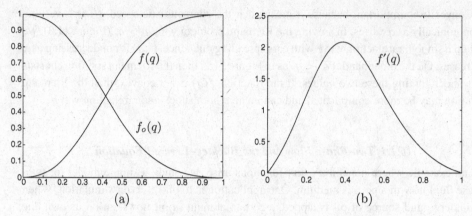

Fig. 16.1. (a) Flux functions (16.2) for the Buckley–Leverett equation, as functions of the water saturation q. (b) The characteristic speed $f'(q)$. Shown for $a = 1/2$ in (16.2).

Buckley and Leverett proposed a simple model for this complex situation in which the flux of water and oil are each given by expressions that depend on the saturation of the fluid:

$$\text{flux of water:} \quad f(q) = \frac{q^2}{q^2 + a(1-q)^2};$$

$$\text{flux of oil:} \quad f_o(q) = \frac{a(1-q)^2}{q^2 + a(1-q)^2}. \tag{16.2}$$

Here $a < 1$ is a constant. Each of the fluxes can be viewed as the product of the saturation of the phase with the average velocity of the phase. In each case the idea is that the average velocity approaches 0 as the saturation goes to 0 (the few molecules present are bound to the substrate) and approaches 1 as the saturation goes to 1 (since the fluid as a whole is flowing at this rate). The fact that $a < 1$ arises from the fact that oil flows less easily than water. So if $q = 1 - q = 0.5$, for example, we expect more water than oil to be flowing. We take $a = 1/2$ in the figures here. Note that $f(q) + f_o(q) \equiv 1$, so that the total fluid flux is the same everywhere, as required by incompressibility. Figure 16.1(a) shows these fluxes as a function of q.

We only need to solve for q, the water saturation, and so we can solve a scalar conservation law

$$q_t + f(q)_x = 0$$

with the flux $f(q)$ given by (16.2). Note that this flux is nonconvex, with a single inflection point. Figure 16.1(b) shows the characteristic speed

$$f'(q) = \frac{2aq(1-q)}{[q^2 + a(1-q)^2]^2},$$

which has a maximum at the point of inflection.

Now consider the Riemann problem with initial states $q_l = 1$ and $q_r = 0$. By following characteristics, we can construct the triple-valued solution shown in Figure 16.2(a). Note that the characteristic velocities are $f'(q)$, so that the profile of this bulge, seen here at time t, is simply the graph of $tf'(q)$ turned sideways.

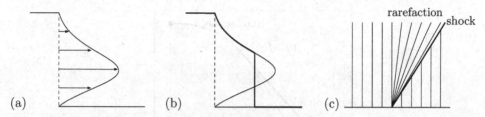

Fig. 16.2. Riemann solution for the Buckley–Leverett equation. (a) Triple-valued solution obtained by following characteristics. (b) Insertion of an area-preserving shock. (c) Structure in the x–t plane.

We can now use the equal-area rule to replace this triple-valued solution by a correct shock. The resulting weak solution is shown in Figure 16.2(b), along with the characteristics in Figure 16.2(c). The postshock value q^* is constant in time (Exercise 16.2), and so is the shock speed

$$s = \frac{f(q^*) - f(q_r)}{q^* - q_r} = \frac{f(q^*)}{q^*}. \tag{16.3}$$

This is to be expected, since the solution to the Riemann problem is still self-similar in the nonconvex case.

Note the physical interpretation of the solution shown in Figure 16.2. As the water moves in, it displaces a certain fraction q^* of the oil immediately. Behind the shock, there is a mixture of oil and water, with less and less oil as time goes on. At a production well (at the point $x = 1$, say), one obtains pure oil until the shock arrives, followed by a mixture of oil and water with diminishing returns as time goes on. It is impossible to recover all of the oil in finite time by this technique.

Note that the Riemann solution involves both a shock and a rarefaction wave and is called a *compound wave*. If $f(q)$ had more inflection points, the solution might involve several shocks and rarefactions (as in Example 16.1 below).

Here we only consider scalar nonconvex problems. For a nonlinear system of equations, similar behavior can arise in any nonlinear field that fails to be genuinely nonlinear (see Section 13.8.4). This arises, for example, in the one-dimensional elasticity equations (2.97) with a nonconvex stress–strain relation. See, e.g., [483], [484].

16.1.2 Solving Nonconvex Riemann Problems

To determine the correct weak solution to a nonconvex scalar conservation law, we need an admissibility criterion for shock waves that applies in this case. A more general form of the entropy condition (11.40), due to Oleinik [347], applies also to nonconvex scalar flux functions f:

Entropy Condition 16.1 (Oleinik). A weak solution $q(x, t)$ is the vanishing-viscosity solution to a general scalar conservation law if all discontinuities have the property that

$$\frac{f(q) - f(q_l)}{q - q_l} \geq s \geq \frac{f(q) - f(q_r)}{q - q_r} \tag{16.4}$$

for all q between q_l and q_r.

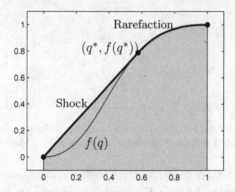

Fig. 16.3. Convex-hull construction of the Riemann solution for the Buckley–Leverett equation with $q_l = 1$ and $q_r = 0$. The shaded region is the convex hull of the set of points lying below the graph of $f(q)$.

For convex f, this requirement reduces to (11.40). (Note that this may not always be the correct admissibility condition to apply; see [285] for a nonconvex traffic-flow example where the vanishing-viscosity criterion would select the wrong solution.)

Convex-Hull Construction

The entropy-satisfying solution to a nonconvex Riemann problem can be determined from the graph of $f(q)$ in a simple manner. If $q_r < q_l$, then construct the *convex hull* of the set $\{(q, y) : q_r \le q \le q_l \text{ and } y \le f(q)\}$. The convex hull is the smallest convex set containing the original set. This is shown in Figure 16.3 for the case $q_l = 1, q_r = 0$.

If we look at the upper boundary of this set, we see that it is composed of a straight line segment from $(0, 0)$ to $(q^*, f(q^*))$ and then follows $y = f(q)$ up to $(1, 1)$. The point of tangency q^* is precisely the postshock value. The straight line represents a shock jumping from $q = 0$ to $q = q^*$, and the segment where the boundary follows $f(q)$ is the rarefaction wave. This works more generally for any two states (provided $q_l > q_r$) and for any f.

Note that the slope of the line segment is equal to the shock speed (16.3). The fact that this line is tangent to the curve $f(q)$ at q^* means that $s = f'(q^*)$, the shock moves at the same speed as the characteristics at this edge of the rarefaction fan, as seen in Figure 16.2(c).

If the shock were connected to some point $\hat{q} < q^*$, then the shock speed $f(\hat{q})/\hat{q}$ would be less than $f'(\hat{q})$, leading to a triple-valued solution. On the other hand, if the shock were connected to some point above q^*, then the entropy condition (16.4) would be violated. This explains the tangency requirement, which comes out naturally from the convex-hull construction. The same construction works for any $q_r < q_l$ lying in $[0, 1]$.

If $q_l < q_r$, then the same idea works, but we look instead at the convex hull of the set of points *above* the graph, $\{(q, y) : q_l \le q \le q_r \text{ and } y \ge f(q)\}$, as illustrated in Example 16.1.

Note that if f is convex, then the convex hull construction gives either a single line segment (single shock) if $q_l > q_r$ or the function f itself (single rarefaction) if $q_l < q_r$.

Example 16.1. Figure 16.4(a) shows another example for the flux function $f(q) = \sin(q)$ with $q_l = \pi/4$ and $q_r = 15\pi/4$. The shaded region is the convex hull of the set of points *above* this curve, since $q_l < q_r$. The lower boundary of this set shows the structure of

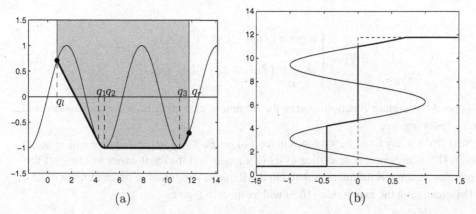

Fig. 16.4. Solving the scalar Riemann problem with a nonconvex flux function $f(q) = \sin(q)$. (a) The flux function $f(q)$. The lower boundary of the convex hull determines the structure of the Riemann solution. (b) The solution $q(x, t)$ as a function of x at time $t = 1$ (heavy line) together with the initial data (dashed line) and the multivalued solution that would be obtained by following characteristics (light line).

the Riemann solution: a shock from q_l to some value q_1, a rarefaction wave from q_1 to $q_2 = 3\pi/2$, a stationary shock from q_2 to $q_3 = 7\pi/2$, and finally a rarefaction from q_3 to q_r. The value of $q_1 \approx 4.2316$ can be found by solving the nonlinear equation

$$\frac{\sin(q_1) - \sin(\pi/4)}{q_1 - \pi/4} = \cos(q_1),$$

since the slope of the line segment from q_l to q_1 must match $f'(q_1)$. Figure 16.4(b) shows the Riemann solution together with the multivalued solution that would be obtained by following characteristics. Note that the equal-area rule of Section 11.7 again applies to the shocks.

Osher's Solution

Osher [349] found a simple representation for the entropy-satisfying similarity solution $q(x, t) = \tilde{q}(x/t)$ for a general nonconvex scalar Riemann problem with arbitrary data q_l and q_r. Let $\xi = x/t$, and set

$$G(\xi) = \begin{cases} \min_{q_l \leq q \leq q_r} (f(q) - \xi q) & \text{if } q_l \leq q_r, \\ \max_{q_r \leq q \leq q_l} (f(q) - \xi q) & \text{if } q_r \leq q_l. \end{cases} \tag{16.5}$$

Then it can be shown that $\tilde{q}(\xi)$ satisfies the equation

$$f(\tilde{q}(\xi)) - \xi \tilde{q}(\xi) = G(\xi). \tag{16.6}$$

In other words, for any given value of ξ, $\tilde{q}(\xi)$ is the value of q for which $f(q) - \xi q$ is minimized or maximized, depending on whether $q_l \leq q_r$ or $q_r \leq q_l$. We can also write this as

$$\tilde{q}(\xi) = \begin{cases} \displaystyle\operatorname*{argmin}_{q_l \leq q \leq q_r} [f(q) - \xi q] & \text{if } q_l \leq q_r, \\[2mm] \displaystyle\operatorname*{argmax}_{q_r \leq q \leq q_l} [f(q) - \xi q] & \text{if } q_r \leq q_l. \end{cases} \tag{16.7}$$

In general the argmin function returns the argument that minimizes the expression, and similarly for argmax.

Note that for any fixed q_0 we can replace $f(q) - \xi q$ by $[f(q) - f(q_0)] - \xi(q - q_0)$ in (16.7). Often an appropriate choice of q_0 (e.g., q_l or q_r) makes it easier to interpret this expression, since it is intimately related to the Rankine–Hugoniot jump condition (11.21).

Differentiating the expression (16.6) with respect to ξ gives

$$[f'(\tilde{q}(\xi)) - \xi]\tilde{q}'(\xi) - \tilde{q}(\xi) = G'(\xi). \tag{16.8}$$

Along every ray $x/t = \xi$ in the Riemann solution we have either $\tilde{q}'(\xi) = 0$ or else $f'(\tilde{q}(\xi)) = \xi$ (in a rarefaction wave), and hence (16.8) reduces to an expression for $\tilde{q}(\xi)$:

$$\tilde{q}(\xi) = -G'(\xi). \tag{16.9}$$

This gives the general solution to the Riemann problem.

The equation (16.6) is particularly useful in the case $\xi = 0$, for which it yields the value $f(\tilde{q}(0))$ along $x/t = 0$. This is the flux value $f(q^{\vee}(q_l, q_r))$ needed in implementing Godunov's method and generalizations. When $\xi = 0$, (16.6) reduces to

$$f(q^{\vee}(q_l, q_r)) = f(\tilde{q}(0)) = G(0) = \begin{cases} \displaystyle\min_{q_l \leq q \leq q_r} f(q) & \text{if } q_l \leq q_r, \\[2mm] \displaystyle\max_{q_r \leq q \leq q_l} f(q) & \text{if } q_r \leq q_l. \end{cases} \tag{16.10}$$

We have already seen this formula for the special case of a convex flux function in (12.4).

16.1.3 Finite Volume Methods for Nonconvex Problems

Finite volume methods for nonconvex problems (see Section 16.1) can be developed using the same approach as for convex problems. To apply Godunov's method to a scalar nonconvex problem, we must compute the flux $F_{i-1/2} = f(Q^{\vee}_{i-1/2})$, where $Q^{\vee}_{i-1/2}$ is the value along $x/t = 0$ in the entropy-satisfying solution to the Riemann problem between states Q_{i-1} and Q_i. This is easily computed using the general formula (16.10), resulting in (12.4). As usual, this flux can be converted into fluctuations $\mathcal{A}^{\pm}\Delta Q_{i-1/2}$ using (12.6).

To apply a high-resolution method, we also need to define one or more waves and corresponding speeds. Since the Riemann solution may consist of several waves in the nonconvex case, one might think it necessary to handle each wave separately. In fact it appears to be sufficient to use a single wave and the corresponding Rankine–Hugoniot speed,

$$\mathcal{W}_{i-1/2} = Q_i - Q_{i-1}, \qquad s_{i-1/2} = \frac{f(Q_i) - f(Q_{i-1})}{Q_i - Q_{i-1}}.$$

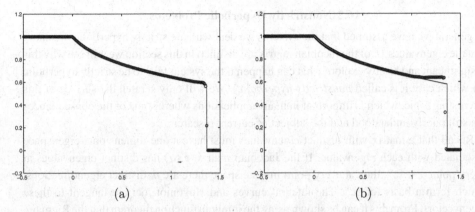

Fig. 16.5. (a) Computed solution to the Buckley–Leverett problem at time $t = 1$ using CLAWPACK with sufficiently small time steps that the Courant number is less than 1. (b) Entropy-violating solution obtained with CLAWPACK when the true Courant number is greater than 1. [claw/book/chap16/bucklev]

In computing the fluctuations an entropy fix must be included, as in the convex case, if the solution contains a transonic rarefaction. It is also important to insure that the Courant number is less than 1. Note that in computing the Courant number we must use the formula

$$\text{Courant number} = \max \left| \frac{\Delta t}{\Delta x} f'(q) \right|,$$

where we maximize over the entire range of q that appears in the solution, e.g., over $[0, 1]$ for the Buckley–Leverett example used above. For nonconvex problems this may be larger than values of $|s_{i-1/2}| \Delta t / \Delta x$ that arise in the course of solving the problem numerically (which is how CLAWPACK estimates the Courant number). This is because the steepest part of the flux function may be in regions that are embedded in shocks and not sampled by the cell averages arising in the problem.

Figure 16.5(a) shows a computed solution to the Buckley–Leverett equation using a high-resolution method of CLAWPACK with a time step that satisfies the CFL condition. Figure 16.5(b) shows a CLAWPACK computation with the parameter values

```
cflv(1) = 1.0,
cflv(2) = 0.8.
```

This aims for a Courant number of 0.8 and forces a step to be retaken whenever the observed Courant number is greater than 1.0. The method fails to produce the correct solution, because the actual Courant number is greater than the estimate used in CLAWPACK, and the time steps being used do not in fact satisfy the CFL condition. The solution obtained does approximate a weak solution, but one in which the shock jumps from $q = 0$ to a value greater than the q^* shown in Figure 16.3, and hence propagates at a speed that is slower than the characteristic speed at some points in between.

See [claw/book/chap16/bucklev] for the CLAWPACK implementation of the Buckley–Leverett equation, and [claw/book/chap16/fsin] for another example, the nonconvex flux function shown in Figure 16.4(a).

16.2 Nonstrictly Hyperbolic Problems

In general we have assumed that the systems we deal with are strictly hyperbolic, meaning that the eigenvalues λ^p of the Jacobian matrix are distinct. In this section we will see why this is significant and briefly explore what can happen if the system fails to be strictly hyperbolic (in which case it is called *nonstrictly hyperbolic*). We will only scratch the surface of this interesting topic, which is important in many applications where some of the consequences are still poorly understood and the subject of current research.

Recall that a matrix with distinct eigenvalues must have a one-dimensional eigenspace associated with each eigenvalue. If the Jacobian matrix $f'(q)$ has distinct eigenvalues at every point $q \in \mathbb{R}^m$, then at every point in state space there are m distinct eigendirections, which form a basis for $\mathbb{R}^m$. The integral curves and Hugoniot loci are tangent to these eigenvectors. From this it can be shown using the implicit-function theorem that the Riemann problem can be uniquely solved for any two states q_l and q_r that are sufficiently close together (see e.g., [263]). This does not necessarily mean that the Riemann problem can be solved globally for any q_l and q_r, but at least locally these curves foliate state space in an organized manner.

Now suppose the system is not strictly hyperbolic and at some point q_0 the Jacobian $f'(q_0)$ has a repeated eigenvalue. The *algebraic multiplicity* of this eigenvalue is the number of times it is repeated as a zero of the characteristic polynomial. The *geometric multiplicity* is the dimension of the linear space of eigenvectors associated with this eigenvalue, which is never greater than the algebraic multiplicity but could be less. In order for the system to be hyperbolic, the geometric multiplicity must be equal to the algebraic multiplicity, since only in this case is the Jacobian matrix diagonalizable. Otherwise the matrix is *defective*, a case discussed in Section 16.3.1.

If the system is hyperbolic but not strictly hyperbolic at q_0, then for some eigenvalue the algebraic and geometric multiplicities are equal but greater than 1. In this multidimensional eigenspace there are infinitely many directions that are eigenvectors, and this infinitude of eigendirections leads to some of the difficulties with nonstrictly hyperbolic systems. Even if this occurs only at a single point in state space, it can complicate the solution to all Riemann problems. It is possible for infinitely many integral curves or Hugoniot loci to coalesce at such a point, which is often called an *umbilic point* because of this behavior. Here we only consider some trivial examples to give an indication of how things can change. Much more interesting examples arise in various applications such as nonlinear elasticity, flow in porous media, and magnetohydrodynamics. For some examples see [51], [142], [146], [209], [210], [227], [228], [235], [237], [238], [240], [334], [335], [372], [397], [398], [460].

16.2.1 Uncoupled Advection Equations

First it is important to note that equal eigenvalues do not always lead to difficulties. In fact some physical systems (such as one-dimensional Riemann problems arising from the multidimensional Euler equations) always have repeated eigenvalues at every point in state space and yet require no special treatment.

As a simple example, suppose we model two distinct tracers being advected in a fluid moving at constant velocity $\bar{u}$, with concentrations denoted by q^1 and q^2. Then we obtain two uncoupled advection equations

$$q_t^1 + \bar{u}q_x^1 = 0,$$
$$q_t^2 + \bar{u}q_x^2 = 0,$$

(16.11)

which are easily solved independently. If we view this as a system, then the coefficient matrix is diagonal:

$$A = \begin{bmatrix} \bar{u} & 0 \\ 0 & \bar{u} \end{bmatrix}.$$

(16.12)

This matrix has eigenvalues $\lambda^1 = \lambda^2 = \bar{u}$ with all of $\mathbb{R}^2$ as the two-dimensional eigenspace; every direction is an eigendirection. Note that the solution to the Riemann problem between arbitrary states q_l and q_r consists of a single wave $\mathcal{W}^1 = q_r - q_l$ with speed $s^1 = \bar{u}$, since the Rankine–Hugoniot jump condition $A(q_r - q_l) = \bar{u}(q_r - q_l)$ is satisfied for every vector $q_r - q_l$. Every curve in state space is an integral curve of this system, but this causes no real problems.

We should note, however, that computationally it is better to use two waves

$$\mathcal{W}^1 = \begin{bmatrix} q_r^1 - q_l^1 \\ 0 \end{bmatrix}, \qquad \mathcal{W}^2 = \begin{bmatrix} 0 \\ q_r^2 - q_l^2 \end{bmatrix},$$

with equal speeds $s^1 = s^2 = \bar{u}$ when using a high-resolution method. This way the limiter is applied separately to each component of the system. Otherwise a discontinuity in q^1 in a region where the concentration q^2 is smooth would lead to a loss of accuracy in the approximation to q^2.

A similar sort of benign nonstrict hyperbolicity is seen in the two-dimensional Euler equations, as discussed in Section 18.8. If we solve a one-dimensional Riemann problem in any direction then the solution consists of four waves: two nonlinear acoustic waves and two linearly degenerate waves moving at the intermediate fluid velocity u^*. These latter waves form the contact discontinuity, across which the two initial gases are in contact. One of these waves carries a jump in density (as in the one-dimensional Euler equations), and the other carries a jump in the transverse velocity (see Section 18.8). As in the advection example above, we could view these as a single wave carrying jumps in both quantities, though computationally it may be best to keep them distinct. If we couple the Euler equations with the transport of different chemical species in the gases, then additional linearly degenerate waves will arise, moving at the same speed. The fact that these waves are linearly degenerate and the same eigenspace arises at every point in state space means that this loss of strict hyperbolicity does not lead to any mathematical or computational difficulties.

16.2.2 Uncoupled Burgers' Equations

Now consider a pair of Burgers' equations,

$$u_t + \frac{1}{2}(u^2)_x = 0,$$

$$v_t + \frac{1}{2}(v^2)_x = 0,$$
(16.13)

for which

$$q = \begin{bmatrix} u \\ v \end{bmatrix}, \qquad f(q) = \frac{1}{2}\begin{bmatrix} u^2 \\ v^2 \end{bmatrix}, \qquad f'(q) = \begin{bmatrix} u & 0 \\ 0 & v \end{bmatrix}.$$
(16.14)

Again this system is easy to solve for any initial data, since the two equations are uncoupled. However, if we attempt to apply the theory developed in Chapter 13 to this system, we see that the state-space structure has become complicated by the fact that this system fails to be strictly hyperbolic along the line $u = v$. We must be careful in labeling the eigenvalues as λ^1 and λ^2, since by convention we assume $\lambda^1 \le \lambda^2$. This requires setting

$$\lambda^1 = \min(u, v), \qquad \lambda^2 = \max(u, v),$$
(16.15)

and hence there is a jump discontinuity in the directions r^1 and r^2 along $u = v$ as shown in Figure 16.6:

$$r^1 = \begin{cases} (1, 0)^T & \text{if } u < v, \\ (0, 1)^T & \text{if } u > v, \end{cases} \qquad r^2 = \begin{cases} (0, 1)^T & \text{if } u < v, \\ (1, 0)^T & \text{if } u > v. \end{cases}$$
(16.16)

For states q on the curve $u = v$, every direction is an eigendirection. This allows r^1 and r^2 to rotate by 90° as we cross this curve. While it may seem that this discontinuity is simply a result of our pedantic insistence on the ordering $\lambda^1 \le \lambda^2$, recall that in solving the Riemann problem it is crucial that we first move from q_l along an integral curve or Hugoniot locus

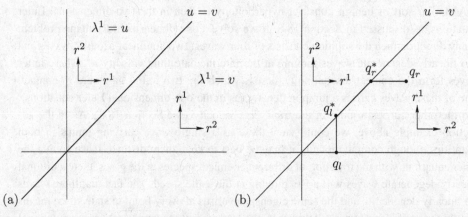

Fig. 16.6. (a) The phase plane for a pair of Burgers equations viewed as a nonstrictly hyperbolic system of equations. The eigenvalues coalesce along $u = v$. (b) Solution to a Riemann problem consisting of three segments.

corresponding to the slower speed, and then along a curve corresponding to the faster speed to reach q_r. Otherwise we cannot patch these waves together into a physically meaningful single-valued solution. For this reason the labeling is important, and the discontinuity in eigendirections across $u = v$ has consequences.

To illustrate the effect this can have on the Riemann solution, consider the data

$$q_l = \begin{bmatrix} 1 \\ 0 \end{bmatrix}, \qquad q_r = \begin{bmatrix} 3 \\ 2 \end{bmatrix}.$$

By solving the uncoupled Burgers equations, we see that $u(x, t)$ consists of a rarefaction fan for $1 \le x/t \le 3$ joining $u_l = 1$ to $u_r = 3$, while $v(x, t)$ is a rarefaction fan for $0 \le x/t \le 2$ joining $v_l = 0$ to $v_r = 2$. These two rarefaction fans partially overlap, so that the solution consists of three distinct regions:

- for $0 \le x/t \le 1$, u is constant while v is varying,
- for $1 \le x/t \le 2$, both u and v vary together,
- for $2 \le x/t \le 3$, only u varies.

The structure of this Riemann solution in state space is shown in Figure 16.6(b). The three distinct waves are clear in this picture as well. We are following an integral curve of r^1 from q_l to $q_r^* = (2, 2)^T$, and an integral curve of r^2 from $q_l^* = (1, 1)^T$ to q_r. The portion between q_l^* and q_r^* is an integral curve of both eigenvector fields.

16.2.3 Undercompressive and Overcompressive Shocks

Recall the Lax Entropy Condition 13.1. For a strictly hyperbolic system of m equations, a shock in the p-family will have $m + 1$ characteristics impinging on it: $m - p + 1$ from the left, since $\lambda^m(q_l) > \lambda^{m-1}(q_l) > \cdots > \lambda^p(q_l) > s$, and p from the right, since $s < \lambda^p(q_r) < \lambda^{p-1}(q_r) < \cdots < \lambda^1(q_r)$. This property can fail for a nonstrictly hyperbolic system. For example, consider the pair of Burgers equations (16.13) with data

$$q_l = \begin{bmatrix} 2 \\ 0 \end{bmatrix}, \qquad q_r = \begin{bmatrix} 0 \\ 2 \end{bmatrix}. \tag{16.17}$$

The solution consists of a shock in u with speed $s = 1$ and a rarefaction wave in v, with speeds ranging from 0 to 2, so that the shock is embedded in the midst of the rarefaction. The states found just to the left and right of the shock are

$$q_l^* = \begin{bmatrix} 2 \\ 1 \end{bmatrix}, \qquad q_r^* = \begin{bmatrix} 0 \\ 1 \end{bmatrix},$$

with eigenvalues $\lambda^1(q_l^*) = 1$, $\lambda^2(q_l^*) = 2$, $\lambda^1(q_r^*) = 0$, and $\lambda^2(q_r^*) = 1$. In this case only the 2-characteristic is impinging on the shock from the left, while only the 1-characteristic is impinging from the right. This shock is said to be *undercompressive*, because it only has two characteristics impinging rather than the three that one would normally expect for a nonlinear system of two equations. Note that it is also ambiguous whether we should call this a 1-shock or a 2-shock, since there is no single value of p for which $\lambda^p(q_l^*) > s > \lambda^p(q_r^*)$ as we normally expect. This is again a consequence of the fact that the labeling of eigenvalues changes as we cross $u = v$, which can only happen where two eigenvalues are equal.

If we instead take the data

$$q_l = \begin{bmatrix} 5 \\ 4 \end{bmatrix}, \qquad q_r = \begin{bmatrix} 1 \\ 2 \end{bmatrix}, \tag{16.18}$$

then the Riemann solution is a shock in both u and v, moving together as a single shock for the system at speed $s = 3$. This is an *overcompressive shock*: in this case all four characteristics impinge on the shock.

16.3 Loss of Hyperbolicity

In order for a first-order system $q_t + f(q)_x = 0$ to be hyperbolic in the sense that we have used the term (more precisely, *strongly hyperbolic*), it is necessary that the Jacobian matrix $f'(q)$ be diagonalizable with real eigenvalues at each q in the physically relevant portion of state space. The system may fail to be hyperbolic at some point if the Jacobian matrix is not diagonalizable, which may happen if the eigenvalues are not distinct. Such a system is called *weakly hyperbolic* if the Jacobian matrix still has real eigenvalues. An example of this, in the context of a linear system, is given in Section 16.3.1. See Section 16.4.2 for a nonlinear example.

It may also happen that the Jacobian matrix is diagonalizable but has complex eigenvalues and is not hyperbolic at all. In some physical problems the equations are hyperbolic for most q but there are some *elliptic regions* of state space where the eigenvalues are complex. This case is discussed in Section 16.3.2.

16.3.1 A Weakly-hyperbolic System

In the examples of Section 16.2, the Jacobian matrix is still diagonalizable even though the eigenvalues are not distinct. When some eigenvalues are equal, it may also happen that the matrix is defective and fails to have a full set of m linearly independent eigenvectors. In this case the system is only weakly hyperbolic and the theory we have developed does not apply directly. However, it is interesting to consider what goes wrong in this case, as it gives additional insight into some aspects of the hyperbolic theory. There are also connections with the theory of hyperbolic equations with singular source terms, which will be studied again in Chapter 17.

Since a small perturbation of a nondiagonalizable matrix can yield one that is diagonalizable, one might wonder what happens if the problem is strongly hyperbolic but the matrix A is *nearly* nondiagonalizable. We should expect to observe some sort of singular behavior as we approach a nondiagonalizable matrix. Consider the system of equations

$$\begin{aligned} q_t^1 + u q_x^1 + \beta q_x^2 &= 0, \\ q_t^2 \qquad\quad + v q_x^2 &= 0, \end{aligned} \tag{16.19}$$

with the coefficient matrix

$$A = \begin{bmatrix} u & \beta \\ 0 & v \end{bmatrix}. \tag{16.20}$$

The eigenvalues are u and v and we assume that these are real numbers, along with the coupling coefficient β. If $u \neq v$, then the matrix is diagonalizable and the problem is hyperbolic. If $u = v$ then the problem is hyperbolic only if $\beta = 0$, in which case the system decouples into independent scalar advection equations for q^1 and q^2.

Suppose that $\beta \neq 0$ and that $v \leq u$. We will study the case $u \to v$. The eigenvalues and eigenvectors of A are

$$\lambda^1 = v, \qquad\qquad \lambda^2 = u,$$

$$r^1 = \begin{bmatrix} \beta \\ v - u \end{bmatrix}, \qquad r^2 = \begin{bmatrix} 1 \\ 0 \end{bmatrix}. \tag{16.21}$$

Note that as $u \to v$, the eigenvector r^1 becomes collinear with r^2 and the eigenvector matrix R becomes singular:

$$R = \begin{bmatrix} \beta & 1 \\ v - u & 0 \end{bmatrix}, \qquad R^{-1} = \frac{1}{u - v} \begin{bmatrix} 0 & -1 \\ u - v & \beta \end{bmatrix} \qquad \text{for } u \neq 0. \tag{16.22}$$

Based on the construction of Riemann solutions presented in Section 3.9, we see that solving the general Riemann problem with arbitrary states q_l and q_r will not be possible in the singular case $u = v$. Only if $q_r - q_l = \alpha^2 r^2$ for some scalar α^2 will we be able to find a solution. In this case $q_l^2 = q_r^2$ and so $q_x^2 \equiv 0$, and the system reduces to an advection equation for q^1 alone.

Now suppose $u = v + \epsilon$ with $\epsilon > 0$, so that the general Riemann problem can be solved by decomposing

$$q_r - q_l = \alpha^1 r^1 + \alpha^2 r^2.$$

Computing $\alpha = R^{-1}(q_r - q_l)$ yields

$$\alpha^1 = -\frac{1}{\epsilon}(q_r^2 - q_l^2),$$

$$\alpha^2 = q_r^1 - q_l^1 + \frac{\beta}{\epsilon}(q_r^2 - q_l^2). \tag{16.23}$$

As $\epsilon \to 0$ the wave strengths blow up unless $q_l^2 = q_r^2$. The intermediate state is

$$q_m = q_l + \alpha^1 r^1 = \begin{bmatrix} q_l^1 - \beta(q_r^2 - q_l^2)/\epsilon \\ q_r^2 \end{bmatrix}, \tag{16.24}$$

as illustrated in Figure 16.7(a). As $\epsilon \to 0$ the eigendirections become collinear and $q_m^1 \to \infty$ (unless $q_r^2 = q_l^2$). Figure 16.7(b) shows the Riemann solution in the x–t plane, with $q = q_m$ only in the narrow wedge $v < x/t < v + \epsilon$. As the wave speeds coalesce, the solution blows

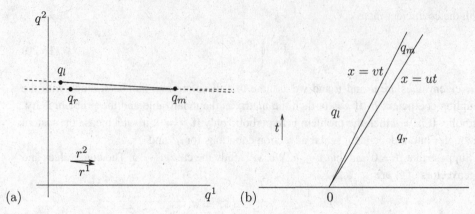

Fig. 16.7. Riemann solution for the case $\epsilon > 0$ but small: (a) in the phase plane, where the eigenvectors are nearly collinear; (b) in the x–t plane, where the wave speeds are nearly equal.

up in the neighborhood of $x = vt$. In the limit the solution can be viewed as consisting of a single jump discontinuity along with a delta-function singularity propagating at the same speed. The exact form of this distribution solution is derived in the next section. Similar phenomena are seen in some nonlinear problems, and are often called *singular shocks* or *delta shocks*. See [236], [239], [366], [441], [440] for some examples.

Singular-Source Interpretation

Another way to solve the Riemann problem for the system (16.19) is to note that the second equation does not depend on q^1 and can be easily solved to yield

$$q^2(x, t) = \overset{\circ}{q}{}^2(x - vt). \tag{16.25}$$

We can then view $\beta q_x^2(x, t) = \beta \overset{\circ}{q}{}_x^2(x - vt)$ as a known source term in the first equation of (16.19), resulting in

$$q_t^1 + u q_x^1 = -\beta \overset{\circ}{q}{}_x^2(x - vt). \tag{16.26}$$

For any known function $\psi(x, t)$, the solution to

$$q_t^1 + u q_x^1 = \psi(x, t) \tag{16.27}$$

is obtained by Duhamel's principle as

$$q^1(x, t) = \overset{\circ}{q}{}^1(x - ut) + \int_0^t \psi(x - u(t - \tau), \tau) \, d\tau. \tag{16.28}$$

Using $\psi(x, t) = -\beta \overset{\circ}{q}{}_x^2(x - vt)$ gives

$$q^1(x, t) = \overset{\circ}{q}{}^1(x - ut) - \beta \int_0^t \overset{\circ}{q}{}_x^2(x - ut + (u - v)\tau) \, d\tau. \tag{16.29}$$

If $u \neq v$ then the integral can be evaluated to give

$$q^1(x, t) = \overset{\circ}{q}{}^1(x - ut) - \frac{\beta}{u - v} [\overset{\circ}{q}{}^2(x - vt) - \overset{\circ}{q}{}^2(x - ut)]. \qquad (16.30)$$

If $u = v$, then (16.29) reduces to

$$q^1(x, t) = \overset{\circ}{q}{}^1(x - ut) - \beta \int_0^t \overset{\circ}{q}{}^2_x(x - vt) d\tau$$

$$= \overset{\circ}{q}{}^1(x - ut) - \beta t \overset{\circ}{q}{}^2_x(x - vt). \qquad (16.31)$$

This result also follows from (16.30) on letting $u \to v$. If $\overset{\circ}{q}{}^2$ is a differentiable function, then this gives the general solution and the differential equation can be solved even in the case $u = v$, but note that in this case the solution grows with t in a manner unlike what we would expect from a hyperbolic equation.

Now consider the Riemann problem again, in which case $\overset{\circ}{q}{}^2$ is not differentiable at $x = 0$. If $u \neq v$, then the formula (16.30) still holds. Along with (16.25) this gives the solution to the Riemann problem, albeit in a different form than previously presented. If $u = v$ then (16.31) still holds if we interpret $\overset{\circ}{q}{}^2_x(x - vt)$ as a delta function. Writing the initial data in terms of the Heaviside function (3.28), we have

$$\overset{\circ}{q}{}^2(x) = q_l^2 + (q_r^2 - q_l^2) H(x) \implies \overset{\circ}{q}{}^2_x(x - vt) = (q_r^2 - q_l^2) \delta(x - vt),$$

and hence

$$q^1(x, t) = q_l^1 + (q_r^1 - q_l^1) H(x - vt) - \beta t (q_r^2 - q_l^2) \delta(x - vt). \qquad (16.32)$$

Note that this same form can be deduced from the Riemann structure shown in Figure 16.7(b), using the formula (16.24) and the observation that the solution q^1 has magnitude $-\beta (q_r^2 - q_l^2)/\epsilon$ over a region of width ϵt. As $\epsilon \to 0$ this approaches the delta function in (16.32).

Physical Interpretation

This Riemann problem has a simple physical interpretation if we take $v = 0$ and $u \geq 0$. Then $q^2(x, t) = \overset{\circ}{q}{}^2(x)$, and the equation (16.26) is an advection equation for q^1 with a delta function source at $x = 0$,

$$q_t^1 + u q_x^1 = D \delta(x), \qquad (16.33)$$

where $D = -\beta(q_r^2 - q_l^2)$. This models the advection of a tracer in a pipe filled with fluid flowing at velocity u, with an isolated source of tracer at $x = 0$ (e.g., a contaminant leaking into the pipe at this point). For $u > 0$ the solution q^1 will have a jump discontinuity at $x = 0$, from q_l^1 to q_m^1, since the tracer advects only downstream and not upstream. The magnitude of the jump will depend on both D (the rate at which tracer is introduced at the source) and u (the rate at which it is carried downstream). For fixed D, larger values of u will lead to a smaller jump since the tracer is carried more rapidly away. As u decreases the magnitude of the jump will rise. As $u \to 0$ we approach the singular solution in which all the tracer introduced at $x = 0$ remains at $x = 0$, yielding a delta function in the solution whose magnitude grows linearly with t, as seen in (16.32).

Similar analysis applies when $v \neq 0$, but now we must view the source location as propagating with velocity v rather than being fixed. In this case it is the relative speed $u - v$ that determines the magnitude of the jump in q^1, as already observed in the general Riemann solution.

16.3.2 Equations of Mixed Type

For some physical systems there are regions of state space where $f'(q)$ has complex eigenvalues. To see what this means, we first consider the linear system

$$\begin{bmatrix} p \\ u \end{bmatrix}_t + \begin{bmatrix} 0 & 1 \\ -1 & 0 \end{bmatrix} \begin{bmatrix} p \\ u \end{bmatrix}_x = 0. \tag{16.34}$$

This looks very similar to the acoustics equations (3.30) (with $u_0 = 0$ and $K_0 = \rho_0 = 1$), but has a crucial change in sign of one of the elements in the coefficient matrix A, resulting in complex eigenvalues $\lambda = \pm i$. The eigenvectors of A are also complex. Evidently we cannot use the techniques of Chapter 3 to solve this system, since it would not make sense physically to decompose the data into complex-valued waves moving at speeds $\pm i$.

We can combine the two equations in (16.34) and eliminate u by differentiating to obtain $p_{tt} = -u_{xt}$ and $u_{tx} = p_{xx}$, and hence

$$p_{tt} + p_{xx} = 0. \tag{16.35}$$

Going through this same process for the acoustics equations (with $+1$ in place of -1 in (16.34)) would result in the second-order wave equation $p_{tt} = p_{xx}$, as derived in Section 2.9.1. Here we instead obtain Laplace's equation (16.35), a second-order elliptic equation. From the theory of elliptic equations, it is known that this equation is well posed only if we specify boundary conditions for p on all boundaries of the region in x–t space where a solution is desired. This means that in order to solve the equation over some domain $[a, b] \times [0, T]$, we would not only have to specify initial data at time $t = 0$ and boundary data at $x = a$ and $x = b$, but also the solution at time T. This does not have the nature of a wave-propagation problem.

For a nonlinear problem it may happen that the Jacobian matrix $f'(q)$ has real eigenvalues at most points in state space but has complex eigenvalues over some relatively small region, called the *elliptic region* of state space. It is then an *equation of mixed type*. In this case it may be that wavelike solutions exist and the initial-value problem is well posed, at least for certain initial data, in spite of the elliptic region.

Problems of this sort often arise when studying wave propagation in materials that can undergo a *phase change*, for example from vapor to liquid. As discussed in Section 14.15, as a gas is compressed it is necessary eventually to consider the attractive intermolecular forces, leading to the van der Waals equation of state (14.66). If we consider the isothermal case where T is held constant, then we obtain the p-system (2.108) with the equation

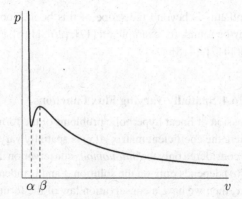

Fig. 16.8. The van der Waals equation of state (16.36).

of state

$$p(v) = \frac{RT}{v - b} - \frac{a}{v^2},$$ (16.36)

where $v = 1/\rho$ is the specific volume. For sufficiently large temperature, this function is monotonically decreasing and the system is hyperbolic. If T is below the critical value $T_c = 8a/(27Rb)$, however, then the function $p(v)$ has the shape indicated in Figure 16.8. As the density increases (v decreases), the intermolecular forces are eventually sufficiently strong that the gas molecules begin to weakly bind to one another, and the gas changes phase and becomes a liquid. When v is sufficiently large it is pure gas (vapor phase) and p increases as v decreases as we expect in an ideal gas. However, as gas molecules begin to bind to one another, the gas pressure is reduced as a result, and so there is a region in which the pressure falls as v is decreased, the region $\alpha < v < \beta$ in Figure 16.8. For sufficiently small v the fluid is entirely liquid and decreasing v further leads to a strong increase in pressure.

Recall that for the p-system (2.108) the Jacobian matrix is

$$f'(q) = \begin{bmatrix} 0 & -1 \\ p'(v) & 0 \end{bmatrix}.$$ (16.37)

The eigenvalues $\pm\sqrt{-p'(v)}$ are real only where $p'(v) < 0$, and so the p–v state space has an elliptic region consisting of the strip $\alpha < v < \beta$, corresponding to the change of phase. We may wish to solve a Riemann problem in which the left and right states are in the two different phases in order to model the motion of the interface between phases. The solution will consist only of states that lie below α or above β, and so the solution essentially remains in the hyperbolic region, but it must include a phase-change jump across the elliptic region.

Even if the left and right states are both in the same phase, it is possible for the Riemann solution to involve states in the other phase. Consider, for example, the case where $v_l, v_r > \beta$ and $u_l > 0$ while $u_r < 0$, so that two gas streams are colliding. Then we expect the specific volume to decrease in the intermediate region after collision, which may lead to the liquid phase forming. In this case the Riemann solution might consist of two phase boundaries moving outward, in addition to shock waves in the gas phases. Similar equations arise in studying solids undergoing a phase transition.

The theory of such problems is beyond the scope of this book. Some other examples and further discussion may be found, for example, in [17], [26], [116], [130], [172], [200], [204], [213], [241], [269], [417], [418].

16.4 Spatially Varying Flux Functions

Chapter 9 contains a discussion of linear hyperbolic problems of the form $q_t + A(x)q_x = 0$ or $q_t + (A(x)q)_x = 0$, where the coefficient matrix $A(x)$ is spatially varying. For nonlinear problems we have thus far considered only the *autonomous* conservation law $q_t + f(q)_x = 0$, where the flux function $f(q)$ depends only on the solution q and is independent of x. If the flux function varies with x, then we have a conservation law of the form

$$q_t + f(q, x)_x = 0. \tag{16.38}$$

A system of this form arises, for example, when studying nonlinear elasticity in a heterogeneous medium where the stress–strain relation $\sigma(\epsilon, x)$ varies with x. See [273] for one application of a wave-propagation algorithm to this problem.

The flux function can be discretized to obtain a flux function $f_i(q)$ associated with the ith grid cell. The Riemann problem at $x_{i-1/2}$ now consists of the problem

$$\begin{aligned} q_t + f_{i-1}(q)_x = 0 \quad &\text{if } x < x_{i-1/2}, \\ q_t + f_i(q)_x = 0 \quad &\text{if } x > x_{i-1/2}, \end{aligned} \tag{16.39}$$

together with the data Q_{i-1} and Q_i. Often the Riemann solution is still a similarity solution of the form $q(x, t) = \tilde{q}(x/t)$ that in general consists of

- left-going waves, which must be shock or rarefaction waves relative to the flux function $f_{i-1}(q)$,
- right-going waves, which must be shock or rarefaction waves relative to the flux function $f_i(q)$,
- a stationary discontinuity at $x/t = 0$ between states $Q_l^{\vee}$ and $Q_r^{\vee}$ just to the left and right of this ray. In order for the solution to be bounded, the physical flux must be continuous across this ray and so

$$f_{i-1}(Q_l^{\vee}) = f_i(Q_r^{\vee}). \tag{16.40}$$

Solving the Riemann problem generally consists in finding states $Q_l^{\vee}$ and $Q_r^{\vee}$ satisfying (16.40) and having the property that Q_{i-1} can be connected to $Q_l^{\vee}$ using only left-going waves, while $Q_r^{\vee}$ can be connected to Q_i using only right-going waves. This cannot always be done, as it requires that there be a sufficient number of outgoing waves available. A simple example where it fails is the variable-coefficient advection equation (9.8) in the case where $u_{i-1} > 0$ and $u_i < 0$. All characteristics are approaching $x_{i-1/2}$, and the Riemann solution contains a delta function at this point rather than consisting only of waves.

For problems where the Riemann problem can be solved in terms of waves, the wave-propagation methods developed previously can be directly applied. Since the flux, unlike q, has no jump across $x_{i-1/2}$, this is often most easily done using the formulation of

Section 15.5. This requires only splitting the flux difference into propagating waves as in (15.74),

$$f_i(Q_i) - f_{i-1}(Q_{i-1}) = \sum_{p=1}^{m} \mathcal{Z}_{i-1/2}^p, \tag{16.41}$$

and often avoids the need to explicitly compute $Q_l^\vee$ and $Q_r^\vee$. For smooth solutions, this can be shown to be second-order accurate following the same approach as in Section 15.6. This method is explored in more detail in [18]. See [147], [148], [163], [243], [270], [288], [317], [361], [453] [460] for other examples of conservation laws with spatially varying fluxes and more discussion of numerical methods.

16.4.1 Traffic Flow with a Varying Speed Limit

The Riemann solution in the spatially varying case can have more complicated structure than in the autonomous case. As a simple scalar example, consider a traffic flow model that incorporates changes in the speed limit or visibility along the length of the highway. An equation of the form developed in Section 11.1 might then be used, but with a flux of the form

$$f(q, x) = u_{\max}(x)q(1 - q) \tag{16.42}$$

in place of (11.6). A linear version of this problem was discussed in Section 9.4.2. As in Chapter 9, a Riemann problem can now be formulated by specifying a piecewise constant function $u_{\max}(x)$ that jumps from $u_{\max,l}$ to $u_{\max,r}$ at $x = 0$, along with piecewise constant data q_l and q_r.

The two flux functions $f_l(q)$ and $f_r(q)$ are distinct quadratic functions as illustrated in Figure 16.9(a) for the case $u_{\max,l} = 2$ and $u_{\max,r} = 1$. This figure also indicates the structure of the Riemann solution for $q_l = 0.13$ and $q_r = 0.1$. The density as a function of x is shown in Figure 16.9(b) at time $t = 40$. The Riemann solution consists of a stationary jump from

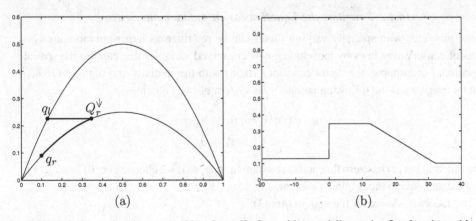

(a)   (b)

Fig. 16.9. Solution to the Riemann problem for traffic flow with a spatially varying flux function, with $q_l = 0.13$. (a) The flux functions $f_l(q) = 2q(1 - q)$ and $f_r(q) = q(1 - q)$ and the states arising in the Riemann solution. In this case $Q_l^\vee = q_l$. (b) The density at $t = 40$. [claw/book/chap16/vctraffic]

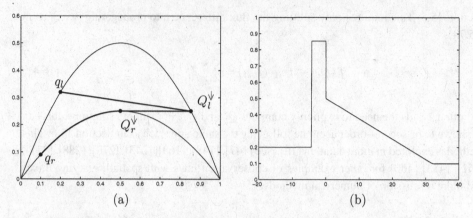

Fig. 16.10. Solution to the Riemann problem for traffic flow with a spatially varying flux function, with $q_l = 0.2$. (a) The flux functions $f_l(q) = 2q(1 - q)$ and $f_r(q) = q(1 - q)$ and the states arising in the Riemann solution. (b) The density at $t = 40$. [claw/book/chap16/vctraffic]

$q_l = Q_l^{\vee}$ to a state $Q_r^{\vee}$ at $x = 0$ along with a rarefaction wave from this state to q_r. In this case there is no left-going wave. The traffic slows down abruptly at the point $x = 0$ where the speed limit u_{max} decreases, and then speeds up again through a rarefaction wave to the state q_r. This is analogous to the example shown in Section 9.4.2, except that the right-going discontinuity becomes a rarefaction wave in the nonlinear case. Note that since we require (16.40), the jump from the flux curve $f_l(q)$ to $f_r(q)$ must occur along a horizontal line in Figure 16.9(a).

If the state q_l is increased above $0.5(1 - \sqrt{0.5}) = 0.1464$, however, then a solution of the form seen in Figure 16.9(a) is no longer possible. The structure illustrated in Figure 16.10 is observed instead, shown here for $q_l = 0.2$. Upstream traffic is now sufficiently heavy that a shock wave forms and moves upstream, so that the Riemann solution contains a left-going shock and a right-going rarefaction wave, along with a stationary jump in density at $x = 0$ from $Q_l^{\vee} = 0.5(1 + \sqrt{0.5})$ to $Q_r^{\vee} = 0.5$.

16.4.2 Rewriting the Equations as an Autonomous System

Some problems with spatially varying fluxes can be rewritten as larger autonomous systems of conservation laws by introducing new conserved variables that capture the spatial variation. For example, the scalar equation (16.38) with the spatially varying flux (16.42) can be rewritten as the following autonomous system of two equations:

$$q_t + (vq(1 - q))_x = 0,$$
$$v_t = 0. \tag{16.43}$$

The function $v(x, t)$ has zero flux and is constant in time, so if we choose $v(x, 0) = u_{max}(x)$, then we are solving the original problem.

The Jacobian matrix for the system (16.43) is

$$\begin{bmatrix} v(1 - 2q) & q(1 - q) \\ 0 & 0 \end{bmatrix}. \tag{16.44}$$

The eigenvalues are $\lambda^1 = v(1 - 2q)$ (the characteristic speed expected from the original scalar equation) and $\lambda^2 = 0$ (the speed of stationary discontinuities at $x = 0$). But notice that this system is only weakly hyperbolic at $q = 0.5$, since the Jacobian is not diagonalizable at that point. This is related to the sudden change in structure of the Riemann solution shown in Figures 16.9 and 16.10 as $Q_r^{\vee}$ reaches the value $q = 0.5$ and $Q_l^{\vee}$ jumps from one root of a quadratic to the other. Note that the weak hyperbolicity at the one point $q = 0.5$ can lead to a Riemann solution that involves three waves (Figure 16.10) even though we have an autonomous system (16.43) of only two equations. This is another example of the sort of difficulty that can be caused by loss of strong hyperbolicity. See Exercise 13.11 for another example of this type of system.

16.5 Nonconservative Nonlinear Hyperbolic Equations

In Chapter 9 we considered nonconservative linear hyperbolic systems of the form $q_t + A(x)q_x = 0$, in the context of variable coefficient advection (the color equation, Section 9.3) and acoustics in a heterogeneous medium (Section 9.6). Solving the Riemann problem at the cell interface $x_{i-1/2}$ with matrices A_{i-1} and A_i yields waves and wave speeds that can be used in the wave-propagation form of the high-resolution algorithms, even though the equation is not in conservation form. In principle the same can be done for a general quasilinear hyperbolic equation of the form

$$q_t + A(q, x)q_x = 0. \tag{16.45}$$

If A depends explicitly on x, then we can first discretize $A(q, x)$ as we did for spatially varying flux functions in the previous section, to obtain a coefficient matrix $A_i(q)$ in the ith cell. Then the Riemann problem at $x_{i-1/2}$ consists of the data Q_{i-1}, Q_i and the equations

$$\begin{aligned} q_t + A_{i-1}(q)q_x = 0 \quad \text{if } x < x_{i-1/2}, \\ q_t + A_i(q)q_x = 0 \quad \text{if } x > x_{i-1/2}. \end{aligned} \tag{16.46}$$

If we can solve this Riemann problem to obtain physically meaningful waves and wave speeds, then the wave-propagation algorithm can be applied as usual. However, in the nonlinear case it may be difficult to determine the correct Riemann solution. This is true even in the autonomous case where there is no explicit dependence on x and the equation is simply

$$q_t + A(q)q_x = 0. \tag{16.47}$$

This differential equation only makes sense where q is differentiable. At discontinuities in q we have previously relied on the integral form of the conservation law to determine the resulting wave structure, using the Rankine–Hugoniot jump conditions. If the equation is not in conservation form, then we cannot use this for guidance.

Even for the simple Burgers equation, the quasilinear form $u_t + uu_x = 0$ is compatible with many different conservation laws, as discussed in Section 11.12. If we were only given the quasilinear equation to solve, it would not be clear what the correct wave speed is for a shock wave. The nonconservative upwind method (12.25) can be viewed a a first-order wave-propagation method for the quasilinear problem using the wave speed $s_{i-1/2}^n = U_i^n$. This gives very different results (see Figure 12.5) than are obtained using $s_{i-1/2}^n = \frac{1}{2}(U_{i-1}^n + U_i^n)$, which yields the method (12.24).

If the quasilinear equation (16.47) can be rewritten as a physically meaningful conservation law (in particular, if q itself should be conserved and $A(q)$ is the Jacobian matrix of some flux function $f(q)$), then that conservation law should generally be solved rather than working with the quasilinear form numerically. However, in some applications a nonconservative equation is naturally obtained that does not have this property. In this case a detailed understanding of the physical problem is generally required in order to determine the proper structure of the Riemann solution.

Notice that if q is discontinuous at a point, then $A(q)$ is typically discontinuous there as well, while q_x contains a delta function singularity centered at this point. The theory of distributions can often be used to study equations involving delta functions and Heaviside functions, but the classical theory only allows these distributions to be multiplied by smooth functions. In (16.47) we have the product of a delta function q_x with a Heaviside function $A(q)$. Such products are generally ambiguous. If the delta function and the Heaviside function are each smoothed out slightly over width ϵ, for example by adding viscosity or diffusion to the problem, then we have ordinary continuous functions for which the product makes sense. But the limiting behavior as $\epsilon \to 0$ depends strongly on how each distribution is smoothed out. This general theory is beyond the scope of the present book. See [86], [87], [101], [194], [267], [268], [344], [345] for some further discussion and examples.

16.6 Nonconservative Transport Equations

One special nonconservative equation will be considered further here since it is easy to handle and often arises in practice. Consider a transport equation (or color equation) of the form

$$\phi_t + u\phi_x = 0 \qquad (16.48)$$

for a tracer $\phi(x, t)$, and suppose that this equation is now coupled with a nonlinear system of m conservation laws that determines the velocity u. Then the full set of $m + 1$ equations is nonlinear, but is not in conservation form, because of (16.48). An example is given in Section 13.12.1, where the shallow water equations are considered along with a tracer ϕ used to distinguish fluid lying to the left and right of a contact discontinuity. In that case the equation (16.48) was rewritten in the conservation form (13.62) and a system of conservation laws obtained. However, in some applications it may be preferable to leave the transport equation in the nonconservative form (16.48). Otherwise one must recover ϕ by taking the ratio of the two conserved quantities $h\phi$ and h. This can lead to some numerical difficulties. In particular, if ϕ is expected to be piecewise constant whereas h and $h\phi$ are both spatially varying, then numerically the ratio $h\phi/h$ may not remain constant in regions where ϕ should be constant.

With the wave-propagation algorithms it is not necessary to put (16.48) into conservation form. In addition to the waves determined from solving the conservation laws for the fluid motion, we can simply add another wave $\mathcal{W}_{i-1/2}^{m+1}$ carrying the jump $\phi_i - \phi_{i-1}$ at speed

$s_{i-1/2}^{m+1} = u_{i-1}^- + u_i^+$ (as motivated by (9.17)). With this formulation, in regions where ϕ is constant we have $\mathcal{W}_{i-1/2}^{m+1} = 0$ and so ϕ will remain constant.

In the example of Section 13.12.1, the tracer is entirely passive and does not affect the fluid dynamics. It is often convenient to add a tracer of this form simply to track the motion of some portion of the fluid relative to the rest of the fluid. This can be very useful in flow visualization. Problems of this form also arise in tracking the movement of pollutants that are present in small quantities and do not affect the flow.

In other problems the value of ϕ may feed back into the fluid dynamics equations, leading to additional coupling between the equations. For example, consider a shock tube problem in which two different ideal polytropic gases (i.e., gamma-law gases with different values γ_l and γ_r) are initially separated by a membrane. Breaking the membrane gives a Riemann problem of the type mentioned in Section 14.15. To solve this problem numerically we must track the constituent gases. After the first time step there will generally be a grid cell that contains a mixture of the two gases (and many such cells in a multidimensional problem). In future time steps this mixing region will be further smeared out due to numerical diffusion in the method. If we let ϕ_i^n be the volume fraction of cell i that contains the left gas at time t_n, then ϕ satisfies the nonconservative transport equation (16.48). This equation must be coupled to the Euler equations (14.8), along with the equation of state (14.23),

$$E = \frac{p}{\gamma - 1} + \frac{1}{2}\rho u^2. \tag{16.49}$$

But now γ depends on ϕ. A cell that has volume fraction ϕ_i of gas l and volume fraction $1 - \phi_i$ of gas r has an effective value of γ_i that is related to γ_l and γ_r by

$$\frac{1}{\gamma_i - 1} = \frac{\phi_i}{\gamma_l - 1} + \frac{1 - \phi_i}{\gamma_r - 1}. \tag{16.50}$$

Rather than solving equation (16.48) for the volume fraction and then computing γ_i from (16.50), one can instead introduce a new variable

$$G = \frac{1}{\gamma - 1} \tag{16.51}$$

and couple the Euler equations to a transport equation for G,

$$G_t + uG_x = 0. \tag{16.52}$$

The equation of state then becomes

$$E = Gp + \frac{1}{2}\rho u^2. \tag{16.53}$$

Note that by using the nonconservative equation (16.52), we insure that G and hence γ will remain constant numerically in regions where there is only a single gas present. Using the continuity equation $\rho_t + (\rho u)_x = 0$ would allow us to rewrite (16.52) as

$$(\rho G)_t + (\rho u G)_x = 0, \tag{16.54}$$

but using this form introduces the possibility that $G = (\rho G)/\rho$ will vary numerically due to variations in ρ and the fact that ρ and ρG both contain numerical errors.

It is also important to transport a quantity such as ϕ or G for which the transport equation continues to be valid in the context of cell averages and numerical diffusion. It is tempting to simply use the transport equation

$$\gamma_t + u\gamma_x = 0 \tag{16.55}$$

to advect γ with the flow. This equation does hold for diffusionless gases where no mixing occurs, since γ will then always be either γ_l or γ_r and the contact discontinuity is simply advected with the flow. But numerically the gases do mix, and so (16.52) must be used instead.

If (16.55) is used numerically, then pressure oscillations will typically develop near the interface between gases, in spite of the fact that the pressure should be constant across the contact discontinuity. This results from using the wrong equation of state (i.e., the wrong value of γ) in the mixture of gases generated by numerical diffusion. For more discussion of these issues and some numerical examples, see [4], [5], [71], [131], [132], [218], [231], [259], [362], [394], [388], [395], [412], [452].

Exercises

16.1.   Determine the entropy-satisfying solution to the Riemann problem for the scalar conservation law $q_t + f(q)_x = 0$ for the following nonconvex flux functions and data. In each case also sketch the characteristics and the structure of the Riemann solution in the x–t plane.
   (a) $f(q) = q^3$, $q_l = 0$, $q_r = 2$,
   (b) $f(q) = q^3$, $q_l = 2$, $q_r = -1$,

16.2.   Use the equal area rule to find an expression for the shock location in the Buckley–Leverett equation, as a function of t. Verify that the Rankine–Hugoniot condition is always satisfied and that the shock propagates at a constant speed that agrees with the speed $f(q^*)/q^*$ determined from the convex-hull construction.

16.3.   For the Buckley–Leverett equation, show that (16.4) is violated if the shock goes above q^*.

16.4.   Explain why it is impossible to have a Riemann solution involving both a shock and a rarefaction when f is convex or concave.

16.5.   For the nonstrictly hyperbolic system of equations (16.13) with data (16.17), plot the following:
   (a) the solution to the Riemann problem in state space,
   (b) the 1-characteristics and 2-characteristics in the x–t plane for this solution, as in the plots of Figure 13.6, for example.

16.6.   Repeat Exercise 16.5 for the data (16.18).

16.7.   (a) Determine the eigenvectors of the Jacobian matrix (16.44), and sketch the integral curves of each eigenvector in state space (the q–v plane).
   (b) Sketch the solutions to the Riemann problems shown in Figures 16.9 and 16.10 in state space.

17

Source Terms and Balance Laws

So far, we have only considered homogeneous conservation laws of the form $q_t + f(q)_x = 0$. As mentioned briefly in Section 2.5, there are many situations in which *source terms* also appear in the equations, so that we wish to solve the system

$$q_t + f(q)_x = \psi(q). \tag{17.1}$$

Note that these are generally called source terms even if physically they represent a sink rather than a source (i.e., net loss rather than gain of the quantity q). The equation (17.1) is also often called a *balance law* rather than a conservation law. We start with a few examples showing how source terms can arise. Others will be encountered later.

Reacting Flow

Fluid dynamics problems often involve chemically reacting fluids or gases. An example was mentioned in Section 2.5. An even simpler example is studied below in Section 17.2. In these examples the reacting species are assumed to represent a small fraction of the volume, and the chemical reactions have no effect on the fluid dynamics. More interesting problems arise when the reactions affect the fluid motion, as in combustion problems where the heat released by the reactions has a pronounced effect on the dynamics of the flow. Often the chemical reactions occur on much faster time scales than the fastest wave speeds in the gas, resulting in problems with *stiff source terms*. Some of the issues that arise in this case are discussed in Section 17.10.

External Forces Such as Gravity

In introducing gas dynamics in Section 2.6, we ignored external forces acting on the gas, such as gravity. External forces give source terms in the momentum equation, and we then do not expect conservation of the initial momentum, since this force will lead to an acceleration of the fluid and a change in its net momentum.

As an example, consider the equations of one-dimensional isentropic gas dynamics in the presence of a gravitational field, pointing in the negative x-direction (so x now measures distance above the earth, say, in a column of gas). The gravitational force acting on the gas causes an acceleration, and hence this force enters into the integral equation for the

375

time-derivative of momentum. The equation (2.33) is replaced by

$$\frac{d}{dt} \int_{x_1}^{x_2} \rho(x,t)u(x,t)\,dx = [\rho(x_1,t)u^2(x_1,t) + p(x_1,t)] - [\rho(x_2,t)u^2(x_2,t) + p(x_2,t)]$$

$$- \int_{x_1}^{x_2} g\rho(x,t)\,dx. \tag{17.2}$$

The differential equation (2.34) becomes

$$(\rho u)_t + (\rho u^2 + p)_x = -g\rho. \tag{17.3}$$

In this case the system (2.39) with q and $f(q)$ given by (2.40) is augmented by a source term with

$$\psi(q) = \begin{bmatrix} 0 \\ -gq^1 \end{bmatrix}.$$

Geometric Source Terms

Often a physical problem in three space dimensions can be reduced to a mathematical problem in one or two dimensions by taking advantage of known symmetries in the solution. For example, if we wish to compute a spherically expanding acoustic wave arising from a pressure perturbation at one point in space, then we can solve a one-dimensional problem in r (distance from the source) and time. However, the homogeneous conservation law in three dimensions may acquire source terms when we reduce the dimension. This follows from the fact that the interval $[r_1, r_2]$ now corresponds to a spherical shell, whose volume varies with r like r^2. Hence a substance whose total mass is fixed but that is spreading out radially will have a density (in mass per unit volume) that is decreasing as the substance spreads out over larger and larger spheres.

As an example, in Section 18.9 we will see that with radial symmetry, the three-dimensional acoustics equations can be reduced to the one-dimensional system

$$p_t + K_0 u_r = -\frac{2K_0 u}{r},$$

$$\rho_0 u_t + p_r = 0, \tag{17.4}$$

where u is now the velocity in the radial direction r. This has the same form as the one-dimensional equations of acoustics but with a geometric source term. Similar geometric source terms arise if we consider flow in a channel or nozzle with varying cross-sectional area. If the variation is relatively slow, then we may be able to model this with a one-dimensional system of equations that includes source terms to model the area variation. This was discussed for a simple advection equation in Section 9.1. More generally this leads to *quasi-one-dimensional* models.

Higher-Order Derivatives

Our focus is on developing methods for first-order hyperbolic systems, but many practical problems also involve higher-order derivatives such as small viscous or diffusive terms. Examples include the advection–diffusion equation $q_t + \bar{u}q_x = \mu q_{xx}$, the Navier–Stokes equations in which viscous terms are added to the Euler equations, or the shallow water equations with bottom friction. Other problems may instead (or in addition) include dispersive terms involving q_{xxx} or other odd-order derivatives. An example is the *Korteweg–de Vries* (KdV) *equation* $q_t + qq_x = q_{xxx}$. Then the equations may be of the form (17.1) with $\psi(q)$ replaced by $\psi(q, q_{xx}, q_{xxx}, \dots)$. We can still view ψ as a source term and apply some of the techniques developed in this chapter. In particular, fractional-step methods are often used to incorporate viscous terms in fluid dynamics problems. This is discussed briefly in the next section and further in Section 17.7.

17.1 Fractional-Step Methods

We will primarily study problems where the homogeneous equation

$$q_t + f(q)_x = 0 \tag{17.5}$$

is hyperbolic and the source terms depend only on q (and perhaps on x) but not on derivatives of q. In this case the equations

$$q_t = \psi(q) \tag{17.6}$$

reduce to independent systems of ODEs at each point x.

One standard approach for such problems is to use a *fractional-step* or *operator-splitting* method, in which we somehow alternate between solving the simpler problems (17.5) and (17.6) in order to approximate the solution to the full problem (17.1). This approach is quite simple to use and is implemented in the CLAWPACK software (see Section 5.4.6). It allows us to use high-resolution methods for (17.5) without change, coupling these methods with standard ODE solvers for the equations (17.6). This approach is described in more detail and analyzed in this chapter. There are situations where a fractional-step method is not adequate, and the analysis presented in this chapter will shed some light on the errors this splitting introduces and when it can be successfully used.

There are also many situations in which the hyperbolic equation is coupled with other terms that involve derivatives of q. For example, the advection–diffusion equation

$$q_t + \bar{u}q_x = \mu q_{xx}$$

can be viewed as an equation of the general form (17.1) in which ψ depends on q_{xx}. This equation should more properly be viewed as

$$q_t + (\bar{u}q - \mu q_x)_x = 0,$$

in conservation form with the flux function $\bar{u}q - \mu q_x$ (see Section 2.2). But in practice it is often simplest and most efficient to use high-resolution explicit methods for the advection part and implicit methods for the diffusion equation $q_t = \mu q_{xx}$, such as the Crank–Nicolson method (4.13). These two approaches can be most easily combined by using a fractional-step method. See [claw/book/chap17/advdiff] for an example.

In Section 19.5 we will also see that it is possible to solve a two-dimensional hyperbolic equation of the form

$$q_t + f(q)_x + g(q)_y = 0$$

by splitting it into two one-dimensional problems $q_t + f(q)_x = 0$ and $q_t + g(q)_y = 0$ and using one-dimensional high-resolution methods for each piece. The same idea extends to three space dimensions. In this context the fractional-step approach is called *dimensional splitting*. The theory developed in this chapter is also useful in analyzing these methods.

17.2 An Advection–Reaction Equation

To illustrate, we begin with a simple advection–reaction equation.

Example 17.1. Consider the linear equation

$$q_t + \bar{u} q_x = -\beta q, \tag{17.7}$$

with data $q(x, 0) = \overset{\circ}{q}(x)$. This would model, for example, the transport of a radioactive material in a fluid flowing at constant speed $\bar{u}$ down a pipe. The material decays as it flows along, at rate β. We can easily compute the exact solution of (17.7), since along the characteristic $dx/dt = \bar{u}$ we have $dq/dt = -\beta q$, and hence

$$q(x, t) = e^{-\beta t} \overset{\circ}{q}(x - \bar{u}t). \tag{17.8}$$

17.2.1 An Unsplit Method

Before discussing fractional-step methods in more detail, we first present an *unsplit* method for (17.7), which more clearly models the correct equation. An obvious extension of the upwind method for advection would be (assuming $\bar{u} > 0$)

$$Q_i^{n+1} = Q_i^n - \frac{\bar{u}\,\Delta t}{\Delta x}\left(Q_i^n - Q_{i-1}^n\right) - \Delta t\,\beta Q_i^n. \tag{17.9}$$

This method is first-order accurate and stable for $0 < \bar{u}\,\Delta t/\Delta x \le 1$; see Exercise 8.3.

A second-order Lax–Wendroff-style method can be developed by using the Taylor series

$$q(x, t + \Delta t) \approx q(x, t) + \Delta t\, q_t(x, t) + \frac{1}{2}\Delta t^2 q_{tt}(x, t). \tag{17.10}$$

As in the derivation of the Lax–Wendroff method in Section 6.1, we must compute q_{tt} from the PDE. Differentiating q_t gives

$$q_{tt} = -\bar{u} q_{xt} - \beta q_t, \qquad q_{tx} = -\bar{u} q_{xx} - \beta q_x,$$

and combining these, we obtain

$$q_{tt} = \bar{u}^2 q_{xx} + 2\bar{u}\beta q_x + \beta^2 q. \tag{17.11}$$

Note that this is more easily obtained by using

$$\partial_t q = (-\bar{u}\partial_x - \beta)q,$$

and hence

$$\partial_t^2 q = (-\bar{u}\partial_x - \beta)^2 q = (\bar{u}^2\partial_x^2 + 2\bar{u}\beta\partial_x + \beta^2)q. \tag{17.12}$$

Using this expression for q_{tt} in (17.10) gives

$$q(x, t + \Delta t) \approx q - \Delta t\,(\bar{u}q_x + \beta q) + \frac{1}{2}\Delta t^2(\bar{u}^2 q_{xx} + 2\bar{u}\beta q_x + \beta^2 q)$$

$$= \left(1 - \Delta t\,\beta + \frac{1}{2}\Delta t^2\beta^2\right)q - \Delta t\,\bar{u}\,(1 - \Delta t\,\beta)\,q_x + \frac{1}{2}\Delta t^2\,\bar{u}^2 q_{xx}. \tag{17.13}$$

We can now approximate x-derivatives by finite differences to obtain the *second-order method*

$$Q_i^{n+1} = \left(1 - \Delta t\,\beta + \frac{1}{2}\Delta t^2\beta^2\right)Q_i^n - \frac{\bar{u}\,\Delta t}{2\,\Delta x}(1 - \Delta t\,\beta)\left(Q_{i+1}^n - Q_{i-1}^n\right)$$

$$+ \frac{\bar{u}^2\Delta t^2}{2\,\Delta x^2}\left(Q_{i-1}^n - 2Q_i^n + Q_{i+1}^n\right). \tag{17.14}$$

In order to model the equation (17.7) correctly to second-order accuracy, we must properly model the interaction between the $\bar{u}q_x$ and the βq terms, which brings in the mixed term $\frac{1}{2}\Delta t^2\,\bar{u}\beta q_x$ in the Taylor series expansion.

For future use we also note that for (17.7) the full Taylor series expansion can be written as

$$q(x, t + \Delta t) = \sum_{j=0}^{\infty} \frac{(\Delta t)^j}{j!}\partial_t^j q(x, t) = \sum_{j=0}^{\infty} \frac{(\Delta t)^j}{j!}(-\bar{u}\partial_x - \beta)^j q(x, t), \tag{17.15}$$

which can be written formally as

$$q(x, t + \Delta t) = e^{-\Delta t\,(\bar{u}\partial_x + \beta)}q(x, t). \tag{17.16}$$

The operator $e^{-\Delta t\,(\bar{u}\partial_x + \beta)}$, which is defined via the Taylor series in (17.15), is called the *solution operator* for the equation (17.7) over a time step of length Δt. Applying this operator to any function of x gives the evolution of this data after time Δt has elapsed.

Note that the second derivative q_{tt} might be harder to compute for a more complicated problem, making it harder to develop second-order numerical methods. It is also not clear how to introduce limiters effectively into the unsplit method (17.14), as might be desirable in solving problems with discontinuous solutions. In some situations, it is possible to use the ideas of high-resolution methods based on Riemann solvers and also include the effects of source terms in the process of solving the Riemann problem. One approach of this form is presented in Section 17.14, and some others can be found in [34], [127], [128], [151], [155], [161], [162], [167], [168], [482], [216], [217], [284], [325], [381], [471]. In many cases, however, the simple fractional-step approach can often be effectively used, as described in the next section.

17.2.2 A Fractional-Step Method

A fractional-step method for (17.7) is applied by first splitting the equation into two *subproblems* that can be solved independently. For the advection–reaction problem (17.7) we would take these to be:

$$\text{Problem A:} \quad q_t + \bar{u} q_x = 0, \tag{17.17}$$

$$\text{Problem B:} \quad q_t = -\beta q. \tag{17.18}$$

The idea of the fractional-step method is to combine these by applying the two methods in an alternating manner. For more complicated problems this has great advantage over attempting to derive an unsplit method. If we split the general problem $q_t + f(q)_x = \psi(q)$ into the homogeneous conservation law and a simple ODE, then we can use standard methods for each. In particular, the high-resolution shock-capturing methods already developed can be used directly for the homogeneous conservation law, whereas trying to derive an unsplit method based on the same ideas while incorporating the source term directly can be more difficult.

As a simple example of the fractional-step procedure, suppose we use the upwind method for the A-step and the forward Euler for the ODE in the B-step for the advection–reaction problem. Then the simplest fractional-step method over one time step would consist of the following two stages:

$$\text{A-step:} \quad Q_i^* = Q_i^n - \frac{\bar{u}\,\Delta t}{\Delta x}\left(Q_i^n - Q_{i-1}^n\right), \tag{17.19}$$

$$\text{B-step:} \quad Q_i^{n+1} = Q_i^* - \beta\,\Delta t\, Q_i^*. \tag{17.20}$$

Note that we first take a time step of length Δt with upwind, starting with initial data Q_i^n to obtain the intermediate value Q_i^*. Then we take a time step of length Δt using the forward Euler method, starting with the data Q^* obtained from the first stage.

It may seem that we have advanced the solution by time $2\,\Delta t$ after taking these two steps of length Δt. However, in each stage we used only some of the terms in the original PDE, and the two stages combined give a consistent approximation to solving the original equation (17.7) over a single time step of length Δt. To check this consistency, we can combine the two stages by eliminating Q^* to obtain a method in a more familiar form:

$$
\begin{aligned}
Q_i^{n+1} &= (1 - \beta\,\Delta t) Q_i^* \\
&= (1 - \beta\,\Delta t)\left[Q_i^n - \frac{\bar{u}\,\Delta t}{\Delta x}\left(Q_i^n - Q_{i-1}^n\right)\right] \\
&= Q_i^n - \frac{\bar{u}\,\Delta t}{\Delta x}\left(Q_i^n - Q_{i-1}^n\right) - \beta\,\Delta t\, Q_i^n + \frac{\bar{u}\beta\,\Delta t^2}{\Delta x}\left(Q_i^n - Q_{i-1}^n\right).
\end{aligned} \tag{17.21}
$$

The first three terms on the right-hand side agree with the unsplit method (17.9). The final term is $\mathcal{O}(\Delta t^2)$ (since $(Q_i^n - Q_{i-1}^n)/\Delta x \approx q_x = \mathcal{O}(1)$), and so a local truncation error analysis will show that this method, though slightly different from (17.9), is also consistent and first-order accurate on the original equation (17.7).

A natural question is whether we could improve the accuracy by using a more accurate method in each step. For example, suppose we use the Lax–Wendroff method in the A-step and the trapezoidal method, or the two-stage Runge–Kutta method, in the B-step. Would we then obtain a second-order accurate method for the original equation? For this particular equation, the answer is yes. In fact if we use pth-order accurate methods for each step, the result will be a pth-order accurate method for the full original equation. But this equation is very special in this regard, and this claim should seem surprising. One would think that splitting the equation into pieces in this manner would introduce some error that depends on the size of the time step Δt and is independent of how well we then approximate the subproblem in each step. In general this is true – there is a "splitting error" that in general would be $\mathcal{O}(\Delta t)$ for the type of splitting used above, and so the resulting fractional-step method will be only first-order accurate, no matter how well we then approximate each step. This will be analyzed in more detail below.

For the case of equation (17.7) there is no splitting error. This follows from the observation that we can solve (17.7) over any time period Δt by first solving the equation (17.17) over time Δt, and then using the result as data to solve the equation (17.18) over time Δt. To verify this, let $u^*(x, \Delta t)$ be the exact solution to the A-problem,

$$q_t^* + \bar{u} q_x^* = 0,$$

$$q^*(x, 0) = \overset{\circ}{q}(x). \tag{17.22}$$

We use a different symbol $q^*(x, t)$ for the solution to this problem rather than $q(x, t)$, which we reserve for the exact solution to the original problem.

Then we have

$$q^*(x, \Delta t) = \overset{\circ}{q}(x - \bar{u}\,\Delta t).$$

If we now use this as data in solving the B-problem (17.18), we will be solving a different equation,

$$q_t^{**} = -\beta q^{**} \tag{17.23}$$

with initial data

$$q^{**}(x, 0) = q^*(x, \Delta t) = \overset{\circ}{q}(x - \bar{u}\,\Delta t).$$

This is just an ODE at each point x, and the solution is

$$q^{**}(x, \Delta t) = e^{-\beta \Delta t}\,\overset{\circ}{q}(x - \bar{u}\,\Delta t).$$

Comparing this with (17.8), we see that we have indeed recovered the solution to the original problem by this two-stage procedure.

Physically we can interpret this as follows. Think of the original equation as modeling a radioactive tracer that is advecting with constant speed $\bar{u}$ (carried along in a fluid, say) and also decaying with rate β. Since the decay properties are independent of the position x, we

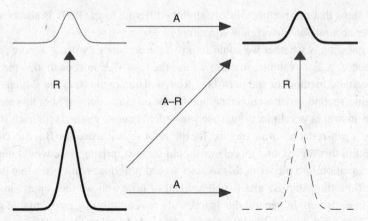

Fig. 17.1. Illustration of a fractional-step procedure for the advection–reaction equation (17.7) when there is no splitting error. The pulse shown in the lower left simultaneously advects and decays as indicated by the diagonal arrow labeled A-R. The same result is obtained if the pulse first advects to the right following the arrow A and then is allowed to decay via the reaction term, following the arrow R, or if the reaction and advection are performed in the opposite order.

can think of first advecting the tracer over time Δt without allowing any decay, and then holding the fluid and tracer stationary while we allow it to decay for time Δt. We will get the same result, and this is what we have done in the fractional-step method. Figure 17.1 illustrates this.

We would also get the same result if we first allowed the tracer to decay at the initial location and then advected the decayed profile, which amounts to switching the order in which the two subproblems are solved (see Figure 17.1). We say that the solution operators for the two subproblems *commute*, since we can apply them in either order and get the same result. In general if these operators commute, then there is no splitting error, a fact that we will investigate more formally in Section 17.3. (Here we are only discussing the Cauchy problem. Boundary conditions can further complicate the situation; see Section 17.9).

Another way to examine the splitting error, which must be used more generally when we do not know the exact solution to the equations involved, is to use Taylor series expansions. (This approach can be used also for nonlinear problems.) If we look at a time step of length Δt, then solving the A-equation gives

$$q^*(x, \Delta t) = q^*(x, 0) + \Delta t\, q_t^*(x, 0) + \frac{1}{2}\Delta t^2 q_{tt}^*(x, 0) + \cdots$$

$$= q^*(x, 0) - \bar{u}\, \Delta t\, q_x^*(x, 0) + \frac{1}{2}\bar{u}^2\Delta t^2 q_{xx}^*(x, 0) - \cdots$$

$$= q(x, 0) - \bar{u}\, \Delta t\, q_x(x, 0) + \frac{1}{2}\bar{u}^2\Delta t^2 q_{xx}(x, 0) - \cdots. \qquad (17.24)$$

Similarly, if we solve the ODE problem (17.23) with general initial data $q^{**}(x, 0)$, we obtain

$$q^{**}(x, \Delta t) = q^{**}(x, 0) + \Delta t\, q_t^{**}(x, 0) + \frac{1}{2}\Delta t^2 q_{tt}^{**}(x, 0) + \cdots$$

$$= \left(1 - \beta\, \Delta t + \frac{1}{2}\beta^2\Delta t^2 + \cdots\right) q^{**}(x, 0). \qquad (17.25)$$

If we now use the result from (17.24) as the initial data in (17.25), we obtain

$$q^{**}(x, \Delta t) = \left(1 - \Delta t\,\beta + \frac{1}{2}\Delta t^2\beta^2 - \cdots\right)\left(q(x, 0) - \bar{u}\,\Delta t\, q_x(x, 0)\right.$$

$$\left. + \frac{1}{2}\bar{u}^2\Delta t^2 q_{xx}(x, 0) + \cdots\right)$$

$$= q - \Delta t\,(\bar{u}q_x + \beta q) + \frac{1}{2}\Delta t^2(\bar{u}^2 q_{xx} + 2\bar{u}\beta q_x + \beta^2 q) + \cdots. \quad (17.26)$$

Comparing this with the Taylor series expansion (17.13) that we used in deriving the unsplit Lax–Wendroff method shows that this agrees with $q(x, \Delta t)$, at least for the three terms shown, and in fact to all orders.

Note that the mixed term $\bar{u}\beta\,\Delta t^2 q_x$ needed in the q_{tt}-term from (17.11) now arises naturally from taking the product of the two Taylor series (17.24) and (17.25). In fact, we see that for this simple equation we can write (17.25) as

$$q^{**}(x, \Delta t) = e^{-\beta\,\Delta t}q^{**}(x, 0),$$

while (17.24) can be written formally as

$$q^*(x, \Delta t) = e^{-\bar{u}\,\Delta t\,\partial_x}\overset{\circ}{q}(x).$$

If we now use $q^*(x, \Delta t)$ as the data $q^{**}(x, 0)$ as we do in the fractional-step method, we obtain

$$q^{**}(x, \Delta t) = e^{-\beta\,\Delta t}e^{-\bar{u}\,\Delta t\,\partial_x}\overset{\circ}{q}(x).$$

Multiplying out the Taylor series as we did in (17.26) verifies that these exponentials satisfy the usual rule, so that to compute the product we need only add the exponents, i.e.,

$$q^{**}(x, \Delta t) = e^{-\Delta t\,(\bar{u}\partial_x + \beta)}\overset{\circ}{q}(x).$$

The exponential appearing here is exactly the solution operator for the original equation, as in (17.16), and so again we see that $q^{**}(x, \Delta t) = q(x, \Delta t)$ and there is no splitting error.

The fact that there is no splitting error for the problem (17.7) is a reflection of the fact that, for this problem, the solution operator for the full problem is exactly equal to the product of the solution operators of the two subproblems (17.17) and (17.18). This is not generally true for other problems.

Example 17.2. Suppose we modify the equation slightly so that the decay rate β depends on x,

$$q_t + \bar{u}q_x = -\beta(x)q. \quad (17.27)$$

Then our previous argument for the lack of a splitting error breaks down – advecting the tracer a distance $\bar{u}\,\Delta t$ and then allowing it to decay, with rates given by the values of β

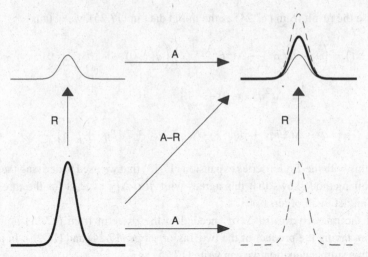

Fig. 17.2. Illustration of a fractional-step procedure for the advection–reaction equation (17.27), where there is a splitting error because the decay rate β depends on x. The pulse shown in the lower left simultaneously advects and decays, as indicated by the diagonal arrow labeled A-R. Different results are obtained if the pulse first advects to the right following the arrow A and then is allowed to decay via the reaction term evaluated at the final position, following the arrow R, or if the reaction and advection are performed in the opposite order.

at the final positions, will not in general give the same result as when the decay occurs continuously as it advects, using the instantaneous rate given by $\beta(x)$ at each point passed.

Figure 17.2 illustrates the fact that solution operators for the two subproblems do not commute, shown for the case $\beta(x) = 1 - x$ over $0 \leq x \leq 1$, so that the decay rate is smaller for larger x. First advecting and then reacting gives too little decay, while first reacting and then advecting gives too much decay. Note that this is shown for very large Δt in Figure 17.2 in order to illustrate the effect clearly. With a numerical fractional-step method we would be using much smaller time steps to solve the problem over this time period, in each step advecting by a small amount and then reacting, so that reasonable results could still be obtained, though formally only first-order accurate as the time step is reduced. See Section 17.5 for some numerical results.

The accuracy of the fractional-step method on (17.27) can be analyzed formally using Taylor series expansions again. Rather than developing this expansion for this particular example, we will first examine the more general case and then apply it to that case.

17.3 General Formulation of Fractional-Step Methods for Linear Problems

Consider a more general linear PDE of the form

$$q_t = (\mathcal{A} + \mathcal{B})q, \tag{17.28}$$

where $\mathcal{A}$ and $\mathcal{B}$ may be differential operators, e.g., $\mathcal{A} = -\bar{u}\partial_x$ and $\mathcal{B} = -\beta(x)$ in Example 17.2. For simplicity suppose that $\mathcal{A}$ and $\mathcal{B}$ do not depend explicitly on t, e.g., $\beta(x)$ is a function of x but not of t. Then we can compute that

$$q_{tt} = (\mathcal{A} + \mathcal{B})q_t = (\mathcal{A} + \mathcal{B})^2 q,$$

and in general

$$\partial_t^j q = (A + B)^j q. \tag{17.29}$$

We have used this idea before in calculating Taylor series, e.g., in (17.12).

Note that if A or B do depend on t, then we would have to use the product rule, e.g.,

$$q_{tt} = (A + B)q_t + (A_t + B_t)q,$$

and everything would become more complicated. Also note that if the problem is nonlinear then the Taylor series expansion can still be used if the solution is smooth, but we don't generally have a simple relation of the form (17.29).

In our simple case we can write the solution at time t using Taylor series as

$$
\begin{aligned}
q(x, \Delta t) &= q(x, 0) + \Delta t (A + B) q(x, 0) + \frac{1}{2} \Delta t^2 (A + B)^2 q(x, 0) + \cdots \\
&= \left(I + \Delta t (A + B) + \frac{1}{2} \Delta t^2 (A + B)^2 + \cdots \right) q(x, 0) \\
&= \sum_{j=0}^{\infty} \frac{\Delta t^j}{j!} (A + B)^j q(x, 0),
\end{aligned}
\tag{17.30}
$$

which formally can be written as

$$q(x, \Delta t) = e^{\Delta t (A + B)} q(x, 0).$$

With the fractional-step method, we instead compute

$$q^*(x, \Delta t) = e^{\Delta t A} q(x, 0),$$

and then

$$q^{**}(x, \Delta t) = e^{\Delta t B} q^*(x, \Delta t) = e^{\Delta t B} e^{\Delta t A} q(x, 0),$$

and so the *splitting error* is

$$q(x, \Delta t) - q^{**}(x, \Delta t) = \left(e^{\Delta t (A+B)} - e^{\Delta t B} e^{\Delta t A} \right) q(x, 0). \tag{17.31}$$

This should be calculated using the Taylor series expansions. We have (17.30) already, while

$$
\begin{aligned}
q^{**}(x, \Delta t) &= \left(I + \Delta t B + \frac{1}{2} \Delta t^2 B^2 + \cdots \right) \left(I + \Delta t A + \frac{1}{2} \Delta t^2 A^2 + \cdots \right) q(x, 0) \\
&= \left(I + \Delta t (A + B) + \frac{1}{2} \Delta t^2 (A^2 + 2BA + B^2) + \cdots \right) q(x, 0). \tag{17.32}
\end{aligned}
$$

The $I + \Delta t (A + B)$ terms agree with (17.30). In the Δt^2 term, however, the term from (17.30) is

$$
\begin{aligned}
(A + B)^2 &= (A + B)(A + B) \\
&= A^2 + AB + BA + B^2. \tag{17.33}
\end{aligned}
$$

In general this is *not* the same as

$$A^2 + 2BA + B^2,$$

and so the splitting error is

$$q(x, \Delta t) - q^{**}(x, \Delta t) = \frac{1}{2}\Delta t^2 (AB - BA)q(x, 0) + \mathcal{O}(\Delta t^3). \qquad (17.34)$$

The splitting error depends on the *commutator* $AB - BA$ and is zero only in the special case when the differential operators A and B commute (in which case it turns out that all the higher-order terms in the splitting error also vanish).

Example 17.3. For the problem considered in Example 17.1,

$$A = -\bar{u}\partial_x \quad \text{and} \quad B = -\beta.$$

We then have $ABq = BAq = \bar{u}\beta q_x$. These operators commute for β constant, and there is no splitting error.

Example 17.4. Now suppose $\beta = \beta(x)$ depends on x as in Example 17.2. Then we have

$$ABq = \bar{u}\partial_x(\beta(x)q) = \bar{u}\beta(x)q_x + \bar{u}\beta'(x)q,$$

while

$$BAq = \beta(x)\bar{u}q_x.$$

These are not the same unless $\beta'(x) = 0$. In general the splitting error will be

$$q(x, \Delta t) - q^{**}(x, \Delta t) = \frac{1}{2}\Delta t^2 \bar{u}\beta'(x)q(x, 0) + \mathcal{O}(\Delta t^3).$$

If we now design a fractional-step method based on this splitting, we will see that the splitting error alone will introduce an $\mathcal{O}(\Delta t^2)$ error in each time step, which can be expected to accumulate to an $\mathcal{O}(\Delta t)$ error after the $T/\Delta t$ time steps needed to reach some fixed time T (in the best case, assuming the method is stable). Hence even if we solve each subproblem *exactly* within the fractional-step method, the resulting method will be only first-order accurate. If the subproblems are actually solved with numerical methods that are sth-order accurate, the solution will still only be first-order accurate no matter how large s is. At least this is true asymptotically as the mesh spacing tends to zero. In practice results that are essentially second-order accurate are observed, for reasons described in Section 17.5.

Of course this order of accuracy can only be obtained for smooth solutions. We are often interested in problems where the solution is not smooth, in which case a lower order of accuracy is generally observed. In this case the ability to easily use high-resolution methods for the hyperbolic portion of the problem is an advantage of the fractional-step approach. On the other hand, it is not so clear that the method even converges in this case, since the above arguments based on Taylor series expansions do not apply directly. For some convergence results, see for example [258], [442], [443].

17.4 Strang Splitting

The above form of fractional-step method, sometimes called the *Godunov splitting*, in general is only first-order accurate formally. It turns out that a slight modification of the splitting idea will yield second-order accuracy quite generally (assuming each subproblem is solved with a method of at least this accuracy). The idea is to solve the first subproblem $q_t = Aq$ over only a half time step of length $\Delta t/2$. Then we use the result as data for a full time step on the second subproblem $q_t = Bq$, and finally take another half time step on $q_t = Aq$. We can equally well reverse the roles of A and B here. This approach is often called *Strang splitting*, as it was popularized in a paper by Strang [426] on solving multidimensional problems.

To analyze the Strang splitting, note that we are now approximating the solution operator $e^{\Delta t (A+B)}$ by $e^{\frac{1}{2}\Delta t A} e^{\Delta t B} e^{\frac{1}{2}\Delta t A}$. Taylor series expansion of this product shows that

$$e^{\frac{1}{2}\Delta t A} e^{\Delta t B} e^{\frac{1}{2}\Delta t A} = \left(I + \frac{1}{2}\Delta t\, A + \frac{1}{8}\Delta t^2 A^2 + \cdots\right)\left(I + \Delta t\, B + \frac{1}{2}\Delta t^2 B^2 + \cdots\right)$$

$$\times \left(I + \frac{1}{2}\Delta t\, A + \frac{1}{8}\Delta t^2 A^2 + \cdots\right)$$

$$= I + \Delta t\,(A + B) + \frac{1}{2}\Delta t^2 (A^2 + AB + BA + B^2) + \mathcal{O}(\Delta t^3).$$

$$\tag{17.35}$$

Comparing with (17.30), we see that the $\mathcal{O}(\Delta t^2)$ term is now captured correctly. The $\mathcal{O}(\Delta t^3)$ term is not correct in general, however, unless $AB = BA$.

Note that over several time steps we can simplify the expression obtained with the Strang splitting. After n steps we have

$$Q^n = \left(e^{\frac{1}{2}\Delta t A} e^{\Delta t B} e^{\frac{1}{2}\Delta t A}\right)\left(e^{\frac{1}{2}\Delta t A} e^{\Delta t B} e^{\frac{1}{2}\Delta t A}\right) \cdots \left(e^{\frac{1}{2}\Delta t A} e^{\Delta t B} e^{\frac{1}{2}\Delta t A}\right) Q^0 \tag{17.36}$$

repeated n times. Dropping the parentheses and noting that $e^{\frac{1}{2}\Delta t A} e^{\frac{1}{2}\Delta t A} = e^{\Delta t A}$, we obtain

$$Q^n = e^{\frac{1}{2}\Delta t A} e^{\Delta t B} e^{\Delta t A} e^{\Delta t B} e^{\Delta t A} \cdots e^{\Delta t B} e^{\frac{1}{2}kA} Q^0. \tag{17.37}$$

This differs from the Godunov splitting only in the fact that we start and end with a half time step on A, rather than starting with a full step on A and ending with B.

Another way to achieve this same effect is to simply take steps of length Δt on each problem, as in the first-order splitting, but to alternate the order of these steps in alternate time steps, e.g.,

$$Q^1 = e^{\Delta t B} e^{\Delta t A} Q^0,$$

$$Q^2 = e^{\Delta t A} e^{\Delta t B} Q^1,$$

$$Q^3 = e^{\Delta t B} e^{\Delta t A} Q^2,$$

$$Q^4 = e^{\Delta t A} e^{\Delta t B} Q^3,$$

$$\vdots$$

If we take an even number of time steps, then we obtain

$$Q^n = (e^{\Delta t\,A}e^{\Delta t\,B})(e^{\Delta t\,B}e^{\Delta t\,A})(e^{\Delta t\,A}e^{\Delta t\,B})(e^{\Delta t\,B}e^{\Delta t\,A})\cdots(e^{\Delta t\,A}e^{\Delta t\,B})(e^{\Delta t\,B}e^{\Delta t\,A})Q^0$$
$$= e^{\Delta t\,A}(e^{\Delta t\,B}e^{\Delta t\,B})(e^{\Delta t\,A}e^{\Delta t\,A})(e^{\Delta t\,B}e^{\Delta t\,B})\cdots(e^{\Delta t\,B}e^{\Delta t\,B})e^{\Delta t\,A}Q^0.$$

Since $e^{\Delta t\,B}e^{\Delta t\,B} = e^{2\,\Delta t\,B}$, this is essentially the same as (17.36) but with $\frac{1}{2}\Delta t$ replaced by Δt. This is generally more efficient than the approach of (17.36), since a single step with the numerical method approximating $e^{\Delta t\,A}$ is typically cheaper than two steps of length $\Delta t/2$. On the other hand, this form is more difficult to implement with variable time steps Δt, as are often used in practice. An even number of steps must be taken and the value of Δt in each pair of steps must be the same, in order to obtain the desired cancellation of errors.

In CLAWPACK, either the Godunov splitting or the Strang splitting (implemented in the form (17.36)) can be selected by setting `method(5) = 1` or 2 respectively. In this case a subroutine `src1.f` must be provided that solves the $q_t = \psi(q)$ subproblem arising from the source terms.

17.5 Accuracy of Godunov and Strang Splittings

The fact that the Strang splitting is so similar to the first-order splitting suggests that the first-order splitting is not really so bad, and in fact it is not. While formally only first-order accurate, the coefficient of the $\mathcal{O}(\Delta t)$ term may be much smaller than coefficients in the second-order terms arising from discretization of $e^{\Delta t\,A}$ and $e^{\Delta t\,B}$.

For this reason the simpler and more efficient Godunov splitting is often sufficient. It is also easier to implement boundary conditions properly with the Godunov splitting, as discussed in Section 17.9.

Example 17.5. Figure 17.3 shows results at time $t = 0.5$ from solving the problem (17.27) with $\bar{u} = 1$, $\beta(x) = 1 - x$, and initial data consisting of a Gaussian pulse centered at $x = 0.25$. The Godunov and Strang splittings are compared, where in each case the

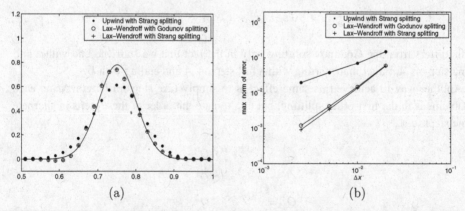

(a)   (b)

Fig. 17.3. Comparison of results with three methods applied to the problem (17.27). (a) Computed and true solution for $\Delta x = 0.02$. (b) Log–log plot of max-norm errors vs. Δx. Note that the Godunov splitting is essentially as accurate as the Strang splitting for this problem. `[claw/book/chap17/nocommute]`

Lax–Wendroff method is used for the advection equation and the second-order two-stage Runge–Kutta method is used for the source term. On this grid with $\Delta x = 0.02$, the results are visually indistinguishable, even though the Godunov splitting is formally only first-order accurate.

For contrast, Figure 17.3(a) also shows the results obtained if the first-order upwind method is used in place of the Lax–Wendroff method in the Strang splitting. This first-order method causes a substantial smearing of the solution.

Figure 17.3(b) shows log–log plots of the error in the max norm for each of these three methods as the grid is refined. For very fine grids the Godunov splitting is slightly less accurate and asymptotically approaches first-order, but even for the finest grid used ($\Delta x = 1/400$) it is only slightly less accurate than the Strang splitting. By contrast the upwind method is much less accurate. As in the discussion of the accuracy of limiters in Section 8.5, it is important to realize that order of accuracy is not the full story.

17.6 Choice of ODE Solver

Consider the Godunov splitting, and suppose we have already obtained Q^* from Q^n by solving the conservation law (17.5). We now wish to advance Q_i^* to Q_i^{n+1} by solving the ODE $q_t = \psi(q)$ over time Δt in each grid cell. In some cases this equation can be solved exactly. For the system (17.4), for example, the source terms alone yield

$$p_t = -\frac{2K_0 u}{r},$$

$$u_t = 0. \tag{17.38}$$

Since u is constant in this system, the value of p_t is constant and this ODE is easily solved exactly, yielding

$$p_i^{n+1} = p_i^* - \Delta t (2K_0 u_i^*)/r_i,$$

$$u_i^{n+1} = u_i^*. \tag{17.39}$$

For more complicated source terms, e.g., those arising from chemical kinetics with many interacting species, it will be necessary to use a numerical ODE solver in each grid cell. We typically want to use a method that is at least second-order accurate to maintain overall accuracy. A wide variety of ODE solvers are available for systems of the general form $y' = \psi(y)$ where $y(t) \in \mathbb{R}^m$. Note, however, that in general we cannot use multistep methods that require more than one level of data (e.g., y^{n-1} as well as y^n) to generate the solution y^{n+1} at the next time level. This is because we only have data Q_i^* to use in computing Q_i^{n+1}. Previous values (e.g., Q_i^n or Q_i^* from the previous time step) are not suitable to use in the context of multistep methods, because Q_i^* is computed from Q_i^n by solving a different equation (the conservation law (17.5)) than the ODE we are now attempting to approximate.

In many cases a simple explicit Runge–Kutta method can be effectively used. These are multistage one-step methods that generate intermediate values as needed to construct higher-order approximations. A simple second-order accurate two-stage method is often sufficient for use with high-resolution methods, for example the classical method

$$Q_i^{**} = Q_i^* + \frac{\Delta t}{2} \psi(Q_i^*),$$

$$Q_i^{n+1} = Q_i^* + \Delta t \, \psi(Q_i^{**}).$$

(17.40)

One must ensure that the explicit method is stable with the time step Δt being used, or perhaps take N time steps of (17.40) using a smaller step size $\Delta t/N$ to advance Q_i^* to Q_i^{n+1} stably.

17.7 Implicit Methods, Viscous Terms, and Higher-Order Derivatives

If the ODEs $q_t = \psi(q)$ are *stiff*, as discussed in Section 17.10, then it may be necessary to use an implicit method in this step in order to use a reasonable time step. In this case other numerical issues arise, and even a stable implicit method may give poor results, as illustrated in Section 17.16.

A natural implicit method to consider is the trapezoidal method, a second-order accurate one-step method that takes the form

$$Q_i^{n+1} = Q_i^* + \frac{\Delta t}{2} [\psi(Q_i^*) + \psi(Q_i^{n+1})].$$

(17.41)

Note, by the way, an advantage of the fractional-step approach for stiff equations. While this is an implicit method, the equations obtained in the ith cell are decoupled from the equations in every other cell, and so these equations can be solved relatively easily. The coupling between grid cells arises only in the hyperbolic part of the equation, which can still be solved with an explicit high-resolution method.

In some cases the source term ψ may depend on derivatives of q as well as the pointwise value, for example if we wish to solve a viscous equation $q_t + f(q)_x = \mu q_{xx}$ by a fractional-step approach. In this case the derivatives will have to be discretized, bringing in values of Q at neighboring grid cells. For example, the term $\psi(Q_i^{n+1})$ in (17.41) would be replaced by

$$\mu(Q_{i-1}^{n+1} - 2Q_i^{n+1} + Q_{i+1}^{n+1})/\Delta x^2,$$

(17.42)

and similarly for $\psi(Q_i^*)$. The trapezoidal method (17.41) would then become the Crank–Nicolson method (4.13), and would require solving a tridiagonal linear system. It is generally necessary to use an implicit method for source terms involving second-order derivatives, since an explicit method would require $\Delta t = \mathcal{O}(\Delta x^2)$. With higher-order derivatives even smaller time steps would be required with an explicit method. The first-order hyperbolic part typically allows $\Delta t = \mathcal{O}(\Delta x)$, based on the CFL condition, and we generally hope to take time steps of this magnitude.

Although the trapezoidal method is second-order accurate and A-stable, it is only marginally stable in the stiff case, and this can lead to problems in the context of stiff hyperbolic equations, as illustrated in Section 17.16. For this reason the so-called *TR-BDF2 method* is generally recommended as a second-order implicit method. This is a two-stage Runge–Kutta method that combines one step of the trapezoidal method over time $\Delta t/2$ with a step of the second-order BDF method, using the intermediate result as another time level. For the ODE $q_t = \psi(q)$ this takes the form

$$Q_i^{**} = Q_i^* + \frac{\Delta t}{4}[\psi(Q_i^*) + \psi(Q_i^{**})],$$

$$Q_i^{n+1} = \frac{1}{3}[4Q_i^{**} - Q_i^* + \Delta t \, \psi(Q_i^{n+1})].$$

(17.43)

If ψ represents viscous terms, say $\psi = \mu q_{xx}$, then again this will have to be discretized using terms of the form (17.42), leading to tridiagonal systems in each stage of the Runge–Kutta method.

An example illustrating the superiority of this method over the Crank–Nicolson method for handling a diffusion term is given in [462], for a reaction–diffusion–advection equation arising in a model of chemotaxis in bacterial growth. Another example of how the trapezoidal method can fail is given in Section 17.16.

17.8 Steady-State Solutions

There are some other potential pitfalls in using a fractional-step method to handle source terms. In this section we consider some of these in relation to computing a steady-state solution, one in which $q_t(x, t) \equiv 0$ and the function $q(x, t)$ is independent of time. For the homogeneous constant-coefficient linear hyperbolic equation $q_t + Aq_x = 0$, if $q_t = 0$ then $q_x = 0$ also and the only steady-state solutions are the constant functions. When a source term is added, there can be more interesting steady-state solutions. Consider the advection–reaction equation (17.7) from Section 17.2, $q_t + \bar{u}q_x = -\beta q$. Setting $q_t = 0$ gives the ODE $q_x = -(\beta/\bar{u})q$, and hence this has the steady-state solution

$$q(x, t) = Ce^{-(\beta/\bar{u})x}.$$

(17.44)

In practice we would have a finite domain and some boundary conditions that must also be satisfied. Consider the same PDE on the domain $0 < x < 1$ with initial data $\overset{\circ}{q}(x)$ and the boundary condition

$$q(0, t) = g_0(t)$$

at the inflow boundary. The general solution is

$$q(x, t) = \begin{cases} e^{-\beta t} \, \overset{\circ}{q}(x - \bar{u}t) & \text{if } t < x/\bar{u}, \\ e^{-(\beta/\bar{u})x} \, g_0(t - x/\bar{u}) & \text{if } t > x/\bar{u}. \end{cases}$$

(17.45)

In the special case $g_0(t) \equiv C$, some constant, the solution $q(x, t)$ will reach the steady state (17.44) for all $t > 1/\bar{u}$, regardless of the initial conditions.

Recall the physical interpretation of this equation as the advection at velocity $\bar{u}$ of a radioactive tracer that decays at rate β. The boundary condition $q(0, t) = C$ corresponds to inflowing fluid having concentration C at all times. Up to time $t = 1/\bar{u}$ the initial data also has an effect on the solution, but after this time all of the fluid (and tracer) initially

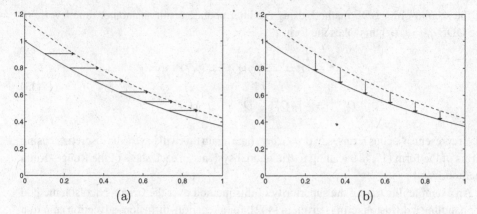

Fig. 17.4. Balance between advection and decay in a steady state solution. (a) The steady-state solution $q(x) = e^{-(\beta/\bar{u})x}$ is shown as the solid line. After advection by a distance $\bar{u}\,\Delta t$ to the right, $q^*(x) = q(x - \bar{u}\,\Delta t)$, the dashed line. (b) The advected solution decays by $e^{-\beta\Delta t}$, reproducing $q(x) = e^{-\beta\Delta t}q^*(x)$.

in the domain $0 < x < 1$ has flowed out the boundary at $x = 1$, and the steady state is reached.

It is important to note that being in a steady state does not in general mean that nothing is happening. In the above example, new tracer is constantly being introduced at $x = 0$, is advected downstream, and decays. The steady state results from a *balance* between the advection and decay processes. The terms $\bar{u}q_x$ and $-\beta q$ are both nonzero but cancel out. This balance is illustrated in Figure 17.4.

This suggests that we may have difficulties with a fractional-step method, where we first solve the advection equation ignoring the reactions and then solve the reaction equation ignoring the advection. Even if we start with the exact steady-state solution, each of these steps can be expected to make a change in the solution. In principle the two effects should exactly cancel out, but numerically they typically will not, since very different numerical techniques are used in each step.

In practice we generally don't know the steady-state solution, so we cannot use this as initial data. Instead we wish to determine the steady state by starting with some arbitrary initial data (perhaps some approximation to the steady state) and then marching forward in time until a steady state is reached. This can be viewed as an iterative method for solving the steady-state problem obtained by setting $q_t = 0$ in the equation, with each time step being one iteration. If all we care about is the steady-state solution, then this may not be a very efficient iterative method to use. A method designed specifically for the steady-state solution may be preferable. Such a method would be designed to converge as rapidly as possible to an accurate steady-state solution without necessarily giving an accurate time-dependent solution along the way. The study of such methods is a major topic in its own right and is not discussed further here.

However, time-marching methods are often used to compute steady-state solutions. For relatively small problems where computational efficiency is not an issue, it may be more cost-effective to use an existing time-marching code than to develop a new code specifically for the steady-state problem. One may also be interested in related time-dependent issues such as how convergence to steady state occurs in the physical system, the dynamic stability of the steady-state solution to small perturbations, or the solution of time-dependent problems that are near a steady state. Such *quasisteady* problems require a time-accurate approach that can also handle steady states well. One approach is presented in Section 17.14.

Fractional-step method can often be used to successfully compute steady-state or quasisteady solutions, but several issues arise. As mentioned above, the steady state results from a balance (cancellation) between two dynamic processes that are a handled separately in a fractional-step method. In some cases the method may not even converge, but instead will oscillate in time near the correct solution. This can happen if a high-resolution method with limiter functions is used for the hyperbolic part, since the limiter depends on the solution and effectively switches between different methods based on the behavior of the solution.

Even when the method converges, the numerical steady state obtained will typically depend on the time step used. This is rather unsatisfying, since the steady solution depends only on x and so we would like the numerical solution generated by a particular method to depend only on Δx. By contrast, unsplit methods can often be developed in which the steady state is independent of Δt. See Exercise 17.4 for one example.

17.9 Boundary Conditions for Fractional-Step Methods

When a fractional-step method is used, we typically need to impose boundary conditions in the hyperbolic step of the procedure. We may also need to impose boundary conditions in the source term step(s) if the source terms involve spatial derivatives of q – for example, if these are diffusion terms, then we are solving the diffusion equation, which requires boundary conditions at each boundary. The boundary conditions for the original PDE must be used to determine any boundary conditions needed for the fractional steps, but the connection between these is often nontrivial.

As a simple example, consider the advection–reaction equation (17.7) with the constant boundary data $q(0, t) = g_0(t) \equiv 1$, which results in the steady-state solution (17.44) for large t (with $C = 1$). Suppose we use a fractional-step method with the Godunov splitting, in which we first solve the advection equation $q_t + \bar{u}q_x = 0$ over time Δt and then the ODE $q_t = -\beta q$ over time Δt. Moreover, suppose we choose Δt so that $\bar{u}\,\Delta t/\Delta x = 1$ and the advection equation is solved exactly via

$$Q_i^* = Q_{i-1}^n, \tag{17.46}$$

and then we also solve the ODE exactly via

$$Q_i^{n+1} = e^{-\beta\,\Delta t} Q_i^*. \tag{17.47}$$

These steps can be combined to yield

$$Q_i^{n+1} = e^{-\beta\,\Delta t} Q_{i-1}^n. \tag{17.48}$$

For this simple problem we observed in Section 17.2.2 that there is no splitting error, and hence this procedure should yield the exact solution. If Q_{i-1}^n is the exact cell average of $q(x, t_n)$ over cell C_{i-1}, then Q_i^{n+1} will be the exact cell average of $q(x, t_{n+1})$ over cell C_i.

But to determine Q_1^{n+1} we must also use the boundary conditions. To implement the step (17.46) at $i = 1$ we must first specify a ghost cell value Q_0^n as described in Chapter 7. (Note that the step (17.47) does not require any boundary data, since we are solving an ODE within each grid cell.)

As a first guess at the value Q_0^n, we might follow the discussion of Section 7.2.2 and use the integral of (7.8). This would give $Q_0^n = 1$, since the specified boundary condition is independent of time. It appears that we have specified the exact ghost-cell value, and we know that the method being used in the interior is exact, and yet this combination will *not* produce the exact solution numerically. The value Q_1^{n+1} will not be the exact cell average of $q(t, t_{n+1})$ over the cell C_1. The computed value will be

$$Q_1^{n+1} = e^{-\beta \Delta t} = e^{-(\beta/\bar{u})\Delta x} = 1 - (\beta/\bar{u})\Delta x + \mathcal{O}(\Delta x^2), \qquad (17.49)$$

while the true cell average of the solution (17.44) is easily computed to be

$$\frac{1}{\Delta x} \int_{C_1} e^{-(\beta/\bar{u})x}\, dx = -\frac{\bar{u}}{\beta \Delta x}\left(e^{-(\beta/\bar{u})\Delta x} - 1\right) = 1 - \frac{1}{2}(\beta/\bar{u})\Delta x + \mathcal{O}(\Delta x^2). \quad (17.50)$$

The numerical value (17.49) is too small by $\mathcal{O}(\Delta x)$. In later time steps this error will propagate downstream, and eventually the entire solution will have an $\mathcal{O}(\Delta x)$ error.

The reason for this error is made clear in Figure 17.5. Figure 17.5(a) shows the steady-state solution, which is used as initial data, along with the function obtained after time Δt if we solve the advection equation alone with the boundary condition $g_0(t) = 1$. Figure 17.5(b) shows the results if we now solve the decay equation using the advected solution as data. Away from the boundary the decay exactly cancels the apparent growth due to the advection, and the exact steady-state solution is recovered. Near the boundary the use of the constant inflow boundary condition $g_0(t) = 1$ leads to the wrong profile.

It is clear how to fix this once we realize that the boundary condition $q(0, t) = 1$ is the proper boundary condition for the full equation $q_t + \bar{u}q_x = -\beta q$, but is *not* the proper boundary condition for the pure advection equation $q_t^* + \bar{u}q_x^* = 0$ that is being solved by

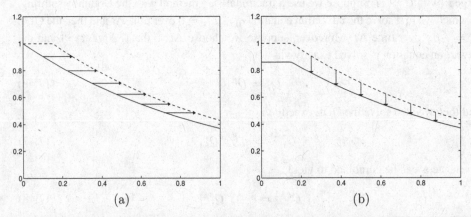

(a)   (b)

Fig. 17.5. The effect of incorrect boundary conditions in a fractional-step method. (a) The steady-state solution $q(x) = e^{-(\beta/\bar{u})x}$ is shown as the solid line. After advection by a distance $\bar{u}\,\Delta t$ to the right with $q^*(0, t) = 1$, the solution is shown as the dashed line. (b) The advected solution decays by $e^{-\beta\Delta t}$ and an error is apparent near the boundary.

the upwind method in the first step of the fractional-step method. If we denote the solution to this equation by $q^*(x, t)$ for $t \geq t_n$, then this function is different from $q(x, t)$ for $t > t_n$ and requires different boundary conditions. Figure 17.4(b) suggests what the correct boundary conditions are. We would like $q^*(x, t_n + \Delta t)$ to be the function $e^{-(\beta/\bar{u})(x - \bar{u}\,\Delta t)} = e^{\beta\,\Delta t} e^{-(\beta/\bar{u})x}$ for all $x \geq 0$, so that after the decay step we will recover $e^{-(\beta/\bar{u})x}$ for all $x \geq 0$. To obtain this we clearly need to impose a boundary condition on $q^*(0, t)$ that is growing with t,

$$q^*(0, t) = e^{\beta(t - t_n)} \equiv g_0^*(t),$$

and it is this function that should be used in determining the ghost-cell value Q_0^n instead of the original boundary condition $g_0(t) = 1$. Note that if we evaluate the integral from (7.8) using this function $g_0^*(t)$, we obtain

$$Q_0^n = \frac{\bar{u}}{\beta\,\Delta x} \left(e^{(\beta/\bar{u})\Delta x} - 1 \right). \tag{17.51}$$

Using this boundary value in the formulas (17.47) and (17.48) results in

$$Q_1^{n+1} = e^{-\beta\,\Delta t} Q_0^n = \frac{\bar{u}}{\beta\,\Delta x} (1 - e^{-\beta\,\Delta x}), \tag{17.52}$$

which is exactly the value (17.50).

For the advection–decay equation with a more general time-dependent boundary condition $q(0, t) = g_0(t)$, the proper boundary condition to use in the advection step is

$$q^*(0, t) = g_0^*(t) = e^{\beta(t - t_n)} g_0(t).$$

The integral (7.8) can perhaps be evaluated exactly using this function in place of g_0, or an approximation such as (7.9) could again be used, which would result in

$$Q_0^n = e^{\beta\,\Delta t/2} g_0 \left(t_n + \frac{\Delta x}{2\bar{u}} \right). \tag{17.53}$$

The proper value for the ghost-cell value Q_{-1}^n can be found similarly if a second ghost cell is needed (for a method with a wider stencil).

For the simple example considered above it was easy to determine the correct boundary conditions for q^* based on our knowledge of the exact solution operators for the advection and decay problems. For other problems it may not be so easy to determine the correct boundary conditions, but an awareness of the issues raised here can often help in deriving better boundary conditions than would be obtained by simply using the boundary conditions from the original equation. Since the required modification to the boundary conditions is typically $\mathcal{O}(\Delta t)$, as in the above example, this can make a significant difference in connection with methods that are second-order accurate or better. A more general procedure for deriving the proper intermediate boundary conditions for a linear hyperbolic equation is discussed in [276].

17.10 Stiff and Singular Source Terms

Most of the source terms that have appeared so far have been bounded functions corresponding to a source that is distributed in space. In some problems source terms naturally arise that are concentrated at a single point, a delta function, or at least are concentrated over a very small region compared to the size of the domain. In some cases this may be an external source depending explicitly on x that is spatially concentrated. An example of this nature was presented in Section 16.3.1, where we considered an advection equation with a delta function source of tracer. We consider another problem of this type in Section 17.11.

In other cases the source $\psi(q)$ depends only on the solution and yet the solution naturally develops structures in which the source terms are nonzero (and very large) only over very small regions in space. For example, this often happens if the source terms model chemical reactions between different species (reacting flow) in cases where the reactions happen on time scales much faster than the fluid dynamic time scales. Then solutions can develop thin *reaction zones* where the chemical-kinetics activity is concentrated. Such problems are said to have *stiff source terms*, in analogy with the classical case of stiff ODEs. Stiffness is common in kinetics problems. Reaction rates often vary by many orders of magnitude, so that some reactions occur on time scales that are very short compared to the time period that must be studied. One classic example of a stiff reacting flow problem is a *detonation wave*; an explosion in which a flammable gas burns over a very thin reaction zone that moves through the unburned gas like a shock wave, but with a more complicated structure. (See, for example, [92], [136], [156].) The thin reaction zone can be idealized as a delta-function source term that moves with the detonation wave. Some simpler examples are studied in this chapter. We will see that singular source terms lead to a modification of the Rankine–Hugoniot jump conditions that determine the structure and speed of propagating discontinuities.

17.11 Linear Traffic Flow with On-Ramps or Exits

As an illustration of a hyperbolic equation with a singular source term, consider traffic flow on a one-lane highway with on-ramps and exits where cars can enter or leave the highway. Then the total number of cars on the highway is not conserved, and instead there are sources and sinks. The corresponding source terms in the equation are delta functions with positive strength at the locations of the on-ramps and negative strength at exits.

As an example, consider a single on-ramp with a flux D at some point x_0, so that the source term is

$$\psi(x) = D\delta(x - x_0). \tag{17.54}$$

To begin with, suppose the traffic is sufficiently light that it moves at some constant speed $\bar{u}$ independent of the density q. Then the traffic flow is modeled by an advection equation as in Section 9.4.2 with the addition of the source term (17.54),

$$q_t + \bar{u}q_x = D\delta(x - x_0). \tag{17.55}$$

This is exactly the problem considered in Section 16.3.1, and the Riemann solution has a jump from q_l to $q_m = q_l + D/\bar{u}$ at $x = x_0$ (resulting from the source) and then a jump from q_m to q_r at $x = \bar{u}t$ (resulting from the initial data, and moving downstream with the traffic).

17.12 Rankine–Hugoniot Jump Conditions at a Singular Source

The jump in q at the on-ramp can be derived from a more general formula, an extension of the Rankine–Hugoniot jump condition to the case where there is a singular source moving with the jump. This formula will be needed to study the nonlinear traffic flow problem.

Consider a general conservation law coupled with a delta-function source term moving at some speed $s(t)$,

$$q_t + f(q)_x = D\,\delta(x - X(t)), \tag{17.56}$$

where $X'(t) = s(t)$. We will compute jump conditions at $X(t)$ using the same procedure as in Section 11.8 for the homogeneous equation. The differential equation (17.56) results from an integral conservation law that, over a small rectangular region such as the one shown in Figure 11.7, has the form (for $s < 0$ as in the figure)

$$\int_{x_1}^{x_1+\Delta x} q(x, t_1 + \Delta t)\,dx - \int_{x_1}^{x_1+\Delta x} q(x, t_1)\,dx$$

$$= \int_{t_1}^{t_1+\Delta t} f(q(x_1, t))\,dt - \int_{t_1}^{t_1+\Delta t} f(q(x_1 + \Delta x, t))\,dt$$

$$+ \int_{t_1}^{t_1+\Delta t} \int_{x_1}^{x_1+\Delta x} D\,\delta(x - X(t))\,dx\,dt. \tag{17.57}$$

Note that over this time interval the point $X(t)$ always lies between x_1 and $x_1 + \Delta x$, so that

$$\int_{x_1}^{x_1+\Delta x} D\,\delta(x - X(t))\,dx = D,$$

and hence the equation (17.57) can be approximated by

$$\Delta x\, q_r - \Delta x\, q_l = \Delta t\, f(q_l) - \Delta t\, f(q_r) + \Delta t\, D + \mathcal{O}(\Delta t^2). \tag{17.58}$$

Using $\Delta x = -s\,\Delta t$, dividing by $-\Delta t$, and taking the limit as $\Delta t \to 0$ gives

$$s(q_r - q_l) = f(q_r) - f(q_l) - D. \tag{17.59}$$

This is identical to the Rankine–Hugoniot jump condition (11.20) but with an additional term resulting from the singular source.

Note that if the source term $\psi(x, t)$ were a bounded function rather than a delta function, then the source term in (17.57) would be

$$\int_{t_1}^{t_1+\Delta t} \int_{x_1}^{x_1+\Delta x} \psi(x, t)\,dx\,dt \approx \Delta t\,\Delta x\, \psi(x, t)$$

$$\approx \Delta t^2 s\, \psi(x_1, t_1). \tag{17.60}$$

After dividing by $-\Delta t$ this would still be $\mathcal{O}(\Delta t)$ and would vanish as $\Delta t \to 0$. Hence a bounded source term does not change the Rankine–Hugoniot jump condition (11.20) at a discontinuity, since its contribution at any single point is negligible. A delta-function source term makes a nontrivial contribution at a single point and hence appears in the jump condition at that point.

Example 17.6. Consider again the on-ramp problem of Section 17.11, modeled by the equation (17.55). In this case $s = 0$ and $f(q) = \bar{u}q$, so that at $x = x_0$, where q jumps from q_l to q_m, the jump condition (17.59) yields

$$q_m - q_l = D/\bar{u}.$$

Note that this makes sense physically: D measures the flux per unit time of cars onto the highway, but the larger $\bar{u}$ is, the more widely spaced these cars are in the existing traffic, and hence the smaller the effect on the density.

Note that this Riemann solution has a similar structure to the Riemann solution illustrated in Figure 9.3 for the the variable-coefficient advection equation

$$q_t + (u(x)q)_x = 0 \qquad\qquad (17.61)$$

in the case where $u(x)$ is discontinuous with a single jump at x_0. In fact there is a connection between the two, since (17.61) can be rewritten as

$$q_t + u(x)q_x = -u'(x)q.$$

This is the color equation with a source term. For the case described in Figure 9.3, $u(x)$ is piecewise constant and hence $u'(x)$ becomes a delta function at x_0.

17.13 Nonlinear Traffic Flow with On-Ramps or Exits

For larger traffic densities a nonlinear traffic model must be used, and again a source term representing incoming cars at an on-ramp can be included. Figure 17.6 shows examples using the traffic flow model of Section 11.1 with velocity function (11.5). The initial density was $q = 0.4$ everywhere and a source of strength D is introduced at $x_0 = 0$ starting at time $t = 0$. In Figure 17.6(a) the source strength is small enough that the structure is essentially the same as it would be in a linear problem: congestion is seen only downstream from the on-ramp, and cars speed up again through a rarefaction wave. Figure 17.6(b) shows the same problem with a slightly larger value of D, in which case the structure is quite different. Since velocity varies with density in the nonlinear model, when the flux of cars onto the highway is too large a traffic-jam shock wave forms and moves upstream from the on-ramp. Note that even if the flux at the on-ramp is now reduced or eliminated this traffic jam will continue to propagate upstream and disrupt traffic. This is one reason that some on-ramps are equipped with traffic lights allowing only "one car per green" to insure that the flux never rises above a certain critical value.

For this model the structure of the Riemann solution can be explicitly determined (see Exercise 17.6). If $q_l > 0.5$, then characteristic signals travel upstream and a traffic-jam shock

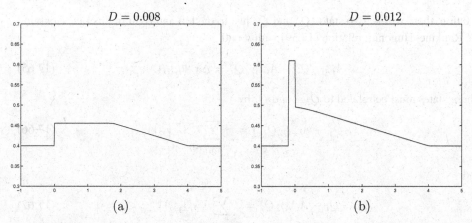

Fig. 17.6. Nonlinear traffic flow with a source term at an on-ramp. (a) The source strength $D = 0.008$ is sufficiently low that upstream traffic is unaffected. (b) For greater source strength $D = 0.012$ a traffic jam moves upstream from the on-ramp. Shown at $t = 20$. [claw/book/chap17/onramp]

will form for any $D > 0$. If $q_l < 0.5$, then a shock forms whenever $D > (1 - 2q_l)^2/4$. For the example in Figure 17.6, this cutoff is at $D = 0.01$.

17.14 Accurate Solution of Quasisteady Problems

In Section 17.8 we observed that fractional-step methods may not be well suited to problems near a steady state, where $f(q)_x$ and $\psi(q)$ are each large in magnitude but nearly cancel out. An alternative unsplit method can be formulated by discretizing the source term as a sum of delta function sources located at the cell interfaces,

$$q_t + f(q)_x = \sum_i \Delta x \, \Psi_{i-1/2}(t)\delta\big(x - x_{i-1/2}\big). \qquad (17.62)$$

At time t_n we then have a Riemann problem at $x_{i-1/2}$ for the equation

$$q_t + f(q)_x = \Delta x \, \Psi_{i-1/2}^n \delta\big(x - x_{i-1/2}\big), \qquad (17.63)$$

with data Q_{i-1} and Q_i. Since the delta-function source is at a fixed location, the solution will consist of propagating waves along with a jump in q at $x_{i-1/2}$, a similar structure to what is observed when for a conservation law with a spatially varying flux function, as discussed in Section 16.4.

Now suppose we use a linearized approximate Riemann solver of the form discussed in Section 15.3 to replace the flux $f(q)$ by $\hat{A}_{i-1/2}q$, where $\hat{A}_{i-1/2}$ is an approximate Jacobian matrix determined by the data Q_{i-1} and Q_i. Then we have a Riemann problem for the equation

$$q_t + \hat{A}_{i-1/2}q_x = \Delta x \, \Psi_{i-1/2}\delta\big(x - x_{i-1/2}\big). \qquad (17.64)$$

The solution will consist of waves propagating with speeds $s_{i-1/2}^p = \hat{\lambda}_{i-1/2}^p$, the eigenvalues of $\hat{A}_{i-1/2}$, and an additional discontinuity in q at $x_{i-1/2}$ (propagating at speed 0) arising from the delta-function source. Instead of a single state $Q_{i-1/2}^\vee$ at $x_{i-1/2}$ in the Riemann solution, there will be two states $Q_l^\vee$ and $Q_r^\vee$ just to the left and right of this ray, satisfying the Rankine–Hugoniot relation (17.59) with $s = 0$,

$$\hat{A}_{i-1/2}Q_r^\vee - \hat{A}_{i-1/2}Q_l^\vee = \Delta x \, \Psi_{i-1/2}. \tag{17.65}$$

These states must be related to Q_{i-1} and Q_i by

$$\hat{A}_{i-1/2}Q_l^\vee - \hat{A}_{i-1/2}Q_{i-1} = \sum_{p:s_{i-1/2}^p < 0} s_{i-1/2}^p \mathcal{W}_{i-1/2}^p \tag{17.66}$$

and

$$\hat{A}_{i-1/2}Q_i - \hat{A}_{i-1/2}Q_r^\vee = \sum_{p:s_{i-1/2}^p > 0} s_{i-1/2}^p \mathcal{W}_{i-1/2}^p, \tag{17.67}$$

where $\mathcal{W}_{i-1/2}^p = \alpha_{i-1/2}^p \hat{r}_{i-1/2}^p$ is the pth wave, which is proportional to the eigenvector $\hat{r}_{i-1/2}^p$ of $\hat{A}_{i-1/2}^p$. Adding (17.66) and (17.67) together and using (17.65) allows us to eliminate $Q_{l,r}^\vee$ and obtain

$$\hat{A}_{i-1/2}(Q_i - Q_{i-1}) - \Delta x \, \Psi_{i-1/2} = \sum_{p=1}^{m} s_{i-1/2}^p \mathcal{W}_{i-1/2}^p. \tag{17.68}$$

If the Roe solver of Section 15.3.2 is used, then the matrix $\hat{A}_{i-1/2}$ satisfies (15.18), and so (17.68) becomes

$$f(Q_i) - f(Q_{i-1}) - \Delta x \, \Psi_{i-1/2} = \sum_{p=1}^{m} s_{i-1/2}^p \mathcal{W}_{i-1/2}^p. \tag{17.69}$$

In order to determine the waves $\mathcal{W}_{i-1/2}^p$ in the Riemann solution, we need only decompose $f(Q_i) - f(Q_{i-1}) - \Delta x \Psi_{i-1/2}$ into eigenvectors as

$$f(Q_i) - f(Q_{i-1}) - \Delta x \, \Psi_{i-1/2} = \sum_{p=1}^{m} \beta_{i-1/2}^p \hat{r}_{i-1/2}^p, \tag{17.70}$$

and then set $\mathcal{W}_{i-1/2}^p = \alpha_{i-1/2}^p \hat{r}_{i-1/2}^p$, where

$$\alpha_{i-1/2}^p = \beta_{i-1/2}^p / s_{i-1/2}^p \quad \text{for } s_{i-1/2}^p \neq 0. \tag{17.71}$$

Godunov's method and related high-resolution methods, when implemented in the wave-propagation form, only require the waves propagating at nonzero speeds, and these can all be obtained by this procedure. Alternatively, the formulation of Section 15.5 can be used to implement the method directly in terms of the waves $\mathcal{Z}_{i-1/2}^p = \beta_{i-1/2}^p \hat{r}_{i-1/2}^p$, with the simple modification that these waves are now defined by solving (17.70) instead of (15.67). As in the discussion of Section 15.5, this procedure can also be used for choices of $\hat{r}^p$ other than the eigenvectors of the Roe matrix.

This method is generally easy to implement by a simple modification of the Riemann solver for the homogeneous equation. It is not formally second-order accurate in general, and may not perform as well as fractional-step methods for some time-dependent problems that are far from steady state.

The method can be greatly advantageous, however, for quasisteady problems, where $f(q)_x \approx \psi(q)$. In this case we expect

$$\frac{f(Q_i) - f(Q_{i-1})}{\Delta x} \approx \Psi_{i-1/2}, \tag{17.72}$$

and hence the left-hand side of (17.69) will be near 0. The waves resulting from the eigen-decomposition will thus have strength near zero, and will cause little change in the solution. These waves will model the deviation from steady state, and it is precisely this information that should be propagated, and to which wave limiters should be applied. Moreover, a numerical steady-state solution computed with this method will satisfy (17.72) with equality. If the source term is appropriately discretized then smooth steady-state solutions will be computed with second-order accuracy even though the transient behavior may not be formally second-order accurate.

A different wave-propagation method with similar features was proposed in [284], but the approach just presented seems to be more robust as well as much easier to implement; see [18] for more discussion. Some other related methods can be found in the references of Section 17.2.1.

17.15 Burgers Equation with a Stiff Source Term

In the remainder of this chapter we consider problems having source terms that do not appear to contain delta functions, but that are typically close to zero over most of the domain while being very large over thin *reaction zones* that dynamically evolve as part of the solution. Such source terms can often be approximated by delta functions, but their location and strength is generally not known *a priori*.

As a simple but illustrative example, consider the Burgers equation with a source term,

$$u_t + \frac{1}{2}(u^2)_x = \psi(u), \tag{17.73}$$

where

$$\psi(u) = \frac{1}{\tau}u(1 - u)(u - \beta), \tag{17.74}$$

with $\tau > 0$ and $0 < \beta < 1$. If we consider the ODE

$$u'(t) = \psi(u(t)) \tag{17.75}$$

alone, we see that $u = 0$, β, 1 are equilibrium points. The middle point $u = \beta$ is an unstable equilibrium, and from any initial data $\mathring{u} \neq \beta$, u asymptotically approaches one of the other equilibria: $u \to 0$ if $\mathring{u} < \beta$ or $u \to 1$ if $\mathring{u} > \beta$. The parameter τ determines the time scale over which u approaches an equilibrium.

For τ very small, any initial data $\mathring{u}(x)$ supplied to the equation (17.73) will rapidly approach a step function with values 0 and 1 (and near-discontinuous behavior where $\mathring{u}(x)$ passes through β) on a much faster time scale than the hyperbolic wave propagation. To see how the solution then evolves, it suffices to consider the case of a Riemann problem with jump from value 0 to 1 or from 1 to 0.

If $u_l = 1$ and $u_r = 0$, then the Burgers equation with no source gives a shock wave moving with the Rankine–Hugoniot speed $1/2$, by (11.23). The source term is then identically zero and has no effect. More general initial data $\mathring{u}(x)$ satisfying $\mathring{u}(x) > \beta$ for $x < 0$ and $\mathring{u}(x) < \beta$ for $x > 0$ would rapidly evolve to this situation and give a shock traveling with speed $1/2$ for any value of $\beta \in (0, 1)$.

The situation is much more interesting in the case where $u_l = 0$ and $u_r = 1$, as studied in [280]. In this case Burgers' equation gives a rarefaction wave that spreads the initial discontinuity out through all intermediate values. The source term opposes this smearing and sharpens all values back towards 0 or 1. These competing effects balance out and result in a smooth solution that rapidly approaches a traveling wave that neither smears nor sharpens further, but simply propagates with some speed s:

$$u(x, t) = w \left(\frac{x - st}{\tau} \right). \tag{17.76}$$

We will see below that $s = \beta$.

The shape of this profile is shown in Figure 17.7(a). The width of the transition zone in $u(x, t)$ is $\mathcal{O}(\tau)$, and for small τ this appears similar to a viscous shock profile with small viscosity, though the competing effects that produce the steady profile are different in that case. As we will see in the next section, computing such a solution on a grid where the structure is not well resolved (e.g., for $\Delta x > \tau$) can be more difficult than correctly approximating a shock wave.

The shape of the profile w and the speed s can be determined by inserting the form (17.76) into (17.73), yielding

$$-sw' + ww' = w(1 - w)(w - \beta), \tag{17.77}$$

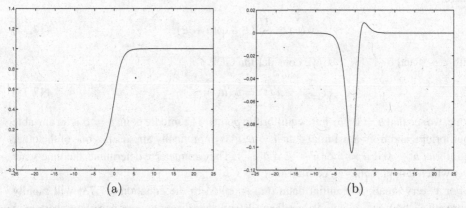

(a)   (b)

Fig. 17.7. (a) Traveling-wave solution $w(\xi)$ to (17.73). (b) The source term $\psi(w(\xi))$ of (17.74). Shown for $\beta = 0.6$.

which gives the ODE

$$w' = \frac{w(1-w)(w-\beta)}{w-s}. \tag{17.78}$$

We also require $w(-\infty) = 0$ and $w(+\infty) = 1$. The equation (17.78) has a solution with these limiting values only if $s = \beta$, since otherwise w cannot cross the unstable equilibrium value β. When $s = \beta$ we can cancel this term in (17.78) and obtain the logistic equation

$$w' = w(1-w), \tag{17.79}$$

with solutions of the form

$$w(\xi) = \frac{e^{\xi}}{1+e^{\xi}} = \frac{1}{2}[1 + \tanh(\xi/2)], \tag{17.80}$$

as shown in Figure 17.7(a).

The propagation speed is $s = \beta$. Unlike the case $u_l = 1$, $u_r = 0$, this speed depends on the structure of the source term. This leads to numerical difficulties if we try to solve the problem on an underresolved grid, as discussed in the next section. More generally, if we replace $f(u) = u^2/2$ by a more general flux function, the propagation speed of the resulting traveling wave will be $s = f'(\beta)$, as shown in [129].

We can relate the result just found to the discussion of singular source terms earlier in this chapter by observing that the source term $\psi(w)$ will be essentially zero except in the transition region, where its magnitude is $\mathcal{O}(1/\tau)$, as seen in Figure 17.7(b). Since this region has width $\mathcal{O}(\tau)$, this suggests that the source terms approximate a delta function as $\tau \to 0$.

The magnitude of the limiting delta function can be found by integrating $\psi(w((x-st)/\tau))$ over all x and taking the limit as $\tau \to 0$, although in fact this value is independent of τ for the wave form w. We can use (17.79) to rewrite $\psi(w(\xi))$ as

$$\psi(w(\xi)) = \frac{1}{\tau}w'(\xi)[w(\xi) - \beta]$$

$$= \frac{1}{\tau}\frac{d}{d\xi}\left(\frac{1}{2}[w(\xi) - \beta]^2\right), \tag{17.81}$$

and hence

$$\int_{-\infty}^{\infty} \psi(w(x-st)/\tau)\,dx = \tau \int_{-\infty}^{\infty} \psi(w(\xi))\,d\xi$$

$$= \frac{1}{2}[w(\xi) - \beta]^2\Big|_{-\infty}^{+\infty}$$

$$= \frac{1}{2} - \beta. \tag{17.82}$$

This value is independent of τ and gives the strength D of the delta-function source observed in the limit $\tau \to 0$,

$$D = \frac{1}{2} - \beta. \tag{17.83}$$

Using the modified Rankine–Hugoniot relation (17.59) results in the speed

$$s = \frac{1}{2} - D = \beta$$

for the limiting jump discontinuity, which is consistent with the speed of the traveling wave, as we should expect. We see that the source term has a nontrivial effect even in the limit $\tau \to 0$ whenever $\beta \neq 1/2$.

Note from Figure 17.7(b) that the source term is negative where $w < \beta$ and positive where $w > \beta$. When $\beta = 1/2$ these two portions exactly cancel and there is no net source term, so the propagation speed agrees with what is expected from the conservation law alone. When $\beta \neq 1/2$ there is a net source or sink of u in the transition zone that affects the speed at which the front propagates. This same effect is seen in many reacting flow problems.

17.16 Numerical Difficulties with Stiff Source Terms

A hyperbolic equation with a source term is said to be *stiff* if the wave-propagation behavior of interest occurs on a much slower time scale than the fastest times scales of the ODE $q_t = \psi(q)$ arising from the source term. An example is a detonation wave arising when gas dynamics is coupled with the chemical kinetics of combustion. Detonation waves can travel rapidly through a combustible gas, but even these fast waves are many orders of magnitude slower than the time scales of some of the chemical reactions appearing in the kinetic source terms. It would often be impossible to simulate the propagation of such waves over distances of physical interest if it were necessary to fully resolve the fastest reactions.

As a simpler example, consider the Burgers equation (17.73) with source (17.74). As we have seen in the previous section, when τ is very small a traveling-wave solution looks essentially like a discontinuity from $u_l = 0$ to $u_r = 1$ propagating at speed $\beta = \mathcal{O}(1)$. Suppose we want to approximate this with $\tau = 10^{-10}$, say, over a domain and time interval that are $\mathcal{O}(1)$. Then the transition from u_l to u_r takes place over a zone of width on the order of 10^{-10}. We will certainly want to use $\Delta x \gg \tau$, and we can't hope to resolve the structure of the traveling wave, only the macroscopic behavior (as in shock capturing). We also do not want to take $\Delta t = \mathcal{O}(\tau)$, but rather want to take $\Delta t = \mathcal{O}(\Delta x)$ based on the CFL restriction for the hyperbolic problem. We hope that if the wave of interest moves less than one grid cell each time step, we will be able to capture it accurately, with little numerical smearing and the physically correct velocity.

The classical theory of numerical methods for stiff *ordinary* differential equations can be found in many sources, e.g., [145], [178], [253]. Recall that an ODE is called stiff if we are trying to compute a particular solution that is varying much more slowly than the fastest time scales inherent in the problem. Typically this means that the solution of interest is evolving on some *slow manifold* in state space and that perturbing the solution slightly off this manifold will result in a rapid transient response, bringing the solution back the manifold, followed by slow evolution once again. Stiff problems often arise in chemical kinetics, where the rates of different reactions can vary by many orders of magnitude. If the fastest reactions are essentially in equilibrium, then the concentrations will vary slowly on time scales governed by slower reactions, and the solution is evolving on a slow manifold.

If the system is perturbed, say by injecting additional reactants, then fast transient behavior may result over a short period of time as the faster reactions reach a new equilibrium.

A simpler example of a stiff ODE is the equation (17.75) with $\tau \ll T$ in (17.74), where T is some reference time over which we are interested in the solution. For most initial data there will be fast transient decay to one of the slow manifolds $u \equiv 0$ or $u \equiv 1$. Perturbing away from one of these values results in a fast transient. An even simpler example is the equation

$$u'(t) = -u(t)/\tau, \qquad (17.84)$$

where $u \equiv 0$ is the slow manifold.

Numerically, stiff ODEs are problematical because we typically want to take "large" steps whenever the solution is "slowly" evolving. However, the numerical method is not exact and hence is constantly perturbing the solution. If these perturbations take the solution off the slow manifold, then in the next time step the solution will have a rapid transient and the desired time step may be much too large. In particular, if an explicit method is used, then the stability restriction of the method will generally require choosing the time step based on the fastest time scale present in the problem, even if we are hoping we do not need to resolve this scale. Luckily many *implicit* methods have much better stability properties and allow one to choose the time step based on the behavior of the solution of interest.

As we will see, solving stiff hyperbolic equations can be even more challenging than solving stiff ODEs. This difficulty arises largely from the fact that in a stiff hyperbolic equation the fastest reactions are often *not* in equilibrium everywhere. In thin regions such as the transition zone of the problem considered in Section 17.15 (see Figure 17.7(b)), or the reaction zone of a detonation wave, there is a fast transient behavior constantly taking place. For some problems it really appears necessary to resolve this zone in order to obtain good solutions, in which case adaptive mesh refinement is often required to efficiently obtain good results. In other cases we can achieve our goal of solving stiff problems on underresolved grids. A number of techniques have been developed for various problems, and here we will primarily illustrate some of the difficulties.

For stiff source terms we will again concentrate on the use of fractional-step methods (though more sophisticated approaches may not use splitting). The *Godunov splitting* applied to $q_t + f(q)_x = \psi(q)$ would simply alternate between solving the following two problems over time step Δt:

$$\text{Problem A:} \quad q_t + f(q)_x = 0, \qquad (17.85)$$

$$\text{Problem B:} \quad q_t = \psi(q). \qquad (17.86)$$

The first thing to observe is that the ODE of (17.86) is going to be stiff, and so we must use an appropriate method for this step in going from Q_i^* to Q_i^{n+1} in the ith cell. One popular class of stiff solvers are the BDF methods (backward differentiation formulas). These are linear multistep methods and require at least two previous time levels to get second-order accuracy or better. As discussed in Section 17.6, this is a problem in the context of a fractional-step method, since we only have one value Q_i^* that we can use.

The trapezoidal method (17.41) can often be effectively used for stiff ODEs, but may fail miserably for hyperbolic equations with stiff source terms. Figure 17.8 shows a sample

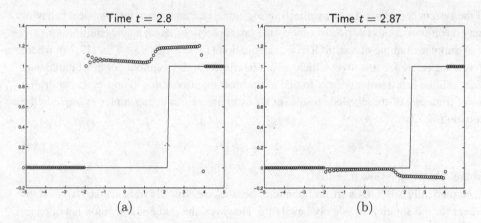

Fig. 17.8. Numerical solution to the Burgers equation with a stiff source term using the trapezoidal method for the source term: (a) after 40 time steps, (b) after 41 time steps. [claw/book/chap17/stiffburgers]

computation on the Burgers-equation example of Section 17.15, with $\beta = 0.8$, $\tau = 10^{-5}$, $\Delta x = 0.1$, and $\Delta t = 0.07$. The initial data was $u = 0$ for $x < 0$ and $u = 1$ for $x > 0$. Figure 17.8(a) shows the solution after 40 time steps, while Figure 17.8(b) shows the solution one time step later. The solution oscillates in time between these two different basic shapes with a set of waves propagating at unphysical speeds.

This behavior arises from the fact that the trapezoidal method is not *L-stable* (see [253]). If we start on the slow manifold, this method does a good job of keeping the numerical solution on the slow manifold even with time steps that are large relative to the faster scales. But for initial data that has an initial fast transient, the method yields oscillations. This is easy to see for the simple ODE (17.84). In this case the "slow manifold" is simply $u \equiv 0$, and starting with any other data gives exponentially fast decay of the true solution towards this state, with rate $1/\tau$. On this problem the trapezoidal method yields

$$U^{n+1} = \left(\frac{1 - \frac{1}{2}\Delta t/\tau}{1 + \frac{1}{2}\Delta t/\tau} \right) U^*. \tag{17.87}$$

If $U^* = 0$ then $U^{n+1} = 0$ also and we stay on the slow manifold. But if $U^* \neq 0$ and $\Delta t/\tau \gg 1$ then $U^{n+1} \approx -U^*$. The "amplification factor" in (17.87) approaches -1 as $-\Delta t/\tau \to -\infty$. Rather than the proper decay, we obtain an oscillation in time unless the transient is well resolved.

For the Burgers equation (17.73) with a stiff source term, nonequilibrium data is constantly being introduced by the averaging process in solving the conservation law near any front. This sets up an oscillation in time, as observed in Figure 17.8.

The trapezoidal method is an A-stable method. In fact, its stability region is exactly the left half plane. The fact that the boundary of the stability region is the imaginary axis, and hence passes through the point at infinity on the Riemann sphere, is responsible for the observed undesirable behavior, since in solving the stiff problem we are interested in letting $-\Delta t/\tau \to -\infty$.

An *L-stable* method is one for which the point at infinity is in the interior of the stability region, and hence the amplification factor approaches something less than 1 in magnitude as $-\Delta t/\tau \to -\infty$. (L-stability is usually defined to also require A-stability, but we use this looser definition for convenience.) The BDF methods have this property. The one-step BDF method is the backward Euler method, which for the equation (17.84) is simply

$$U^{n+1} = U^* - \frac{\Delta t}{\tau}U^{n+1} \implies U^{n+1} = \left(\frac{1}{1 + \Delta t/\tau}\right)U^*. \tag{17.88}$$

Note that $(1 + \Delta t/\tau)^{-1} \to 0$ as $\Delta t/\tau \to -\infty$, and so this method can be used on stiff problems even when the initial data is not on the slow manifold. The backward Euler method is only first-order accurate. In the present context, this does not really matter, since we expect the source terms to be active only over thin regions where there are fast transients that we cannot resolve with any accuracy anyway. What we primarily require is the L-stability property.

In some situations we may want to use a second-order method that is L-stable so that we can obtain good accuracy of source terms in regions where they are smooth (or where transients are well resolved) and also avoid oscillations in regions of stiffness. In the context of a fractional-step method we must use a one-step method, and one possible choice is the TR-BDF2 method (17.43) described in Section 17.6, which is L-stable. Figure 17.9 shows results obtained if we apply the TR-BDF2 method to the stiff source term in the Burgers-equation example. The use of this method eliminates the oscillations and unphysical states that were seen in Figure 17.8.

However, it still produces an incorrect solution in the stiff case, as seen in Figure 17.9(a). The wave now looks reasonable but is traveling at the wrong speed. It behaves fine if the grid is further refined, or equivalently if the value of τ is increased as shown in Figure 17.9(b). But the method still fails on underresolved grids in spite of the good behavior of the ODE solver.

This illustrates a difficulty often seen with stiff source terms. Similar behavior was observed by Colella, Majda & Roytburd [82] for detonation waves, and the problem has been discussed in many papers since. Even the simplest advection equation together with the source term (17.74) gives similar results. An analysis of this model problem, presented in LeVeque & Yee [293], carries over to the Burgers-equation example.

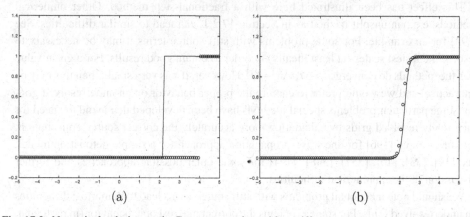

(a)   (b)

Fig. 17.9. Numerical solution to the Burgers equation with a stiff source term using the TR-BDF2 method for the source term: (a) the stiff case $\tau = 10^{-5}$; (b) the nonstiff case $\tau = 0.1$. [claw/book/chap17/stiffburgers]

Suppose

$$U_i^n = \begin{cases} 0 & \text{if } i < I, \\ 1 & \text{if } i \geq I, \end{cases}$$

as suggested by the result plotted in Figure 17.9(a). Then a high-resolution method for the Burgers equation will reduce to Godunov's method (since all slopes will be limited to 0), and we will obtain

$$U_i^* = \begin{cases} 0 & \text{if } i < I, \\ 1 - \Delta t/(2\,\Delta x) & \text{if } i = I, \\ 1 & \text{if } i > I, \end{cases}$$

since the jump propagates at speed $1/2$. For the calculation shown in Figure 17.9(a), $U_I^* = 1 - \Delta t/(2\,\Delta x) = 0.65$. We now solve the stiff ODE, and since $U_I^* < \beta = 0.8$, the solution decays rapidly towards the equilibrium at 0, and so $U_I^{n+1} \approx 0$ after time Δt. In all other cells the source term is zero. We thus obtain

$$U_i^{n+1} \approx \begin{cases} 0 & \text{if } i < I+1, \\ 1 & \text{if } i \geq I+1, \end{cases}$$

and the wave has shifted over by one grid cell. The wave thus propagates at a speed of one grid cell per time step, which is a purely numerical artifact and not the correct physical speed.

Note that with a smaller time step we would have $U_I^* > \beta$, in which case $U_I^{n+1} \approx 1$, and so $U^{n+1} \approx U^n$. Now the numerical solution remains stationary (propagates with speed 0), again not the correct solution.

This difficulty arises from the fact that the numerical method for the homogeneous hyperbolic equation introduces numerical diffusion, leading to a smearing of the discontinuity or steep gradient that should be observed physically. The source term is then active over the entire region where the solution is smeared, leading to a larger contribution from this term than is physical. As we saw in Section 17.12, the speed at which the jump propagates is directly related to the strength of the source concentrated at the jump, so an incorrect net source term leads to incorrect propagation speeds.

This effect has been illustrated here with a fractional-step method. Other numerical methods, e.g., an unsplit method as in Section 17.2.1, can lead to similar difficulties. See [293] for an example. For some problems with stiff source terms it may be necessary to resolve the fastest scales (at least locally) in order to obtain good results. For a given value of τ the methods do converge as $\Delta t, \Delta x \to 0$. However, if τ is very small, then this may be impractical and we would prefer to capture the proper behavior on an underresolved grid. For some particular problems special methods have been developed that avoid the need for such finely resolved grids by calculating more accurately the correct source contribution; see for example [196] for one wave-propagation approach for a simple detonation model. See [19], [33], [40], [357], [348], [494] for some other possible approaches and further discussion of these numerical difficulties.

We should note that not all problems with stiff source terms lead to numerical difficulties on underresolved grids. For some problems the correct macroscopic behavior (in particular, correct propagation speeds) is observed even if the fast time scales are not well resolved.

To understand why, it is useful to consider the equation

$$q_t + f(q)_x = \psi(q) + \epsilon q_{xx}, \tag{17.89}$$

in which a diffusive or viscous term has been added in addition to the source term, which we assume is stiff with some fast time scale $\tau \ll 1$. Denote the solution to this equation by $q^{\tau,\epsilon}$. In practice there is typically physical dissipation present, but (as usual with the vanishing-viscosity approach) we assume this is very small, and we wish to find the limit $\lim_{\epsilon \to 0} q^{\tau,\epsilon}$ where τ is fixed at some physically correct value. On an underresolved grid we cannot hope to capture the detailed behavior on this fast time scale, and so we really seek to compute an approximation to

$$\lim_{\tau \to 0} \left(\lim_{\epsilon \to 0} q^{\tau,\epsilon} \right). \tag{17.90}$$

However, the numerical method will typically introduce dissipation $\epsilon > 0$ that depends on Δx, so that on an underresolved grid we can easily have $\epsilon \gg \tau$. As we refine the grid we are really approximating

$$\lim_{\epsilon \to 0} \left(\lim_{\tau \to 0} q^{\tau,\epsilon} \right), \tag{17.91}$$

at least as long as we remain on underresolved grids. It is only when we reach a grid that resolves the fast scale (which we don't want to do) that we would begin to approximate the correct limit (17.90). Pember [356] has conjectured that the problems leading to numerical difficulties on underresolved grids are precisely those for which the limits $\epsilon \to 0$ and $\tau \to 0$ do not commute, i.e., for which (17.90) and (17.91) give different results.

This is the case for the stiff Burgers-equation example of Section 17.15, for example. Adding a viscous term to (17.73) gives

$$u_t + \frac{1}{2}(u^2)_x = \frac{1}{\tau}u(1-u)(u-\beta) + \epsilon u_{xx}. \tag{17.92}$$

This equation has an exact traveling-wave solution that generalizes (17.80),

$$u(x,t) = \frac{1}{2}(1 + \tanh(\mu(x - st))), \tag{17.93}$$

where the values μ and s are given by

$$\mu = \frac{1}{\tau}\left(1 + \sqrt{1 + 8\epsilon/\tau}\right)^{-1}$$

and

$$s = \left(\frac{\beta}{2} - \frac{1}{4}\right)\left(1 + \sqrt{1 + 8\epsilon/\tau}\right) + \frac{1}{2}.$$

This solution can be found using results in [254], where a more general equation (with a cubic source term) is studied. For fixed τ, as $\epsilon \to 0$ we recover $\mu = 1/2\tau$ and $s = \beta$, so that (17.93) agrees with (17.80). On the other hand, if τ is smaller than ϵ, then this solution is quite different and the limits $\epsilon \to 0$ and $\tau \to 0$ do not commute.

In some physical problems adding additional dissipation does not substantially change the solution, and for such problems good results can often be obtained on underresolved grids. In the next section we study one class of problems for which this is true.

17.17 Relaxation Systems

In many physical problems there is an equilibrium relationship between the variables that is essentially maintained at all times. If the solution is perturbed away from this equilibrium, then it rapidly *relaxes* back towards the equilibrium. The problem considered in Section 17.15 has this flavor, but in that case there is a single variable u and two possible stable equilibria $u = 0$ and $u = 1$, leading to numerical difficulties. In this section we consider the situation where there are several variables and perhaps many different equilibrium states, but there is a unique equilibrium relationship between the variables.

A simple model problem is the system of two equations

$$u_t + v_x = 0,$$
$$v_t + au_x = \frac{f(u) - v}{\tau} \tag{17.94}$$

for $0 < \tau \ll 1$, where $a > 0$ and $f(u)$ is a given function. The equilibrium states are those for which $v = f(u)$. If $v \neq f(u)$, then in the second equation the right-hand side dominates the term au_x, and v is rapidly driven back towards equilibrium (except perhaps in narrow reaction zones where u has a very steep gradient).

Note that if we simply assume $v \equiv f(u)$ in (17.94), then we can discard the second equation. Inserting this equilibrium assumption into the first equation, we then obtain the scalar conservation law

$$u_t + f(u)_x = 0. \tag{17.95}$$

Therefore we might expect solutions to the system (17.94) to be well modeled by the *reduced equation* (17.95) for small τ. In fact this is true, provided that a so-called *subcharacteristic condition* is satisfied. Observe that the homogeneous part of the system (17.94) is a linear hyperbolic system. The coefficient matrix has eigenvalues $\lambda^{1,2} = \pm\sqrt{a}$. Hence information can propagate no faster than speed $\sqrt{a}$, and adding the source term in (17.94) does not change this fact. On the other hand, the reduced equation (17.95) has characteristic speed $f'(u)$. If $|f'(u)| > \sqrt{a}$, then we cannot expect the solution of (17.94) to behave well in the limit $\tau \to 0$. The subcharacteristic condition for this problem is the requirement

$$-\sqrt{a} \le f'(u) \le \sqrt{a}. \tag{17.96}$$

Similar conditions hold for larger systems involving relaxation, and generally state that the characteristic speed of the reduced equation must fall within the range spanned by the characteristic speeds of the homogeneous part of the original system. This terminology was introduced by Liu [310], who studied more general relaxation systems in which au is replaced by a possibly nonlinear function $\sigma(u)$ in (17.94). Such problems arise in nonlinear elasticity (see Section 2.12.4), in which case $\sigma(u)$ represents the stress as a function of strain.

Many physical systems contain rapid relaxation processes that maintain an equilibrium, and often we solve the reduced equations based on this equilibrium. In fact the Euler equations of gas dynamics can be viewed as the reduced equations for more accurate models of the physics that include relaxation of vibrational or chemical nonequilibrium states that arise from intermolecular collisions. For many practical purposes the Euler equations are sufficient, though for some problems it is necessary to consider nonequilibrium flows; see for example [69], [70], [474], or [497]. A related example of the Euler equations relaxing towards isothermal flow is given in Section 17.17.3.

Relaxation systems are also used in modeling traffic flow. In Section 11.1 we derived a scalar conservation law by specifying the velocity of cars as a function of density $U(\rho)$. This assumes that drivers can react infinitely quickly to changes in density and are always driving at the resulting *equilibrium velocity*. In fact the velocity should relax towards this value at some rate depending on the drivers' reaction time. This can be modeled with a system of two equations, one for the density and a second for the velocity. Such models are often called *second-order models* in traffic flow (the scalar equation is the first-order model) and were first studied by Payne [354] and Whitham [486]. For more recent work, see for example [16], [99], [260], [296], [400], [498].

On the theoretical side there is considerable interest in the study of quite general relaxation systems and the conditions under which convergence of solutions occurs as $\tau \to 0$. See for example [64], [62], [305], [310], [336], [337], [491], [495]. Numerically it is useful to note that these relaxation problems can typically be solved on underresolved grids, as the next example shows. (This fact is the basis for the class of numerical *relaxation schemes* discussed in Section 17.18, in which stiff relaxation terms are intentionally introduced.)

Example 17.7. Figure 17.10 shows some solutions to the relaxation system (17.94) for $f(u) = \frac{1}{2}u^2$, $a = 1$, and the Riemann data

$$u_l = 1, \quad u_r = 0, \quad v_l = f(u_l) = \frac{1}{2}, \quad v_r = f(u_r) = 0. \quad (17.97)$$

The reduced equation is Burgers' equation, and we expect a shock traveling with speed $1/2$ in the limit as $\tau \to 0$, so it should be at $x = 0.4$ at the time $t = 0.8$ shown in all figures. Numerical solutions to the system (17.94) for various values of τ are shown, obtained using a fractional-step method. Solving the Riemann problem for the homogeneous linear system gives two waves with speeds ± 1. The state between these waves is not in equilibrium even if the Riemann data is, and the source term relaxes this state towards equilibrium. When τ is large (slow relaxation), the structure of the linear system is clearly visible, but as $\tau \to 0$ we observe convergence to the solution of Burgers' equation.

Note that for this problem the numerical fractional-step method behaves well in the limit $\tau \to 0$. The steep gradient near $x/t = 0.4$ (where the source term is active) is not well resolved on the grid for the smallest value of τ used in Figure 17.10, and yet the correct macroscopic behavior is observed. This is a reflection of the fact that if we add $\mathcal{O}(\epsilon)$ viscosity to the system (17.94), then the limits $\epsilon \to 0$ and $\tau \to 0$ commute, as described in Section 17.16. The basic reason for this is that the parameter τ also acts like a viscosity in the system (though this is not clear from the form of (17.94)), and so adding additional viscosity does not change the nature of the solution.

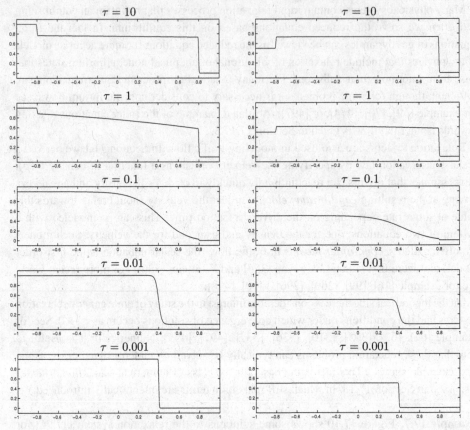

Fig. 17.10. Solution to the relaxation system (17.94) for five different values of τ, all shown at time $t = 0.8$. The left column shows u, and the right column shows v.

17.17.1 Chapman–Enskog Expansion

To see that τ is like a viscosity in the relaxation system (17.94), we use a so-called *Chapman–Enskog expansion*,

$$v(x, t) = f(u(x, t)) + \tau v_1(x, t) + \tau^2 v_2(x, t) + \cdots . \qquad (17.98)$$

The form of this expansion is motivated by the fact that $v \to f(u)$ as $\tau \to 0$. Inserting this in the first equation of (17.94) gives

$$u_t + [f(u) + \tau v_1 + \tau^2 v_2 + \cdots]_x = 0,$$

or

$$u_t + f(u)_x = -\tau v_{1x} + \cdots . \qquad (17.99)$$

To determine $v_1(x, t)$ we insert (17.98) in the second equation of (17.94), yielding

$$[f'(u)u_t + \tau v_{1t} + \tau^2 v_{2t} + \cdots] + au_x = -(v_1 + \tau v_2 + \cdots),$$

or by (17.99),

$$[f'(u)(-f(u)_x - \tau v_{1x} + \cdots) + \tau v_{1t} + \tau^2 v_{2t} + \cdots] + au_x = -(v_1 + \tau v_2 + \cdots).$$

Equating the $\mathcal{O}(1)$ terms for $\tau \ll 1$ gives

$$-f'(u)f(u)_x + au_x = -v_1$$

and hence

$$v_1 = -[a - f'(u)^2]u_x.$$

Using this in (17.99) gives

$$u_t + f(u)_x = \tau(\beta(u)u_x)_x + \mathcal{O}(\tau^2), \tag{17.100}$$

where

$$\beta(u) = a - [f'(u)]^2.$$

The equation (17.100) is a refined version of the reduced equation (17.95). For $\tau > 0$ we see that this parameter plays the role of a viscosity provided that $\beta(u) > 0$, which is true exactly when the subcharacteristic condition (17.96) is satisfied.

17.17.2 Violating the Subcharacteristic Condition

It is interesting to ask what happens if the subcharacteristic condition is violated. If $f(u) = bu$ is a linear function and $|b| > \sqrt{a}$, then the solution will blow up along the characteristic $\text{sgn}(b)\sqrt{a}$ as $\tau \to 0$. This case has been studied in [292]. Adding some viscosity to the system can stabilize the solution, since then the system is parabolic and allows arbitrary propagation speeds. If f is nonlinear, then the nonlinearity may also stabilize the solution. There are some physical problems where this is of interest, notably the case of *roll waves* in shallow water theory on a sloping surface [223], [486]. However, for the vast majority of physical problems involving relaxation the appropriate subcharacteristic condition is satisfied.

17.17.3 Thermal Relaxation and Isothermal Flow

As an example of a relaxation system we consider the Euler equations with a relaxation term driving the temperature towards a constant value. The equations of isothermal flow were introduced in Section 14.6, for gas in a one-dimensional tube surrounded by a bath at constant temperature. We assume that heat flows in or out of the bath instantaneously, so that a constant temperature is maintained in the gas (so heat is extracted from the gas just behind a shock wave, and flows into the gas in a rarefaction wave).

This is obviously not a perfect model of the physical situation. A better model would be the relaxation system

$$
\begin{bmatrix} \rho \\ \rho u \\ E \end{bmatrix}_t + \begin{bmatrix} \rho u \\ \rho u^2 + p \\ (E+p)u \end{bmatrix}_x = \begin{bmatrix} 0 \\ 0 \\ -[E - \bar{E}(\rho, \rho u)]/\tau \end{bmatrix} \tag{17.101}
$$

together with an appropriate equation of state for the gas, e.g., (14.23) for a polytropic ideal gas. Here $\bar{E}(\rho, \rho u)$ is the energy in the gas that results if we bring T to the bath temperature $\bar{T}$ without changing the density or momentum. According to the ideal gas law (14.9), we then have $p = R\rho\bar{T} = a^2\rho$, where $a = \sqrt{R\bar{T}}$ is the isothermal sound speed. Using this in the equation of state (14.23) gives

$$
\bar{E}(\rho, \rho u) = \frac{a^2\rho}{\gamma - 1} + \frac{1}{2}\rho u^2. \tag{17.102}
$$

The quantity $\tau > 0$ is the time scale over which the energy $E - \bar{E}$ flows into the tube, the reciprocal of the *relaxation rate*. The isothermal equations (14.34),

$$
\begin{bmatrix} \rho \\ \rho u \end{bmatrix}_t + \begin{bmatrix} \rho u \\ \rho u^2 + a^2\rho \end{bmatrix}_x = 0, \tag{17.103}
$$

result from letting $\tau \to 0$. These are the reduced equations corresponding to the relaxation system (17.101).

Note that the subcharacteristic condition is satisfied for this system provided that $a \leq c$, where a is the isothermal sound speed and $c = \sqrt{\gamma p/\rho}$ is the sound speed for the full polytropic Euler equations. Near equilibrium we have $p \approx (\gamma - 1)(\bar{E} - \frac{1}{2}\rho u^2) = R\bar{T}\rho$, and so $c = \sqrt{\gamma R\bar{T}}$. Since $\gamma > 1$ we have $c > a$.

The system (17.101) has a structure similar to (17.94). We can divide the variables $(\rho, \rho u, E)$ into *reduced variables* $(\rho, \rho u)$ and the *relaxation variable* E, which relaxes quickly to an equilibrium state $\bar{E}$, a unique value for any given $(\rho, \rho u)$. The reduced equation is obtained by assuming $E \equiv \bar{E}(\rho, \rho u)$ and eliminating the relaxation variable. The system (17.94) has the same structure, with u being the reduced variable and v the relaxation variable. This structure is important in that it allows us to perform a Chapman–Enskog expansion as in Section 17.17.1 to show that the relaxation time τ plays the role of a viscosity in the reduced equation. This in turn suggests that numerical methods will be successful on these relaxation systems even on underresolved grids.

In particular, it is important here that there is a unique equilibrium state $\bar{E}$ corresponding to any given values of the reduced variables $(\rho, \rho u)$. The relaxation variable E converges to this value regardless of the initial value of E. By contrast, in the example of Burgers' equation with a stiff source term, (17.73), there is no separation into reduced and relaxation variables, since this is a scalar equation and u plays both roles. There are two possible equilibrium states $u = 0$ and $u = 1$, and which one we relax towards depends on the initial value of u. This means that a smearing of u can lead to the calculation of an incorrect equilibrium state, resulting in incorrect wave speeds, as was illustrated in Section 17.16. Note also that the example of Section 17.16 suggests that τ does not play the role of a viscosity in the stiff Burgers example. Adding viscosity to a rarefaction wave should leave it essentially unchanged, whereas the stiff source term of (17.73) converts the rarefaction wave into a thin reaction zone that approaches a discontinuity as $\tau \to 0$.

17.18 Relaxation Schemes

In the example of the previous section, the relaxation system is the "correct" physical model and the reduced system is an approximation valid in the limit $\tau \to 0$. In practice one might want to use the reduced system (e.g., the isothermal equations) instead of solving the more complicated relaxation system, which involves stiff source terms as well as being a larger system of equations. Similarly, the scalar conservation law (17.95) can be viewed as an approximation to the relaxation system (17.94).

In some situations we may wish to turn this viewpoint around and instead view a relaxation system such as (17.94) as an approximation to the conservation law (17.95). Suppose we wish to solve the nonlinear conservation law $u_t + f(u)_x = 0$ numerically but for some reason don't wish to write a nonlinear Riemann solver for $f(u)$. Then we can artificially introduce a relaxation variable v and instead solve the system (17.94) for some a and τ, using a fractional-step method. This only requires solving a *linear* hyperbolic system with simple characteristic structure in each time step. For a scalar problem this may be pointless, since the nonlinear Riemann problem is typically easy to solve, but for a nonlinear system of m conservation laws the same idea can be applied, by using the relaxation system

$$u_t + v_x = 0,$$
$$v_t + Au_x = \frac{f(u) - v}{\tau}, \qquad (17.104)$$

where $u, v \in \mathbb{R}^m$ and A is some $m \times m$ matrix. The original nonlinear system of m equations is converted into a linear system of $2m$ equations with the nonlinearity concentrated in the source terms.

The coefficient matrix of (17.104) has the form

$$B = \begin{bmatrix} 0 & I \\ A & 0 \end{bmatrix}, \qquad (17.105)$$

with eigenvalues $\pm\sqrt{\lambda}$, where λ is an eigenvalue of A. In order for the relaxation system to be strictly hyperbolic, we require $\lambda > 0$ for each eigenvalue of A, i.e., A must be positive definite. We also need the subcharacteristic condition to hold, which requires that the eigenvalues of the Jacobian matrix $f'(u)$ should always lie within the range of eigenvalues of B, i.e., within the interval $\pm \max \sqrt{\lambda^p}$. This is a generalization of the subcharacteristic condition (17.96) for the case $m = 1$.

This numerical approach was introduced by Jin and Xin [225]. Methods of this type are generally called *relaxation schemes*, and have since been extensively studied, e.g., [12], [60], [64], [164], [222], [226], [233], [261], [306], [439]. Jin and Xin present some theory and numerical results for the Euler equations, taking A to be a diagonal matrix so that the Riemann problem for (17.104) is easy to solve. The choice of elements of A must be based on estimates of the minimum and maximum wave speed that will arise in the problem so that the subcharacteristic condition is satisfied. In [225] the stiff source term is solved using an implicit Runge–Kutta method with a small value of τ. In practice it seems to work as well to simply consider the limit $\tau \to 0$, called the *relaxed scheme* in [225]. We can then implement a single step of the relaxation scheme using a fractional-step method of the following form:

1. U^n, V^n are used as data for the homogeneous version of (17.104). Solving this linear hyperbolic system over time Δt produces values U^*, V^*.
2. These values are used as data for the system with only the source terms, so $u_t = 0$ and $v_t = [f(u) - v]/\tau$. The solution of this in the limit $\tau \to 0$ has u constant and $v \to f(u)$, so we simply set

$$U^{n+1} = U^*,$$
$$V^{n+1} = f(U^{n+1}).$$
$$(17.106)$$

With this approach we do not need to choose a specific value of τ or solve the ODE; we simply solve the linear hyperbolic system and then reset V to $f(U)$. Note that the success of relaxation schemes depends on the fact that, for relaxation systems, the stiff source terms do not cause numerical difficulties on underresolved grids.

The advantage of a relaxation scheme is that it avoids the need for a Riemann solver for the nonlinear flux function $f(u)$, since the hyperbolic system to be solved now involves the linear coefficient matrix (17.104) instead. The relaxation scheme can alternatively be viewed as a particular way of defining an approximate Riemann solver for the nonlinear problem, as shown in [288], which results in some close connections with other approximate Riemann solvers introduced in Section 15.3. A particularly simple choice of A leads to the Lax–Friedrichs method (4.20); see Exercise 17.9.

Exercises

17.1. Suppose $\beta(x)$ varies with x in the problem of Section 17.2. Derive an unsplit second-order accurate method for this problem.

17.2. Compute the $\mathcal{O}(\Delta t^3)$ term in the splitting error for the Strang splitting (17.35).

17.3. Determine the splitting error for the Godunov splitting applied to the system (17.4).

17.4. Suppose we wish to numerically approximate the steady-state solution to (17.7) for the boundary-value problem with $q(0, t) = C$. The exact solution is given by (17.44).
 (a) Consider the unsplit method (17.9), and suppose we have reached a numerical steady state so that $Q_i^{n+1} = Q_i^n$ for all i. Show that this numerical steady-state solution satisfies

$$Q_i = \frac{Q_{i-1}}{1 + \beta \, \Delta x / \bar{u}}.$$

Hence the numerical steady-state solution is accurate to $\mathcal{O}(\Delta x)$ and is independent of Δt.
 (b) Now consider the fractional-step method (17.21), and show that the numerical

steady-state solution satisfies

$$Q_i = \frac{Q_{i-1}}{1 + (\beta \, \Delta x / \bar{u})(1 - \beta \, \Delta t)}.$$

Again this agrees with the true steady-state solution to $\mathcal{O}(\Delta x) = \mathcal{O}(\Delta t)$, but now the numerical steady state obtained depends on the time step Δt.

17.5. Apply the approach of Section 17.14 to the advection–reaction problem (17.7), with $\Psi_{i-1/2} = -\frac{\beta}{2}(Q_{i-1} + Q_i)$. Use the resulting fluctuations in (4.43) to determine a first-order accurate unsplit method for this equation. How does this method compare to (17.9)? Is the resulting numerical steady state independent of Δt?

17.6. Determine the structure of the on-ramp Riemann problem of Section 17.13. In particular, show that a shock forms whenever $D > (1 - 2q_l)^2/4$.

17.7. Solve the relaxation system (17.101) numerically using CLAWPACK and the Godunov splitting for the source terms. Compare the results on a shock-tube problem with results obtained by solving the isothermal equations. Try various values of τ and both resolved and underresolved grids.

17.8. Determine the eigenvectors of the matrix B in (17.105) in terms of the eigenvectors of A.

17.9. Take $A = (\Delta x / \Delta t)^2 I$ in the relaxation system (17.104). Show that the relaxed scheme described at the end of Section 17.18 then reduces to the Lax–Friedrichs method (4.20). How does the subcharacteristic condition relate to the CFL condition in this case?

Part three
Multidimensional Problems

Multidimensional Hyperbolic Problems

Practical problems involving conservation laws and hyperbolic systems must frequently be solved in more than one space dimension. Most of this book has been devoted to the one-dimensional theory and algorithms, but in the remaining chapters we will see that this forms the basis for understanding and solving multidimensional problems as well.

In two dimensions a conservation law takes the form

$$q_t + f(q)_x + g(q)_y = 0, \tag{18.1}$$

where $q(x, y, t)$ is a vector of m conserved quantities, and $f(q)$ and $g(q)$ are flux functions in the x- and y-directions, as described below. More generally, a quasilinear hyperbolic system has the form

$$q_t + A(q, x, y, t)q_x + B(q, x, y, t)q_y = 0, \tag{18.2}$$

where the matrices A and B satisfy certain conditions given below in Section 18.5. In three dimensions a third term would be added to each of these equations:

$$q_t + f(q)_x + g(q)_y + h(q)_z = 0, \tag{18.3}$$

and

$$q_t + A(q, x, y, z, t)q_x + B(q, x, y, z, t)q_y + C(q, x, y, z, t)q_z = 0, \tag{18.4}$$

respectively.

18.1 Derivation of Conservation Laws

We begin by deriving the conservation law (18.1) in two dimensions from the more fundamental integral form. Again the integral form can be used directly in the development of finite volume methods, as we will see beginning in Chapter 19.

As in one dimension (see Chapter 2), we derive the conservation law by considering an arbitrary spatial domain Ω over which q is assumed to be conserved, so that the integral of q over Ω varies only due to flux across the boundary of Ω. This boundary is denoted by

$\partial\Omega$. We thus have

$$\frac{d}{dt}\iint_\Omega q(x, y, t)\, dx\, dy = \text{net flux across } \partial\Omega. \qquad (18.5)$$

The net flux is determined by integrating the flux of q normal to $\partial\Omega$ around this boundary.

Let $f(q)$ represent the flux of q in the x-direction (per unit length in y, per unit time). This means that the total flux through an interval from (x_0, y_0) to $(x_0, y_0 + \Delta y)$ over time Δt is roughly $\Delta t\, \Delta y\, f(q(x_0, y_0))$, for Δt and Δy sufficiently small.

Similarly, let $g(q)$ be the flux in the y-direction, and let $\vec{f}(q) = (f(q), g(q))$ be the flux vector. Finally, let $\vec{n}(s) = (n^x(s), n^y(s))$ be the outward-pointing unit normal vector to $\partial\Omega$ at a point $(x(s), y(s))$ on $\partial\Omega$, where s is the arclength parameterization of $\partial\Omega$. Then the flux at $\vec{x}(s) = (x(s), y(s))$ in the direction $\vec{n}(s)$ is

$$\vec{n}(s) \cdot \vec{f}(q(x(s), y(s), t)) = n^x(s)f(q) + n^y(s)g(q), \qquad (18.6)$$

and (18.5) becomes

$$\frac{d}{dt}\iint_\Omega q(x, y, t)\, dx\, dy = -\int_{\partial\Omega} \vec{n} \cdot \vec{f}(q)\, ds. \qquad (18.7)$$

If q is smooth then we can use the divergence theorem to rewrite this as

$$\frac{d}{dt}\iint_\Omega q(x, y, t)\, dx\, dy = -\iint_\Omega \vec{\nabla} \cdot \vec{f}(q)\, dx\, dy, \qquad (18.8)$$

where the divergence of $\vec{f}$ is

$$\vec{\nabla} \cdot \vec{f}(q) = f(q)_x + g(q)_y.$$

This leads to

$$\iint_\Omega [q_t + \vec{\nabla} \cdot \vec{f}(q)]\, dx\, dy = 0. \qquad (18.9)$$

Since this must hold over any arbitrary region Ω, the integrand must be zero, and we obtain the conservation law (18.1) in differential form. Note that this argument is exactly analogous to what we did in Chapter 2 in one dimension. As in the one-dimensional case, this derivation assumes q is smooth and so the differential form holds only for smooth solutions. To properly compute discontinuous solutions we will again use finite volume methods based on the integral form.

The same argument extends to three dimensions with the flux vector

$$\vec{f}(q) = [f(q), g(q), h(q)]. \qquad (18.10)$$

If we integrate over an arbitrary volume Ω, so that $\partial\Omega$ is the surface bounding this volume, then we obtain (18.3) for smooth solutions.

In the notation above we have assumed q is a scalar. If q, $f(q)$, $g(q)$, $h(q) \in \mathbb{R}^m$, then the vector $\vec{f}$ of (18.10) is a vector in $\mathbb{R}^{3m}$, while $\vec{n}(s) = (n^x, n^y, n^z)$ is in $\mathbb{R}^3$, and we interpret dot products by the formula

$$\check{f}(q) \equiv \vec{n} \cdot \vec{f}(q) = n^x f(q) + n^y g(q) + n^z h(q). \tag{18.11}$$

In general we use an arrow on a symbol to denote a spatial vector with components corresponding to each spatial dimension. In addition to $\vec{n}$ and $\vec{f}$, we also use $\vec{u} = (u, v, w)$ for the velocity vector and $\vec{A} = (A, B, C)$ as a vector of matrices. We will see below that we need to investigate the linear combination

$$\check{A} \equiv \vec{n} \cdot \vec{A} = n^x A + n^y B + n^z C \tag{18.12}$$

to determine whether the system (18.4) is hyperbolic. We use the breve accent ˘ to denote a quantity that has been restricted to a particular direction specified by $\vec{n}$. As a mnemonic device, the circular arc of the breve accent can be thought of as indicating rotation to the desired direction. This will be heavily used in Chapter 23, where we discuss numerical methods on general quadrilateral grids. One-dimensional Riemann problems will be solved in the direction normal to each cell edge in order to compute the normal fluxes, and doing so requires rotating the flux function to that direction.

For simplicity we will mostly restrict our attention to the case of two dimensions, but the essential ideas extend directly to three dimensions, and this case is briefly discussed as we go along.

18.2 Advection

As a simple example, suppose a fluid is flowing with a known velocity $\vec{u} = (u(x, y, t), v(x, y, t))$ in the plane, and let the scalar $q(x, y, t)$ represent the concentration of a tracer, measured in units of mass per unit area in the plane (see Section 9.1). Then the flux functions are

$$\begin{aligned} f &= u(x, y, t) q(x, y, t), \\ g &= v(x, y, t) q(x, y, t), \end{aligned} \tag{18.13}$$

so that $\vec{f}(q) = \vec{u}q$ and we obtain the conservation law

$$q_t + (uq)_x + (vq)_y = 0. \tag{18.14}$$

Note that in this case f and g may depend explicitly on (x, y, t) as well as on the value of q, and that the derivation above carries over to this situation. If in fact $(u(x, y, t), v(x, y, t)) = (\bar{u}, \bar{v})$ is constant in space and time, so the fluid is simply translating at constant speed in a fixed direction, then (18.14) reduces to

$$q_t + \bar{u}q_x + \bar{v}q_y = 0.$$

The solution is then easily seen to be

$$q(x, y, t) = \overset{\circ}{q}(x - \bar{u}t, y - \bar{v}t), \tag{18.15}$$

so that the initial density simply translates at this velocity. The solution to the more general variable-coefficient advection equation is discussed in Section 20.5.

18.3 Compressible Flow

In Section 2.6 we saw the equations of compressible gas dynamics in one space dimension, in the simplest case where the equation of state relates the pressure directly to the density,

$$p = P(\rho). \tag{18.16}$$

These equations are easily extended to two space dimensions. The conservation-of-mass equation (continuity equation) for the density $\rho(x, y, t)$ is identical to the advection equation (18.14) derived in the previous section:

$$\rho_t + (\rho u)_x + (\rho v)_y = 0. \tag{18.17}$$

But now the velocities (u, v) are not known *a priori*. Instead the continuity equation must be coupled with equations for the conservation of x-momentum ρu and y-momentum ρv. Each of these momenta advects with the fluid motion, giving fluxes analogous to (18.14) with q replaced by ρu or ρv respectively. In addition, pressure variations lead to acceleration of the fluid. Variation in the x-direction, measured by p_x, accelerates the fluid in that direction and appears in the equation for $(\rho u)_t$, while p_y appears in the equation for $(\rho v)_t$. The conservation-of-momentum equations are

$$(\rho u)_t + (\rho u^2 + p)_x + (\rho u v)_y = 0,$$
$$(\rho v)_t + (\rho u v)_x + (\rho v^2 + p)_y = 0. \tag{18.18}$$

These equations (18.17) and (18.18), together with the equation of state (18.16), give a closed system of three conservation laws for mass and momentum. If the equation of state is more complicated, then these equations will generally also have to be coupled with the equation for the conservation of energy, as discussed in Chapter 14. The simple case will suffice for our purposes now. In particular, from these equations we can derive the linearized equations of acoustics just as we did in Section 2.8 in one space dimension.

These gas dynamics equations can be written as a system of conservation laws of the form (18.1) with

$$q = \begin{bmatrix} \rho \\ \rho u \\ \rho v \end{bmatrix}, \quad f(q) = \begin{bmatrix} \rho u \\ \rho u^2 + p \\ \rho u v \end{bmatrix} = \begin{bmatrix} q^2 \\ (q^2)^2/q^1 + P(q^1) \\ q^2 q^3/q^1 \end{bmatrix},$$
$$g(q) = \begin{bmatrix} \rho v \\ \rho u v \\ \rho v^2 + p \end{bmatrix} = \begin{bmatrix} q^3 \\ q^2 q^3/q^1 \\ (q^3)^2/q^1 + P(q^1) \end{bmatrix}. \tag{18.19}$$

These equations can also be written in quasilinear form

$$q_t + f'(q)q_x + g'(q)q_y = 0, \tag{18.20}$$

in terms of the Jacobian matrices

$$f'(q) = \begin{bmatrix} 0 & 1 & 0 \\ -u^2 + P'(\rho) & 2u & 0 \\ -uv & v & u \end{bmatrix}, \quad g'(q) = \begin{bmatrix} 0 & 0 & 1 \\ -uv & v & u \\ -v^2 + P'(\rho) & 0 & 2v \end{bmatrix}. \tag{18.21}$$

18.4 Acoustics

Linearizing the equations derived in the previous section about a constant state $q_0 = (\rho_0, u_0, v_0)$ gives

$$q_t + f'(q_0)q_x + g'(q_0)q_y = 0,$$

where q now represents perturbations from the constant state q_0. In particular, if we wish to study acoustics in a stationary gas, then we can take $u_0 = v_0 = 0$ and the Jacobian matrices simplify considerably:

$$f'(q_0) = \begin{bmatrix} 0 & 1 & 0 \\ P'(\rho_0) & 0 & 0 \\ 0 & 0 & 0 \end{bmatrix}, \qquad g'(q_0) = \begin{bmatrix} 0 & 0 & 1 \\ 0 & 0 & 0 \\ P'(\rho_0) & 0 & 0 \end{bmatrix}. \qquad (18.22)$$

As in one dimension, we can now manipulate these equations to derive an equivalent linear system in terms of perturbations in pressure and velocity, a linear system of the form

$$q_t + Aq_x + Bq_y = 0, \qquad (18.23)$$

where (again for $u_0 = v_0 = 0$)

$$q = \begin{bmatrix} p \\ u \\ v \end{bmatrix}, \qquad A = \begin{bmatrix} 0 & K_0 & 0 \\ 1/\rho_0 & 0 & 0 \\ 0 & 0 & 0 \end{bmatrix}, \qquad B = \begin{bmatrix} 0 & 0 & K_0 \\ 0 & 0 & 0 \\ 1/\rho_0 & 0 & 0 \end{bmatrix}. \qquad (18.24)$$

These are the *equations of acoustics* in two space dimensions, where again $K_0 = \rho_0 P'(\rho_0)$ is the bulk modulus of compressibility. More generally, for acoustics against a background flow with constant velocity $\vec{u}_0 = (u_0, v_0)$, the coefficient matrices are

$$A = \begin{bmatrix} u_0 & K_0 & 0 \\ 1/\rho_0 & u_0 & 0 \\ 0 & 0 & u_0 \end{bmatrix}, \qquad B = \begin{bmatrix} v_0 & 0 & K_0 \\ 0 & v_0 & 0 \\ 1/\rho_0 & 0 & v_0 \end{bmatrix}. \qquad (18.25)$$

18.5 Hyperbolicity

Recall that in one space dimension the linear system $q_t + Aq_x = 0$ is said to be *hyperbolic* if the matrix A is diagonalizable with real eigenvalues. In two space dimensions we need this condition to hold for each of the coefficient matrices A and B, but we also need something more: the same property should hold for any linear combination of these matrices. This is formalized in Definition 18.1 below after some motivation.

The essence of hyperbolicity is that wavelike solutions should exist. In one space dimension a linear system of m equations generally gives rise to m waves moving at constant speeds and unchanged shape. In two dimensions we should see this same behavior if we take special initial data that varies only in one direction – not just the x-, or y-direction, but any arbitrary direction specified by a unit vector $\vec{n} = (n^x, n^y)$, so that the data has the form

$$q(x, y, 0) = \mathring{q}(\vec{n} \cdot \vec{x}) = \mathring{q}(n^x x + n^y y). \qquad (18.26)$$

The contour lines of $q(x, y, 0)$ are straight lines, and we expect wave motion in the direction normal to these lines, which is the direction given by $\vec{n}$.

In particular, there should be special initial data of this form that yields a *single* wave propagating at some constant speed s, a *plane wave* of the form

$$q(x, y, t) = \breve{q}(\vec{n} \cdot \vec{x} - st).$$

This is the multidimensional analogue of the simple wave discussed in Section 3.5. If we compute $q_t, q_x,$ and q_y for this *Ansatz* and insert them into the equation $q_t + Aq_x + Bq_y = 0$, we find that

$$\breve{A} \, \breve{q}'(\vec{n} \cdot \vec{x} - st) = s \, \breve{q}'(\vec{n} \cdot \vec{x} - st),$$

where

$$\breve{A} = \vec{n} \cdot \vec{A} = n^x A + n^y B. \tag{18.27}$$

Except for the trivial case $\breve{q} \equiv$ constant, (18.27) can only hold if s is an eigenvalue of the matrix $\breve{A}$, with $\breve{q}'(\xi)$ a corresponding eigenvector of this matrix for each value of ξ. This leads to our definition of hyperbolicity in two space dimensions.

Definition 18.1. *The constant-coefficient system $q_t + Aq_x + Bq_y = 0$ is (strongly) hyperbolic if, for every choice of $\vec{n}$, the matrix $\breve{A} = \vec{n} \cdot \vec{A}$ is diagonalizable with real eigenvalues. The quasilinear system (18.20) is hyperbolic in some region of state space if the Jacobian matrix $\breve{f}'(q) = \vec{n} \cdot \vec{f}'(q) = n^x f'(q) + n^y g'(q)$ is diagonalizable with real eigenvalues for every $\vec{n}$, for all q in this region.*

Note in particular that for $\vec{n} = (1, 0)$ or $\vec{n} = (0, 1)$ we have propagation in the x- or y-direction respectively. In these cases we obtain the usual one-dimensional conditions on the matrices A and B separately. The obvious three-dimensional extension of this definition is given in Section 18.6.

For the acoustics equations with (18.24), we have

$$\breve{A} = \begin{bmatrix} 0 & n^x K_0 & n^y K_0 \\ n^x/\rho_0 & 0 & 0 \\ n^y/\rho_0 & 0 & 0 \end{bmatrix}. \tag{18.28}$$

This matrix has eigenvalues that are independent of $\vec{n}$:

$$\breve{\lambda}^1 = -c_0, \qquad \breve{\lambda}^2 = 0, \qquad \breve{\lambda}^3 = +c_0,$$

where $c_0 = \sqrt{K_0/\rho_0}$ is the speed of sound. This is exactly what we should expect, since sound waves can propagate in any direction at the same speed (for the uniform isotropic medium we are considering here, with ρ_0 and K_0 constant).

For acoustics against a constant background flow, we expect sound waves to propagate at speed c_0 relative to the moving fluid. For the matrices A and B of (18.25) we have

$$\breve{A} = \begin{bmatrix} \breve{u}_0 & n^x K_0 & n^y K_0 \\ n^x/\rho_0 & \breve{u}_0 & 0 \\ n^y/\rho_0 & 0 & \breve{u}_0 \end{bmatrix}, \qquad (18.29)$$

where $\breve{u}_0 = \vec{n} \cdot \vec{u}_0$ is the fluid velocity in the $\vec{n}$-direction. Since this differs from (18.28) only by a multiple of the identity matrix, it has the same eigenvectors (given below in (18.33)), and the eigenvalues are simply shifted by $\breve{u}_0$:

$$\breve{\lambda}^1 = \breve{u}_0 - c_0, \qquad \breve{\lambda}^2 = \breve{u}_0, \qquad \breve{\lambda}^3 = \breve{u}_0 + c_0, \qquad (18.30)$$

exactly as we expected.

In one space dimension we can diagonalize a general linear hyperbolic equation using the matrix of eigenvectors, decoupling it into independent scalar advection equations for each characteristic variable. For a linear system in more dimensions, we can do this in general only for the special case of a plane-wave solution. The full system $q_t + Aq_x + Bq_y = 0$ with arbitrary data can be diagonalized only if the coefficient matrices commute, e.g., if $AB = BA$ in (18.23), in which case the matrices have the *same* eigenvectors. Then A and B can be *simultaneously diagonalized* by a common eigenvector matrix R:

$$A = R\Lambda^x R^{-1}, \qquad B = R\Lambda^y R^{-1},$$

where $\Lambda^x = \text{diag}(\lambda^{x1}, \ldots, \lambda^{xm})$ and $\Lambda^y = \text{diag}(\lambda^{y1}, \ldots, \lambda^{ym})$ contain the eigenvalues, which may be different. The system in (18.23) can then be diagonalized by setting $w = R^{-1}q$ to obtain

$$w_t + \Lambda^x w_x + \Lambda^y w_y = 0,$$

yielding m independent advection equations. Note that in this case there are only m distinct directions in which information propagates. The pth characteristic variable w^p propagates with velocity $(\lambda^{xp}, \lambda^{yp})$. This is not what we would expect in acoustics, for example, since sound waves can propagate in any direction.

If $AB \neq BA$, then there is no single transformation that will simultaneously diagonalize A and B. If the system is hyperbolic, then we can diagonalize each matrix separately,

$$A = R^x \Lambda^x (R^x)^{-1}, \qquad B = R^y \Lambda^y (R^y)^{-1},$$

but the two matrices have different eigenvectors R^x and R^y, respectively. In this case the equations are more intricately coupled. This is the usual situation physically. In the case of acoustics, for example, the matrices A and B of (18.24) or (18.25) are not simultaneously diagonalizable. The matrix A of (18.25) has right eigenvectors

$$r^{x1} = \begin{bmatrix} -Z_0 \\ 1 \\ 0 \end{bmatrix}, \quad r^{x2} = \begin{bmatrix} 0 \\ 0 \\ 1 \end{bmatrix}, \quad r^{x3} = \begin{bmatrix} Z_0 \\ 1 \\ 0 \end{bmatrix}, \quad \text{so that} \quad R^x = \begin{bmatrix} -Z_0 & 0 & Z_0 \\ 1 & 0 & 1 \\ 0 & 1 & 0 \end{bmatrix},$$

$$(18.31)$$

while the matrix B has right eigenvectors

$$r^{y1} = \begin{bmatrix} -Z_0 \\ 0 \\ 1 \end{bmatrix}, \quad r^{y2} = \begin{bmatrix} 0 \\ 1 \\ 0 \end{bmatrix}, \quad r^{y3} = \begin{bmatrix} Z_0 \\ 0 \\ 1 \end{bmatrix}, \quad \text{so that} \quad R^y = \begin{bmatrix} -Z_0 & 0 & Z_0 \\ 0 & 1 & 0 \\ 1 & 0 & 1 \end{bmatrix},$$

(18.32)

where $Z_0 = q_0 c_0$ is again the impedance.

Note that for a plane wave moving in the x-direction these acoustics equations reduce to $q_t + Aq_x = 0$ with A given by (18.25). This system has exactly the same eigenstructure as the coupled acoustics–advection system of Section 3.10. In this case the y-component of the velocity perturbation, v, is simply advected with the background velocity u_0 and does not affect the acoustics. (Recall that in such a plane wave we assume the variables only vary with x. Variations of v in the y-direction would of course generate acoustic signals.)

A plane wave in the y-direction gives a similar structure, but now the eigenvectors r^{y1} and r^{y3} corresponding to acoustic waves involving the pressure p and vertical velocity perturbation v, while the x-component of the velocity perturbations, u, is simply advected at the background speed v_0.

The more general matrix $\breve{A}$ from (18.29) has eigenvalues (18.30) and the eigenvectors

$$\breve{r}^1 = \begin{bmatrix} -Z_0 \\ n^x \\ n^y \end{bmatrix}, \quad \breve{r}^2 = \begin{bmatrix} 0 \\ -n^y \\ n^x \end{bmatrix}, \quad \breve{r}^3 = \begin{bmatrix} Z_0 \\ n^x \\ n^y \end{bmatrix},$$

(18.33)

which reduce to (18.31) or (18.32) when $\vec{n}$ is in the x- or y-direction. Note that more generally the acoustic waves $\breve{r}^1$ and $\breve{r}^3$ have velocity components in the $\vec{n}$-direction, as we expect for these compressional waves. The 2-wave carries velocity perturbations in the orthogonal direction (a shear wave), which are simply advected with the flow. In Section 22.1 we consider elastic waves in a solid that resists shear motion, in which case shear waves have more interesting structure.

In one space dimension we could diagonalize the acoustics equations to obtain a coupled pair of advection equations. Solutions consist simply of two waves advecting with velocities $-c_0$ and $+c_0$ in the two possible directions. In two dimensions, even though the structure of each matrix is that of one-dimensional acoustics, the nondiagonalizable coupling between them leads to a much richer structure. In general we obtain waves propagating in all of the infinitely many possible directions in the plane.

18.6 Three-Dimensional Systems

The three-dimensional linear system

$$q_t + Aq_x + Bq_y + Cq_z = 0 \tag{18.34}$$

is hyperbolic provided that, for any direction defined by the unit vector $\vec{n} = (n^x, n^y, n^z)$, the matrix $\breve{A}$ given by (18.12) is diagonalizable with real eigenvalues and a complete set of eigenvectors. The eigenvalues have the interpretation of physical propagation velocities for plane waves in this direction.

The acoustics equations in three dimensions are a linear hyperbolic system for perturbations in the pressure and three velocity components (u, v, w). Rather than displaying the coefficient matrices A, B, and C separately, it is more compact to just display $\breve{A}$ in an arbitrary direction $\vec{n}$. We have

$$
q = \begin{bmatrix} p \\ u \\ v \\ w \end{bmatrix}, \qquad
\breve{A} = \begin{bmatrix}
\breve{u}_0 & n^x K_0 & n^y K_0 & n^z K_0 \\
n^x/\rho_0 & \breve{u}_0 & 0 & 0 \\
n^y/\rho_0 & 0 & \breve{u}_0 & 0 \\
n^z/\rho_0 & 0 & 0 & \breve{u}_0
\end{bmatrix}, \tag{18.35}
$$

where

$$
\breve{u}_0 = n^x u_0 + n^y v_0 + n^z w_0 \tag{18.36}
$$

is the component of the background velocity in the direction $\vec{n}$. For any choice of direction, the eigenvalues of $\breve{A}$ are

$$
\breve{\lambda}^1 = \breve{u}_0 - c_0, \qquad \breve{\lambda}^2 = \breve{\lambda}^3 = \breve{u}_0, \qquad \breve{\lambda}^4 = \breve{u}_0 + c_0. \tag{18.37}
$$

Note that there is a two-dimensional eigenspace corresponding to the eigenvalue $\breve{\lambda}^2 = \breve{\lambda}^3 = \breve{u}_0$, since shear waves can now carry an arbitrary jump in each of the two velocity components orthogonal to $\vec{n}$.

18.7 Shallow Water Equations

In two space dimensions the shallow water equations take the form

$$
h_t + (hu)_x + (hv)_y = 0,
$$

$$
(hu)_t + \left(hu^2 + \frac{1}{2}gh^2\right)_x + (huv)_y = 0, \tag{18.38}
$$

$$
(hv)_t + (huv)_x + \left(hv^2 + \frac{1}{2}gh^2\right)_y = 0,
$$

where h is the depth and (u, v) the velocity vector, so that hu and hv are the momenta in the two directions. These are a natural generalization of the one-dimensional equations (13.5) and are identical to the two-dimensional compressible flow equations derived in Section 18.3 if we replace ρ by h there and use the hydrostatic equation of state

$$
p = P(h) = \frac{1}{2}gh^2 \tag{18.39}
$$

as derived in (13.3) (taking $\bar{\rho} = 1$). From (18.21), the flux Jacobian matrices are thus

$$
f'(q) = \begin{bmatrix} 0 & 1 & 0 \\ -u^2 + gh & 2u & 0 \\ -uv & v & u \end{bmatrix}, \qquad
g'(q) = \begin{bmatrix} 0 & 0 & 1 \\ -uv & v & u \\ -v^2 + gh & 0 & 2v \end{bmatrix}. \tag{18.40}
$$

Let $c = \sqrt{gh}$ be the speed of gravity waves. Then the matrix $f'(q)$ has eigenvalues and eigenvectors

$$\lambda^{x1} = u - c, \qquad \lambda^{x2} = u, \qquad \lambda^{x3} = u + c,$$

$$r^{x1} = \begin{bmatrix} 1 \\ u - c \\ v \end{bmatrix}, \quad r^{x2} = \begin{bmatrix} 0 \\ 0 \\ 1 \end{bmatrix}, \quad r^{x3} = \begin{bmatrix} 1 \\ u + c \\ v \end{bmatrix}. \tag{18.41}$$

The Jacobian $g'(q)$ has a similar set of eigenvalues and eigenvectors,

$$\lambda^{y1} = v - c, \qquad \lambda^{y2} = v, \qquad \lambda^{y3} = v + c,$$

$$r^{y1} = \begin{bmatrix} 1 \\ u \\ v - c \end{bmatrix}, \quad r^{y2} = \begin{bmatrix} 0 \\ -1 \\ 0 \end{bmatrix}, \quad r^{y3} = \begin{bmatrix} 1 \\ u \\ v + c \end{bmatrix} \tag{18.42}$$

in which the roles of u and v are switched along with x and y.

In each case the 1-wave and 3-wave are nonlinear gravity waves, while the 2-wave is linearly degenerate. Compare these with (13.64) and (13.65), the Jacobian matrix and eigenstructure for the one-dimensional shallow water equations augmented by a passive tracer.

18.7.1 The Plane-Wave Riemann Problem

Consider a two-dimensional Riemann problem for the shallow water equations with variation only in the x-direction. In this case the velocity v plays no dynamic role in the gravity waves, and any jump in v is simply carried along passively at the fluid velocity u_m that arises between the two nonlinear waves. This is again a contact discontinuity that lies at the interface between the two original fluids. The fluid to the left always has y-velocity v_l while the one to the right has y-velocity v_r. Figure 13.20 illustrates this, if we let the dark and light regions now represent different velocities v. Figure 18.1 gives another illustration of this, showing a top view of a two-dimensional version of Figure 13.20. The contact discontinuity is also called a *shear wave* in this context.

One should recall that we are ignoring fluid viscosity with this hyperbolic model. In reality a jump discontinuity in shear velocity would be smeared out due to frictional forces (diffusion of the momentum hv in the x-direction) and may lead to Kelvin–Helmholtz instabilities along such an interface.

The true solution to this Riemann problem is easily computed using the one-dimensional theory. We simply solve the one-dimensional problem ignoring v, and then introduce a jump in v at the contact surface.

To verify that the two-dimensional shallow water equations are hyperbolic, we compute the Jacobian matrix $\breve{f}'(q) = \vec{n} \cdot \vec{f}'(q)$ in an arbitrary direction $\vec{n}$,

$$\breve{f}'(q) = \begin{bmatrix} 0 & n^x & n^y \\ n^x gh - u\breve{u} & \breve{u} + n^x u & n^y u \\ n^y gh - v\breve{u} & n^x v & \breve{u} + vn^y \end{bmatrix}, \tag{18.43}$$

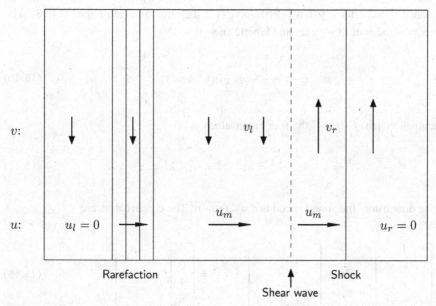

Fig. 18.1. Solution to a Riemann problem in the x-direction for the two-dimensional shallow water equations. The depth h and normal velocity u are as shown in Figure 13.20, and the shading in that figure represents v.

where $\breve{u} = \vec{n} \cdot \vec{u}$. This matrix has eigenvalues and eigenvectors given by

$$\breve{\lambda}^1 = \breve{u} - c, \qquad \breve{\lambda}^2 = \breve{u}, \qquad \breve{\lambda}^3 = \breve{u} + c,$$

$$\breve{r}^1 = \begin{bmatrix} 1 \\ u - n^x c \\ v - n^y c \end{bmatrix}, \quad \breve{r}^2 = \begin{bmatrix} 0 \\ -n^y \\ n^x \end{bmatrix}, \quad \breve{r}^3 = \begin{bmatrix} 1 \\ u + n^x c \\ v + n^y c \end{bmatrix}. \tag{18.44}$$

The expressions (18.41) and (18.42) are special cases of this. For any direction $\vec{n}$ the eigenvalues are real and correspond to wave speeds 0, $\pm c$ relative to the moving fluid.

18.8 Euler Equations

The two-dimensional Euler equations have the same form as the compressible flow equations presented in Section 18.3, but with the addition of an energy equation for the general case where the equation of state is more complicated than (18.16):

$$q = \begin{bmatrix} \rho \\ \rho u \\ \rho v \\ E \end{bmatrix}, \qquad f(q) = \begin{bmatrix} \rho u \\ \rho u^2 + p \\ \rho u v \\ (E + p)u \end{bmatrix}, \qquad g(q) = \begin{bmatrix} \rho v \\ \rho u v \\ \rho v^2 + p \\ (E + p)v \end{bmatrix}. \tag{18.45}$$

The equation of state for a γ-law polytropic gas is the obvious extension of (14.23). The total energy is the sum of internal and kinetic energy,

$$E = \frac{p}{\gamma - 1} + \frac{1}{2}\rho(u^2 + v^2). \tag{18.46}$$

The Jacobian matrix $f'(q)$ has the four eigenvalues

$$\lambda^1 = u - c, \quad \lambda^2 = u, \quad \lambda^3 = u, \quad \lambda^4 = u + c. \tag{18.47}$$

As in one dimension, the sound speed is $c = \sqrt{\gamma p / \rho}$. The eigenvectors are

$$r^{x1} = \begin{bmatrix} 1 \\ u - c \\ v \\ H - uc \end{bmatrix}, \quad r^{x2} = \begin{bmatrix} 1 \\ u \\ v \\ \frac{1}{2}(u^2 + v^2) \end{bmatrix}, \quad r^{x3} = \begin{bmatrix} 0 \\ 0 \\ 1 \\ v \end{bmatrix}, \quad r^{x4} = \begin{bmatrix} 1 \\ u + c \\ v \\ H + uc \end{bmatrix}. \tag{18.48}$$

The eigenvalues and eigenvectors in the y-direction are similar, with the roles of u and v reversed.

18.8.1 The Plane-Wave Riemann Problem

Consider a Riemann problem in which the data varies only in x. For the one-dimensional Euler equations the density can be discontinuous across the contact discontinuity, as illustrated in Figure 14.1. In the two-dimensional extension, there can also be a jump in the transverse velocity v across the contact discontinuity, exactly as was illustrated for the two-dimensional shallow water equations in Figure 18.1. The jump in density and the jump in shear velocity are carried by two independent linearly degenerate waves that both travel at the same velocity. These two waves correspond to the two eigenvalues $\lambda^2 = \lambda^3 = u$ of the Jacobian matrix. (The two vectors r^{x2} and r^{x3} in (18.48) are just one possible basis for this two-diemensional eigenspace.)

  This two-dimensional Riemann problem is easily solved based on the one-dimensional theory, just as in the case of the shallow water equations. We can follow the procedure of Section 14.11, ignoring the transverse velocity v, since the primitive variables u and p are still continuous across the contact discontinuity. We then introduce a jump in v from v_l to v_r at the contact discontinuity. Note that this also gives a jump in E in the eigenvector r^{x3}, since v comes into the equation of state (18.46).

18.8.2 Three-Dimensional Euler Equations

In three space dimensions the Euler equations are similar, but with the addition of a fifth equation for the conservation of momentum ρw in the z-direction, where w is the z-component of velocity. The conserved quantities and fluxes are then

$$
q = \begin{bmatrix} \rho \\ \rho u \\ \rho v \\ \rho w \\ E \end{bmatrix}, \quad
f(q) = \begin{bmatrix} \rho u \\ \rho u^2 + p \\ \rho u v \\ \rho u w \\ (E+p)u \end{bmatrix}, \quad
g(q) = \begin{bmatrix} \rho v \\ \rho u v \\ \rho v^2 + p \\ \rho v w \\ (E+p)v \end{bmatrix}, \quad
h(q) = \begin{bmatrix} \rho w \\ \rho u w \\ \rho v w \\ \rho w^2 + p \\ (E+p)w \end{bmatrix}.
$$

$$(18.49)$$

The equation of state now includes the kinetic energy $\frac{1}{2}\rho(u^2 + v^2 + w^2) = \frac{1}{2}\rho\vec{u} \cdot \vec{u}$, where $\vec{u} = (u, v, w)$ is the velocity vector. In any arbitrary direction $\vec{n}$ there are two nonlinear acoustic fields with eigenvalues $(\vec{n}\cdot\vec{u})\pm c$, and three linearly degenerate fields with eigenvalue $\vec{n} \cdot \vec{u}$. These three fields correspond to jumps in the density (entropy waves), and jumps in the two transverse velocities (shear waves). For example, if $\vec{n} = (1, 0, 0)$, then we are looking in the x-direction and arbitrary jumps in ρ, v, and w across the contact discontinuity can all propagate with speed u.

18.9 Symmetry and Reduction of Dimension

For some problems we may be able to reduce the complexity of the numerical problem substantially by taking advantage of symmetry. For example, if we are solving a problem where the solution is known to be radially symmetric, then we should be able to rewrite the equations in polar or spherical coordinates, obtaining a system that reduces to a problem in the single space variable r. The transformed equations will typically involve *geometric source terms*.

For example, when rewritten in polar r–θ coordinates, the compressible flow equations (18.19) take the form

$$
\frac{\partial}{\partial t}\begin{bmatrix} r\rho \\ r\rho U \\ r\rho V \end{bmatrix}
+ \frac{\partial}{\partial r}\begin{bmatrix} r\rho U \\ r\rho U^2 + p \\ r\rho U V \end{bmatrix}
+ \frac{1}{r}\frac{\partial}{\partial \theta}\begin{bmatrix} r\rho V \\ r\rho U V \\ r\rho V^2 + p \end{bmatrix} = 0, \qquad (18.50)
$$

where $U(r, \theta, t)$ is the velocity in the radial direction and $V(r, \theta, t)$ is the velocity in the θ-direction. If we assume that $V(r, \theta, t) \equiv 0$ and there is no variation in the θ-direction, then these equations reduce to the two equations

$$
(r\rho)_t + (r\rho U)_r = 0,
$$
$$
(r\rho U)_t + (r\rho U^2 + p)_r = 0. \qquad (18.51)
$$

This system can be rewritten as

$$
\rho_t + (\rho U)_r = -(\rho U)/r,
$$
$$
(\rho U)_t + (\rho U^2 + p)_r = -(\rho U^2)/r, \qquad (18.52)
$$

which has exactly the same form as the one-dimensional system of equations (2.38), but with the addition of a geometric source term on the right-hand side.

The full two- or three-dimensional Euler equations with radial symmetry yield

$$\rho_t + (\rho U)_r = -\frac{\alpha}{r}(\rho U),$$

$$(\rho U)_t + (\rho U^2 + p)_r = -\frac{\alpha}{r}(\rho U^2), \qquad (18.53)$$

$$E_t + ((E + p)U)_r = -\frac{\alpha}{r}((E + p)U),$$

where $\alpha = 1$ in two dimensions and $\alpha = 2$ in three dimensions.

Even if the real problems of interest must be studied multidimensionally, radially symmetric solutions are very valuable in testing and validating numerical codes. A highly accurate solution to the one-dimensional problem can be computed on a fine grid and used to test solutions computed with the multidimensional solver. This is useful not only in checking that the code gives essentially the correct answer in at least some special cases, but also in determining whether the numerical method is isotropic or suffers from *grid-orientation effects* that lead to the results being better resolved in some directions than in others. See Section 21.7.1 for one such example.

The D-dimensional acoustics equations with radial symmetry reduce to

$$p_t + K_0 U_r = -\frac{\alpha}{r}(K_0 U),$$

$$\rho_0 U_t + p_r = 0, \qquad (18.54)$$

where again $\alpha = D - 1$.

Exercises

18.1. Consider the system $q_t + Aq_x + Bq_y = 0$ with

$$A = \begin{bmatrix} 3 & 1 \\ 1 & 3 \end{bmatrix}, \qquad B = \begin{bmatrix} 0 & 2 \\ 2 & 0 \end{bmatrix}.$$

Show that these matrices are simultaneously diagonalizable, and determine the general solution to this system with arbitrary initial data. In particular, sketch how the solution evolves in the x–y plane with data

$$q^1(x, y, 0) = \begin{cases} 1 & \text{if } x^2 + y^2 \le 1, \\ 0 & \text{otherwise,} \end{cases} \qquad q^2(x, y, 0) \equiv 0.$$

18.2. (a) Suppose that A and B are both symmetric matrices. Show that the system $q_t + Aq_x + Bq_y = 0$ must then be hyperbolic.

   (b) The matrices A and B are *simultaneously symmetrizable* if there is an invertible matrix M such that $M^{-1}AM$ and $M^{-1}BM$ are both symmetric. Show that in this case the system $q_t + Aq_x + Bq_y = 0$ must be hyperbolic.

   (c) Show that the matrices in (18.25) for the linearized acoustics equations are simultaneously symmetrizable with a matrix M of the form $M = \text{diag}(d_1, 1, 1)$.

18.3. Consider the two-dimensional system $q_t + Aq_x + Bq_y = 0$ with matrices

$$A = \begin{bmatrix} 1 & 10 \\ 0 & 2 \end{bmatrix}, \qquad B = \begin{bmatrix} 2 & 0 \\ 10 & 1 \end{bmatrix},$$

each of which is diagonalizable with real eigenvalues. Show, however, that this system is not hyperbolic. (See also Exercise 19.1.)

18.4. Determine the eigenvectors of the three-dimensional acoustics matrix $\breve{A}$ from (18.35).

18.5. Show that the three-dimensional system (2.115) of Maxwell's equations is hyperbolic.

Multidimensional Numerical Methods

In the remainder of the book we concentrate on finite volume methods for multidimensional hyperbolic equations. We will begin by considering uniform Cartesian grids in two dimensions, using the notation illustrated in Figure 19.1(a). The value Q_{ij}^n represents a cell average over the (i, j) grid cell at time t_n,

$$Q_{ij}^n \approx \frac{1}{\Delta x \, \Delta y} \int_{y_{j-1/2}}^{y_{j+1/2}} \int_{x_{i-1/2}}^{x_{i+1/2}} q(x, y, t_n) \, dx \, dy. \tag{19.1}$$

As in one dimension, we can use the integral form of the equations to determine how this cell average varies with time, and develop finite volume methods based on numerical approximations to the fluxes at each cell edge. Various approaches to doing this are summarized starting in Section 19.2.

19.1 Finite Difference Methods

Rather than working with the cell averages and the integral form of the equations, one could instead view Q_{ij}^n as a pointwise approximation to the value of $q(x_i, y_j, t_n)$ at the point indicated in Figure 19.1(b). Discretizing the differential equations by finite differences then gives a finite difference method. As in one dimension, this approach often gives methods that look very similar to related finite volume methods. We will concentrate on finite volume methods, since this viewpoint allows the derivation of methods that are more robust when discontinuities are present, as well as being exactly conservative. However, it is sometimes useful to think of the methods in terms of their finite difference interpretation, in particular when computing the local truncation error by comparing with a Taylor series expansion of the true solution at the point (x_i, y_j). The flux differences arising in a finite volume method are often seen to give approximations to terms in this Taylor expansion, at least when applied to smooth solutions.

19.1.1 Taylor Series Expansion of the Exact Solution

We develop the Taylor series expansion of the exact solution at a point after a single time step for the constant-coefficient linear hyperbolic system

$$q_t + A q_x + B q_y = 0. \tag{19.2}$$

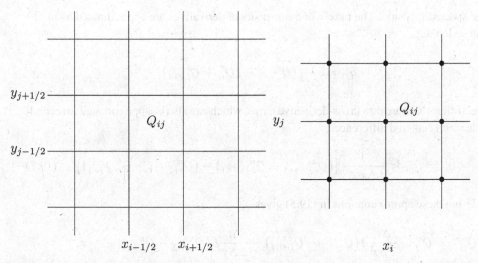

Fig. 19.1. (a) Finite volume grid in two space dimensions, where Q_{ij} represents a cell average. (b) Finite difference grid in two dimensions, where Q_{ij} represents a pointwise value.

We will require higher-order time derivatives in this expansion, which are most easily found from the expression

$$\partial_t^j q = [-(A\partial_x + B\partial_y)]^j q. \tag{19.3}$$

So, in particular,

$$q_{tt} = A^2 q_{xx} + AB q_{yx} + BA q_{xy} + B^2 q_{yy}. \tag{19.4}$$

Note that although $q_{yx} = q_{xy}$, in general $AB \neq BA$, and so we cannot combine the middle two terms.

The Taylor series expansion at (x_i, y_j) after time Δt can be written as

$$
\begin{aligned}
q(x_i, y_j, t_n + \Delta t) &= q + \Delta t\, q_t + \frac{1}{2}\Delta t^2 q_{tt} + \cdots \\
&= q - \Delta t\,(Aq_x + Bq_y) \\
&\quad + \frac{1}{2}\Delta t^2 (A^2 q_{xx} + AB q_{yx} + BA q_{xy} + B^2 q_{yy}) + \cdots .
\end{aligned} \tag{19.5}
$$

Note that if A and B vary with x and y, then (19.3) still holds, but (19.4) becomes

$$q_{tt} = A(Aq_x)_x + A(Bq_y)_x + B(Aq_x)_y + B(Bq_y)_y. \tag{19.6}$$

The expansion becomes somewhat more complicated, as would the Lax–Wendroff method developed in the next subsection.

19.1.2 The Lax–Wendroff Method

The Lax–Wendroff method for the equation (19.2) is obtained by keeping only the terms shown in (19.5) and replacing the derivatives by centered finite differences, just as in one space dimension. The purely one-dimensional derivatives are approximated as in one dimension, e.g.,

$$q_{yy} = \frac{1}{\Delta y^2}\left(Q_{i,j-1}^n - 2Q_{ij}^n + Q_{i,j+1}^n\right).$$

In addition, there are now cross-derivative terms, which can also be approximated to second-order with centered differences,

$$q_{xy} = q_{yx} \approx \frac{1}{4\,\Delta x\,\Delta y}\left[\left(Q_{i+1,j+1}^n - Q_{i-1,j+1}^n\right) - \left(Q_{i+1,j-1}^n - Q_{i-1,j-1}^n\right)\right]. \quad (19.7)$$

Using these approximations in (19.5) gives

$$
\begin{aligned}
Q_{ij}^{n+1} = {}& Q_{ij}^n - \frac{\Delta t}{2\,\Delta x}A\left(Q_{i+1,j}^n - Q_{i-1,j}^n\right) - \frac{\Delta t}{2\,\Delta y}B\left(Q_{i,j+1}^n - Q_{i,j-1}^n\right) \\
& + \frac{\Delta t^2}{2\,\Delta x^2}A^2\left(Q_{i+1,j}^n - 2Q_{ij}^n + Q_{i-1,j}^n\right) + \frac{\Delta t^2}{2\,\Delta y^2}B^2\left(Q_{i,j+1}^n - 2Q_{ij}^n + Q_{i,j-1}^n\right) \\
& + \frac{\Delta t^2}{8\,\Delta x\,\Delta y}(AB + BA)\left[\left(Q_{i+1,j+1}^n - Q_{i-1,j+1}^n\right) - \left(Q_{i+1,j-1}^n - Q_{i-1,j-1}^n\right)\right].
\end{aligned}
$$

$$(19.8)$$

This has a nine-point stencil involving all nine of the points in Figure 19.1(b) in the update for Q_{ij}. In Section 19.3.1 we will see that this same method can be viewed as a finite volume method for updating the cell average Q_{ij} shown in Figure 19.1(a), resulting from defining numerical fluxes at the four edges of the cell in a natural way, based on the nine nearby cell values.

19.2 Finite Volume Methods and Approaches to Discretization

As in one space dimension, the two-dimensional Lax–Wendroff method suffers from problems with numerical dispersion, leading to phase errors and to unphysical oscillations in problems with discontinuities or steep gradients. By reinterpreting this method as a finite volume method and introducing upwind biasing and flux limiting, we will see that it can be greatly improved. We will also write the resulting methods in a more general form that applies also to nonlinear conservation laws and to variable-coefficient problems, by using the waves resulting from solving Riemann problems at each cell edge as the basis for upwinding and limiting.

This goal is pursued starting in the next chapter, where we begin to focus on this approach to developing multidimensional high-resolution methods. This is certainly not the only approach, however, and a wide variety of other multidimensional algorithms have been developed and successfully used. The remainder of this chapter is devoted to introducing some general notions of multidimensional finite volume methods.

Three of the most popular general approaches to obtaining multidimensional methods are:

- *Fully discrete flux-differencing methods.* In Section 19.3 we will see that the Lax–Wendroff method can be rewritten as a flux-differencing method. A numerical flux at each edge of the grid cell is defined, based on the data at the beginning of the time step. Differencing these fluxes gives the update to the cell average over a time step. To obtain better than first-order accuracy it is necessary to use the Taylor series expansion developed above in defining these fluxes. To obtain high-resolution nonoscillatory results it is also necessary to introduce limiters. There are many ways in which this can be done, and one particular method of this form is developed in Chapters 20 and 21.

- *Semidiscrete methods with Runge–Kutta time stepping.* Rather than using the Taylor series to replace time derivatives by spatial derivatives, we can focus on obtaining good accuracy of the flux at one instant in time and then use a Runge–Kutta method to perform the time stepping. This approach was introduced in one space dimension in Section 10.4. The two-dimensional extension is briefly discussed in Section 19.4.

- *Dimensional splitting.* By far the simplest approach to obtaining a multidimensional method is to apply a fractional-step method to split a multidimensional problem up into a sequence of one-dimensional problems. To solve $q_t + f(q)_x + g(q)_y = 0$, for example, we might alternate between solving $q_t + f(q)_x = 0$ and $q_t + g(q)_y = 0$, similarly to the way fractional-step methods are used for handling source terms, as discussed in Chapter 17. This approach, which is often surprisingly effective, is discussed in Section 19.5.

We will concentrate on methods that use one-dimensional Riemann solvers as a basic tool in the determination of interface fluxes. This is possible because the local problem at the edge of a grid cell is essentially one-dimensional in the direction normal to the edge. In order to obtain better than first-order accuracy it is necessary to bring in more multidimensional information, as is clear from the Taylor series expansion (19.5), but this can be done in various ways while still only using one-dimensional Riemann solutions.

One particular approach is developed starting in Chapter 20, in the form of the wave-propagation algorithms developed in [282], [283], and [257]. These algorithms take the form of fully discrete flux-differencing methods and are implemented in CLAWPACK in such a way that they can be applied to a wide variety of hyperbolic problems. Some related methods of this form can be found, for example, in [22], [24], [80], [100], [389].

Other methods have been developed that are based on a full decomposition of the data into multidimensional waves, rather than relying on one-dimensional Riemann solvers. For some examples, see [3], [52], [105], [106], [134], [135], [199], [342], [414], [415], [428].

19.3 Fully Discrete Flux-Differencing Methods

In deriving the two-dimensional conservation law $q_t + f(q)_x + g(q)_y = 0$ in Section 18.1, we considered an arbitrary region Ω. Now consider the special case where Ω is a rectangular grid cell of the form $\mathcal{C}_{ij} = [x_{i-1/2}, x_{i+1/2}] \times [y_{j-1/2}, y_{j+1/2}]$, as shown in Figure 19.1, where $x_{i+1/2} - x_{i-1/2} = \Delta x$ and $y_{j+1/2} - y_{j-1/2} = \Delta y$. In this special case, the formula (18.7) simplifies, since the normal vector always points in either the x- or the y-direction. The normal flux is given by $f(q)$ along the left and right edges and by $g(q)$ along the top and bottom. Integrating around the edges as required in (18.7) then gives

$$\frac{d}{dt} \iint_{C_{ij}} q(x, y, t)\, dx\, dy = \int_{y_{j-1/2}}^{y_{j+1/2}} f\big(q(x_{i+1/2}, y, t)\big)\, dy - \int_{y_{j-1/2}}^{y_{j+1/2}} f\big(q(x_{i-1/2}, y, t)\big)\, dy$$

$$+ \int_{x_{i-1/2}}^{x_{i+1/2}} g\big(q(x, y_{j+1/2}, t)\big)\, dx - \int_{x_{i-1/2}}^{x_{i+1/2}} g\big(q(x, y_{j-1/2}, t)\big)\, dx.$$

$$(19.9)$$

If we integrate this expression from t_n to t_{n+1} and divide by the cell area $\Delta x\, \Delta y$, we are led to a fully discrete flux-differencing method of the form

$$Q_{ij}^{n+1} = Q_{ij}^n - \frac{\Delta t}{\Delta x}\big[F_{i+1/2,j}^n - F_{i-1/2,j}^n\big] - \frac{\Delta t}{\Delta y}\big[G_{i,j+1/2}^n - G_{i,j-1/2}^n\big], \quad (19.10)$$

where

$$F_{i-1/2,j}^n \approx \frac{1}{\Delta t\, \Delta y} \int_{t_n}^{t_{n+1}} \int_{y_{j-1/2}}^{y_{j+1/2}} f\big(q(x_{i-1/2}, y, t)\big)\, dy\, dt,$$

$$(19.11)$$

$$G_{i,j-1/2}^n \approx \frac{1}{\Delta t\, \Delta x} \int_{t_n}^{t_{n+1}} \int_{x_{i-1/2}}^{x_{i+1/2}} g\big(q(x, y_{j-1/2}, t)\big)\, dx\, dt.$$

The numerical fluxes F^n and G^n at each edge are typically computed from the data Q^n at the initial time. (As in one dimension, these methods can also be extended to nonconservative hyperbolic systems, see Section 19.3.3.)

For the linear system $q_t + Aq_x + Bq_y = 0$ we can obtain approximations to these interface fluxes by using the Taylor expansion (19.5), which can be rewritten as

$$q(x, y, t_n + \Delta t) = q - \Delta t \left(Aq - \frac{\Delta t}{2} A^2 q_x - \frac{\Delta t}{2} ABq_y \right)_x$$

$$- \Delta t \left(Bq - \frac{\Delta t}{2} B^2 q_y - \frac{\Delta t}{2} BAq_x \right)_y + \cdots. \quad (19.12)$$

This suggests that we need

$$F_{i-1/2,j} \approx Aq(x_{i-1/2}, y_j, t_n) - \frac{\Delta t}{2} A^2 q_x(x_{i-1/2}, y_j, t_n) - \frac{\Delta t}{2} ABq_y(x_{i-1/2}, y_j, t_n),$$

$$G_{i,j-1/2} \approx Bq(x_i, y_{j-1/2}, t_n) - \frac{\Delta t}{2} B^2 q_y(x_i, y_{j-1/2}, t_n) - \frac{\Delta t}{2} BAq_x(x_i, y_{j-1/2}, t_n).$$

$$(19.13)$$

It can be shown that for this problem these expressions agree with the integrals in (19.11) to $\mathcal{O}(\Delta t^2)$.

19.3.1 Flux-Differencing Form of the Lax–Wendroff Method

The Lax–Wendroff method (19.8) for the constant-coefficient linear system $q_t + Aq_x + Bq_y = 0$ can be interpreted as a method of the form (19.10), where the fluxes are given by

$$
\begin{aligned}
F_{i-1/2,j} = {}& \frac{1}{2}A(Q_{i-1,j} + Q_{ij}) - \frac{\Delta t}{2\,\Delta x}A^2(Q_{ij} - Q_{i-1,j}) \\
& - \frac{\Delta t}{8\,\Delta y}AB[(Q_{i,j+1} - Q_{ij}) + (Q_{i-1,j+1} - Q_{i-1,j}) \\
& \qquad\qquad + (Q_{ij} - Q_{i,j-1}) + (Q_{i-1,j} - Q_{i-1,j-1})], \\
G_{i,j-1/2} = {}& \frac{1}{2}B(Q_{i,j-1} + Q_{ij}) - \frac{\Delta t}{2\,\Delta y}B^2(Q_{ij} - Q_{i,j-1}) \\
& - \frac{\Delta t}{8\,\Delta x}BA[(Q_{i+1,j} - Q_{ij}) + (Q_{i+1,j-1} - Q_{i,j-1}) \\
& \qquad\qquad + (Q_{ij} - Q_{i-1,j}) + (Q_{i,j-1} - Q_{i-1,j-1})].
\end{aligned}
\tag{19.14}
$$

These fluxes relate directly to (19.13). Note in particular that the expression ABq_y in (19.13), for example, is approximated by

$$
\begin{aligned}
ABq_y(x_{i-1/2}, y_j, t_n) \approx {}& \frac{1}{4\,\Delta y}[AB(Q_{i,j+1} - Q_{ij}) + AB(Q_{i-1,j+1} - Q_{i-1,j}) \\
& \qquad + AB(Q_{ij} - Q_{i,j-1}) + AB(Q_{i-1,j} - Q_{i-1,j-1})]. \quad (19.15)
\end{aligned}
$$

In Chapters 20 and 21 we will see how this method can be greatly improved by introducing an upwind bias and flux limiting into the formulas.

19.3.2 Godunov's Method

For a general conservation law, the simplest flux-differencing method of the form (19.10) is Godunov's method. A natural two-dimensional generalization of the method developed in Section 15.1 is obtained by simply solving the normal Riemann problem at each cell edge to find the value $Q^\vee$ that propagates with speed 0, and then evaluating the appropriate flux function at this value to obtain the numerical flux at this edge. This gives

$$
\begin{aligned}
F_{i-1/2,j} &= f(Q^\vee_{i-1/2,j}), \\
G_{i,j-1/2} &= g(Q^\vee_{i,j-1/2}),
\end{aligned}
\tag{19.16}
$$

where $Q^\vee_{i-1/2,j}$ is obtained by solving the Riemann problem for $q_t + f(q)_x = 0$ with data $Q_{i-1,j}$ and Q_{ij}, while $Q^\vee_{i,j-1/2}$ is obtained by solving the Riemann problem for $q_t + g(q)_y = 0$ with data $Q_{i,j-1}$ and Q_{ij}. As in one dimension, approximate Riemann solvers can be used in place of the exact Riemann solution.

For a linear system of equations with $f(q) = Aq$ and $g(q) = Bq$, we denote the eigenvector matrices for A and B by R^x and R^y, respectively, and the eigenvalue matrices by Λ^x and Λ^y, as in Section 18.5. We can then define matrices $A^\pm$ and $B^\pm$ analogous to

(4.45) by

$$A^{\pm} = R^x(\Lambda^x)^{\pm}(R^x)^{-1}, \qquad B^{\pm} = R^y(\Lambda^y)^{\pm}(R^y)^{-1}. \tag{19.17}$$

In terms of this notation, we find that the Godunov fluxes for a linear problem are the natural generalization of the one-dimensional flux (4.56),

$$F_{i-1/2,j} = A^+ Q_{i-1,j} + A^- Q_{ij},$$
$$G_{i,j-1/2} = B^+ Q_{i,j-1} + B^- Q_{ij}. \tag{19.18}$$

This amounts to using only the first terms in the fluxes (19.13), and an upwind approximation to these. Of course, this method is only first-order accurate and moreover is typically stable only for Courant number up to $1/2$ in two dimensions. This is illustrated for the advection equation in Section 20.4.

19.3.3 Fluctuation Form

As in one space dimension, we will develop finite volume methods in a more general form than the flux-differencing formula (19.10), so that they are also applicable to hyperbolic equations that are not in conservation form. To make this extension we will rewrite the method as follows, motivated by the one-dimensional method (15.62):

$$Q_{ij}^{n+1} = Q_{ij} - \frac{\Delta t}{\Delta x}\left(\mathcal{A}^+ \Delta Q_{i-1/2,j} + \mathcal{A}^- \Delta Q_{i+1/2,j}\right)$$

$$- \frac{\Delta t}{\Delta y}\left(\mathcal{B}^+ \Delta Q_{i,j-1/2} + \mathcal{B}^- \Delta Q_{i,j+1/2}\right)$$

$$- \frac{\Delta t}{\Delta x}\left(\tilde{F}_{i+1/2,j} - \tilde{F}_{i-1/2,j}\right) - \frac{\Delta t}{\Delta y}\left(\tilde{G}_{i,j+1/2} - \tilde{G}_{i,j-1/2}\right). \tag{19.19}$$

The term $\mathcal{A}^+ \Delta Q_{i-1/2,j}$, for example, represents the first-order Godunov update to the cell value Q_{ij} resulting from the Riemann problem at the edge $(i - 1/2, j)$. The other three similar terms are the Godunov updates resulting from the Riemann problems at the other three edges. For the linear system discussed above, these fluctuations are simply given by

$$\mathcal{A}^{\pm} \Delta Q_{i-1/2,j} = A^{\pm}(Q_{ij} - Q_{i-1,j}),$$
$$\mathcal{B}^{\pm} \Delta Q_{i,j-1/2} = B^{\pm}(Q_{ij} - Q_{i,j-1}). \tag{19.20}$$

For Godunov's method we take $\tilde{F} = \tilde{G} = 0$ everywhere. Later the fluxes $\tilde{F}$ and $\tilde{G}$ will be used for correction terms, both those arising from introducing slopes as in one dimension to model the A^2 and B^2 terms in (19.13), and also new ones modeling the cross-derivative terms involving AB and BA in (19.13).

For a general nonlinear conservation law where the Godunov fluxes are defined by (19.16), we can set

$$\mathcal{A}^+ \Delta Q_{i-1/2, j} = f(Q_{ij}) - f(Q_{i-1/2, j}^{\vee}),$$

$$\mathcal{A}^- \Delta Q_{i-1/2, j} = f(Q_{i-1/2, j}^{\vee}) - f(Q_{i-1, j}),$$

$$\mathcal{B}^+ \Delta Q_{i, j-1/2} = g(Q_{ij}) - g(Q_{i, j-1/2}^{\vee}),$$

$$\mathcal{B}^- \Delta Q_{i, j-1/2} = g(Q_{i, j-1/2}^{\vee}) - g(Q_{i, j-1}).$$

(19.21)

Godunov's method results from using these formulas in (19.19) and setting all $\tilde{F} = \tilde{G} = 0$. As in one dimension, the fluctuations $\mathcal{A}^{\pm} \Delta Q$ and $\mathcal{B}^{\pm} \Delta Q$ can also be computed in terms of the waves and speeds arising in the Riemann solution, using formulas analogous to (4.42).

19.4 Semidiscrete Methods with Runge–Kutta Time Stepping

As we see already from the Lax–Wendroff method with fluxes (19.14), obtaining even second-order accuracy with a flux-differencing method based on the time-integrated fluxes (19.11) can lead to complicated formulas. As discussed in Section 10.3 for one-dimensional problems, this Taylor series approach is not easily extended to obtain higher-order methods. For this reason, another popular approach is to proceed as in Section 10.4 and use the expression (19.9) to derive ordinary differential equations for the evolution of the cell averages

$$Q_{ij}(t) = \iint_{\mathcal{C}_{ij}} q(x, y, t) \, dx \, dy.$$

(19.22)

This is accomplished by defining numerical flux functions

$$F_{i-1/2, j}(Q(t)) \approx \frac{1}{\Delta y} \int_{y_{j-1/2}}^{y_{j+1/2}} f(q(x_{i-1/2}, y, t)) \, dy,$$

$$G_{i, j-1/2}(Q(t)) \approx \frac{1}{\Delta x} \int_{x_{i-1/2}}^{x_{i+1/2}} g(q(x, y_{j-1/2}, t)) \, dx$$

(19.23)

by some procedure based on the nearby cell averages at this instant in time. Then the system of ODEs

$$\frac{d}{dt} Q_{ij}(t) = -\frac{1}{\Delta x} \left[F_{i+1/2, j}(Q(t)) - F_{i-1/2, j}(Q(t)) \right]$$

$$- \frac{1}{\Delta y} \left[G_{i, j+1/2}(Q(t)) - G_{i, j-1/2}(Q(t)) \right]$$

(19.24)

is solved by an ODE method, typically a multistage Runge–Kutta method. In order to achieve high-order accuracy it is still necessary to use information from several grid cells nearby in defining the fluxes (19.23), typically by some multidimensional interpolation method. However, since we do not attempt to also approximate the time-derivative terms, this may be relatively simple. To obtain a high-resolution method, it is necessary to include some form of upwinding and/or limiting in the process of approximating the flux, e.g., by a multidimensional version of the ENO method described in Section 10.4.4.

19.5 Dimensional Splitting

The easiest way to extend one-dimensional numerical methods to more space dimensions is to use *dimensional splitting*, an application of the fractional-step procedure discussed in Chapter 17. A multidimensional problem is simply split into a sequence of one-dimensional problems. This is easy to apply on a Cartesian grid aligned with the coordinate axes as shown in Figure 19.1.

For example, the two-dimensional linear problem

$$q_t + Aq_x + Bq_y = 0$$

can be split into

$$x\text{-sweeps}: \quad q_t + Aq_x = 0, \tag{19.25}$$
$$y\text{-sweeps}: \quad q_t + Bq_y = 0. \tag{19.26}$$

In the x-sweeps we start with cell averages Q_{ij}^n at time t_n and solve one-dimensional problems $q_t + Aq_x = 0$ along each row of cells $\mathcal{C}_{ij}$ with j fixed, updating Q_{ij}^n to Q_{ij}^*:

$$Q_{ij}^* = Q_{ij}^n - \frac{\Delta t}{\Delta x}\left(F_{i+1/2,j}^n - F_{i-1/2,j}^n\right), \tag{19.27}$$

where $F_{i-1/2,j}^n$ is an appropriate numerical flux for the one-dimensional problem between cells $\mathcal{C}_{i-1,j}$ and $\mathcal{C}_{ij}$. In the y-sweeps we then use the Q_{ij}^* values as data for solving $q_t + Bq_y = 0$ along each column of cells with i fixed, which results in Q_{ij}^{n+1}:

$$Q_{ij}^{n+1} = Q_{ij}^* - \frac{\Delta t}{\Delta x}\left(G_{i,j+1/2}^* - G_{i,j-1/2}^*\right). \tag{19.28}$$

Note that there will generally be a splitting error (see Section 17.3) unless the operators $\mathcal{A} = A\partial_x$ and $\mathcal{B} = B\partial_y$ commute, i.e., unless $AB = BA$. Only in the case where the multidimensional problem decouples into scalar advection equations can we use dimensional splitting with no splitting error. Even in this case we must be careful with boundary conditions (see Section 17.9).

However, the splitting error is often no worse than the errors introduced by the numerical methods in each sweep, and dimensional splitting can be a very effective approach. It gives a simple and relatively inexpensive way to extend one-dimensional high-resolution methods to two or three dimensions.

Note that with the dimensional-splitting approach we do not explicitly model the cross-derivative terms involving q_{xy} in the Taylor series expansion (19.5). In each sweep we only model second derivatives in each coordinate direction, q_{xx} and q_{yy}, which appear in the one-dimensional algorithm. The q_{xy} term arises automatically through the fractional-step procedure. The intermediate solution q^* resulting from x-sweeps involves terms modeling Aq_x. In the y-sweeps we compute terms modeling Bq_y^*, which thus model $B(Aq_x)_y$.

Instead of this *Godunov splitting*, one might instead use the *Strang splitting*

$$Q_{ij}^* = Q_{ij}^n - \frac{\Delta t}{2\,\Delta x}\left(F_{i+1/2,j}^n - F_{i-1/2,j}^n\right),$$

$$Q_{ij}^{**} = Q_{ij}^* - \frac{\Delta t}{\Delta x}\left(G_{i,j+1/2}^* - G_{i,j-1/2}^*\right), \tag{19.29}$$

$$Q_{ij}^{n+1} = Q_{ij}^n - \frac{\Delta t}{2\,\Delta x}\left(F_{i+1/2,j}^{**} - F_{i-1/2,j}^{**}\right),$$

as discussed in Section 17.4. With the Strang splitting we also obtain terms modeling $A(Bq_y)_x$ from the second x-sweep, which are also needed in the Taylor series expansion. Only in the constant-coefficient case with $AB = BA$ does the Godunov splitting give a fully second-order accurate method. However, in practice there is often very little difference in results obtained with the two approaches, as is also the case for other fractional-step methods as discussed in Section 17.5.

In fact, if the Strang splitting is implemented as in (19.29), it may give worse results than the Godunov splitting, because the x-sweeps are taken with time step $\Delta t/2$ and hence will typically have Courant number less than $1/2$. This introduces more numerical smearing as well as more work. If instead the Strang splitting is implemented by simply alternating the order in which x-sweeps and y-sweeps are performed, then this is avoided (see Section 17.4). However, this is somewhat harder to implement in connection with variable-size time steps, and in CLAWPACK the form (19.29) is implemented as the Strang splitting, though the Godunov splitting is generally recommended instead.

19.5.1 CLAWPACK *Implementation*

In order to apply the fractional-step approach, we need to be able to solve each of the one-dimensional equations (19.25) and (19.26), and hence must have two different Riemann solvers available. For many physical systems the equations and solution procedure are essentially the same in each direction. For example, the acoustics equations (18.25) in an isotropic medium have exactly the same form when restricted to plane waves in x or y, but with the roles of u and v reversed. The same is true for the shallow water equations (18.38) and the Euler equations (18.45). For this reason it is often simplest to write a single Riemann solver with a flag indicating the desired direction. This convention is used in CLAWPACK, where a single subroutine `rpn2` must be provided that solves the Riemann problem normal to edges of grid cells along one slice of the domain. The flag `ixy` indicates whether the slice is in the x-direction or the y-direction. The other parameters of this routine are identical to the parameters appearing in the one-dimensional Riemann solver `rp1`.

The two-dimensional CLAWPACK code also allows a second Riemann solver `rpt2` to be provided. This must solve a different sort of Riemann problem in the transverse direction, as described in Section 21.3, and is not used in the dimensional-splitting algorithm.

Dimensional splitting is invoked in CLAWPACK by setting `method(3) = -1` for the Godunov splitting, which is generally recommended, or `method(3) = -2` for Strang splitting. Positive values of method(3) instead invoke the unsplit methods described in later chapters. (See Section 21.3.)

Exercise

19.1. The system given in Exercise 18.3 is hyperbolic in x and y separately, and so we can apply dimensional splitting to attempt to solve this nonhyperbolic system. Implement this in CLAWPACK, and analyze what happens.

20

Multidimensional Scalar Equations

In this chapter we will develop high-resolution methods for scalar hyperbolic equations in two space dimensions. We begin by considering the constant-coefficient advection equation and show how the waves obtained by solving the Riemann problem at each cell interface can be naturally used to define high-resolution fluxes. We then extend these methods to variable-coefficient advection and nonlinear scalar conservation laws. In the next chapter they are extended further to hyperbolic systems of equations.

We first consider the scalar advection equation

$$q_t + uq_x + vq_y = 0, \tag{20.1}$$

with u and v constant. In the figures illustrating these methods we will generally assume $u > 0$ and $v > 0$, and this case will sometimes be assumed when we wish to be specific, but most of the formulas will be presented in a manner that applies for flow in any direction. The notation $u^\pm$ and $v^\pm$ meaning the positive or negative part of the velocity will frequently be used, with the definition (4.40).

The true solution for this equation is simply $q(x, y, t) = \overset{\circ}{q}(x - ut, y - vt)$, but for our present purposes the Taylor series expansion is more illuminating:

$$q(x, y, t_{n+1}) = q(x, y, t_n) + \Delta t\, q_t(x, y, t_n) + \frac{1}{2}(\Delta t)^2 q_{tt}(x, y, t_n) + \cdots$$
$$= q(x, y, t_n) - u\,\Delta t\, q_x - v\,\Delta t\, q_y$$
$$+ \frac{1}{2}(\Delta t)^2[u^2 q_{xx} + vu q_{xy} + uv q_{yx} + v^2 q_{yy}] + \cdots. \tag{20.2}$$

This comes from (19.5) with $A = u$ and $B = v$, and will be useful in identifying terms arising in finite volume approximations to the advection equation.

20.1 The Donor-Cell Upwind Method for Advection

The simplest finite volume method for the advection equation is the first-order upwind method, which takes the general form

$$Q_{ij}^{n+1} = Q_{ij} - \frac{\Delta t}{\Delta x}[u^+(Q_{ij} - Q_{i-1,j}) + u^-(Q_{i+1,j} - Q_{ij})]$$
$$- \frac{\Delta t}{\Delta y}[v^+(Q_{ij} - Q_{i,j-1}) + v^-(Q_{i,j+1} - Q_{ij})]. \tag{20.3}$$

447

This uses an upwind approximation to the derivatives q_x and q_y in the $\mathcal{O}(\Delta t)$ terms of the Taylor series expansion (20.2). This method has the form (19.19) with $\tilde{F} = \tilde{G} = 0$ and fluctuations

$$
\begin{aligned}
\mathcal{A}^\pm \Delta Q_{i-1/2,j} &= u^\pm (Q_{ij} - Q_{i-1,j}), \\
\mathcal{B}^\pm \Delta Q_{i,j-1/2} &= v^\pm (Q_{ij} - Q_{i,j-1}),
\end{aligned}
\tag{20.4}
$$

which are a special case of (19.20).

This upwind method agrees with Godunov's method as described in Section 19.3.2 for this scalar equation. The fluxes for the method (20.3) are

$$
\begin{aligned}
F_{i-1/2,j} &= u^+ Q_{i-1,j} + u^- Q_{ij}, \\
G_{i,j-1/2} &= v^+ Q_{i,j-1} + v^- Q_{ij},
\end{aligned}
\tag{20.5}
$$

which agree with the Godunov fluxes (19.18). In this case each Riemann solution consists of a single wave carrying the jump in Q between the neighboring two grid cells, propagating at speed u horizontally or at speed v vertically depending on the orientation of the two cells. The value of $Q^\vee$ at each interface depends on whether the relevant velocity is positive or negative.

The first-order accurate method (20.3) for the advection equation is often called the *donor-cell upwind* (DCU) method. Each flux in (20.5) approximates the amount of q flowing normal to the edge, assuming that the only contribution to this flux is from the adjacent cell on the upwind side (the donor cell). This is indicated schematically in Figure 20.1(a) for the case $u, v > 0$.

Note that the updated value Q_{ij}^{n+1} depends on only the three values Q_{ij}, $Q_{i-1,j}$, and $Q_{i,j-1}$. This is clearly not correct, since the flow is really at an angle to the grid, as indicated in Figure 20.1(b), and the value $Q_{i-1,j-1}$ should also affect Q_{ij}^{n+1}. The CFL condition (Section 4.4) suggests that this may cause stability problems, and indeed this method does not have the best possible stability properties. It will be shown in Section 20.4 that this

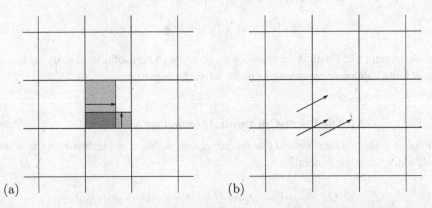

(a) (b)

Fig. 20.1. (a) Waves moving normal to the cell interfaces give the updates for the DCU method. (b) The true velocity (u, v) is at an angle to the grid, and information from cell $(i-1, j-1)$ should also affect the new value in cell (i, j). This corner coupling is missing in the DCU method.

method is stable only for Δt small enough that

$$\left| \frac{u\, \Delta t}{\Delta x} \right| + \left| \frac{v\, \Delta t}{\Delta y} \right| \le 1. \tag{20.6}$$

An improved upwind method is developed in Section 20.2, which takes account of the flow direction more fully and has the stability bound

$$\max \left(\left| \frac{u\Delta t}{\Delta x} \right|, \left| \frac{v\Delta t}{\Delta y} \right| \right) \le 1. \tag{20.7}$$

This is better than (20.6) whenever u and v are both nonzero, i.e., when flow is at an angle to the grid.

20.2 The Corner-Transport Upwind Method for Advection

For the advection equation, a better first-order accurate upwind method can be derived by taking the reconstruct–evolve–average approach of Algorithm 4.1, extended in the obvious way to two space dimensions:

- View the cell averages at time t_n as defining a piecewise constant function $\tilde{q}^n(x, y, t_n)$ with constant value Q_{ij}^n in cell C_{ij},
- Evolve the advection equation exactly with this data over time Δt,
- Average the resulting solution $\tilde{q}^n(x, y, t_{n+1})$ back onto the grid.

For the constant-coefficient advection equation this is easily done, since

$$\tilde{q}^n(x, y, t_{n+1}) = \tilde{q}^n(x - u\, \Delta t, y - v\, \Delta t, t_n).$$

The exact solution is the same piecewise constant function, simply shifted by $(u\, \Delta t, v\, \Delta t)$. So we find that

$$Q_{ij}^{n+1} = \frac{1}{\Delta x\, \Delta y} \int_{x_{i-1/2}}^{x_{i+1/2}} \int_{y_{j-1/2}}^{y_{j+1/2}} \tilde{q}^n(x - u\, \Delta t, y - v\, \Delta t, t_n)\, dx\, dy$$

$$= \frac{1}{\Delta x\, \Delta y} \int_{x_{i-1/2}-u\, \Delta t}^{x_{i+1/2}-u\, \Delta t} \int_{y_{j-1/2}-v\, \Delta t}^{y_{j+1/2}-v\, \Delta t} \tilde{q}^n(x, y, t_n)\, dx\, dy. \tag{20.8}$$

The new cell average Q_{ij}^{n+1} is given by the cell average of $\tilde{q}^n(x, y, t_n)$ over the shaded region shown in Figure 20.2. Since $\tilde{q}^n(x, y, t_n)$ is constant in each grid cell, this reduces to a simple convex combination of four cell values:

$$Q_{ij}^{n+1} = \frac{1}{\Delta x\, \Delta y} \big[(\Delta x - u\, \Delta t)(\Delta y - v\, \Delta t)Q_{ij}^n + (\Delta x - u\, \Delta t)(v\, \Delta t)Q_{i,j-1}^n$$

$$+ (\Delta y - v\, \Delta t)(u\, \Delta t)Q_{i-1,j}^n + (u\, \Delta t)(v\, \Delta t)Q_{i-1,j-1}^n \big]. \tag{20.9}$$

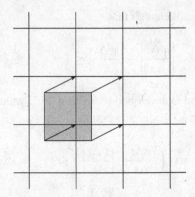

Fig. 20.2. The corner-transport upwind method is obtained by shifting the piecewise constant data by distance $(u\,\Delta t, v\,\Delta t)$ and averaging back on the grid. Alternatively, the new value Q_{ij}^{n+1} is determined by averaging the piecewise constant function over the shaded region shown in the figure.

This can be rearranged to yield

$$Q_{ij}^{n+1} = Q_{ij} - \frac{u\,\Delta t}{\Delta x}(Q_{ij} - Q_{i-1,j}) - \frac{v\,\Delta t}{\Delta y}(Q_{ij} - Q_{i,j-1})$$

$$+ \frac{1}{2}(\Delta t)^2 \left\{ \frac{u}{\Delta x}\left[\frac{v}{\Delta y}(Q_{ij} - Q_{i,j-1}) - \frac{v}{\Delta y}(Q_{i-1,j} - Q_{i-1,j-1}) \right] \right.$$

$$\left. + \frac{v}{\Delta y}\left[\frac{u}{\Delta x}(Q_{ij} - Q_{i-1,j}) - \frac{u}{\Delta x}(Q_{i,j-1} - Q_{i-1,j-1}) \right] \right\}. \quad (20.10)$$

The top line of this expression corresponds to the donor-cell upwind method. The additional terms can be arranged in several different ways. They have been displayed here in a manner that relates directly to the Taylor series expansion (20.2). We see that the final term in (20.10) models the cross-derivative terms $uvq_{yx} + vuq_{xy}$ in the $\mathcal{O}((\Delta t)^2)$ term of that expansion.

The method (20.10) is often called *corner-transport upwind* (CTU) method (following Colella [80]), since it includes the proper transport across the corner from cell $C_{i-1,j-1}$ to C_{ij}. It is still only first-order accurate, for two reasons: It is missing approximations to the q_{xx} and q_{yy} terms in (20.2), and the approximations to uq_x and vq_y terms are only first-order one-sided approximations. Both of these deficiencies can be addressed by introducing slopes in the x- and y-directions separately, just as we did in one dimension. Consequently a high-resolution version of this algorithm is easy to construct, as we will do in Section 20.6. See [25], [470] for discussion of some related algorithms.

20.3 Wave-Propagation Implementation of the CTU Method

Before discussing high-resolution corrections, we will develop a different implementation of the CTU method that will be much easier to extend to variable-coefficient advection and to other hyperbolic systems. Figure 20.3 shows the basis for a wave-propagation view of this method, in which all waves propagate at velocity (u, v) in the correct physical direction. To be specific we will continue to assume $u > 0$ and $v > 0$ in the figures and formulas in this section. More general formulas are given in the next section.

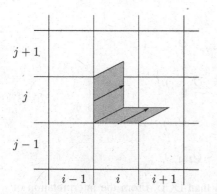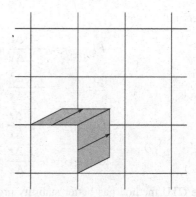

Fig. 20.3. (a) Transverse propagation affecting the fluxes $\tilde{F}_{i+1/2,j}$ and $\tilde{G}_{i,j+1/2}$. (b) Transverse propagation affecting the fluxes $\tilde{F}_{i-1/2,j}$ and $\tilde{G}_{i,j-1/2}$.

From the interface between cells $\mathcal{C}_{i-1,j}$ and $\mathcal{C}_{ij}$, for example, there is a wave that propagates into cells $\mathcal{C}_{ij}$ and $\mathcal{C}_{i,j+1}$. The jump across this wave is $Q_{ij} - Q_{i-1,j}$, and this increment affects both cell averages Q_{ij} and $Q_{i,j+1}$. As Figure 20.3 shows, there are four distinct waves that affect Q_{ij}^n. The effect of each wave is to modify the cell average by the jump across the wave multiplied by the fraction of the cell covered by the wave. These fractions are easily worked out, noting that the small triangular portion of each wave moving transversely into neighboring cells has area $\frac{1}{2}(u\,\Delta t)(v\,\Delta t) = \frac{1}{2}uv(\Delta t)^2$. The wave from interface $(i-1/2,\,j)$, for example, modifies Q_{ij} by

$$\left(\frac{u\,\Delta t\,\Delta y - \frac{1}{2}uv(\Delta t)^2}{\Delta x\,\Delta y} \right)(Q_{ij} - Q_{i-1,j}) \qquad (20.11)$$

and modifies $Q_{i,j+1}$ by

$$\left(\frac{\frac{1}{2}uv(\Delta t)^2}{\Delta x\,\Delta y} \right)(Q_{ij} - Q_{i-1,j}). \qquad (20.12)$$

Note that the update (20.11) is present in the formula (20.10), split into two parts. The latter part, corresponding to the triangular piece, is grouped with three other term corresponding to the triangular pieces of the other three waves shown in Figure 20.3 that affect this cell.

Thus we can view (20.10) as consisting of the DCU method (the first line), in which waves simply move normal to the cell interfaces as shown in Figure 20.1, combined with a set of corrections for the transverse propagation of the waves. These corrections can be viewed as fluxes through edges of the cell. The triangular region in cell $(i, j+1)$ of Figure 20.3(a) has been transferred from cell (i, j) and hence corresponds to a flux through edge $(i, j+1/2)$. This transfer can be represented by a flux $\tilde{G}_{i,j+1/2}$. Taking this viewpoint, we find that we can rewrite (20.10) in the form (19.19) by defining $\mathcal{A}^{\pm}\Delta Q$ and $\mathcal{B}^{\pm}\Delta Q$ as in (20.4) and the correction fluxes as

$$\tilde{F}_{i-1/2,j} = -\frac{1}{2}\frac{\Delta t}{\Delta y}\, uv(Q_{i-1,j} - Q_{i-1,j-1}),$$

$$\tilde{F}_{i+1/2,j} = -\frac{1}{2}\frac{\Delta t}{\Delta y}\, uv(Q_{ij} - Q_{i,j-1}),$$

$$\tilde{G}_{i,j-1/2} = -\frac{1}{2}\frac{\Delta t}{\Delta x}\, uv(Q_{i,j-1} - Q_{i-1,j-1}),$$ (20.13)

$$\tilde{G}_{i,j+1/2} = -\frac{1}{2}\frac{\Delta t}{\Delta x}\, uv(Q_{ij} - Q_{i-1,j}).$$

The CTU method has better stability properties than DCU. From the interpretation of the algorithm given at the beginning of Section 20.2, we expect the method to be stable for any time step for which the piecewise constant function does not shift more than one grid cell in the time step. This gives the stability bound (20.7), and the method is stable for Courant numbers up to 1. This can be shown formally in the 1-norm by extending the proof of Section 8.3.4 to two dimensions using the convex combination (20.9). Stability in the 2-norm is demonstrated using von Neumann analysis in the next section.

20.4 von Neumann Stability Analysis

For constant-coefficient linear equations, von Neumann analysis is often the easiest way to determine stability bounds, as discussed in Section 8.3.3 in one space dimension. As two-dimensional examples, in this section we consider the DCU method (20.3) and the CTU method (20.10). Similar analysis can also be performed for the Lax–Wendroff method or the wave-propagation version (introduced in Section 20.6) provided that no limiter function is applied. (Applying limiters makes the method nonlinear and von Neumann analysis can no longer be used.) See [282] for stability analyses of these cases and three-dimensional generalizations, and also [202].

To be specific we will assume u, $v > 0$, although similar analysis applies to other choices of signs. Then the DCU method (20.3) becomes

$$Q_{IJ}^{n+1} = Q_{IJ}^{n} - v^x(Q_{IJ}^{n} - Q_{I-1,J}^{n}) - v^y(Q_{IJ}^{n} - Q_{I,J-1}^{n}),$$ (20.14)

where $v^x = u\,\Delta t/\Delta x$ and $v^y = v\,\Delta t/\Delta y$. (We use I, J as the grid indices in this section, so that $i = \sqrt{-1}$ can be used in the complex exponential.) As in one dimension, Fourier analysis decouples the constant-coefficient linear difference equation into separate equations for each mode, so it suffices to consider data consisting of a single arbitrary Fourier mode

$$Q_{IJ}^{n} = e^{i(\xi I\,\Delta x + \eta J\,\Delta y)},$$ (20.15)

where ξ and η are the wave numbers in x and y. Inserting this into (20.14) gives

$$Q_{IJ}^{n+1} = g(\xi, \eta, \Delta x, \Delta y, \Delta t)\, Q_{IJ}^{n}$$ (20.16)

with amplification factor

$$g(\xi, \eta, \Delta x, \Delta y, \Delta t) = (1 - v^x - v^y) + v^x e^{-i\xi\,\Delta x} + v^y e^{-i\eta\,\Delta y}.$$ (20.17)

The method is stable in the 2-norm provided that $|g| \leq 1$ for all choices of ξ and η. Values of g lie a distance at most $\nu^x + \nu^y$ from the point $1 - \nu^x - \nu^y$ in the complex plane, and hence the method is stable for $0 \leq \nu^x + \nu^y \leq 1$. By considering other choices for the sign of u and v we find that in general the stability limit (20.6) is required for the DCU method.

We now turn to the CTU method, which has the form (20.10) when u, $v > 0$. With our current notation this becomes

$$Q_{IJ}^{n+1} = Q_{IJ}^{n} - \nu^x \left(Q_{IJ}^{n} - Q_{I-1,J}^{n} \right) - \nu^y \left(Q_{IJ}^{n} - Q_{I,J-1}^{n} \right)$$
$$+ \frac{1}{2} \nu^x \nu^y \left[\left(Q_{IJ}^{n} - Q_{I,J-1}^{n} \right) - \left(Q_{I-1,J}^{n} - Q_{I-1,J-1}^{n} \right) \right.$$
$$\left. + \left(Q_{IJ}^{n} - Q_{I-1,J}^{n} \right) - \left(Q_{I,J-1}^{n} - Q_{I-1,J-1}^{n} \right) \right]. \qquad (20.18)$$

Inserting the Fourier mode (20.15) into this again gives an expression of the form (20.16) with amplification factor

$$g(\xi, \eta, \Delta x, \Delta y, \Delta t) = 1 - \nu^x (1 - e^{-i\xi \, \Delta x}) - \nu^y (1 - e^{-i\eta \, \Delta y})$$
$$+ \frac{1}{2} \nu^x \nu^y [(1 - e^{-i\eta \, \Delta y}) - e^{-i\xi \, \Delta x} (1 - e^{-i\eta \, \Delta y})$$
$$+ (1 - e^{-i\xi \, \Delta x}) - e^{-i\eta \, \Delta y} (1 - e^{-i\xi \, \Delta x})]$$
$$= [1 - \nu^x (1 - e^{-i\xi \, \Delta x})][1 - \nu^y (1 - e^{-i\eta \, \Delta y})]. \qquad (20.19)$$

Now g is the product of two one-dimensional terms. The method is stable if and only if both terms lie in the unit circle for all choices of ξ and η, and so the method is stable provided $\max(\nu^x, \nu^y) \leq 1$. By considering other choices for the sign of u and v we find that in general the stability limit (20.7) is required for the CTU method.

20.5 The CTU Method for Variable-Coefficient Advection

The formulas (20.13) are for the advection equation in the special case u, $v > 0$. For different directions of flow the fluxes must be properly specified to reflect the propagation directions. In this section we will give the general formulas based on a simple wave-propagation procedure. We also now consider the more general context where we allow the velocities to vary in space, since this is equally easy to handle with the wave-propagation approach.

Here we consider the *color-equation* form of the advection equation,

$$q_t + u(x, y)q_x + v(x, y)q_y = 0, \qquad (20.20)$$

as discussed in Section 9.3. The conservative form of the two-dimensional advection equation can be solved by extensions of the approach developed in Section 9.5.2.

We assume that the velocities are specified at cell edges (see Section 9.5 for the one-dimensional case) with $u_{i-1/2,j}$ specified at the edge between cells $(i - 1, j)$ and (i, j) and $v_{i,j-1/2}$ at the edge between cells $(i, j - 1)$ and (i, j). Only the normal velocity is needed at each edge in order to determine the normal flux through that portion of the cell boundary. Ideally these should be averages of the true normal velocity along the corresponding edge of the cell (see Section 20.8).

The Riemann problem at each interface leads to a single wave with speed given by the edge velocity. In the x-direction we have

$$\mathcal{W}_{i-1/2,j} = Q_{ij} - Q_{i-1,j},$$
$$s_{i-1/2,j} = u_{i-1/2,j}, \tag{20.21}$$

and in the y-direction

$$\mathcal{W}_{i,j-1/2} = Q_{ij} - Q_{i,j-1},$$
$$s_{i,j-1/2} = v_{i,j-1/2}. \tag{20.22}$$

The fluctuations needed for the DCU algorithm are the natural generalizations of (20.4),

$$\mathcal{A}^{\pm}\Delta Q_{i-1/2,j} = s^{\pm}_{i-1/2,j}\mathcal{W}_{i-1/2,j} = u^{\pm}_{i-1/2,j}(Q_{ij} - Q_{i-1,j}),$$
$$\mathcal{B}^{\pm}\Delta Q_{i,j-1/2} = s^{\pm}_{i,j-1/2}\mathcal{W}_{i,j-1/2} = v^{\pm}_{i,j-1/2}(Q_{ij} - Q_{i,j-1}), \tag{20.23}$$

To compute the correction fluxes needed for the CTU method, we view each wave as potentially propagating transversely into each of the neighboring cells (see Figure 20.4). Rather than giving a single expression for each correction flux, we will build up these fluxes by adding in any transverse terms arising from each Riemann problem as it is solved.

At the beginning of each time step we set

$$\tilde{F}_{i-1/2,j} := 0 \quad \text{and} \quad \tilde{G}_{i,j-1/2} := 0 \qquad \forall i, j.$$

After solving each Riemann problem in the x-direction, at interface $(i - 1/2, j)$, we set

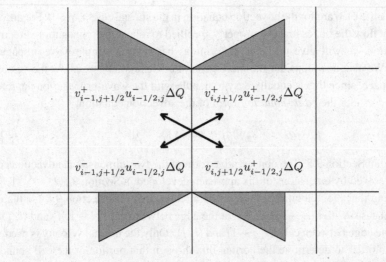

Fig. 20.4. The jump $\Delta Q_{i-1/2,j}$ may propagate in any of four directions, depending on the velocities, and so four neighboring $\tilde{G}$-fluxes may need to be updated.

$\mathcal{A}^{\pm}\Delta Q_{i-1/2,j}$ as in (20.23) and then update the nearby correction fluxes by

$$
\tilde{G}_{i-1,j-1/2} := \tilde{G}_{i-1,j-1/2} - \frac{1}{2}\frac{\Delta t}{\Delta x} v_{i-1,j-1/2}^{-}u_{i-1/2,j}^{-}(Q_{ij} - Q_{i-1,j}),
$$

$$
\tilde{G}_{i-1,j+1/2} := \tilde{G}_{i-1,j+1/2} - \frac{1}{2}\frac{\Delta t}{\Delta x} v_{i-1,j+1/2}^{+}u_{i-1/2,j}^{-}(Q_{ij} - Q_{i-1,j}),
$$

$$
\tilde{G}_{i,j-1/2} := \tilde{G}_{i,j-1/2} - \frac{1}{2}\frac{\Delta t}{\Delta x} v_{i,j-1/2}^{-}u_{i-1/2,j}^{+}(Q_{ij} - Q_{i-1,j}), \tag{20.24}
$$

$$
\tilde{G}_{i,j+1/2} := \tilde{G}_{i,j+1/2} - \frac{1}{2}\frac{\Delta t}{\Delta x} v_{i,j+1/2}^{+}u_{i-1/2,j}^{+}(Q_{ij} - Q_{i-1,j}).
$$

This takes into account all possible triangular regions. Normally three out of these four updates will be zero, as in the case where u and v are constant. At least two will always be zero, since $u_{i-1/2,j}$ can't be both positive and negative. But if, for example, $u_{i-1/2,j} > 0$ and $v_{i,j+1/2} > 0$ while $v_{i,j-1/2} < 0$, then both $\tilde{G}_{i,j+1/2}$ and $\tilde{G}_{i,j-1/2}$ will be updated, since the wave is evidently flowing transversely into both the cell above and the cell below in this case.

It may also happen that a single interface flux, say $\tilde{G}_{i,j+1/2}$, will be updated by more than one flux correction arising from different Riemann problems, for example if $v_{i,j+1/2} > 0$ and $u_{i-1/2,j} > 0$ while $u_{i+1/2,j} < 0$.

The unsplit algorithms are implemented in CLAWPACK using this same approach. The correction fluxes are all initialized to zero. A Riemann problem is solved at each cell interface, and in addition to determining the fluctuations, the appropriate nearby correction fluxes are updated.

This is implemented by means of a second *transverse Riemann solver* that takes the fluctuation $\mathcal{A}^{+}\Delta Q_{i-1/2,j} = u_{i-1/2,j}^{+}(Q_{ij} - Q_{i-1,j})$, for example, and produces a down-going transverse fluctuation

$$
\mathcal{B}^{-}\mathcal{A}^{+}\Delta Q_{i-1/2,j} = v_{i,j-1/2}^{-}u_{i-1/2,j}^{+}(Q_{ij} - Q_{i-1,j}) \tag{20.25}
$$

and an up-going transverse fluctuation

$$
\mathcal{B}^{+}\mathcal{A}^{+}\Delta Q_{i-1/2,j} = v_{i,j+1/2}^{+}u_{i-1/2,j}^{+}(Q_{ij} - Q_{i-1,j}). \tag{20.26}
$$

These are used to update the correction fluxes $\tilde{G}_{i,j-1/2}$ and $\tilde{G}_{i,j+1/2}$, respectively, as in (20.24). The left-going fluctuation $\mathcal{A}^{-}\Delta Q_{i-1/2,j} = u_{i-1/2,j}^{-}(Q_{ij} - Q_{i-1,j})$ results in the transverse fluctuations

$$
\mathcal{B}^{-}\mathcal{A}^{-}\Delta Q_{i-1/2,j} = v_{i,j-1/2}^{-}u_{i-1/2,j}^{-}(Q_{ij} - Q_{i-1,j}) \tag{20.27}
$$

and

$$
\mathcal{B}^{+}\mathcal{A}^{-}\Delta Q_{i-1/2,j} = v_{i,j+1/2}^{+}u_{i-1/2,j}^{-}(Q_{ij} - Q_{i-1,j}). \tag{20.28}
$$

that are used to update $\tilde{G}_{i-1,j-1/2}$ and $\tilde{G}_{i-1,j+1/2}$, respectively. This approach generalizes quite naturally to hyperbolic systems of equations and is presented in Section 21.2. The form of the Riemann solvers required in CLAWPACK is discussed in more detail in Section 21.3.

A similar approach is taken when sweeping in the y-direction. After solving the Riemann problem at interface $(i, j - 1/2)$, we set $\mathcal{B}^{\pm} \Delta Q_{i,j-1/2}$ as in (20.23) and then update the nearby correction fluxes by

$$\tilde{F}_{i-1/2,j-1} := \tilde{F}_{i-1/2,j-1} - \frac{1}{2} \frac{\Delta t}{\Delta y} u_{i-1/2,j-1}^- v_{i,j-1/2}^- (Q_{ij} - Q_{i,j-1}),$$

$$\tilde{F}_{i+1/2,j-1} := \tilde{F}_{i+1/2,j-1} - \frac{1}{2} \frac{\Delta t}{\Delta y} u_{i+1/2,j-1}^+ v_{i,j-1/2}^- (Q_{ij} - Q_{i,j-1}),$$

$$(20.29)$$

$$\tilde{F}_{i-1/2,j} := \tilde{F}_{i-1/2,j} - \frac{1}{2} \frac{\Delta t}{\Delta y} u_{i-1/2,j}^- v_{i,j-1/2}^+ (Q_{ij} - Q_{i,j-1}),$$

$$\tilde{F}_{i+1/2,j} := \tilde{F}_{i+1/2,j} - \frac{1}{2} \frac{\Delta t}{\Delta y} u_{i+1/2,j}^+ v_{i,j-1/2}^+ (Q_{ij} - Q_{i,j-1}).$$

20.6 High-Resolution Correction Terms

As noted at the end of Section 20.2, the CTU method fails to be second-order accurate because it is based on first-order accurate approximations to the q_x and q_y terms in the Taylor series expansion, and is missing the q_{xx} and q_{yy} terms altogether. Both of these problems can be fixed by adding in additional correction fluxes based entirely on the one-dimensional theory. In each direction we wish to replace the first-order upwind approximation by a Lax–Wendroff approximation in that direction. (In two dimensions we also need the cross-derivative terms in the Taylor series expansion (20.2), but these have already been included via the transverse terms in the CTU method.) To improve the method we make the following updates to the correction fluxes already defined:

$$\tilde{F}_{i-1/2,j} := \tilde{F}_{i-1/2,j} + \frac{1}{2} |u_{i-1/2,j}| \left(1 - \frac{\Delta t}{\Delta x} |u_{i-1/2,j}|\right) \widetilde{\mathcal{W}}_{i-1/2,j},$$

$$(20.30)$$

$$\tilde{G}_{i,j-1/2} := \tilde{G}_{i,j-1/2} + \frac{1}{2} |v_{i,j-1/2}| \left(1 - \frac{\Delta t}{\Delta y} |v_{i,j-1/2}|\right) \widetilde{\mathcal{W}}_{i,j-1/2}.$$

These have exactly the same form as the one-dimensional correction flux (6.56). In the present case there is only a single wave $\mathcal{W}_{i-1/2,j} = Q_{ij} - Q_{i-1,j}$, and as usual $\widetilde{\mathcal{W}}_{i-1/2,j}$ represents a limited version of this wave, obtained by comparing this wave with the wave in the upwind direction. If $u_{i-1/2,j} > 0$ and $v_{i,j-1/2} < 0$, for example, then $\mathcal{W}_{i-1/2,j}$ is compared to $\mathcal{W}_{i-3/2,j}$ while $\mathcal{W}_{i,j-1/2}$ is compared to $\mathcal{W}_{i,j+1/2}$.

20.7 Relation to the Lax–Wendroff Method

Suppose we apply the method just derived to the constant-coefficient advection equation (20.1) with no limiters. We might suspect this should reduce to the Lax–Wendroff method for the advection equation, since this is what happens in this situation in one dimension. Indeed, the pure x- and y-derivatives will be approximated as in the Lax–Wendroff method, but the cross-derivative terms are not. Instead, combining the previous expressions yields a flux-differencing method of the form (19.10) with fluxes

$$F_{i-1/2,j} = \frac{1}{2}u(Q_{i-1,j} + Q_{ij}) - \frac{\Delta t}{2\,\Delta x}u^2(Q_{ij} - Q_{i-1,j})$$

$$- \frac{\Delta t}{2\,\Delta y}[u^- v^-(Q_{i,j+1} - Q_{ij}) + u^+ v^-(Q_{i-1,j+1} - Q_{i-1,j})$$

$$+ u^- v^+(Q_{ij} - Q_{i,j-1}) + u^+ v^+(Q_{i-1,j} - Q_{i-1,j-1})],$$

(20.31)

$$G_{i,j-1/2} = \frac{1}{2}v(Q_{i,j-1} + Q_{ij}) - \frac{\Delta t}{2\,\Delta y}v^2(Q_{ij} - Q_{i,j-1})$$

$$- \frac{\Delta t}{2\,\Delta x}[v^- u^-(Q_{i+1,j} - Q_{ij}) + v^+ u^-(Q_{i+1,j-1} - Q_{i,j-1})$$

$$+ v^- u^+(Q_{ij} - Q_{i-1,j}) + v^+ u^+(Q_{i,j-1} - Q_{i-1,j-1})].$$

Compare this with the Lax–Wendroff method (19.14) for the case $A = u$, $B = v$. Instead of approximating the cross-derivative term with a simple average of four nearby fluxes, as is done in (19.15) for Lax–Wendroff, the wave-propagation algorithm uses

$$uvq_y(x_{i-1/2}, y_j) \approx \frac{1}{\Delta y}[u^- v^-(Q_{i,j+1} - Q_{ij}) + u^+ v^-(Q_{i-1,j+1} - Q_{i-1,j})$$

$$+ u^- v^+(Q_{ij} - Q_{i,j-1}) + u^+ v^+(Q_{i-1,j} - Q_{i-1,j-1})].$$

(20.32)

In this constant-coefficient case only one of these four terms will be nonzero. Rather than averaging four nearby approximations to q_y, only one is used, taken from the upwind direction.

This leads to an improvement in the stability of the method. The Lax–Wendroff method is generally stable only if

$$\frac{\Delta t}{\Delta x}\sqrt{u^2 + v^2} \le 1,$$

(20.33)

whereas the wave-propagation version is stable up to Courant number 1 in the sense of (20.7). If $u = v$, then this is better by a factor of $\sqrt{2}$. In Chapter 21 we will see that similar improvements can be made in the Lax–Wendroff method for systems of equations, by generalizing (20.32) to systems using the matrices $A^\pm$ and $B^\pm$.

20.8 Divergence-Free Velocity Fields

Note that the conservative advection equation

$$q_t + (u(x, y)q)_x + (v(x, y)q)_y = 0$$

(20.34)

and the color equation

$$q_t + u(x, y)q_x + v(x, y)q_y = 0$$

(20.35)

are mathematically equivalent if the velocity field is divergence-free,

$$u_x(x, y) + v_y(x, y) = 0,$$

(20.36)

a case that arises in many applications. In this case (20.35) is often called the *advective form* of the equation, while (20.34) is the *conservative form*.

The constraint (20.36) holds, for example, for two-dimensional models of incompressible flow and more generally for any flow in which the net flux through the boundary of any arbitrary region Ω should be zero. For then we have

$$0 = \int_{\partial\Omega} \vec{n}(s) \cdot \vec{u}(s)\, ds = \iint_{\Omega} \vec{\nabla} \cdot \vec{u}(x, y)\, dx\, dy, \qquad (20.37)$$

where, as in Section 18.1, $\vec{n} \cdot \vec{u}$ is the normal velocity.

If $q(x, y, t)$ measures the density of a conserved tracer in a divergence-free flow, then we expect the integral of q to be conserved and generally hope to achieve this numerically even when discretizing the color equation. Note that numerical conservation would be guaranteed if the conservative equations (20.34) were used instead, but there are other potential disadvantages in using this form.

The methods developed in Sections 20.5 and 20.6 will be conservative on the color equation provided that the edge velocities satisfy

$$\frac{1}{\Delta x}\left(u_{i+1/2, j} - u_{i-1/2, j}\right) + \frac{1}{\Delta y}\left(v_{i, j+1/2} - v_{i, j-1/2}\right) = 0, \qquad (20.38)$$

as will be verified below. This is a natural discrete version of (20.36) across a grid cell. More fundamental is the integral interpretation of this condition. Since $\Delta y\, u_{i\pm1/2, j}$ and $\Delta x\, v_{i, j\pm1/2}$ are supposed to approximate integrals of the normal velocity along the four sides of the grid cell, we see that multiplying (20.38) by $\Delta x\, \Delta y$ gives a discrete form of the requirement (20.37). In particular, the discrete divergence-free condition (20.38) will be satisfied if we determine the edge velocities by computing exact averages of the normal velocities,

$$u_{i-1/2, j} = \frac{1}{\Delta y} \int_{y_{j-1/2}}^{y_{j+1/2}} u\left(x_{i-1/2}, y\right) dy,$$

$$v_{i, j-1/2} = -\frac{1}{\Delta x} \int_{x_{i-1/2}}^{x_{i+1/2}} v\left(x, y_{j-1/2}\right) dx. \qquad (20.39)$$

In this case (20.38) follows immediately as a special case of (20.37) for $\Omega = \mathcal{C}_{ij}$, the (i, j) grid cell.

Unfortunately the integrals in (20.39) may be hard to evaluate exactly. If the velocities are smooth, then simply evaluating the normal velocity at the midpoint of each edge will give values that are second-order accurate, but probably will not exactly satisfy the condition (20.38). In Section 20.8.1 we will see an approach to specifying these values using a stream function that is often quite simple to implement.

First we verify that the condition (20.38) does lead to discrete conservation when the color equation is solved using the method developed above. Recall that the wave-propagation method for the color equation can be written in the form (19.19) with the fluctuations (20.23) and correction fluxes $\tilde{F}$ and $\tilde{G}$ arising from both transverse propagation and high-resolution corrections, if these are included. Since these corrections are implemented by differencing the fluxes $\tilde{F}$ and $\tilde{G}$, they will maintain conservation. Thus we only need to worry about the fluctuations in verifying that the full method is conservative, and it is enough to consider the DCU method

$$Q_{ij}^{n+1} = Q_{ij} - \frac{\Delta t}{\Delta x}\left[u_{i-1/2,j}^{+}(Q_{ij} - Q_{i-1,j}) + u_{i+1/2,j}^{-}(Q_{i+1,j} - Q_{ij})\right]$$

$$- \frac{\Delta t}{\Delta y}\left[v_{i,j-1/2}^{+}(Q_{ij} - Q_{i,j-1}) + v_{i,j+1/2}^{-}(Q_{i,j+1} - Q_{ij})\right]. \tag{20.40}$$

Summing this equation over all i and j and rearranging the sum on the right to collect together all terms involving $Q_{ij} = Q_{ij}^n$, we obtain

$$\sum_{i,j} Q_{ij}^{n+1} = \sum_{i,j} Q_{ij}^n\left[1 + \frac{\Delta t}{\Delta x}\left(u_{i+1/2,j} - u_{i-1/2,j}\right) + \frac{\Delta t}{\Delta y}\left(v_{i,j+1/2} - v_{i,j-1/2}\right)\right]. \tag{20.41}$$

We see that the method is conservative provided that (20.38) is satisfied, in the sense that $\sum Q_{ij}^{n+1} = \sum Q_{ij}^n$ up to boundary fluxes.

20.8.1 Stream-Function Specification of Velocities

It is often easiest to define a two-dimensional divergence-free velocity field in terms of a *stream function* $\psi(x, y)$. Any continuous and piecewise differential scalar function $\psi(x, y)$ can be used to define a velocity field via

$$\begin{aligned} u(x, y) &= \psi_y(x, y), \\ v(x, y) &= -\psi_x(x, y). \end{aligned} \tag{20.42}$$

This velocity field will be divergence-free, since

$$u_x + v_y = \psi_{yx} - \psi_{xy} = 0.$$

Note that the velocity field $\vec{u} = (u, v)$ is orthogonal to $\vec{\nabla}\psi = (\psi_x, \psi_y)$, and hence contour lines of ψ in the x–y plane are *streamlines* of the flow, and are simply particle paths of the flow in the case we are considering, where ψ and hence $\vec{u}$ is independent of t.

If we know the stream function ψ for a velocity field, then it is easy to compute the edge velocities (20.39) by integrating (20.42), yielding the simple formulas

$$u_{i-1/2,j} = \frac{1}{\Delta y}\left[\psi\left(x_{i-1/2}, y_{j+1/2}\right) - \psi\left(x_{i-1/2}, y_{j-1/2}\right)\right],$$

$$v_{i,j-1/2} = -\frac{1}{\Delta x}\left[\psi\left(x_{i+1/2}, y_{j-1/2}\right) - \psi\left(x_{i-1/2}, y_{j-1/2}\right)\right]. \tag{20.43}$$

We simply difference ψ between two corners to determine the total flow normal to that edge. More generally, differencing ψ between any two points in the plane gives the total flow normal to the line between those points, a fact that is useful in defining edge velocities on more general curvilinear grids (see Section 23.5.2).

The expressions (20.43) might also be interpreted as centered approximations to the derivatives in (20.42), but since they are exactly equal to the integrals (20.39), the discrete divergence-free condition (20.38) will be satisfied. This fact can also be verified directly from the formulas (20.43), since differencing the edge velocities as in (20.38) using the expressions (20.43) leads to a complete cancellation of the four corner values of ψ.

20.8.2 Solid-Body Rotation

As an example with circular streamlines, consider the stream function

$$\psi(x, y) = x^2 + y^2. \tag{20.44}$$

The resulting velocity field

$$u(x, y) = 2y, \qquad v(x, y) = -2x \tag{20.45}$$

corresponds to solid-body rotation. This is a nice test problem for two-dimensional advection algorithms, since the true solution is easily found for any initial data. In particular, the solution at time $t = N\pi$ agrees with the initial data for any integer N, since the flow has then made N complete rotations.

Example 20.1. Figure 20.5 shows the results of solid-body rotation on a 80×80 grid with data $q = 0$ except in a square region where $q = 1$ and a circular region where q is *cone-shaped*, growing to a value 1 at the center:

$$q(x, y, 0) = \begin{cases} 1 & \text{if } 0.1 < x < 0.6 \quad \text{and} \quad -0.25 < y < 0.25, \\ 1 - r/0.35 & \text{if } r \equiv \sqrt{(x + 0.45)^2 + y^2} < 0.35, \\ 0 & \text{otherwise.} \end{cases} \tag{20.46}$$

The results are not perfect, of course. The discontinuity in q is smeared out, and the peak of the cone is chopped off. However, this unsplit high-resolution method (using the MC limiter) gives much better results than would be obtained with more classical methods. For example, Figure 20.6 shows what we would obtain with the first-order CTU method or the second-order method with no limiter. This figure also shows results obtained using dimensional splitting with one-dimensional high-resolution methods, which compare very well with the unsplit results of Figure 20.5.

20.9 Nonlinear Scalar Conservation Laws

The methods developed in Sections 20.5 and 20.6 extend easily to nonlinear scalar conservation laws $q_t + f(q)_x + g(q)_y = 0$. The one-dimensional Riemann problem normal to each cell edge is solved as in one dimension, resulting in waves, speeds, and fluctuations. In the x-direction we have

$$\mathcal{W}_{i-1/2,j} = Q_{ij} - Q_{i-1,j},$$

$$s_{i-1/2,j} = \begin{cases} [f(Q_{ij}) - f(Q_{i-1,j})]/(Q_{ij} - Q_{i-1,j}) & \text{if } Q_{i-1,j} \neq Q_{ij}, \\ f'(Q_{ij}) & \text{if } Q_{i-1,j} = Q_{ij}, \end{cases} \tag{20.47}$$

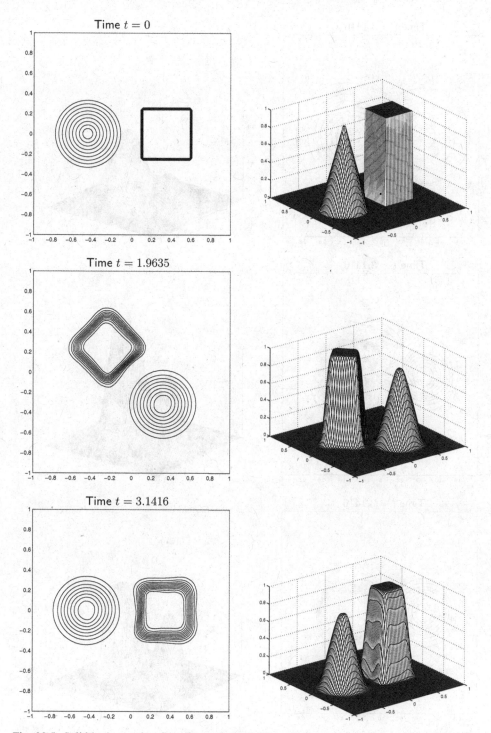

Fig. 20.5. Solid-body rotation from Example 20.1. The solution computed on an 80×80 grid is shown at three different times. Top: $t = 0$; middle: $t = 5\pi/8$; bottom: $t = \pi$. At each time the solution is shown as a contour plot (left) and a mesh plot (right). Contour lines are at the values $q = 0.05, 0.15, 0.25, \ldots, 0.95$. [claw/book/chap20/rotate]

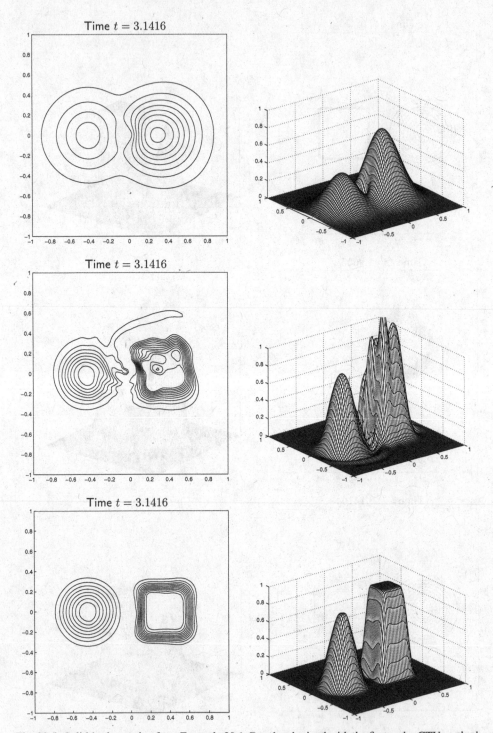

Fig. 20.6. Solid-body rotation from Example 20.1. Results obtained with the first-order CTU method (top), the second-order "Lax–Wendroff" method with no limiter (middle), and dimensional splitting with the high-resolution method in each direction (bottom). Each is shown only at the final time $t = \pi$. [claw/book/chap20/rotate]

and in the y-direction

$$\mathcal{W}_{i,j-1/2} = Q_{ij} - Q_{i,j-1},$$

$$s_{i,j-1/2} = \begin{cases} [g(Q_{ij}) - g(Q_{i,j-1})]/(Q_{ij} - Q_{i,j-1}) & \text{if } Q_{i,j-1} \neq Q_{ij}, \\ g'(Q_{ij}) & \text{if } Q_{i,j-1} = Q_{ij}. \end{cases} \tag{20.48}$$

The fluctuations can be defined simply as

$$\mathcal{A}^{\pm}\Delta Q_{i-1/2,j} = s_{i-1/2,j}^{\pm}\mathcal{W}_{i-1/2,j},$$
$$\mathcal{B}^{\pm}\Delta Q_{i,j-1/2} = s_{i,j-1/2}^{\pm}\mathcal{W}_{i,j-1/2}, \tag{20.49}$$

except in the case of transonic rarefactions, where these must be modified as in Section 12.3. The wave and speed can be used to compute second-order correction terms as in (20.30),

$$\tilde{F}_{i-1/2,j} := \tilde{F}_{i-1/2,j} + \frac{1}{2}|s_{i-1/2,j}|\left(1 - \frac{\Delta t}{\Delta x}|s_{i-1/2,j}|\right)\tilde{\mathcal{W}}_{i-1/2,j},$$

$$\tilde{G}_{i,j-1/2} := \tilde{G}_{i,j-1/2} + \frac{1}{2}|s_{i,j-1/2}|\left(1 - \frac{\Delta t}{\Delta y}|s_{i,j-1/2}|\right)\tilde{\mathcal{W}}_{i,j-1/2}. \tag{20.50}$$

The only subtle point is the determination of transverse velocities for the CTU terms corresponding to the correction fluxes developed in Section 20.5 for the advection equation. We no longer have velocities specified at nearby cell edges. Instead, the transverse velocity must be determined by approximating $g'(q)$ based on the data $Q_{i-1,j}$ and Q_{ij} (or other data nearby). One natural approach that generalizes quite easily to systems of equations, as we will see in the next chapter, is to choose the transverse velocity to be

$$\hat{v}_{i-1/2,j} = \begin{cases} [g(Q_{ij}) - g(Q_{i-1,j})]/(Q_{ij} - Q_{i-1,j}) & \text{if } Q_{i-1,j} \neq Q_{ij}, \\ g'(Q_{ij}) & \text{if } Q_{i-1,j} = Q_{ij}. \end{cases} \tag{20.51}$$

We then set

$$\tilde{G}_{i-1,j-1/2} := \tilde{G}_{i-1,j-1/2} - \frac{1}{2}\frac{\Delta t}{\Delta x}\hat{v}_{i-1/2,j}^{-}s_{i-1/2,j}^{-}(Q_{ij} - Q_{i-1,j}),$$

$$\tilde{G}_{i-1,j+1/2} := \tilde{G}_{i-1,j+1/2} - \frac{1}{2}\frac{\Delta t}{\Delta x}\hat{v}_{i-1/2,j}^{+}s_{i-1/2,j}^{-}(Q_{ij} - Q_{i-1,j}),$$

$$\tilde{G}_{i,j-1/2} := \tilde{G}_{i,j-1/2} - \frac{1}{2}\frac{\Delta t}{\Delta x}\hat{v}_{i-1/2,j}^{-}s_{i-1/2,j}^{+}(Q_{ij} - Q_{i-1,j}),$$

$$\tilde{G}_{i,j+1/2} := \tilde{G}_{i,j+1/2} - \frac{1}{2}\frac{\Delta t}{\Delta x}\hat{v}_{i-1/2,j}^{+}s_{i-1/2,j}^{+}(Q_{ij} - Q_{i-1,j}). \tag{20.52}$$

Note that this is somewhat different from (20.24) in that a single transverse velocity $\hat{v}_{i-1/2,j}$ is used based on the data $Q_{i-1,j}$ and Q_{ij} rather than the four edge values $v_{i,j\pm1/2}$ and $v_{i-1,j\pm1/2}$ appearing in (20.24). Similarly, a transverse velocity $\hat{u}_{i,j-1/2}$ is defined in the course of solving Riemann problems in the y-direction and is used to update nearby $\tilde{F}$-fluxes.

20.9.1 Burgers Equation

The inviscid Burgers equation (11.13) can be generalized to two space dimensions as

$$u_t + n^x \left(\frac{1}{2}u^2\right)_x + n^y \left(\frac{1}{2}u^2\right)_y = 0 \tag{20.53}$$

where $\vec{n} = (n^x, n^y)$ is an arbitrary unit vector. For $\vec{n} = (1, 0)$ or $(0, 1)$, this is just the one-dimensional Burgers equation in x or y respectively. More generally this can be reduced to a one-dimensional Burgers equation at angle $\theta = \tan^{-1}(n^y/n^x)$ to the x-axis. If we introduce new coordinates ξ in this direction and η in the orthogonal direction, then (20.53) reduces to

$$u_t + \left(\frac{1}{2}u^2\right)_\xi = 0. \tag{20.54}$$

Along each slice in the ξ-direction, we can solve this one-dimensional equation to obtain the solution $u(x, y, t)$ along this slice.

Figure 20.7 shows some sample results, using the same initial data (20.46) as for the solid-body rotation example shown in Figure 20.5. Two different angles, $\theta = 0$ and $\theta = \pi/4$, are illustrated. These were computed on a 300×300 grid using the high-resolution wave-propagation algorithm with the MC limiter.

20.10 Convergence

Convergence theory for multidimensional numerical methods is even more difficult than for one-dimensional problems. For scalar problems several results are known, however, and some of these are summarized in this section. See [156], [245] for more detailed discussions.

20.10.1 Convergence of Dimensional Splitting

In considering the convergence of dimensionally split methods, the first natural question is whether convergence occurs when the exact solution operator is used in each one-dimensional sweep (19.25) and (19.26). For nonlinear conservation laws this was shown by Crandall & Majda [95], even when the solution contains shock waves. Let $\mathcal{S}(t)$ represent the true solution operator of the full equation $q_t + f(q)_x + g(q)_y = 0$ over time t, so $\mathcal{S}(t)\overset{\circ}{q}$ is the (unique) entropy-satisfying solution at time t, $(\mathcal{S}(t)\overset{\circ}{q})(x, t) = q(x, y, t)$. Similarly, let $\mathcal{S}^x(t)$ and $\mathcal{S}^y(t)$ be the solution operators for the one-dimensional problems $q_t + f(q)_x = 0$ and $q_t + g(q)_y = 0$, respectively. Then convergence in the 1-norm of both the Godunov and Strang splitting is guaranteed by the following theorem.

Theorem 20.1 (Crandall & Majda [95]). *If the exact solution operator is used in each step of the fractional-step procedure, then the method converges to the weak solution of the two-dimensional scalar conservation law, i.e.,*

$$\|\mathcal{S}(T)\overset{\circ}{q} - [\mathcal{S}^y(\Delta t)\mathcal{S}^x(\Delta t)]^n \overset{\circ}{q}\|_1 \to 0$$

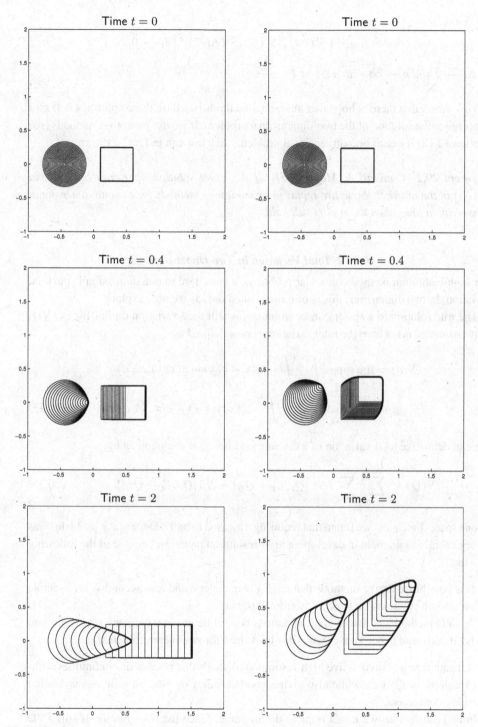

Fig. 20.7. Solution of the two-dimensional Burgers equation (20.53) at angle θ to the x-axis. Left column: $\theta = 0$. Right column: $\theta = \pi/4$. Contour lines are at $u = 0.05 : 0.05 : 0.95$. [claw/book/chap20/burgers]

and

$$\|\mathcal{S}(T)\overset{\circ}{q} - [\mathcal{S}^x(\Delta t/2)\mathcal{S}^y(\Delta t)\mathcal{S}^x(\Delta t/2)]^n \overset{\circ}{q}\|_1 \to 0$$

as $\Delta t \to 0$ and $n \to \infty$ with $n\,\Delta t = T$.

This shows that there is hope that numerical methods based on these splittings will also converge to the solution of the two-dimensional problem. If we use monotone methods (see Section 12.12) for each one-dimensional problem, then this can in fact be shown:

Theorem 20.2 (Crandall & Majda [95]). *If the exact solution operators $\mathcal{S}^x(\Delta t)$ and $\mathcal{S}^y(\Delta t)$ in the above theorem are replaced by monotone methods for the one-dimensional conservation laws, then the results still hold.*

20.10.2 Total Variation in Two Dimensions

For high-resolution methods on scalar problems, a basic tool in one dimension is the total variation. In two dimensions this is of more limited use, as we now explore.

The true solution to a scalar conservation law is still total variation diminishing (TVD) in two dimensions, where the total variation is now defined as

$$TV(q) = \limsup_{\epsilon \to 0} \frac{1}{\epsilon} \int_{-\infty}^{\infty} \int_{-\infty}^{\infty} |q(x + \epsilon, y) - q(x, y)|\, dx\, dy$$

$$+ \limsup_{\epsilon \to 0} \frac{1}{\epsilon} \int_{-\infty}^{\infty} \int_{-\infty}^{\infty} |q(x, y + \epsilon) - u(x, y)|\, dx\, dy. \qquad (20.55)$$

We can define the total variation of a discrete grid function analogously by

$$TV(Q) = \sum_{i=-\infty}^{\infty} \sum_{j=-\infty}^{\infty} [\Delta y\, |Q_{i+1,j} - Q_{ij}| + \Delta x\, |Q_{i,j+1} - Q_{ij}|]. \qquad (20.56)$$

In one space dimension we found that requiring a method to be TVD for scalar problems was a very useful requirement in developing high-resolution methods, because of the following two facts:

1. It is possible to derive methods that are TVD in general and also second-order accurate on smooth solutions (at least away from extrema).
2. A TVD method guarantees approximations that all lie in some compact set (as the grid is refined), and hence convergence can be proved for nonlinear problems.

This meant that we could derive high-resolution methods that resolve discontinuities without spurious oscillations while also giving good accuracy on smooth solutions and being provably convergent.

In two dimensions we might hope to do the same. Since the true solution is still TVD and we again wish to avoid spurious oscillations, we might try to require that the numerical method be TVD. With the variation defined as in (20.56), this would guarantee that all approximate solutions lie in an appropriate compact set and allow us to prove convergence, just as in one dimension. It is then natural to look for conditions similar to Harten's conditions of Theorem 6.1 that might guarantee the solution is TVD and also be loose enough to allow second-order accuracy. Unfortunately, an attempt to do this resulted instead in the following negative result.

Theorem 20.3 (Goodman & LeVeque [159]). *Except in certain trivial cases, any method that is TVD in two space dimensions is at most first-order accurate.*

This does not mean, however, that it is impossible to achieve high-resolution results in two dimensions. In fact the method described in Section 20.9 works very well in practice and gives results that are typically as sharp and accurate as one would expect based on one-dimensional experience.

Also, dimensional splitting often works very well when one-dimensional high-resolution methods are applied in each direction separately. Note that if the second-order Strang splitting is used, then this method is "second-order accurate" to the extent that the one-dimensional high-resolution method is. Moreover, in each sweep limiters are applied that keep the one-dimensional variation along that row of cells from increasing, and thus they do a good job of insuring that no spurious oscillations arise. The problem is that this is not enough to prove that the two-dimensional variation defined by (20.56) does not increase. A rather pathological example constructed in [159] shows that the two-dimensional variation may in fact increase. In practice this is not generally an issue, however, and Theorem 20.3 simply means that the TVD notion is not as useful for proving convergence of high-resolution methods in two dimensions as it is in one dimension.

A number of other techniques have instead been introduced for proving convergence of numerical methods for nonlinear scalar conservation laws in more than one dimension. One approach that has been quite successful is to use the theory of measure-valued solutions of conservation laws introduced by DiPerna [110]. This requires a weaker condition than uniformly bounded variation of the approximate solutions. Szepessy [430] used this to prove convergence of a finite element method, and Coquel & Le Floch [89], [90] applied a similar approach to finite volume methods, as did Kröner & Rokyta [247]. See [74], [75], [76], [246], [251], [473], [485] for some other work on convergence and error estimates for multidimensional scalar problems.

Exercises

20.1. Consider the advection equation $q_t + q_x + q_y = 0$ with initial data

$$Q_{ij}^0 = \begin{cases} 1 & \text{if } i + j \leq 0, \\ 0 & \text{if } i + j > 0 \end{cases}$$

for the Cauchy problem ($-\infty < i, j < \infty$). Suppose $\Delta t = \Delta x = \Delta y$, so that the Courant number is 1. Determine the solution Q_{ij}^1 and Q_{ij}^2 after 1 and 2 time steps when each of the following algorithms is applied:

(a) The DCU algorithm of Section 20.1. Observe that there is an exponentially growing oscillation and the method is unstable at this Courant number.

(b) The CTU algorithm of Section 20.2. Observe that the method is stable and produces the exact solution at this Courant number.

20.2. Consider the Cauchy problem for the constant-coefficient advection equation with $u, v > 0$. Suppose we apply dimensional splitting using the Godunov splitting defined by (19.27) and (19.28) with the one-dimensional first-order upwind algorithm in each sweep. Eliminate Q^* to determine how Q^{n+1} is defined in terms of Q^n, and show that this is equivalent to the CTU algorithm of Section 20.2. Would the same be true for the variable-coefficient advection equation?

20.3. Verify that (20.41) follows from (20.40).

20.4. Show that computing the discrete divergence (20.38) using the edge velocities (20.43) leads to a complete cancellation of the four corner values of ψ, verifying that use of the stream function gives a divergence-free discrete velocity field.

20.5. Use CLAWPACK to solve the solid-body rotation problem of Section 20.8.2 on the domain $[-1, 1] \times [0, 1]$ with $\overset{\circ}{q}(x, y) \equiv 0$ and boundary conditions

$$q(x, 0, t) = \begin{cases} 0 & \text{if } x < -0.8 \text{ or } -0.2 < x < 0, \\ 1 & \text{if } -0.8 \leq x \leq -0.2, \end{cases}$$

and extrapolation boundary conditions along the remainder of the boundary. Observe that the solution should reach a steady state after $t = \pi$ with the profile specified along the inflow boundary being reproduced at the outflow boundary. Compare how well the various methods implemented in CLAWPACK perform on this problem. You might also try other choices of the inflow boundary conditions, e.g., a smooth function of x.

20.6. In one space dimension the true solution to the variable-coefficient color equation $q_t + u(x)q_x = 0$ is TVD. Is the same true in two dimensions for the equation $q_t + u(x, y)q_x + v(x, y)q_y = 0$ with respect to the total variation (20.56)? Hint: Consider the special case $q_t + u(y)q_x = 0$.

21

Multidimensional Systems

In this chapter the high-resolution wave-propagation algorithms developed in Chapter 20 for scalar problems are extended to hyperbolic systems. We start with constant-coefficient linear systems, where the essential ingredients are most easily seen. A Riemann problem is first solved normal to each cell edge (a simple eigendecomposition in the linear case). The resulting waves are used to update cell averages on either side. The addition of correction terms using wave limiters (just as in one dimension) gives high-resolution terms modeling the pure x- and y-derivative terms in the Taylor series expansion (19.5). The cross-derivative terms are handled by simple extension of the corner-transport upwind (CTU) idea presented for the advection equation in Sections 20.2 through 20.5. In general this requires solving a second set of Riemann problems transverse to the interface. For a linear system this means performing a second eigendecomposition using the coefficient matrix in the transverse direction. Extending the methods to variable-coefficient or nonlinear systems is then easy, using ideas that are already familiar from one space dimension. The solutions (or approximate solutions) to the more general Riemann problems are used in place of the eigendecompositions, and the method is implemented in a wave-propagation form that applies very generally.

21.1 Constant-Coefficient Linear Systems

We again consider the constant-coefficient linear system $q_t + Aq_x + Bq_y = 0$ discussed in Chapter 19, where in particular the Lax–Wendroff and Godunov methods for this system were presented. The numerical fluxes for these two methods are given by (19.14) and (19.18) respectively. Our goal is to create a high-resolution version of the Lax–Wendroff method that incorporates upwinding and limiting. In addition to upwinding the first-order term (to rewrite it as the Godunov flux plus a high-resolution correction), we also wish to upwind the cross-derivative terms, as motivated by the CTU method for advection presented in Section 20.2. This can be accomplished by replacing the approximation (19.15) to ABq_y used in the Lax-Wendroff method with

$$ABq_y\left(x_{i-1/2}, y_j\right) \approx \frac{1}{\Delta y}[A^- B^-(Q_{i,j+1} - Q_{ij}) + A^+ B^-(Q_{i-1,j+1} - Q_{i-1,j})$$

$$+ A^- B^+(Q_{ij} - Q_{i,j-1}) + A^+ B^+(Q_{i-1,j} - Q_{i-1,j-1})], \quad (21.1)$$

469

generalizing the expression (20.28) for the CTU method on the advection equation. A similar approximation is used for the BAq_x term in the G-fluxes. Here $A^\pm$ and $B^\pm$ are defined as usual by (19.17). Note the important fact that

$$A^- B^- + A^+ B^- + A^- B^+ + A^+ B^+ = (A^- + A^+)(B^- + B^+) = AB \qquad (21.2)$$

for arbitrary matrices A and B, which ensures that we are still using a consistent approximation to the cross-derivative terms. Rather than multiplying each of the four jumps in Q by $\frac{1}{4}AB$ as in the Lax–Wendroff method, the product AB is split into four unequal pieces based on upwinding.

In practice we do not compute these matrices or the matrix products indicated above. Instead these terms in the flux are computed by solving Riemann problems and accumulating contributions to the fluxes based on the direction of wave propagation, exactly as was done for the advection equation in Section 20.2. This approach makes it easy to extend the method to variable-coefficient or nonlinear problems.

In spite of the fact that we do not actually compute the fluxes in terms of these matrices, it is useful to display them in this form for comparison with the Lax–Wendroff fluxes (19.14). We have

$$F_{i-1/2,j} = A^+ Q_{i-1,j} + A^- Q_{ij} + \frac{1}{2}\sum_{p=1}^{m} |\lambda^{xp}| \left(1 - \frac{\Delta t}{\Delta x}|\lambda^{xp}|\right) \widetilde{\mathcal{W}}^p_{i-1/2,j}$$

$$- \frac{\Delta t}{2\Delta y}[A^- B^-(Q_{i,j+1} - Q_{ij}) + A^+ B^-(Q_{i-1,j+1} - Q_{i-1,j})$$

$$+ A^- B^+(Q_{ij} - Q_{i,j-1}) + A^+ B^+(Q_{i-1,j} - Q_{i-1,j-1})], \qquad (21.3)$$

and a similar expression for $G_{i,j-1/2}$. Here $\mathcal{W}^p_{i-1/2,j} = \alpha^p_{i-1/2,j} r^{xp}$ is the pth wave in the Riemann solution, with $\alpha^p_{i-1/2,j} = (R^x)^{-1}(Q_{ij} - Q_{i-1,j})$. The limited version $\widetilde{\mathcal{W}}^p_{i-1/2,j}$ is obtained by comparing this wave with $\widetilde{\mathcal{W}}^p_{I-1/2,j}$, where

$$I = \begin{cases} i-1 & \text{if } \lambda^{xp} > 0, \\ i+1 & \text{if } \lambda^{xp} < 0. \end{cases}$$

If no limiter is used, then, as in one dimension,

$$\sum_{p=1}^{m} |\lambda^{xp}| \left(1 - \frac{\Delta t}{\Delta x}|\lambda^{xp}|\right) \mathcal{W}^p_{i-1/2,j} = |A| \left(I - \frac{\Delta t}{\Delta x}|A|\right)(Q_{ij} - Q_{i-1,j})$$

and

$$A^+ Q_{i-1,j} + A^- Q_{ij} + \frac{1}{2}\sum_{p=1}^{m} |\lambda^{xp}| \left(1 - \frac{\Delta t}{\Delta x}|\lambda^{xp}|\right) \mathcal{W}^p_{i-1/2,j}$$

$$= A^+ Q_{i-1,j} + A^- Q_{ij} + \frac{1}{2}|A| \left(I - \frac{\Delta t}{\Delta x}|A|\right)(Q_{ij} - Q_{i-1,j})$$

$$= \frac{1}{2}A(Q_{i-1,j} + Q_{ij}) - \frac{1}{2}\frac{\Delta t}{\Delta x}A^2(Q_{ij} - Q_{i-1,j}), \qquad (21.4)$$

which agrees with the corresponding terms in the Lax–Wendroff flux (19.14). On the other hand, if all waves are fully limited so that $\widetilde{\mathcal{W}}^p_{i-1/2,j} = 0$, then these terms in the flux reduce to the Godunov flux (19.18).

21.2 The Wave-Propagation Approach to Accumulating Fluxes

To implement the method described above, we take an approach very similar to what was done in Sections 20.5 and 20.6 for the advection equation. We use the form (19.19), which means we need fluctuations and correction fluxes. We present the algorithm for computing each of these in a framework that easily extends to nonlinear systems of equations by using approximate Riemann solvers:

1. Initialize $\tilde{F}_{i-1/2,j} = 0$ and $\tilde{G}_{i,j-1/2} = 0$ at each interface.
2. Sweep through the grid, solving each Riemann problem in x. At the interface between cells $\mathcal{C}_{i-1,j}$ and $\mathcal{C}_{ij}$ we use data $Q_{i-1,j}$ and Q_{ij} to compute waves $\mathcal{W}^p_{i-1/2,j}$ and speeds $s^p_{i-1/2,j}$. We also compute fluctuations $\mathcal{A}^-\Delta Q_{i-1/2,j}$ and $\mathcal{A}^+\Delta Q_{i-1/2,j}$ exactly as in one space dimension. For the constant-coefficient linear case the $\mathcal{W}$ and s will be eigenvectors and eigenvalues of A and we will have

$$\mathcal{A}^-\Delta Q_{i-1/2,j} = \sum_{p=1}^{m} \left(s^p_{i-1/2,j}\right)^- \mathcal{W}^p_{i-1/2,j} = A^-\Delta Q_{i-1/2,j},$$

$$\mathcal{A}^+\Delta Q_{i-1/2,j} = \sum_{p=1}^{m} \left(s^p_{i-1/2,j}\right)^+ \mathcal{W}^p_{i-1/2,j} = A^+\Delta Q_{i-1/2,j}. \tag{21.5}$$

3. The waves are limited to obtain $\widetilde{\mathcal{W}}^p_{i-1/2,j}$ and these are used to update the correction fluxes at this interface:

$$\tilde{F}_{i-1/2,j} := \tilde{F}_{i-1/2,j} + \frac{1}{2}\sum_{p=1}^{m} \left|s^p_{i-1/2,j}\right| \left(1 - \frac{\Delta t}{\Delta x}\left|s^p_{i-1/2,j}\right|\right)\widetilde{\mathcal{W}}^p_{i-1/2,j}. \tag{21.6}$$

4. The right-going fluctuation $\mathcal{A}^+\Delta Q_{i-1/2,j}$ is used to compute an up-going transverse fluctuation $\mathcal{B}^+\mathcal{A}^+\Delta Q_{i-1/2,j}$ and a down-going transverse fluctuation $\mathcal{B}^-\mathcal{A}^+\Delta Q_{i-1/2,j}$ by solving a *transverse Riemann problem*. We have seen an exmple of this for the advection equation $q_t + uq_x + vq_y = 0$ in Section 20.5, where $\mathcal{A}^+\Delta Q_{i-1/2,j} = u^+_{i-1/2,j}(Q_{ij} - Q_{i-1,j})$ and the transverse fluctuations are defined by (20.25) and (20.26),

$$\mathcal{B}^\pm\mathcal{A}^+\Delta Q_{i-1/2,j} = v^\pm_{i,j\pm1/2}u^+_{i-1/2,j}(Q_{ij} - Q_{i-1,j}). \tag{21.7}$$

In general the symbols $\mathcal{B}^+\mathcal{A}^+\Delta Q$ and $\mathcal{B}^-\mathcal{A}^+\Delta Q$ each represent a single m-vector obtained by some decomposition of the fluctuation $\mathcal{A}^+\Delta Q$. The notation is motivated by the linear case, in which case we want

$$\mathcal{B}^\pm\mathcal{A}^+\Delta Q_{i-1/2,j} = B^\pm A^+(Q_{ij} - Q_{i-1,j}). \tag{21.8}$$

In the linear system case these are computed by decomposing the fluctuation $\mathcal{A}^+\Delta Q_{i-1/2,j}$ into eigenvectors of B,

$$\mathcal{A}^+\Delta Q_{i-1/2,j} = \sum_{p=1}^{m} \beta^p r^{yp},$$

and then setting

$$\mathcal{B}^{\pm}\mathcal{A}^+\Delta Q_{i-1/2,j} = \sum_{p=1}^{m}(\lambda^{yp})^{\pm}\beta^p r^{yp}. \tag{21.9}$$

This *wave decomposition* of $\mathcal{A}^+\Delta Q_{i-1/2,j}$ can be viewed as solving a second Riemann problem in the transverse direction, even though it is not based on left and right states as we normally interpret a Riemann solver. The net contribution of all right-going waves is split up into up-going and down-going parts based on the eigenvectors corresponding to plane waves in the y-direction.

5. These fluctuations $\mathcal{B}^{\pm}\mathcal{A}^+\Delta Q_{i-1/2,j}$ are used to update the correction fluxes above and below cell $\mathcal{C}_{ij}$:

$$\tilde{G}_{i,j+1/2} := \tilde{G}_{i,j+1/2} - \frac{\Delta t}{2\,\Delta x}\mathcal{B}^+\mathcal{A}^+\Delta Q_{i-1/2,j},$$

$$\tilde{G}_{i,j-1/2} := \tilde{G}_{i,j-1/2} - \frac{\Delta t}{2\,\Delta x}\mathcal{B}^-\mathcal{A}^+\Delta Q_{i-1/2,j}. \tag{21.10}$$

6. In a similar manner, the left-going fluctuation $\mathcal{A}^-\Delta Q_{i-1/2,j}$ is split into transverse fluctuations $\mathcal{B}^{\pm}\mathcal{A}^-\Delta Q_{i-1/2,j}$, which are then used to update the fluxes above and below cell $\mathcal{C}_{i-1,j}$:

$$\tilde{G}_{i-1,j+1/2} := \tilde{G}_{i-1,j+1/2} - \frac{\Delta t}{2\,\Delta x}\mathcal{B}^+\mathcal{A}^-\Delta Q_{i-1/2,j},$$

$$\tilde{G}_{i-1,j-1/2} := \tilde{G}_{i-1,j-1/2} - \frac{\Delta t}{2\,\Delta x}\mathcal{B}^-\mathcal{A}^-\Delta Q_{i-1/2,j}. \tag{21.11}$$

Note that these updates to nearby $\tilde{G}$ fluxes are exactly analogous to what was done in (20.24) for the scalar advection equation.

7. Steps 2–6 are now repeated for each Riemann problem in y, at interfaces between cells $\mathcal{C}_{i,j-1}$ and $\mathcal{C}_{ij}$. The resulting waves $\mathcal{W}_{i,j-1/2}$ are limited by comparisons in the y-direction and used to update $\tilde{G}_{i,j-1/2}$. In solving these Riemann problems we also compute fluctuations $\mathcal{B}^{\pm}\Delta Q_{i,j-1/2}$, which are then split transversely into $\mathcal{A}^{\pm}\mathcal{B}^+\Delta Q_{i,j-1/2}$ and $\mathcal{A}^{\pm}\mathcal{B}^-\Delta Q_{i,j-1/2}$. These four transverse fluctuations are used to modify four nearby $\tilde{F}$ fluxes, as was done in (20.25) for the advection equation.

8. Finally, the updating formula (19.19) is applied to advance by time Δt,

$$Q_{ij}^{n+1} = Q_{ij} - \frac{\Delta t}{\Delta x}\left(\mathcal{A}^+\Delta Q_{i-1/2,j} + \mathcal{A}^-\Delta Q_{i+1/2,j}\right)$$

$$- \frac{\Delta t}{\Delta y}\left(\mathcal{B}^+\Delta Q_{i,j-1/2} + \mathcal{B}^-\Delta Q_{i,j+1/2}\right)$$

$$- \frac{\Delta t}{\Delta x}\left(\tilde{F}_{i+1/2,j} - \tilde{F}_{i-1/2,j}\right) - \frac{\Delta t}{\Delta y}\left(\tilde{G}_{i,j+1/2} - \tilde{G}_{i,j-1/2}\right). \tag{21.12}$$

21.3 CLAWPACK **Implementation**

This wave-propagation algorithm is implemented in CLAWPACK by assuming that the user has provided two Riemann solvers. One, called rpn2, solves the Riemann problem normal to any cell interface and is similar to the one-dimensional Riemann solver rp1. For dimensional splitting only this one Riemann solver is needed. The new Riemann solver needed for the unsplit algorithm is called rpt2 and solves Riemann problems of the sort just described in the transverse direction.

Each Riemann solver must be capable of solving the appropriate Riemann problem in either the x-direction or the y-direction, depending on which edge of the cell we are working on. The computations are organized by first taking sweeps in the x-direction along each row of the grid and then sweeps in the y-direction along each column. In each sweep a one-dimensional slice of the data is passed into the Riemann solver rpn2, so that ql and qr in this routine are exactly analogous to ql and qr in the one-dimensional Riemann solver rp1. A flag ixy is also passed in to indicate whether this is an x-slice (if ixy=1) or a y-slice (if ixy=2) of the data. The corresponding one-dimensional slice of the auxiliary array is also passed in.

The subroutine returns vectors of fluctuations (amdq, apdq) and waves and speeds (wave, s) obtained by solving the one-dimensional Riemann problem at each interface along the slice, as in rp1. In order to use the dimensional-splitting methods described in Section 19.5, only this Riemann solver rpn2 is required.

To perform the transverse Riemann solves required in the multidimensional wave-propagation algorithms, each of the fluctuations amdq ($= \mathcal{A}^-\Delta Q$) and apdq ($= \mathcal{A}^+\Delta Q$) must be passed into the transverse solver rpt2, so this routine is called twice. Within the subroutine this parameter is called asdq ($= \mathcal{A}^*\Delta Q$), and a parameter imp indicates which fluctuation this is (imp=1 if asdq $= \mathcal{A}^-\Delta Q$, and imp=2 if asdq $= \mathcal{A}^+\Delta Q$). For many problems the subroutine's action may be independent of the value of imp. In particular, for a constant-coefficient linear system the vector $\mathcal{A}^*\Delta Q$ is simply decomposed into eigenvectors of B in either case. For a variable-coefficient problem, however, the matrix B may be different to the left and right of the interface, and so the decomposition may depend on which direction the fluctuation is propagating.

The routine rpt2 returns bmasdq ($= \mathcal{B}^-\mathcal{A}^*\Delta Q$) and bpasdq ($= \mathcal{B}^+\mathcal{A}^*\Delta Q$), the splitting of this fluctuation in the transverse direction. These terms are used to update the correction fluxes $\tilde{G}$ nearby.

The same routine is used during the y-sweeps to split $\mathcal{B}^\pm\Delta Q$ into $\mathcal{A}^\pm\mathcal{B}^\pm\Delta Q$, so when ixy=2 it is important to realize that the input parameter asdq represents either $\mathcal{B}^-\Delta Q$ or $\mathcal{B}^+\Delta Q$, while the outputs bmasdq and bpasdq now represent $\mathcal{A}^\pm\mathcal{B}^*\Delta Q$. Similarly, in rpn2 the parameter asdq represents $\mathcal{A}^*\Delta Q$ in the x-sweeps, as described above, and represents $\mathcal{B}^*\Delta q$ in the y-sweeps. It may be easiest to simply remember that in these routines "a" always refers to the normal direction and "b" to the transverse direction.

For many systems of equations the Riemann solver for the x-sweeps and y-sweeps take a very similar form, especially if the equations are isotropic and have exactly the same form in any direction (as is the case for many physical systems such as acoustics, shallow water, or gas dynamics). Then the cases ixy=1 and ixy=2 may be distinguished only by which component of q represents the normal velocity and which is the transverse velocity. This is the reason that a single Riemann solver rpn2 with a flag ixy is required rather than separate Riemann solvers in the x- and y-directions.

The parameter values method(2) and method(3) determine what method is used in CLAWPACK. If method(2)=1 then the first-order updates are used but the second-order corrections based on limited waves are not used, i.e., step 3 in the above algorithm is skipped. If method(3)=0 then no transverse propagation is done, i.e., steps 4–6 are skipped. If method(3)=1 then these steps are performed. If method(3)=2 then an additional improvement is made to the algorithm, in which the correction terms from step 3 are also split in the transverse direction. This is accomplished by applying the transverse solver rpt2 to the vectors

$$\mathcal{A}^- \Delta Q_{i-1/2,j} + \sum_{p=1}^{m} |s_{i-1/2,j}^p| \left(1 - \frac{\Delta t}{\Delta x}|s_{i-1/2,j}^p|\right) \widetilde{\mathcal{W}}_{i-1/2,j}^p$$

and

$$\mathcal{A}^+ \Delta Q_{i-1/2,j} - \sum_{p=1}^{m} |s_{i-1/2,j}^p| \left(1 - \frac{\Delta t}{\Delta x}|s_{i-1/2,j}^p|\right) \widetilde{\mathcal{W}}_{i-1/2,j}^p$$

instead of to $\mathcal{A}^- \Delta Q_{i-1/2,j}$ and $\mathcal{A}^+ \Delta Q_{i-1/2,j}$. The rationale for this is explained in [283].

If method(3)<0 is specified, then dimensional splitting is used instead of this unsplit algorithm, as has already been described in Section 19.5.1.

See the CLAWPACK *User Guide* and sample programs for more description of the normal and transverse Riemann solvers. As an example we consider the acoustics equations in the next section. The CLAWPACK Riemann solver for this system may be found in [claw/book/chap21/acoustics].

21.4 Acoustics

As an example, consider the two-dimensional acoustics equations (18.24) with no background flow ($u_0 = v_0 = 0$). In this case the eigenvectors of A and B are given in (18.31) and (18.32), and the eigenvalues of each are $\lambda^{x1} = -c_0$, $\lambda^{x2} = 0$, and $\lambda^{x3} = c_0$.

The Riemann solver rpn2 must solve the Riemann problem $q_t + Aq_x = 0$ in the x-direction when ixy=1 or $q_t + Bq_y = 0$ in the y-direction when ixy=2.

If ixy=1, then we decompose $\Delta Q_{i-1/2,j} = Q_{ij} - Q_{i-1,j}$ as

$$\Delta Q = \alpha^1 r^{x1} + \alpha^2 r^{x2} + \alpha^3 r^{x3}, \tag{21.13}$$

where the eigenvectors are given in (18.31). For clarity the subscript $i - 1/2, j$ has been dropped from ΔQ and also from the coefficients α, which are different at each interface of course.

Solving the linear system (21.13) for α yields

$$\alpha^1 = \frac{-\Delta Q^1 + Z_0 \Delta Q^2}{2Z_0},$$

$$\alpha^2 = \Delta Q^3, \tag{21.14}$$

$$\alpha^3 = \frac{\Delta Q^1 + Z_0 \Delta Q^2}{2Z_0}.$$

The waves are then given by

$$\mathcal{W}^1 = \alpha^1 \begin{bmatrix} -Z_0 \\ 1 \\ 0 \end{bmatrix}, \qquad \mathcal{W}^2 = \alpha^2 \begin{bmatrix} 0 \\ 0 \\ 1 \end{bmatrix}, \qquad \mathcal{W}^3 = \alpha^3 \begin{bmatrix} Z_0 \\ 1 \\ 0 \end{bmatrix}, \qquad (21.15)$$

and the corresponding wave speeds are $s^1 = -c_0$, $s^2 = 0$, and $s^3 = c_0$.

The fluctuations are then given by

$$\mathcal{A}^- \Delta Q = s^1 \mathcal{W}^1, \qquad \mathcal{A}^+ \Delta Q = s^3 \mathcal{W}^3.$$

Note that the 2-wave makes no contribution to these fluctuations or to the second-order correction terms, where the contribution is also weighted by s^2, so the implementation can be made slightly more efficient by propagating only the 1-wave and 3-wave. (This is done in [claw/book/chap21/acoustics], where mwaves=2 is used.) If there were a nonzero background flow (u_0, v_0), then the wave speeds would be $s^1 = u_0 - c_0$, $s^2 = u_0$, and $s^3 = u_0 + c_0$. In this case it would be necessary to use all three waves and consider the sign of each s^p in computing the fluctuations.

If ixy=2 then we are sweeping in the y-direction. We then need to decompose $\Delta Q = \Delta Q_{i,j-1/2} = Q_{ij} - Q_{i,j-1}$ as

$$\Delta Q = \alpha^1 r^{y1} + \alpha^2 r^{y2} + \alpha^3 r^{y3},$$

where the eigenvectors are given in (18.32). This yields

$$\alpha^1 = \frac{-\Delta Q^1 + Z_0 \Delta Q^3}{2Z_0},$$

$$\alpha^2 = \Delta Q^2, \qquad (21.16)$$

$$\alpha^3 = \frac{\Delta Q^1 + Z_0 \Delta Q^3}{2Z_0}.$$

The waves are

$$\mathcal{W}^1 = \alpha^1 \begin{bmatrix} -Z_0 \\ 0 \\ 1 \end{bmatrix}, \qquad \mathcal{W}^2 = \alpha^2 \begin{bmatrix} 0 \\ 1 \\ 0 \end{bmatrix}, \qquad \mathcal{W}^3 = \alpha^3 \begin{bmatrix} Z_0 \\ 0 \\ 1 \end{bmatrix}, \qquad (21.17)$$

and the corresponding wave speeds are $s^1 = -c_0$, $s^2 = 0$, and $s^3 = c_0$. The fluctuations are then

$$\mathcal{B}^- \Delta Q = s^1 \mathcal{W}^1, \qquad \mathcal{B}^+ \Delta Q = s^3 \mathcal{W}^3,$$

but recall that in the CLAWPACK Riemann solver these are again denoted by amdq and apdq.

Note that these formulas are essentially the same for each value of ixy, except that the roles of the second and third components of Q are switched, depending on which velocity u or v is the velocity normal to the interface. In the CLAWPACK Riemann solver [claw/book/chap21/acoustics/rpn2ac.f], this is easily accomplished by using indices

$$
\text{mu} = \begin{cases} 2 & \text{if ixy} = 1, \\ 3 & \text{if ixy} = 2, \end{cases} \qquad \text{mv} = \begin{cases} 3 & \text{if ixy} = 1, \\ 2 & \text{if ixy} = 2 \end{cases} \tag{21.18}
$$

for the normal (mu) and transverse (mv) components of Q.

In fact, these formulas are easily generalized to solve a Riemann problem at any angle to the x and y axes. This is discussed in Section 23.6, where acoustics on a general quadrilateral grid is discussed.

The transverse Riemann solver rpt2 must take a fluctuation $\mathcal{A}^* \Delta Q$ and split it into $\mathcal{B}^- \mathcal{A}^* \Delta Q$ and $\mathcal{B}^+ \mathcal{A}^* \Delta Q$, or take a fluctuation $\mathcal{B}^* \Delta q$ and split it into $\mathcal{A}^- \mathcal{B}^* \Delta Q$ and $\mathcal{A}^+ \mathcal{B}^* \Delta Q$. This requires another splitting into eigenvectors of these matrices and multiplication by the corresponding eigenvalues. This is described in the next section for the more general problem of acoustics in heterogeneous media. For the constant-coefficient case, see also the simpler transverse Riemann solver [claw/book/chap21/acoustics/rpt2ac.f].

21.5 Acoustics in Heterogeneous Media

In Section 21.4 the normal and transverse Riemann solvers for acoustics in a homogeneous material were discussed. In this section we extend this to the case of a heterogeneous material, where the density $\rho(x, y)$ and the bulk modulus $K(x, y)$ may vary in space. The one-dimensional case has been studied in Section 9.6, and here we develop the two-dimensional generalization.

As in one dimension, the linear hyperbolic system can be solved in the nonconservative form

$$
q_t + A(x, y)q_x + B(x, y)q_y = 0, \tag{21.19}
$$

where

$$
q = \begin{bmatrix} p \\ u \\ v \end{bmatrix}, \quad A = \begin{bmatrix} 0 & K(x, y) & 0 \\ 1/\rho(x, y) & 0 & 0 \\ 0 & 0 & 0 \end{bmatrix}, \quad B = \begin{bmatrix} 0 & 0 & K(x, y) \\ 0 & 0 & 0 \\ 1/\rho(x, y) & 0 & 0 \end{bmatrix}.
$$

$$\tag{21.20}$$

This equation is not in conservation form, but can still be handled by high-resolution methods if we use the fluctuation form. The solution to the Riemann problem normal to each cell interface is computed exactly as in the one-dimensional case of Section 9.6. Let ρ_{ij} and c_{ij} be the density and sound speed in the (i, j) cell, where $c_{ij} = \sqrt{K_{ij}/\rho_{ij}}$. Then the Riemann problem at the $(i - 1/2, j)$ edge, for example, gives

$$
\mathcal{W}^1 = \alpha^1 \begin{bmatrix} -Z_{i-1,j} \\ 1 \\ 0 \end{bmatrix}, \quad \mathcal{W}^2 = \alpha^2 \begin{bmatrix} 0 \\ 0 \\ 1 \end{bmatrix}, \quad \mathcal{W}^3 = \alpha^3 \begin{bmatrix} Z_{ij} \\ 1 \\ 0 \end{bmatrix},
$$

where $Z_{ij} = \rho_{ij} c_{ij}$ is the impedance in the (i, j) cell, and

$$\alpha^1 = \frac{-\Delta Q^1 + Z_{ij} \Delta Q^2}{Z_{i-1,j} + Z_{ij}},$$

$$\alpha^2 = \Delta Q^3, \qquad (21.21)$$

$$\alpha^3 = \frac{\Delta Q^1 + Z_{i-1,j} \Delta Q^2}{Z_{i-1,j} + Z_{ij}}.$$

The subscript $i - 1/2, j$ has been omitted from α and ΔQ here for clarity.

These reduce to (21.14) in the case of constant impedance. As usual, the fluctuations $\mathcal{A}^- \Delta Q$ and $\mathcal{A}^+ \Delta Q$ are given by the product of the waves and wave speeds,

$$\mathcal{A}^- \Delta Q_{i-1/2,j} = s^1_{i-1/2,j} \mathcal{W}^1_{i-1/2,j}, \qquad \mathcal{A}^+ \Delta Q_{i-1/2,j} = s^3_{i-1/2,j} \mathcal{W}^3_{i-1/2,j},$$

where $s^1_{i-1/2,j} = -c_{i-1,j}$ and $s^3_{i-1/2,j} = c_{ij}$ are the appropriate wave speeds.

21.5.1 Transverse Propagation

The right-going fluctuation $\mathcal{A}^+ \Delta Q$ is split into up-going and down-going fluctuations $\mathcal{B}^+ \mathcal{A}^+ \Delta Q$ and $\mathcal{B}^- \mathcal{A}^+ \Delta Q$ that modify the fluxes $\tilde{G}_{i,j+1/2}$ and $\tilde{G}_{i,j-1/2}$ above and below the cell (i, j), respectively. To compute the down-going fluctuation $\mathcal{B}^- \mathcal{A}^+ \Delta Q$, for example, we need to decompose the vector $\mathcal{A}^+ \Delta Q$ into eigenvectors corresponding to up-going and down-going waves arising from the interface at $(i, j - 1/2)$,

$$\mathcal{A}^+ \Delta Q_{i-1/2,j} = \beta^1 \begin{bmatrix} -Z_{i,j-1} \\ 0 \\ 1 \end{bmatrix} + \beta^2 \begin{bmatrix} 0 \\ -1 \\ 0 \end{bmatrix} + \beta^3 \begin{bmatrix} Z_{ij} \\ 0 \\ 1 \end{bmatrix}, \qquad (21.22)$$

with speeds $-c_{i,j-1}, 0, c_{ij}$ respectively. Solving this linear system gives

$$\beta^1 = \frac{-\left(\mathcal{A}^+ \Delta Q_{i-1/2,j}\right)^1 + \left(\mathcal{A}^+ \Delta Q_{i-1/2,j}\right)^3 Z_{ij}}{Z_{i,j-1} + Z_{ij}}, \qquad (21.23)$$

where $(\mathcal{A}^+ \Delta Q_{i-1/2,j})^p$ is the pth element of the vector $\mathcal{A}^+ \Delta Q_{i-1/2,j}$. The coefficient β^1 is the only one needed to compute the down-going fluctuation, which is obtained by multiplying the first wave in (21.22) by the speed of this down-going wave,

$$\mathcal{B}^- \mathcal{A}^+ \Delta Q_{i-1/2,j} = -c_{i,j-1} \beta^1 \begin{bmatrix} -Z_{i,j-1} \\ 0 \\ 1 \end{bmatrix}. \qquad (21.24)$$

To compute the up-going fluctuation $\mathcal{B}^- \mathcal{A}^+ \Delta Q$, we instead decompose the vector $\mathcal{A}^+ \Delta Q$ into eigenvectors corresponding to up-going and down-going waves arising from the interface at $(i, j + 1/2)$,

$$\mathcal{A}^+\Delta Q_{i-1/2,j} = \beta^1 \begin{bmatrix} -Z_{ij} \\ 0 \\ 1 \end{bmatrix} + \beta^2 \begin{bmatrix} 0 \\ -1 \\ 0 \end{bmatrix} + \beta^3 \begin{bmatrix} Z_{i,j+1} \\ 0 \\ 1 \end{bmatrix}, \qquad (21.25)$$

with speeds $-c_{ij}, 0, c_{i,j+1}$ respectively. Solving this linear system gives

$$\beta^3 = \frac{\left(\mathcal{A}^+\Delta Q_{i-1/2,j}\right)^1 + \left(\mathcal{A}^+\Delta Q_{i-1/2,j}\right)^3 Z_{ij}}{Z_{ij} + Z_{i,j+1}}. \qquad (21.26)$$

The coefficient β^3 is the only one needed to compute the up-going fluctuation, which is obtained by multiplying the third wave in (21.25) by the speed of this up-going wave,

$$\mathcal{B}^+\mathcal{A}^+\Delta Q_{i-1/2,j} = c_{i,j+1}\beta^3 \begin{bmatrix} Z_{i,j+1} \\ 0 \\ 1 \end{bmatrix}. \qquad (21.27)$$

The left-going fluctuation $\mathcal{A}^-\Delta Q_{i-1/2,j}$ must similarly be decomposed in two different ways to compute the transverse fluctuations $\mathcal{B}^-\mathcal{A}^-\Delta Q_{i-1/2,j}$ and $\mathcal{B}^+\mathcal{A}^-\Delta Q_{i-1/2,j}$. The formulas are quite similar with i replaced by $i - 1$ in the sound speeds c and impedances Z above. See [claw/book/chap21/corner/rpt2acv.f].

Example 21.1. Figure 21.1(a) shows a heterogeneous medium with piecewise constant density and bulk modulus. Figure 21.2 shows the calculation of an acoustic pulse propagating in this medium. The pulse is initially a square rightward-propagating plane-wave pulse in pressure, as indicated in Figure 21.1(b). The pressure perturbation is nonzero only for

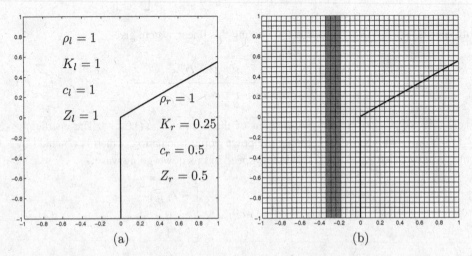

Fig. 21.1. (a) Piecewise-constant heterogeneous material for Example 21.1. (b) Illustration of how the interface cuts through a Cartesian grid, and the initial pressure pulse.

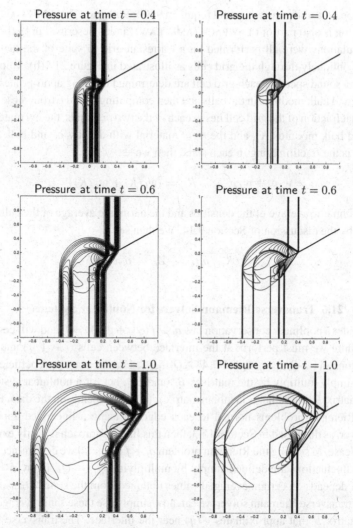

Fig. 21.2. Contours of pressure for an acoustic pulse propagating in the material shown in Figure 21.1, at three different times (from bottom to top). The calculation on the left was done on a 100×100 uniform Cartesian grid. [claw/book/chap21/corner] The calculation on the right is highly resolved using adaptive mesh refinement. [claw/book/chap21/corner/amr]

$-0.35 < x < -0.2$. When the pulse hits the interface, it is partially reflected and partially transmitted. As the pulse moves up the ramp portion of the interface, observe that the usual law of reflection is satisfied: the angle of incidence of the original pulse is equal to the angle of reflection. The transmitted wave is also oblique to the grid, at an angle determined by Snell's law that depends on the difference in wave speeds between the two media.

Two calculations are shown in Figure 21.2: a coarse-grid calculation on a 100×100 grid on the left, and a highly refined adaptive mesh refinement (AMR) calculation on the right. Fine grids are used only where required near the discontinuities in pressure. To obtain the same resolution on a uniform grid would require a 960×960 grid. The AMR code used for this computation is also part of CLAWPACK (AMRCLAW), and is described in [32].

These calculations were all performed on a Cartesian grid in spite of the fact that the interface cuts obliquely through the grid cells as illustrated in Figure 21.1(b). Values of the impedance and sound speed in each grid cell are determined by using appropriate averages of the density and bulk modulus in the cells and then computing Z and c from these. We first determine what fraction of the grid cell lies in each of the two materials, the *left* material with density ρ_l and bulk modulus K_l, and the *right* material with density ρ_r and bulk modulus K_r. If w_l, w_r is the fraction lying in each state, then we set

$$\rho_{ij} = w_l\rho_l + w_r\rho_r, \qquad K_{ij} = (w_l/K_l + w_r/K_r)^{-1}. \tag{21.28}$$

We use the arithmetic average of the densities and the harmonic average of the bulk moduli, as suggested by the discussion of Section 9.14. We then set

$$c_{ij} = \sqrt{K_{ij}/\rho_{ij}}, \qquad Z_{ij} = \rho_{ij}c_{ij}. \tag{21.29}$$

21.6 Transverse Riemann Solvers for Nonlinear Systems

We now consider a nonlinear conservation law $q_t + f(q)_x + g(q)_y = 0$ and will concentrate on the procedure we must perform at the interface between cells $(i - 1, j)$ and (i, j) to split fluctuations $\mathcal{A}^{\pm}\Delta Q_{i-1/2,j}$ into $\mathcal{B}^{\pm}\mathcal{A}^{\pm}\Delta Q_{i-1/2,j}$. For the constant-coefficient linear problem we simply multiply by the matrices B^- and B^+, but for a nonlinear system there is no single matrix B, but rather a Jacobian matrix $g'(q)$ that depends on the data. However, if we solve Riemann problems normal to each edge by using a linearized approximate Riemann solver, as discussed in Section 15.3, then this linear approach is easily extended to the nonlinear case. In solving the Riemann problem $q_t + f(q)_x = 0$ we determined a matrix $\hat{A}$ so that the fluctuations are defined simply by multiplying $Q_{ij} - Q_{i-1,j}$ by $\hat{A}^-$ and $\hat{A}^+$. The matrix $\hat{A}$ depends on certain averaged values obtained from the states $Q_{i-1,j}$ and Q_{ij}. To define the transverse Riemann solver we can now simply use these same averaged values to define a matrix $\hat{B}$ that approximates $g'(q)$ near the interface. The transverse Riemann solver then returns

$$\begin{aligned} \mathcal{B}^-\mathcal{A}^*\Delta Q &= \hat{B}^-(\mathcal{A}^*\Delta Q), \\ \mathcal{B}^+\mathcal{A}^*\Delta Q &= \hat{B}^+(\mathcal{A}^*\Delta Q). \end{aligned} \tag{21.30}$$

This is illustrated in the next section for the shallow water equations. Examples for the Euler equations can be found on the webpage [claw/book/chap21/euler].

21.7 Shallow Water Equations

The numerical methods developed above for multidimensional acoustics can be extended easily to nonlinear systems such as the shallow water equations. For the dimensional-splitting method we only need a normal Riemann solver (rpn2 in CLAWPACK). This is essentially identical to the one-dimensional Riemann problem for the shallow water equations with a passive tracer as discussed in Section 13.12.1. In practice an approximate

Riemann solver is typically used, e.g., the Roe solver developed in Section 15.3.3. This is very easily extended to the two-dimensional case. In the x-direction, for example, the velocity v does not affect the nonlinear waves, and so the Roe averages $\bar{h}$ and $\hat{u}$ are computed as in (15.32) and (15.35) respectively,

$$\bar{h} = \frac{1}{2}(h_l + h_r), \qquad \hat{u} = \frac{\sqrt{h_l}\, u_l + \sqrt{h_r}\, u_r}{\sqrt{h_l} + \sqrt{h_r}}, \qquad (21.31)$$

where $h_l = h_{i-1,j}$, $h_r = h_{ij}$, etc. We also need an average value $\hat{v}$, discussed below. The Roe matrix is then

$$\hat{A} = \begin{bmatrix} 0 & 1 & 0 \\ -\hat{u}^2 + g\bar{h} & 2\hat{u} & 0 \\ -\hat{u}\hat{v} & \hat{v} & \hat{u} \end{bmatrix}, \qquad (21.32)$$

with eigenvalues and eigenvectors

$$\hat{\lambda}^{x1} = \hat{u} - \hat{c}, \qquad \hat{\lambda}^{x2} = \hat{u}, \qquad \hat{\lambda}^{x3} = \hat{u} + \hat{c},$$

$$\hat{r}^{x1} = \begin{bmatrix} 1 \\ \hat{u} - \hat{c} \\ \hat{v} \end{bmatrix}, \qquad \hat{r}^{x2} = \begin{bmatrix} 0 \\ 0 \\ 1 \end{bmatrix}, \qquad \hat{r}^{x3} = \begin{bmatrix} 1 \\ \hat{u} + \hat{c} \\ \hat{v} \end{bmatrix}, \qquad (21.33)$$

where $\hat{c} = \sqrt{g\bar{h}}$. Entropy fixes and limiters are applied just as in one dimension.

The average velocity $\hat{v}$ is given by the Roe average

$$\hat{v} = \frac{\sqrt{h_l}\, v_l + \sqrt{h_r}\, v_r}{\sqrt{h_l} + \sqrt{h_r}} \qquad (21.34)$$

With this choice the resulting matrix $\hat{A}$ in (21.32) satisfies the usual requirement

$$\hat{A}(Q_r - Q_l) = f(Q_r) - f(Q_l) \qquad (21.35)$$

for the Roe matrix. In the first printing of this book it was incorrectly stated that with this choice of $\hat{v}$ the equation (21.35) is not satisfied, because it is not obvious that the third equation of this system holds,

$$-\hat{u}\hat{v}\delta^1 + \hat{v}\delta^2 + \hat{u}\delta^3 = h_r u_r v_r - h_l u_l v_l, \qquad (21.36)$$

where $\delta = Q_r - Q_l$, but in fact it does. Note that solving (21.36) for $\hat{v}$ gives

$$\hat{v} = \frac{(h_r u_r v_r - h_l u_l v_l) - \hat{u}(h_r v_r - h_l v_l)}{(h_r u_r - h_l u_l) - \hat{u}(h_r - h_l)}$$

$$= \frac{a_l v_l + a_r v_r}{a_l + a_r},$$

where

$$a_l = h_l(\hat{u} - u_l), \qquad a_r = h_r(u_r - \hat{u}).$$

This appears to be different weighting of v_l and v_r than what is given by (21.34), but it can be shown that these are actually equivalent. The formulation (21.34) is preferable since it does not require dividing by a quantity that will be zero when $u_l = u_r$.

To use the method developed in Section 21.2, we must also provide a transverse Riemann solver, similar to the one developed in Section 18.4 for the two-dimensional acoustics equations. For the nonlinear shallow water equations we wish to take the right-going flux $\mathcal{A}^+ \Delta Q_{i-1/2,j}$, for example, and split it into an up-going part $\mathcal{B}^+ \mathcal{A}^+ \Delta Q_{i-1/2,j}$ and a down-going part $\mathcal{B}^- \mathcal{A}^+ \Delta Q_{i-1/2,j}$. As developed in Sections 21.1 through 21.6, the basic idea is to split the vector $\mathcal{A}^+ \Delta Q_{i-1/2,j}$ into eigenvectors of a matrix B approximating $g'(q)$. For this nonlinear system the Jacobian varies with q. However, we can use the Roe-averaged quantities $\bar{h}, \hat{u}$, and $\hat{v}$ to define a natural approximate matrix $\hat{B}$ to use for this decomposition. The eigenvalues and eigenvectors are as in (18.42) but with (h, u, v) replaced by the Roe averages. The formula (21.30) is then used to define the transverse fluctuations.

Normal and transverse Riemann solvers for the shallow water equations can be found in the directory [claw/book/chap21/radialdam]. This approach has been used in the example shown below.

21.7.1 A Radial Dam-Break Problem

Figure 21.3 shows a radial dam-break problem for the two-dimensional shallow water equations. The depth is initially $h = 2$ inside a circular dam and $h = 1$ outside. When the dam is removed, a shock wave travels radially outwards while a rarefaction wave moves inwards. This is similar to the structure of the one-dimensional dam-break Riemann problem. The fluid itself is moving outwards, and is accelerated either abruptly through the shock wave or smoothly through the rarefaction wave. Figure 21.4 shows the time evolution of both the depth and the radial momentum as a function of r, the distance from the origin, over a longer time period. This was computed by solving the one-dimensional equations

$$h_t + (hU)_r = -\frac{hU}{r},$$

$$(hU)_t + \left(hU^2 + \frac{1}{2}gh^2\right)_r = -\frac{hU^2}{r},$$

(21.37)

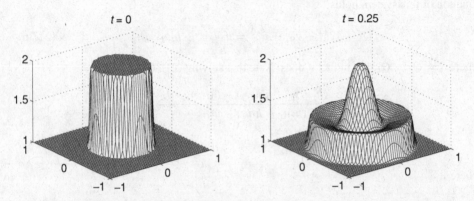

Fig. 21.3. Depth of water h for a radial dam-break problem, as computed on a 50×50 grid.

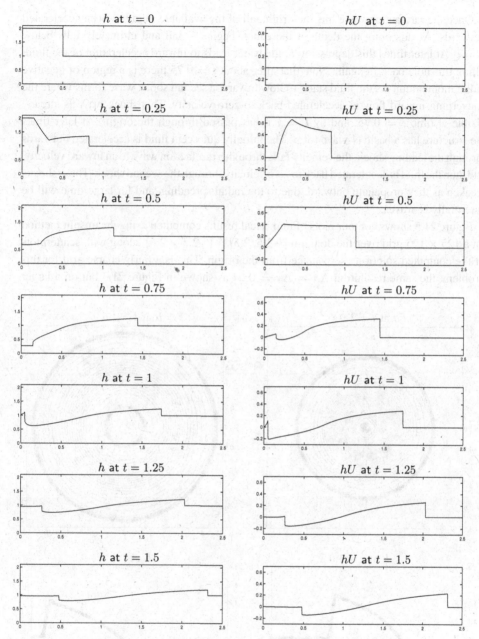

Fig. 21.4. Solution to the radial dam-break problem as a function of r. Left: depth h. Right: radial momentum hU. [claw/book/chap21/radialdam/1drad]

where $U(r, t)$ is the radial velocity. These follow from (18.52) with the hydrostatic pressure (18.39). Note that at time $t = 0.25$ the depth and momentum are no longer constant between the shock and rarefaction wave as in the one-dimensional Riemann problem. This is due to the source terms in (21.37), which physically arise from the fact that the fluid is spreading out and it is impossible to have constant depth and constant nonzero radial velocity.

Once the rarefaction wave hits the origin, all of the available fluid has been accelerated outwards. At this point the depth at the center begins to fall and ultimately falls below $h = 1$. At later times this depression in the water leads to inward acceleration of the fluid, filling this hole back in again. Note that after about $t = 0.75$ there is a region of negative radial momentum. As the fluid starts to flow inward, a second shock wave forms where the converging inward flow is decelerated back to zero velocity. This shock wave is already visible at time $t = 0.75$, and by $t = 1$ it has passed through the origin. At later times the structure has a basic N-wave form. The initially quiescent fluid is accelerated outwards through the leading shock, the velocity falls through a rarefaction wave to an inward velocity, and then the fluid is decelerated back to zero velocity through the second shock. These shocks weaken as they propagate outward, due to the radial spreading, and for large time will be essentially N-waves.

Figure 21.5 shows contour plots of numerical results computed using the unsplit method on a 125×125 grid over the domaim $[-2.5, 2.5] \times [-2.5 \times 2.5]$, along with scatterplots of the computed solution vs. distance from the origin. This is a fairly coarse grid for this problem, the same resolution $\Delta x = \Delta y = 0.04$ as shown in Figure 21.3 but on a larger

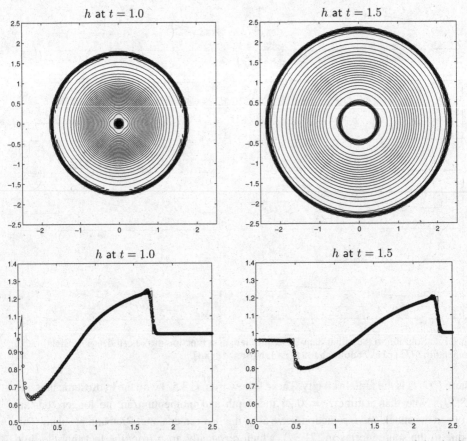

Fig. 21.5. Computed solutions for the radial dam-break problem. Top: contour plots of the depth at two times. Bottom: scatterplots of the depth vs. distance from the origin at the same two times. Contour levels are $0.61 : 0.02 : 1.31$. [claw/book/chap21/radialdam]

domain. The depth is shown at two different times. At $t = 1.0$ the shock near the origin cannot be resolved on this grid. The structure looks correct elsewhere, however. At $t = 1.5$ the basic structure is captured well everywhere. Note in particular that the depth near the origin stabilizes at a value $h \approx 0.96$ that is well captured also in the two-dimensional results.

21.8 Boundary Conditions

Boundary conditions in two (or three) space dimensions can be handled in much the same way as in one dimension. The grid on the desired computational domain consists of *interior cells* that we will label by $i = 1, 2, \ldots, m_x$ and $j = 1, 2, \ldots, m_y$. This grid is extended by introducing a set of *ghost cells* on all sides, for $i = 1 - m_{BC}, \ldots, 0$ and $i = m_x + 1, \ldots, m_x + m_{BC}$, and for $j = 1 - m_{BC}, \ldots, 0$ and $j = m_y + 1, \ldots, m_y + m_{BC}$. Figure 21.6 shows a portion of such an extended grid for the case $m_{BC} = 2$. At the beginning of each time step the ghost-cell values are filled, based on data in the interior cells and the given boundary conditions, and then the algorithm of choice is applied over the extended domain. How many rows of ghost cells are needed depends on the stencil of the algorithm. The high-resolution algorithms presented in Chapters 20 and 21 generally require two rows of ghost cells as shown in the figure. This allows us to solve a Riemann problem at the the boundary of the original domain and also one additional Riemann problem outside the domain. The waves from this Riemann problem do not enter the domain and so do not affect the solution directly, but are used to limit the waves arising from the original boundary. In the discussion below we assume $m_{BC} = 2$, but it should be clear how to extend each of the boundary conditions for larger values of m_{BC}.

The manner in which ghost cells are filled depends on the nature of the given boundary conditions. Periodic boundary conditions, for example, are easy to apply (as in the one-dimensional case of Section 7.1) by simply copying data from the opposite side of the grid. Below we will consider other standard cases, extending the approaches that were developed

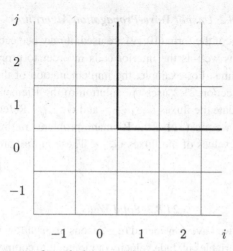

Fig. 21.6. The lower left corner of a typical computational domain is shown as the dark line. Interior grid cells are labeled with $i = 1, 2, \ldots$ and $j = 1, 2, \ldots$. The domain is extended with $m_{BC} = 2$ rows of ghost cells on each side.

in Chapter 7 to multidimensional problems. In particular, we consider solid-wall boundary conditions and the use of extrapolation at outflow boundaries. All of these are implemented in the default CLAWPACK routine [claw/clawpack/2d/lib/bc2.f].

21.8.1 Dimensional Splitting

We will primarily concentrate on the techniques needed when unsplit methods of the form (19.10) or (19.19) are used, in which the data Q^n is used to compute all fluctuations and fluxes in both the x- and y-directions. If a dimensional-splitting method is used, as described in Section 19.5, then additional issues arise. If we sweep first in the x-direction to obtain Q^* from Q^n, and then sweep in the y-direction to obtain Q^{n+1} from Q^* (using the Godunov splitting), we will need to specify boundary conditions for Q^* that may differ from the physical boundary conditions originally given for q. Some discussion of this point was given in Section 17.9 in the context of fractional-step methods for source terms. For dimensional splitting, appropriate ghost-cell values for Q^* can often be obtained by first extending Q^n to all ghost cells and then sweeping over the rows of ghost cells along the top and bottom of the grid ($j = -1, 0$ and $j = m_y + 1, m_y + 2$) as well as over the rows of interior cells ($j = 1, 2, \ldots, m_y$). This modifies the ghost-cell values by solving the same one-dimensional equation as in the interior, and gives Q^*-values in these cells that can now be used as the ghost-cell values in the y-sweeps.

Note that if we wish to use the Strang splitting (19.29), then the ghost-cell values along the left and right edges ($i = -1, 0, m_x + 1, m_x + 2$) must also be updated to Q^* so that we can apply y-sweeps to these rows of cells as well as in the interior. Then we will have Q^{**}-values in these ghost cells, which are needed in taking the final x-sweep to obtain Q^{n+1} from Q^{**}. To do this requires that we have twice as many ghost cells (at least along the left and right boundaries), so that Q^* can be obtained in the ones where Q^{**} is ultimately needed.

21.8.2 Unsplit Wave-Propagation Algorithms

Even if unsplit algorithms of the form (19.19) are used, it may be necessary to sweep over the rows of ghost cells as well as the interior cells in order to properly implement the multidimensional algorithms. For example, the implementation of the wave-propagation algorithm discussed in Section 21.2 uses the solution to the Riemann problem between cells $\mathcal{C}_{i-1,j}$ and $\mathcal{C}_{ij}$ to update the fluxes $\tilde{G}_{i-1,j+1/2}$ and $\tilde{G}_{i,j+1/2}$. Referring to Figure 21.6, we see that when $j = 0$ we must solve the Riemann problems in this row of ghost cells in order to obtain proper values of the fluxes $\tilde{G}_{i,1/2}$. These in turn are used to update the interior values Q_{i1}.

21.8.3 Solid Walls

In many of the problems we have considered (e.g., acoustics, shallow water equations, gas dynamics), the solution variables include velocity or momentum components in each spatial dimension. A common boundary condition is that the velocity normal to a wall should be zero, corresponding to a solid wall that fluid cannot pass through. As in one dimension, we can implement this boundary condition by a suitable extension of the solution to the ghost cells. Consider the left edge of the domain, for example, where the x-component of the velocity should vanish. We first extrapolate all components in a symmetric manner, setting

$$Q_{0,j} = Q_{1,j}, \qquad Q_{-1,j} = Q_{2,j} \qquad \text{for } j = 1, 2, \ldots, m_y. \qquad (21.38)$$

We then negate the component of Q_{ij} (for $i = -1, 0$) that corresponds to the x-component of velocity or momentum. We perform a similar extension at the right boundary. At the bottom boundary we extrapolate

$$Q_{i,0} = Q_{i,1}, \qquad Q_{i,-1} = Q_{i,2} \qquad \text{for } i = -1, 0, 1, \ldots, m_x + 1, m_x + 2, \qquad (21.39)$$

and then negate the y-component of the velocity or momentum in these ghost cells. Note that we apply this procedure in the rows of ghost cells (e.g., $i = -1, 0$) as well as in the interior cells in order to insure that all the ghost cells shown in Figure 21.6 are filled, including the four corner cells where i and j both have values 0 or -1. One should always insure that these corner cells are properly filled, especially if combinations of different boundary conditions are used at the two adjacent sides.

Note that in the procedure just outlined, the tangential component of velocity is simply extrapolated from the interior. For the problems we have considered (acoustics, shallow water, gas dynamics), it really doesn't matter what value the tangential velocity has in the ghost cells, since any jump in this quantity will propagate with zero normal velocity and will not affect the solution in the interior cells. This is due to the symmetry of the extrapolated data, which results in the contact discontinuity in the Riemann solution having zero velocity, properly mimicking a stationary wall. Hence any tangential velocity is allowed by these boundary conditions.

21.8.4 No-Slip Boundary Condition

In many fluid dynamics problems there is another physical boundary condition we might wish to impose at a solid wall. The *no-slip boundary condition* states that the tangential velocity should also vanish at the wall along with the normal velocity, so that fluid adjacent to the wall is stationary. This is expected due to friction between the wall and fluid molecules, which keeps molecules from slipping freely along the wall. However, this friction is present only in viscous fluids, and hyperbolic equations only model inviscid fluids, so we are not able to impose the no-slip condition in these models. If fluid viscosity is introduced, we obtain a parabolic equation (e.g., the Navier–Stokes equations instead of the inviscid Euler equations) that allows (in fact, requires) more boundary conditions to be specified.

If the physical viscosity of the fluid is very small relative to the typical fluid velocity away from the wall (i.e., if the *Reynolds number* is large), then there will often be a thin *boundary layer* adjacent to the wall in which the tangential velocity rapidly approaches the zero velocity of the wall. The thickness of this layer depends on the magnitude of the viscosity ϵ and often vanishes as $\epsilon \to 0$. As in the case of shock waves, the inviscid hyperbolic equation attempts to model the $\epsilon = 0$ limit.

In some applications this is a suitable approximation. For example, in many aerodynamics problems the thickness of the physical boundary layer on the surface of a body is much smaller than a computational cell. For this reason the inviscid Euler equations are often used rather than the more expensive Navier–Stokes equations. However, caution must be used, since in some problems the viscous effects at the boundary do have a substantial influence on the global solution, even when the viscosity is very small. This is particularly true if the geometry is such that the boundary layer separates from the wall at some point (as must happen at the trailing edge of a wing, for example). Then the vorticity generated by a no-slip boundary will move away from the wall and perhaps lead to large-scale turbulence that persists even when ϵ is extremely small, but would not be seen in an ideal inviscid fluid. The numerical viscosity that is inherent in any "inviscid" algorithm can lead to similar effects, but may mimic flow at the wrong Reynolds number. A full discussion of these issues is beyond the scope of this book.

21.8.5 *Extrapolation and Absorbing Boundary Conditions*

For many problems we must use a computational domain that is smaller than the physical domain, particularly if we must cut off an essentially infinite domain at some point to obtain a finite computational domain, as already discussed in Section 7.3.1. We then wish to impose boundary conditions that allow us to compute on this smaller domain and obtain results that agree well with what would be computed on a larger domain. If the computational domain is large enough for the problem of interest, then we expect that there will only be outgoing waves at the boundary of this domain. There should not be substantial incoming waves unless they have a known form (as from some known external source) that can be imposed as part of the boundary conditions, as was done in Section 7.3.2 in one dimension. Here we will assume there should be no incoming waves, in which case our goal is to impose boundary conditions on the computational domain that are *nonreflecting*, or *absorbing*, and that allow any outgoing waves to disappear without generating spurious incoming waves.

It may be that the outgoing waves should interact outside the computational domain in such a way that incoming waves are generated, which should appear at the boundary at a later time. In general we cannot hope to model such processes via the boundary conditions, and this would be an indication that our computational domain is simply not large enough to capture the full problem.

In one space dimension, we saw in Section 7.3.1 that quite effective absorbing boundary conditions can be obtained simply by using zero-order extrapolation. This idea can be extended easily to more dimensions as well. For example, along the left and bottom edge we would set

$$Q_{0j} = Q_{1j}, \quad Q_{-1,j} = Q_{1j} \qquad \text{for } j = 1, 2, \ldots, m_y, \tag{21.40}$$

and then

$$Q_{i0} = Q_{i1}, \quad Q_{i,-1} = Q_{i1} \qquad \text{for } i = -1, 0, \ldots, m_x + 2. \tag{21.41}$$

Note that we have filled the corner ghost cells in Figure 21.6 as well as the edge ghost cells. The value obtained in all four of the corner cells is Q_{11}. The same value would be obtained if we reversed the order above and first extrapolated in y and then in x.

This simple approach to absorbing boundary conditions often works very well in multi-dimensional problems. As in one dimension, its success rests on the fact that the Riemann problem at the edge of the computational domain has the same data on either side, resulting in zero-strength waves and in particular no incoming waves.

While surprisingly effective, this approach is unfortunately not quite as effective as in one dimension, except in the special case of plane waves exiting normal to the boundary of the domain. An outgoing wave at some angle to the grid can be viewed as a superposition of various waves moving in the x- and y-directions. Some of these waves should be incoming from the perspective of the computational boundary. This is clear from the fact that solving a Riemann problem in x or y in the midst of such an oblique wave will result in nontrivial waves moving with both positive and negative speeds. Using zero-order extrapolation will result in the loss of some of this information. The fact that there are no incoming waves normal to the boundary results in an incorrect representation of the outgoing oblique wave, which appears computationally as an incoming reflected wave. The strength of this reflection

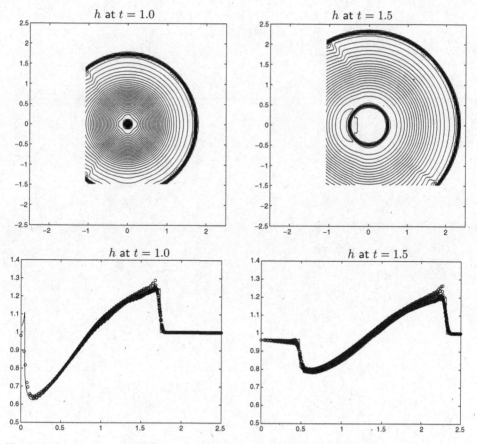

Fig. 21.7. Computed solutions for the radial dam-break problem on a reduced domain $[-1.1, 2.5] \times [-1.5, 2.5]$ with zero-order extrapolation boundary conditions. Top: contour plots of the depth at two times. Bottom: scatterplots of the depth vs. distance from the origin at the same two times. Compare Figure 21.5, where the same resolution has been used on a larger domain. Contour levels are $0.61 : 0.02 : 1.31$. [claw/book/chap21/radialdamabc]

generally depends on the angle of the wave to the boundary and is typically worst in corners. This is illustrated in Example 21.2 below, which shows the effects of this error. Still, these boundary conditions are fairly effective considering their simplicity, and are often good enough in practice.

There is an extensive literature on more sophisticated approaches to specifying absorbing boundary conditions for wave-propagation problems. See, for example, [1], [21], [28], [117], [123], [176], [197], [230].

Example 21.2. As an example, consider the radial dam-break problem for the shallow water equations described in Section 21.7.1. We now solve this same problem on a 90×100 grid in a smaller domain $[-1.1, 2.5] \times [-1.5, 2.5]$ rather than the full domain $[-2.5, 2.5] \times [-2.5, 2.5]$ used to compute the results shown in Figure 21.5. The same mesh size $\Delta x = \Delta y = 0.04$ is used. As seen in the contour plots of Figure 21.7, small-amplitude reflected waves are generated as the shock wave leaves the computational domain. The effect of these errors is also observed in the scatterplots of depth vs. distance from the origin.

Elastic Waves

A brief introduction to one-dimensional elasticity theory and elastic wave propagation was given in Section 2.12. In this chapter we will explore the full three-dimensional elasticity equations in the context of elastic wave propagation, or *elastodynamics*. There are many references available on the basic theory of linear and nonlinear elastodynamics (e.g., [6], [11], [141], [249], [255], [367], [422]), though often not in the first-order hyperbolic form we need. In this chapter the equations, eigenstructure, and Riemann solutions are written out in detail for several different variants of the linear problem.

The notation and terminology for these equations differs widely between different fields of application. Much of the emphasis in the literature is on steady-state problems, or *elastostatics*, in which the goal is to determine the deformation of an object and the internal stresses that result from some applied force. These boundary-value problems are often posed as second-order or fourth-order elliptic equations. We will concentrate instead on the hyperbolic nature of the first-order time-dependent problem, and the eigenstructure of this system. This is important in many wave-propagation applications such as seismic modeling in the earth or the study of ultrasound waves propagating through biological tissue. For small deformations, linear elasticity can generally be used. But even this case can be challenging numerically, since most practical problems involve heterogeneous materials and complicated geometry. High-resolution finite volume methods are well suited to these problems, since interfaces between different materials are handled naturally in the process of solving Riemann problems. This has already been explored in one dimension in Section 9.6. These methods can also be extended to nonlinear elasticity equations by incorporating an appropriate nonlinear Riemann solver, allowing the solution of problems with *finite deformations* (meaning larger than infinitesimal), in which case shock waves can form. For even larger deformations plastic behavior is observed, which can also be modeled with hyperbolic systems in some cases. For some examples of the application of hyperbolic theory and Riemann solvers to elastic and elastic–plastic problems, see for example [8], [30], [55], [83], [85], [164], [273], [327], [360], [406], [454], [455], [456]. Here we restrict our attention to linear elastic problems.

Recall from Section 2.12 that there are generally two basic types of waves that can propagate in an elastic solid, P-waves (pressure waves or primary waves) and S-waves (shear waves or secondary waves). In one dimension the P-waves and S-waves can be modeled separately by disjoint systems of two equations each. In multidimensional problems there is a coupling between these modes and the situation is more complicated. However, we will see that a plane-wave problem in any of the three coordinate directions leads to a system that decouples into the simple structure seen in Section 2.12.4. This means that finite volume methods based on Riemann solvers in the coordinate directions are easy to apply.

22.1 Derivation of the Elasticity Equations

In this section we will informally derive the full three-dimensional elasticity equations. The clear discussion of Davis and Selvadurai [102] has largely motivated the derivation given here, but other derivations and discussion of elastodynamics can be found in many sources, such as those listed above.

We first generalize the notation of Section 2.12.1 from two to three dimensions. The displacement $\vec{\delta}(x, y, z, t)$ now has three components, and $\nabla\vec{\delta}$ is a 3×3 matrix. The strain tensor ϵ is again defined by

$$\epsilon = \frac{1}{2}[\nabla\vec{\delta} + (\nabla\vec{\delta})^T] = \begin{bmatrix} \epsilon^{11} & \epsilon^{12} & \epsilon^{13} \\ \epsilon^{21} & \epsilon^{22} & \epsilon^{23} \\ \epsilon^{31} & \epsilon^{32} & \epsilon^{33} \end{bmatrix}. \tag{22.1}$$

This symmetric matrix has six distinct elements – three extensional strains and three shear strains – given by

$$\epsilon^{11} = \delta_x^1, \qquad \epsilon^{22} = \delta_y^2, \qquad \epsilon^{33} = \delta_z^3$$
$$\epsilon^{12} = \frac{1}{2}(\delta_y^1 + \delta_x^2), \quad \epsilon^{13} = \frac{1}{2}(\delta_z^1 + \delta_x^3), \quad \epsilon^{23} = \frac{1}{2}(\delta_z^2 + \delta_y^3). \tag{22.2}$$

The stress tensor σ is also a 3×3 symmetric matrix with six distinct elements,

$$\sigma = \begin{bmatrix} \sigma^{11} & \sigma^{12} & \sigma^{13} \\ \sigma^{21} & \sigma^{22} & \sigma^{23} \\ \sigma^{31} & \sigma^{32} & \sigma^{33} \end{bmatrix}, \tag{22.3}$$

with all elements varying as functions of space and time. This is a tensorial quantity that is written in matrix form corresponding to x–y coordinates. At any point in space this stress tensor represents the internal forces acting at that point. If we introduce a surface through the point with unit normal vector $\vec{n}$, then the traction (force per unit area) acting on this surface is given by the vector $\sigma \cdot \vec{n}$. In particular, the three columns of σ represent the traction acting on planes normal to the x-, y-, and z-axes respectively. Relative to these planes, the components σ^{11}, σ^{22}, and σ^{33} are the *normal stress* components, while σ^{12}, σ^{13}, and σ^{23} are the *shear stress* components. But it is important to keep in mind that for a plane not aligned with the coordinates, the normal and shear stresses relative to that plane will each in general have values that depend on all components of σ.

We can derive a system of conservation laws governing wave motion as a generalization to the systems (2.91) and (2.98) in the one-dimensional case. These equations have the form

$$\epsilon_t^{11} - u_x = 0,$$

$$\epsilon_t^{22} - v_y = 0,$$

$$\epsilon_t^{33} - w_z = 0,$$

$$\epsilon_t^{12} - \frac{1}{2}(v_x + u_y) = 0,$$

$$\epsilon_t^{23} - \frac{1}{2}(v_z + w_y) = 0, \qquad (22.4)$$

$$\epsilon_t^{13} - \frac{1}{2}(u_z + w_x) = 0,$$

$$\rho u_t - \sigma_x^{11} - \sigma_y^{12} - \sigma_z^{13} = 0,$$

$$\rho v_t - \sigma_x^{12} - \sigma_y^{22} - \sigma_z^{23} = 0,$$

$$\rho w_t - \sigma_x^{13} - \sigma_y^{23} - \sigma_z^{33} = 0.$$

The first six equations follow directly from the definition of ϵ in terms of the spatial gradient of $\vec{\delta}$, whereas the velocity (u, v, w) is the time derivative of $\vec{\delta}$. (The first of these was derived in (2.92).) The final three equations in (22.4) express the dynamic relationship between the acceleration and the net force resulting from all the stresses.

The equations (22.4) must be completed by specifying a constitutive stress–strain relationship between σ and ϵ. In general this might be nonlinear, but for small deformations a linear stress–strain relation can be assumed, leading to the multidimensional equations of linear elasticity derived in the next subsection.

22.1.1 Linear Elasticity

For small deformations the stress and strain can be related by a generalization of Hooke's law, which has the general form

$$\sigma^{ij} = \sum_{k,l} C^{ijkl} \epsilon^{kl}. \qquad (22.5)$$

The tensor C has 81 components, but by symmetry only 21 are independent. We will make a considerable further simplification by assuming that the material is *isotropic*, and hence the material behavior is the same in any direction. In this case the six independent components of σ can be related to those of ϵ by means of a 6×6 matrix, which will be displayed below.

If we apply a small force σ^{11} in the x-direction to an elastic bar, we expect the material to stretch by a linearly proportional small amount,

$$\epsilon^{11} = \frac{1}{E}\sigma^{11}, \qquad (22.6)$$

as in Hooke's law. The parameter E is called *Young's modulus*. In general we also expect the
bar to contract slightly in the y- and z-directions as it is stretched in x. Since the material
is isotropic, we expect the strains ϵ^{22} and ϵ^{33} to be equal to one another and, for small
deformations, linear in ϵ^{11}:

$$\epsilon^{22} = \epsilon^{33} = -\nu\epsilon^{11}.$$

The parameter ν is *Poisson's ratio*. For most materials $0 < \nu < 0.5$, although there are
strange materials for which $\nu < 0$ (e.g., [252]). Thermodynamics requires $-1 \leq \nu \leq 0.5$.
If $\nu = 0.5$, then the material is *incompressible*, a mathematical idealization in that in reality
any material can be compressed if sufficient force is applied. The assumption $\nu < 0.5$ is
required for a hyperbolic formulation.
   Using (22.6), we can write

$$\epsilon^{22} = \epsilon^{33} = -\frac{\nu}{E}\sigma^{11}.$$

Similarly, a force applied in the y- or z-direction will also typically cause strains in all
three directions. More generally we can think of applying normal stresses σ^{11}, σ^{22}, and
σ^{33} simultaneously, resulting in strains that are a linear combination of those obtained from
each stress separately. This leads to the extensional strains

$$
\begin{aligned}
\epsilon^{11} &= \frac{1}{E}\sigma^{11} - \frac{\nu}{E}\sigma^{22} - \frac{\nu}{E}\sigma^{33}, \\
\epsilon^{22} &= \frac{1}{E}\sigma^{22} - \frac{\nu}{E}\sigma^{11} - \frac{\nu}{E}\sigma^{33}, \\
\epsilon^{33} &= \frac{1}{E}\sigma^{33} - \frac{\nu}{E}\sigma^{11} - \frac{\nu}{E}\sigma^{22}.
\end{aligned}
\tag{22.7}
$$

The shear strains and shear stresses are related to one another by the simpler relations

$$\sigma^{12} = 2\mu\epsilon^{12}, \quad \sigma^{13} = 2\mu\epsilon^{13}, \quad \sigma^{23} = 2\mu\epsilon^{23}. \tag{22.8}$$

where $\mu \geq 0$ is the *shear modulus*. For elastic materials the shear modulus can be determined
in terms of E and ν as

$$\mu = \frac{E}{2(1+\nu)}. \tag{22.9}$$

See [102], for example, for a derivation.

Combining (22.7) and (22.8) gives the desired stress–strain relation

$$
\begin{bmatrix} \epsilon^{11} \\ \epsilon^{22} \\ \epsilon^{33} \\ \epsilon^{12} \\ \epsilon^{23} \\ \epsilon^{13} \end{bmatrix}
=
\begin{bmatrix}
1/E & -v/E & -v/E & 0 & 0 & 0 \\
-v/E & 1/E & -v/E & 0 & 0 & 0 \\
-v/E & -v/E & 1/E & 0 & 0 & 0 \\
0 & 0 & 0 & 1/2\mu & 0 & 0 \\
0 & 0 & 0 & 0 & 1/2\mu & 0 \\
0 & 0 & 0 & 0 & 0 & 1/2\mu
\end{bmatrix}
\begin{bmatrix} \sigma^{11} \\ \sigma^{22} \\ \sigma^{33} \\ \sigma^{12} \\ \sigma^{23} \\ \sigma^{13} \end{bmatrix}.
\tag{22.10}
$$

We can invert this matrix to instead determine the stress in terms of the strain,

$$
\begin{bmatrix} \sigma^{11} \\ \sigma^{22} \\ \sigma^{33} \\ \sigma^{12} \\ \sigma^{23} \\ \sigma^{13} \end{bmatrix}
=
\begin{bmatrix}
\lambda + 2\mu & \lambda & \lambda & 0 & 0 & 0 \\
\lambda & \lambda + 2\mu & \lambda & 0 & 0 & 0 \\
\lambda & \lambda & \lambda + 2\mu & 0 & 0 & 0 \\
0 & 0 & 0 & 2\mu & 0 & 0 \\
0 & 0 & 0 & 0 & 2\mu & 0 \\
0 & 0 & 0 & 0 & 0 & 2\mu
\end{bmatrix}
\begin{bmatrix} \epsilon^{11} \\ \epsilon^{22} \\ \epsilon^{33} \\ \epsilon^{12} \\ \epsilon^{23} \\ \epsilon^{13} \end{bmatrix}.
\tag{22.11}
$$

Here we have introduced the parameter λ defined by

$$
\lambda = \frac{vE}{(1+v)(1-2v)}.
\tag{22.12}
$$

This does not have any direct physical interpretation, but is useful in that it appears in the inverse above. The relation (22.9) has also been used to simplify the form of this inverse. The parameter λ should not be confused with an eigenvalue, for which we use the symbol s in this chapter. The parameters λ and μ are often called the *Lamé parameters* for the material. From (22.9) and (22.12) we can also compute E and v from λ and μ, as

$$
E = \frac{\mu(3\lambda + 2\mu)}{\lambda + \mu}, \qquad v = \frac{1}{2}\left(\frac{\lambda}{\lambda + \mu}\right).
\tag{22.13}
$$

The relationship (22.11) can be used to convert (22.4) into a closed system of nine equations for the velocities $\vec{u} = (u, v, w)$ and either σ or ϵ, by eliminating the other set of six parameters (analogously to choosing (2.93) or (2.95) in the one-dimensional case). Either way we obtain a hyperbolic linear system of nine equations. The two alternative systems are similarity transformations of one another and have the same eigenvalues, as they must, since they model the same elastic waves.

We will use $\vec{u}$ and σ, as is more common in linear elasticity. Then we need expressions for the time derivatives of the stresses. These may be obtained by using (22.11) to write, for example,

$$
\sigma_t^{11} = (\lambda + 2\mu)\epsilon_t^{11} + \lambda\epsilon_t^{22} + \lambda\epsilon_t^{33}
$$

and then using the equations of motion (22.4) to evaluate the time derivatives on the right-hand side. We obtain the system

$$\sigma_t^{11} - (\lambda + 2\mu)u_x - \lambda v_y - \lambda w_z = 0,$$
$$\sigma_t^{22} - \lambda u_x - (\lambda + 2\mu)v_y - \lambda w_z = 0,$$
$$\sigma_t^{33} - \lambda u_x - \lambda v_y - (\lambda + 2\mu)w_z = 0,$$
$$\sigma_t^{12} - \mu(v_x + u_y) = 0,$$
$$\sigma_t^{23} - \mu(v_z + w_y) = 0, \qquad (22.14)$$
$$\sigma_t^{13} - \mu(u_z + w_x) = 0,$$
$$\rho u_t - \sigma_x^{11} - \sigma_y^{12} - \sigma_z^{13} = 0,$$
$$\rho v_t - \sigma_x^{12} - \sigma_y^{22} - \sigma_z^{23} = 0,$$
$$\rho w_t - \sigma_x^{13} - \sigma_y^{23} - \sigma_z^{33} = 0.$$

This can be written as

$$q_t + Aq_x + Bq_y + Cq_z = 0, \qquad (22.15)$$

with

$$q = \begin{bmatrix} \sigma^{11} \\ \sigma^{22} \\ \sigma^{33} \\ \sigma^{12} \\ \sigma^{23} \\ \sigma^{13} \\ u \\ v \\ w \end{bmatrix}, \quad A = \begin{bmatrix} 0 & 0 & 0 & 0 & 0 & 0 & -(\lambda+2\mu) & 0 & 0 \\ 0 & 0 & 0 & 0 & 0 & 0 & -\lambda & 0 & 0 \\ 0 & 0 & 0 & 0 & 0 & 0 & -\lambda & 0 & 0 \\ 0 & 0 & 0 & 0 & 0 & 0 & 0 & -\mu & 0 \\ 0 & 0 & 0 & 0 & 0 & 0 & 0 & 0 & 0 \\ 0 & 0 & 0 & 0 & 0 & 0 & 0 & 0 & -\mu \\ -1/\rho & 0 & 0 & 0 & 0 & 0 & 0 & 0 & 0 \\ 0 & 0 & 0 & -1/\rho & 0 & 0 & 0 & 0 & 0 \\ 0 & 0 & 0 & 0 & 0 & -1/\rho & 0 & 0 & 0 \end{bmatrix},$$

$$(22.16)$$

and similar matrices B and C with the nonzero elements shifted to different locations.

The matrices A, B, and C do not commute, and so these equations are generally coupled in the multidimensional case. This is not surprising, since we expect that elastic waves, like acoustic waves, can propagate equally well in any direction. In fact the eigenvalues of $\breve{A} = n^x A + n^y B + n^z C$ are the same for any unit vector $\vec{n}$, and are given by

$$s^1 = -c_p, \quad s^2 = c_p, \quad s^3 = -c_s, \quad s^4 = c_s,$$
$$s^5 = -c_s, \quad s^6 = c_s, \quad s^7 = s^8 = s^9 = 0. \qquad (22.17)$$

These are not ordered monotonically, but instead the three eigenvalues with modulus 0 are grouped last, since in practice we only need to propagate six waves after solving any one-dimensional Riemann problem. The 1- and 2- waves are the P-waves, propagating in the directions $\pm\vec{n}$. There are also two sets of S-waves corresponding to the fact that shear motions are in the two-dimensional plane orthogonal to this direction. The wave speeds are given by

$$c_p = \sqrt{\frac{\lambda + 2\mu}{\rho}}, \qquad c_s = \sqrt{\frac{\mu}{\rho}}. \tag{22.18}$$

The Riemann problem in each coordinate direction is easy to solve. In the x-direction, for example, we have the following eigenvectors of the matrix A of (22.16):

$$r^{1,2} = \begin{bmatrix} \lambda + 2\mu \\ \lambda \\ \lambda \\ 0 \\ 0 \\ 0 \\ \pm c_p \\ 0 \\ 0 \end{bmatrix}, \quad r^{3,4} = \begin{bmatrix} 0 \\ 0 \\ 0 \\ \mu \\ 0 \\ 0 \\ 0 \\ \pm c_s \\ 0 \end{bmatrix}, \quad r^{5,6} = \begin{bmatrix} 0 \\ 0 \\ 0 \\ 0 \\ 0 \\ \mu \\ 0 \\ 0 \\ \pm c_s \end{bmatrix}. \tag{22.19}$$

These correspond to P-waves in x, shear waves with displacement in the y-direction, and shear waves with displacement in the z-direction, respectively. The other three eigenvectors, corresponding to $\lambda^{7,8,9} = 0$, are given by

$$r^7 = \begin{bmatrix} 0 \\ 0 \\ 0 \\ 0 \\ 1 \\ 0 \\ 0 \\ 0 \\ 0 \end{bmatrix}, \quad r^8 = \begin{bmatrix} 0 \\ 1 \\ 0 \\ 0 \\ 0 \\ 0 \\ 0 \\ 0 \\ 0 \end{bmatrix}, \quad r^9 = \begin{bmatrix} 0 \\ 0 \\ 1 \\ 0 \\ 0 \\ 0 \\ 0 \\ 0 \\ 0 \end{bmatrix}. \tag{22.20}$$

These correspond to jumps in σ^{23}, σ^{22}, or σ^{33} alone, each of which causes no wave propagation in x. The matrices B and C have similar eigenvectors, again with the nonzero elements appropriately rearranged.

Note that if we solve a plane-wave problem in which there is only variation in x, then $q_y = q_z = 0$ and the three-dimensional system reduces to $q_t + Aq_x = 0$. In this case the system does decouple into systems of the form discussed in Section 2.12.4. Actually the P-waves given by $r^{1,2}$ in (22.19) carry variation in σ^{22} and σ^{33} as well as in σ^{11} and u. However, this stress in the y- and z-directions exactly balances the stress in x in such a way that the strain is entirely in the x-direction, i.e., $\epsilon^{22} = \epsilon^{33} = 0$, as is clear when the stress components of $r^{1,2}$ are inserted into (22.10). In spite of the fact that the Poisson ratio is typically nonzero, a compressional plane wave in an infinite solid causes no deformation in the orthogonal directions (as indicated in Figure 2.2). Also, since the second and third columns of A in (22.16) are identically zero, we see that the stresses σ^{22} and σ^{33} cause no dynamic effects and we can drop these variables from the system in deriving the one-dimensional equations (2.93). But note that these one-dimensional equations are based on the assumption of a plane wave in an infinite three-dimensional solid, as discussed further in Section 22.3. Other "one-dimensional" situations lead to different equations. For example, the equations modeling longitudinal waves in thin elastic rod are discussed in Section 22.6.

22.1.2 The Bulk Modulus and Acoustics

The *mean stress* in a solid is defined to be one third the trace of the stress tensor,

$$\frac{1}{3} \operatorname{tr}(\sigma) = \frac{1}{3}(\sigma^{11} + \sigma^{22} + \sigma^{33}). \tag{22.21}$$

This is an invariant of the stress tensor, i.e., it has the same value regardless of the choice of coordinate system used. The trace of the strain tensor is also an invariant, and this value

$$e \equiv \operatorname{tr}(\epsilon) = \epsilon^{11} + \epsilon^{22} + \epsilon^{33}, \tag{22.22}$$

is called the *volumetric strain*. It approximates the relative change in volume in the strained solid. By adding together the three equations of (22.7), we find that

$$e = \frac{1 - 2\nu}{E} \operatorname{tr}(\sigma), \tag{22.23}$$

and hence the mean stress is related to the volumetric strain by

$$\frac{1}{3} \operatorname{tr}(\sigma) = Ke, \tag{22.24}$$

where the *bulk modulus of compressibility* K is defined by

$$K = \frac{E}{3(1 - 2\nu)} = \lambda + \frac{2}{3}\mu. \tag{22.25}$$

Averaging the first three equations of (22.14) gives an evolution equation for the mean stress,

$$\frac{1}{3}(\sigma^{11} + \sigma^{22} + \sigma^{33})_t - K(u_x + v_y + w_z) = 0. \tag{22.26}$$

We can relate the elastodynamics equations to the acoustics equations derived earlier for gas dynamics if we make the additional assumption on that the stress is *hydrostatic*, as it is in a fluid. This means that there is no shear stress, $\sigma^{12} = \sigma^{13} = \sigma^{23} = 0$, and the extensional stress components are all equal and negative,

$$\sigma^{11} = \sigma^{22} = \sigma^{33} \equiv -p. \tag{22.27}$$

The value p is called the *hydrostatic pressure*, and has the opposite sign from the stresses as discussed in Section 2.12.4. In this case the stress tensor (22.3) reduces to $-pI$ where I is the identity matrix. Rather than working with this tensor we can reduce the equations and deal only with the scalar pressure p, which satisfies $p = -Ke$ by (22.24), since $p = -\frac{1}{3}\operatorname{tr}(\sigma)$. The equation (22.26) becomes an evolution equation for the hydrostatic pressure,

$$p_t + K(u_x + v_y + w_z) = 0. \tag{22.28}$$

Since we now also assume that $\sigma^{12} = \sigma^{13} = \sigma^{23} = 0$, we can drop the middle three equations of (22.14) and the final three become

$$\rho u_t + p_x = 0,$$
$$\rho v_t + p_y = 0, \tag{22.29}$$
$$\rho w_t + p_z = 0.$$

We recognize (22.28), (22.29) as defining the three-dimensional acoustics equations from Section 18.6. This system has wave speeds given by the "speed of sound"

$$c = \sqrt{\frac{K}{\rho}} = \sqrt{\frac{\lambda + \frac{2}{3}\mu}{\rho}}. \tag{22.30}$$

Note that this is different than the P-wave speed c_p of (22.18), which is the sound speed actually observed in solids. However, since a fluid does not support shear stresses we should set $\mu = 0$, in which case (22.18) and (22.30) do agree. The acoustics equations are sometimes used as an approximate system of equations for modeling P-waves in solids when shear waves are relatively unimportant, particularly in solids where μ is small compared to λ.

22.2 The Plane-Strain Equations of Two-Dimensional Elasticity

We can reduce the three-dimensional equations (22.14) to two space dimensions by setting $q_z \equiv 0$, for example, if we assume there is no variation in the z-direction. Note that the strain $\epsilon^{33} = \delta_z^3$ must be zero in this case, since it is the z-derivative of the z-displacement. (We discuss below when this assumption is reasonable.) The stress σ^{33} will not generally be zero, but can be determined in terms of σ^{11} and σ^{22} as discussed below. Dropping the equation for σ^{33} from (22.14) along with all z-derivative terms, the remaining eight equations reduce to two decoupled systems of equations,

$$\sigma_t^{11} - (\lambda + 2\mu)u_x - \lambda v_y = 0,$$
$$\sigma_t^{22} - \lambda u_x - (\lambda + 2\mu)v_y = 0,$$
$$\sigma_t^{12} - \mu(v_x + u_y) = 0, \tag{22.31}$$
$$\rho u_t - \sigma_x^{11} - \sigma_y^{12} = 0,$$
$$\rho v_t - \sigma_x^{12} - \sigma_y^{22} = 0,$$

and

$$\sigma_t^{23} - \mu w_y = 0,$$

$$\sigma_t^{13} - \mu w_x = 0, \tag{22.32}$$

$$\rho w_t - \sigma_x^{13} - \sigma_y^{23} = 0.$$

The latter system (22.32) models shear waves with motion orthogonal to the x–y plane. Waves modeled by this system have speed c_s, the S-wave speed given in (22.18).

The system (22.31) is the more interesting system, and models both P-waves and S-waves for which the motion is in the x–y plane. The S-waves modeled by this system have material motion orthogonal to the direction the wave is propagating, but still within the x–y plane. This system (22.31) is often called the two-dimensional *plane-strain equations*, since the strain is confined entirely to the x–y plane. This is a reasonable model for plane waves propagating through a three-dimensional elastic body in cases where there is no variation in the z-direction, for example, if the x–y plane is a representative slice through a three-dimensional solid with essentially infinite extent in the z-direction, as might occur in modeling large-scale seismic waves in the earth, for example. If it is correct to assume that there is no variation in the z-direction, then it is also valid to assume that $\epsilon^{33} = 0$. Otherwise, if ϵ^{33} had some nonzero value independent of z, then the displacement δ^3 would have to be of the form $\delta^3 = \epsilon^{33}(z - z_0)$ and grow without bound in z. This is not reasonable for finite-amplitude waves. Of course, as the material is compressed in the x- or y-direction it will *try* to expand in z (when $\nu \neq 0$), but it will be prevented from doing so by the adjacent material, which is trying equally hard to expand in the other direction. The result is a nonzero stress σ^{33} while ϵ^{33} remains zero. Indeed, setting $\epsilon^{33} = 0$ in the system (22.11) yields

$$\sigma^{11} = (\lambda + 2\mu)\epsilon^{11} + \lambda\epsilon^{22}, \tag{22.33}$$

$$\sigma^{22} = \lambda\epsilon^{11} + (\lambda + 2\mu)\epsilon^{22}, \tag{22.34}$$

$$\sigma^{33} = \lambda\epsilon^{11} + \lambda\epsilon^{22}. \tag{22.35}$$

The first two equations of this set are all that are needed for the two-dimensional system (22.31), but the stress σ^{33} can also be computed from (22.35) if desired. Alternatively we can obtain

$$\sigma^{33} = \nu(\sigma^{11} + \sigma^{22}) \tag{22.36}$$

from the third equation of (22.7) by setting $\epsilon^{33} = 0$.

Note that if we invert the stress–strain relation (22.33)–(22.35) to find ϵ^{11} and ϵ^{22} in terms of σ^{11} and σ^{22}, we find that

$$\hat{E}\epsilon^{11} = \sigma^{11} - \hat{\nu}\sigma^{22},$$

$$\hat{E}\epsilon^{22} = \sigma^{22} - \hat{\nu}\sigma^{11}, \tag{22.37}$$

where

$$\hat{E} = \frac{E}{1 - \nu^2}, \qquad \hat{\nu} = \frac{\nu}{1 - \nu}. \tag{22.38}$$

The equations (22.37) have the same form as the three-dimensional stress–strain relations (22.7), but with different effective values for the Young's modulus $\hat{E}$ and Poisson ratio $\hat{\nu}$. These relations can be derived either by inverting the 2×2 system given by (22.33) and (22.34), or from (22.7) by using (22.36).

It is important to note that the plane-strain system (22.31) does *not* in general model elastic waves in a thin plate, in spite of the fact that it might seem natural to view this as a two-dimensional elastic medium. Wave propagation in a plate can be modeled by a two-dimensional hyperbolic system, but (22.31) is not the correct one; see Section 22.5.

We now discuss the eigenstructure of the system (22.31). Rather than displaying the matrices A and B separately in this case, it is more compact and perhaps also more revealing to show the linear combination $\breve{A} = n^x A + n^y B$, where $\vec{n}$ is again a unit vector in an arbitrary direction. The matrix $\breve{A}$ is then the coefficient matrix for the one-dimensional problem modeling the propagation of plane waves in the $\vec{n}$-direction. Setting $\vec{n} = (1, 0)$ or $(0, 1)$ below recovers the matrices A and B separately. We have

$$
q = \begin{bmatrix} \sigma^{11} \\ \sigma^{22} \\ \sigma^{12} \\ u \\ v \end{bmatrix}, \quad
\breve{A} = - \begin{bmatrix} 0 & 0 & 0 & n^x(\lambda + 2\mu) & n^y\lambda \\ 0 & 0 & 0 & n^x\lambda & n^y(\lambda + 2\mu) \\ 0 & 0 & 0 & n^y\mu & n^x\mu \\ n^x/\rho & 0 & n^y/\rho & 0 & 0 \\ 0 & n^y/\rho & n^x/\rho & 0 & 0 \end{bmatrix}. \tag{22.39}
$$

The eigenvalues of $\breve{A}$ are

$$
\breve{s}^1 = -c_p, \quad \breve{s}^2 = c_p, \quad \breve{s}^3 = -c_s, \quad \breve{s}^4 = c_s, \quad \breve{s}^5 = 0, \tag{22.40}
$$

where we use s instead of λ to avoid confusion with the Lamé parameter. The P-wave eigenvectors are

$$
\breve{r}^1 = \begin{bmatrix} \lambda + 2\mu(n^x)^2 \\ \lambda + 2\mu(n^y)^2 \\ 2\mu n^x n^y \\ n^x c_p \\ n^y c_p \end{bmatrix}, \quad
\breve{r}^2 = \begin{bmatrix} \lambda + 2\mu(n^x)^2 \\ \lambda + 2\mu(n^y)^2 \\ 2\mu n^x n^y \\ -n^x c_p \\ -n^y c_p \end{bmatrix}, \tag{22.41}
$$

while the S-wave eigenvectors $r^{3,4}$ and the stationary wave r^5 are

$$
\breve{r}^3 = \begin{bmatrix} -2n^x n^y \mu \\ 2n^x n^y \mu \\ [(n^x)^2 - (n^y)^2]\mu \\ -n^y c_s \\ n^x c_s \end{bmatrix}, \quad
\breve{r}^4 = \begin{bmatrix} -2n^x n^y \mu \\ 2n^x n^y \mu \\ [(n^x)^2 - (n^y)^2]\mu \\ n^y c_s \\ -n^x c_s \end{bmatrix}, \quad
\breve{r}^5 = \begin{bmatrix} (n^y)^2 \\ (n^x)^2 \\ -n^x n^y \\ 0 \\ 0 \end{bmatrix}. \tag{22.42}
$$

Observe that P-waves $\breve{r}^{1,2}$ have velocity components directed in the $\pm\vec{n}$-direction, the direction in which the plane wave propagates. The S-waves, on the other hand, have motion in the orthogonal direction $\pm(-n^y, n^x)$. Note also that a P-wave in the x- or y-direction has $\sigma^{12} = 0$, but that a P-wave propagating in any other direction has $\sigma^{12} \neq 0$. This is because the elements of the stress tensor have been expressed in $x-y$ coordinates, and representing a purely extensional stress in some other direction requires all components of σ to be nonzero.

22.3 One-Dimensional Slices

We can reduce the systems of equations (22.31) and (22.32) even further if we assume that there is no variation in the y-direction. As in our discussion of the plane-strain equations above, this is typically valid if we are considering a one-dimensional slice through an essentially infinite medium in cases where there is variation in only one direction (e.g., a plane wave propagating through the earth). It is not a valid model for waves in a "one-dimensional" thin elastic rod, which is discussed in Section 22.6.

Setting all y-derivatives to zero in (22.31) results in the two decoupled systems

$$\sigma_t^{11} - (\lambda + 2\mu)u_x = 0,$$
$$\rho u_t - \sigma_x^{11} = 0 \qquad\qquad (22.43)$$

and

$$\sigma_t^{12} - \mu v_x = 0,$$
$$\rho v_t - \sigma_x^{12} = 0. \qquad\qquad (22.44)$$

These are the systems (2.95) and (2.100) introduced in Section 2.12.4. They model P-waves with displacement in x and S-waves with displacement in y, respectively, with wave speeds c_p and c_s given by (22.18).

We have dropped the equation for σ^{22}, which is given by

$$\sigma^{22} = \lambda\epsilon^{11} = \hat{\nu}\sigma^{11}. \qquad\qquad (22.45)$$

Note that in this case $\sigma^{33} = \sigma^{22}$ by (22.36).

The equations (22.32) reduce to

$$\sigma_t^{13} - \mu w_x = 0,$$
$$\rho w_t - \sigma_x^{13} = 0. \qquad\qquad (22.46)$$

This system models an independent set of shear waves in which the displacement is in the z-direction rather than in the y-direction (and propagation still in the x-direction).

22.4 Boundary Conditions

Boundary conditions for elastic solids can be imposed in much the same way as for acoustics (see Sections 7.3 and 21.8), but there are now a wider variety of physically meaningful boundary conditions to consider. We will use the two-dimensional plane-strain equations (22.31) for illustration, and consider a point along the left edge of the domain, which we assume is at $x = 0$. We must then determine ghost-cell values Q_{0j} and $Q_{-1,j}$ based on interior values and the physical boundary conditions. It should be clear how to translate this discussion to other boundaries and to three dimensions.

Periodic or extrapolation boundary conditions are easily imposed, as for other equations, as discussed in Section 21.8. The more interesting cases are where either the motion of the boundary or the traction applied to the boundary is specified. These are discussed in the next two subsections.

22.4.1 Specified Motion

Suppose the velocity of the boundary at $x = 0$, $y = y_j$ is known. Call the velocity at this point (U, V) for brevity. Then following the discussion of Sections 7.3.4 and 21.8.3, we can impose this by specifying the ghost-cell values as follows:

$$\text{for } Q_{0j}: \quad \sigma_{0j}^{11} = \sigma_{1j}^{11}, \qquad \sigma_{0j}^{12} = \sigma_{1j}^{12}, \qquad \sigma_{0j}^{22} = \sigma_{1j}^{22},$$

$$u_{0j} = 2U - u_{1j}, \qquad v_{0j} = 2V - v_{1j};$$

$$\text{for } Q_{-1,j}: \quad \sigma_{-1,j}^{11} = \sigma_{2j}^{11}, \qquad \sigma_{-1,j}^{12} = \sigma_{2j}^{12}, \qquad \sigma_{-1,j}^{22} = \sigma_{2j}^{22}, \tag{22.47}$$

$$u_{-1,j} = 2U - u_{2j}, \qquad v_{-1,j} = 2V - v_{2j}.$$

When the Riemann problem is solved at $x = 0$ (i.e., at cell interface $i = 1/2$), this choice insures that the intermediate state $Q_{1/2}^{\vee}$ has velocity (U, V) and satisfies the required physical boundary condition. An important special case is $U = 0$, $V = 0$, in which case the boundary is fixed at this point.

Note that in the case of an elastic solid we must specify both u and v. This differs from two-dimensional acoustics or inviscid fluid dynamics, where only the normal component of velocity is specified as discussed in Sections 21.8.3 and 21.8.4. For an inviscid fluid there can be slip along the boundary, and so the tangential component of velocity cannot be specified. For an elastic solid we must specify both. For a three-dimensional problem we would also have to specify w at the boundary, e.g., $w = 0$ at a fixed boundary. The stresses are not specified and must be free to react as necessary to the imposed motion. This is accomplished by simply reflecting σ from the interior values in (22.47), and the values observed in the resulting Riemann solution $Q_{1/2}^{\vee}$ can be used to obtain the surface traction if this is desired as part of the solution to the problem.

22.4.2 Specified Traction

Often we wish instead to specify the traction at a point on the boundary and compute the resulting motion. At the boundary $x = 0$ this amounts to specifying the values of σ^{11} and σ^{12}, say as $\sigma^{11} = S^{11}$ and $\sigma^{12} = S^{12}$ at the point $x = 0$, $y = y_j$. In particular, if this is a free boundary (the edge of an elastic solid with no external force applied), then we should set $S^{11} = 0$ and $S^{12} = 0$. This is called a *traction-free boundary*. Note that we have no physical control over σ^{22} at this boundary.

The proper ghost cell values for general S^{11} and S^{12} are then given by:

$$\text{for } Q_{0j}: \quad \sigma_{0j}^{11} = 2S^{11} - \sigma_{1j}^{11}, \quad \sigma_{0j}^{12} = 2S^{12} - \sigma_{1j}^{12}, \quad \sigma_{0j}^{22} = \sigma_{1j}^{22},$$

$$u_{0j} = u_{1j}, \qquad\qquad v_{0j} = v_{1j};$$

$$\text{for } Q_{-1,j}: \quad \sigma_{-1,j}^{11} = 2S^{11} - \sigma_{2j}^{11}, \quad \sigma_{-1,j}^{12} = 2S^{12} - \sigma_{2j}^{12}, \quad \sigma_{-1,j}^{22} = \sigma_{2j}^{22},$$

$$u_{-1,j} = u_{2j}, \qquad\qquad v_{-1,j} = v_{2j}.$$

$$(22.48)$$

For a three-dimensional problem we would also need to specify $\sigma^{13} = S^{13}$, while σ^{23} and σ^{33} would simply be reflected from the interior, as is done for σ^{22} in (22.48).

22.5 The Plane-Stress Equations and Two-Dimensional Plates

In Section 22.2 we derived a two-dimensional hyperbolic system by using the plane-strain assumption that all displacements are confined to the x–y plane. As discussed there, this is valid if we assume the material is essentially infinite in the z-direction and there is no variation of the solution in that direction. We now consider a different situation, in which the three-dimensional domain is a thin plate bounded by the planes $z = \pm h$ for some small h. We wish to derive two-dimensional equations that model waves whose wavelength is long relative to the thickness of the plate. Each surface of the plate is a free boundary and must be traction-free (see Section 22.4.2), so we have

$$\sigma^{13} \equiv 0, \qquad \sigma^{23} \equiv 0, \qquad \sigma^{33} \equiv 0 \tag{22.49}$$

on the boundary planes $z = \pm h$. To derive a two-dimensional system of equations we will assume that these are identically zero throughout the plate. Then from (22.8) we also have

$$\epsilon^{13} \equiv 0, \qquad \epsilon^{23} \equiv 0. \tag{22.50}$$

Note, however, that we cannot assume $\epsilon^{33} = 0$. Instead, from (22.7) with $\sigma^{33} = 0$ we obtain

$$\epsilon^{33} = -\frac{\nu}{E}(\sigma^{11} + \sigma^{22}), \tag{22.51}$$

along with

$$\epsilon^{11} = \frac{1}{E}(\sigma^{11} - \nu\sigma^{22}),$$

$$\epsilon^{22} = \frac{1}{E}(-\nu\sigma^{11} + \sigma^{22}). \tag{22.52}$$

Adding these last two equations together gives

$$\epsilon^{11} + \epsilon^{22} = \frac{1 - \nu}{E}(\sigma^{11} + \sigma^{22}). \tag{22.53}$$

Since $\epsilon^{11} + \epsilon^{22} = \delta_x^1 + \delta_y^2$ is the x–y divergence of the displacement, we see that if $\epsilon^{11} + \epsilon^{22} \neq 0$, then there is compression or expansion in the x–y plane, and in this case from (22.51) there must be compensating motion in the z-direction whenever the Poisson ratio v is nonzero. No matter how thin the plate may appear, it is still three-dimensional, and stretching or compressing it in the x–y plane leads to motion in z of the same order of magnitude.

To derive two-dimensional equations we will assume, however, that σ^{11} and σ^{22}, and hence ϵ^{33}, are independent of z. Recall that $\epsilon^{33} = \delta_z^3$, so $\epsilon^{33} > 0$ corresponds to the plate bulging out, while $\epsilon^{33} < 0$ corresponds to the plate becoming thinner.

As the plate thickens or thins, the z-velocity w must be nonzero. Moreover, it is clearly not valid to assume that the value of w is independent of z. For example, if the plate is bulging out, then we must have $w > 0$ for $0 < z < h$ and $w < 0$ for $-h < z < 0$. Since $w = \delta_t^3$, we have $w_z = \delta_{zt}^3 = \epsilon_t^{33}$, and the assumption that ϵ^{33} is independent of z means that w varies linearly in z. Although w is not zero, it is close to zero and symmetric about $z = 0$. The plate equations are most rigorously defined by integrating the three-dimensional equations in z from $-h$ to h and the integral of w then reduces to zero, to leading order. This justifies ignoring the effects of w in deriving the two-dimensional plate equations.

Using the assumptions (22.49) and (22.50) in the system (22.4) gives the following system of equations (after dropping the equations for ϵ^{13}, ϵ^{23}, ϵ^{33}, and w):

$$\epsilon_t^{11} - u_x = 0,$$

$$\epsilon_t^{22} - v_y = 0,$$

$$\epsilon_t^{12} - \frac{1}{2}(v_x + u_y) = 0, \tag{22.54}$$

$$\rho u_t - \sigma_x^{11} - \sigma_y^{12} = 0,$$

$$\rho v_t - \sigma_x^{12} - \sigma_y^{22} = 0.$$

To close this system we need the constitutive relations relating σ to ϵ, which are given by (22.52) along with $\sigma^{12} = 2\mu\epsilon^{12}$ coming from (22.8). We can write these either as

$$\epsilon^{11} = \frac{1}{E}(\sigma^{11} - v\sigma^{22}),$$

$$\epsilon^{22} = \frac{1}{E}(-v\sigma^{11} + \sigma^{22}), \tag{22.55}$$

$$\epsilon^{12} = \frac{1}{2\mu}\sigma^{12}$$

or, by inverting the system, as

$$\sigma^{11} = \left(\frac{2\mu}{1-v}\right)(\epsilon^{11} + v\epsilon^{22}),$$

$$\sigma^{22} = \left(\frac{2\mu}{1-v}\right)(v\epsilon^{11} + \epsilon^{22}), \tag{22.56}$$

$$\sigma^{12} = 2\mu\epsilon^{12}.$$

As usual, we can eliminate either ϵ or σ from (22.54). If we eliminate ϵ, then we obtain the system

$$\sigma_t^{11} - \left(\frac{2\mu}{1-\nu}\right) u_x - \left(\frac{2\mu\nu}{1-\nu}\right) v_y = 0,$$

$$\sigma_t^{22} - \left(\frac{2\mu\nu}{1-\nu}\right) u_x - \left(\frac{2\mu}{1-\nu}\right) v_y = 0,$$

$$\sigma_t^{12} - \mu(v_x + u_y) = 0, \tag{22.57}$$

$$\rho u_t - \sigma_x^{11} - \sigma_y^{12} = 0,$$

$$\rho v_t - \sigma_x^{12} - \sigma_y^{22} = 0.$$

Note that this system has different coefficients than the plane-strain system (22.31). However, if we define

$$\hat{\lambda} = \frac{2\mu\nu}{1-\nu} = \frac{2\mu\lambda}{\lambda + 2\mu} \tag{22.58}$$

as an effective Lamé parameter for the plate, then

$$\frac{2\mu}{1-\nu} = \hat{\lambda} + 2\mu,$$

and the system (22.57) has exactly the same form as (22.31) but with $\hat{\lambda}$ in place of λ. Hence any method developed for the plane-strain case can also be applied to the plane-stress equations simply by changing the value of λ. In particular, the eigenstructure is the same as that developed in Section 22.2 (with $\hat{\lambda}$ in place of λ), and so we see from the eigenvalues (22.40) that the characteristic wave speeds for waves in a plate are

$$\hat{c}_p = \sqrt{\frac{\hat{\lambda} + 2\mu}{\rho}} = \sqrt{\frac{2\mu}{\rho(1-\nu)}} = \sqrt{\frac{E}{\rho(1-\nu^2)}} \tag{22.59}$$

and

$$c_s = \sqrt{\mu/\rho}, \tag{22.60}$$

respectively. The speed c_s is the usual S-wave speed, but the wave speed $\hat{c}_p$ is smaller than c_p from (22.18) when $0 < \nu < 1/2$, since

$$\hat{\lambda} = \lambda\left(1 - \frac{\nu}{1-\nu}\right) < \lambda. \tag{22.61}$$

We will refer to waves propagating at the velocity $\hat{c}_p$ as $\hat{P}$-waves.

Example 22.1. We can investigate wave propagation in a thin plate numerically by solving the three-dimensional elasticity equations in a plate with finite thickness $-h < z < h$ and imposing traction-free boundary conditions $\sigma^{13} = \sigma^{23} = \sigma^{33} = 0$ at $z = \pm h$. To illustrate an isolated $\hat{P}$-wave we can assume the wave is propagating in the x-direction

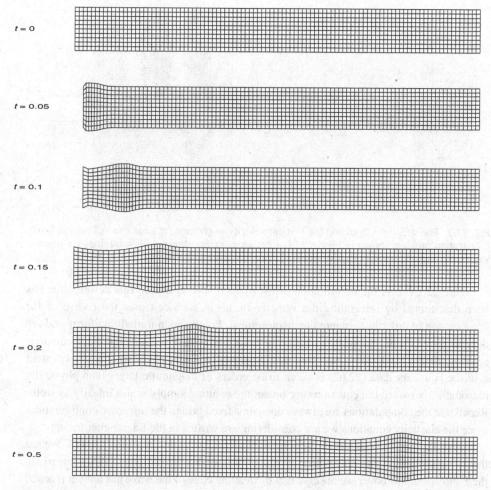

$t = 0$

$t = 0.05$

$t = 0.1$

$t = 0.15$

$t = 0.2$

$t = 0.5$

Fig. 22.1. Side view of a thin plate in the x–z plane. The plate is pushed inwards at the left, giving rise to a $\hat{P}$-wave as described in the text. [`claw/book/chap22/plate`]

and there is no variation in y, and hence no strain in the y-direction. In this case we can solve a two-dimensional problem in the x–z plane. The appropriate equations are now the plane-strain equations of Section 22.2, rewritten in x and z instead of x and y. Figures 22.1 and 22.2 show results of a numerical calculation in which the traction-free boundary conditions $\sigma^{13} = \sigma^{33} = 0$ are imposed at $z = \pm 0.01$ (see Section 22.4.2) and the plate is initially undisturbed. Boundary conditions at $x = 0$ are given by specifying the velocity (see Section 22.4.1) as

$$u(0, z, t) = \begin{cases} a \sin(2\pi t/T) & \text{if } t \le T, \\ 0 & \text{if } t > T, \end{cases}$$

$$w(0, z, t) = 0$$

$$(22.62)$$

with $T = 0.15$. This corresponds to pushing the plate inward and then bringing the edge back to its original location, slowly enough that the wavelength is long relative to the thickness.

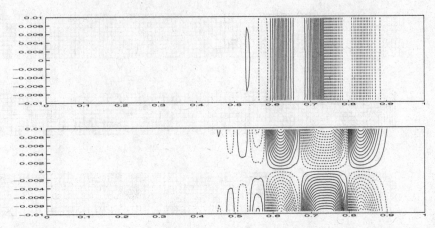

Fig. 22.2. The stress σ^{11} (top) and the vertical velocity w (bottom) at time $t = 0.5$ for the wave-propagation problem shown in Figure 22.1. Contour plots are shown with solid lines for positive values and dashed lines for negative values. [claw/book/chap22/plate]

Figure 22.1 shows the deformation of the plate at several times. This deformation has been determined by integrating the velocity (u, w) in each cell over the course of the computation in order to compute the displacement. Points on an initially uniform grid are then displaced by this amount to illustrate the deformed solid. For clarity these deformations are magnified considerably over what is appropriate for linear elasticity (i.e., the constant a in the boundary data (22.62) is taken to be orders of magnitude larger than physically reasonable, but since the equations are linear, the solution simply scales linearly as well). Recall that the computations are always done on a fixed grid in the reference configuration, since the elasticity equations we are considering are written in the Lagrangian frame.

Figure 22.2 shows contours of the computed σ^{11} and w at the final time $t = 0.5$. We see that σ^{11} essentially varies only with x in spite of the fact that the velocity w varies linearly in the z-direction. Moreover we observe that the leading edge of the wave has not yet reached $x = 0.9$ at this time. The material parameters were chosen to be $\rho = 1$, $\lambda = 2$, and $\mu = 1$ so that $c_p = 2$. Rather than traveling at this P-wave speed, the disturbance is propagating at the velocity $\hat{c}_p = \sqrt{3}$, and at time $t = 0.5$ has reached approximately $0.5\hat{c}_p = 0.866$ rather than $0.5c_p = 1$.

If this same computation were repeated but with $u = w = 0$ specified along $z = \pm h$ instead of $\sigma^{13} = \sigma^{33} = 0$ (plane strain rather than a thin plate), the result would look like the figure on the left of Figure 2.2 rather than Figure 22.1, and the disturbance would have reached $x = 1$ at time $t = 0.5$. In this case $w = 0$ would be exactly maintained and there would be no deformation in the z-direction.

It may seem strange that the speed of a $\hat{P}$-wave in a thin plate is smaller than c_p. After all, the plate is composed of a material characterized by the Lamé parameters λ and μ, and should have a characteristic propagation speed given by c_p. However, a $\hat{P}$-wave is not simply a P-wave propagating along the plate with deformation confined to the x–y plane. Rather, it is a wave that can be viewed as a superposition of many waves propagating transversely through the plate, bouncing back and forth between the top and bottom surfaces and hence moving at a slower effective speed along the plate. This internal reflection at the free surface is not apparent in Figure 22.2, since the zig-zagging waves effectively combine into standing waves in the z-direction. In a thicker plate the transverse wave motion would be more apparent. (A similar effect was seen in Section 9.14 for waves propagating in a layered heterogeneous material.)

22.6 A One-Dimensional Rod

Consider a thin elastic rod that is long in the x-direction and has small cross-sectional area, say $-h \leq y \leq h$ and $-h \leq z \leq h$. If we consider compressional waves propagating down the rod whose wavelength is long compared to h, then these can be modeled with a one-dimensional system of equations. We can start with the plane-stress equations (22.54), which model a thin plate $-h \leq z \leq h$, and now restrict also to $-h \leq y \leq h$ by imposing traction-free boundary conditions on these surfaces: $\sigma^{12} = \sigma^{22} = 0$. As in the derivation of the plane-stress equations, we now assume that in fact $\sigma^{12} = \sigma^{22} = 0$ throughout the rod, and also that $v = 0$ (in addition to the previous assumption that $w = 0$ and $\sigma^{k3} = 0$ for $k = 1, 2, 3$). Then (22.54) reduces to

$$
\epsilon_t^{11} - u_x = 0,
$$
$$
\rho u_t - \sigma_x^{11} = 0. \tag{22.63}
$$

From (22.52) we now have the constitutive relation $\epsilon^{11} = \sigma^{11}/E$, and so the one-dimensional system can be rewritten as

$$
\sigma_t^{11} - E u_x = 0,
$$
$$
\rho u_t - \sigma_x^{11} = 0. \tag{22.64}
$$

This has the same structure as the equations (22.43) derived for a one-dimensional slice of a three-dimensional solid. But the fact that the rod has traction-free boundaries and is free to contract or expand in y and z leads to a different wave speed,

$$
c_p^{\text{rod}} = \sqrt{E/\rho}, \tag{22.65}
$$

as seen by computing the eigenvalues of the coefficient matrix from (22.64).

22.7 Two-Dimensional Elasticity in Heterogeneous Media

The multidimensional elastic wave equations can be solved numerically using essentially the same procedure as for the linear acoustics equations, as was described in Chapter 21. This is also easily implemented for a heterogeneous medium by allowing each grid cell to have distinct values for the density ρ and Lamé parameters λ and μ. For the two-dimensional equations discussed in Section 22.2, the Riemann solvers in [claw/book/chap22/rp] give an implementation of the necessary eigendecompositions. These are based directly on the eigenstructure determined in Section 22.2. Examples will be presented for the plane-strain case described there, but the same solver work also for the plane-stress equations modeling a thin plate if λ is replaced by $\hat{\lambda}$ as described in Section 22.5.

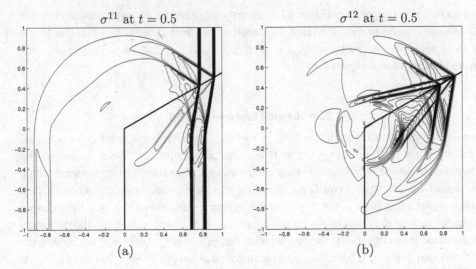

Fig. 22.3. Elastic wave propagation with initial data consisting of a P-wave as shown in Figure 21.1(b): (a) the stress σ^{11}; (b) the shear stress σ^{12}. [`claw/book/chap22/corner`]

Example 22.2. Figure 22.3 shows an example in the same domain indicated in Figure 21.1(a), but with the two regions now containing different elastic materials with parameters

$$
\begin{aligned}
\rho_l &= 1, & \rho_r &= 1, \\
\lambda_l &= 4, & \lambda_r &= 2, \\
\mu_l &= 0.5, & \mu_r &= 1, \\
c_{pl} &= \sqrt{5} \approx 2.2, & c_{pr} &= 2, \\
c_{sl} &= \sqrt{0.5} \approx 0.7, & c_{sr} &= 1.
\end{aligned}
\tag{22.66}
$$

The initial data is zero everywhere, except for a perturbation as in Figure 21.1(b), in which

$$
\sigma^{11} = \lambda_l + 2\mu_l, \quad \sigma^{22} = \lambda_l, \quad \sigma^{12} = 0, \quad u = c_{pl}, \quad v = 0 \tag{22.67}
$$

for $-0.35 < x < -0.2$. This is an eigenvector r^2 from (22.41) (with $\vec{n} = (1, 0)$) and hence is a right-going P-wave. After hitting the interface, the transmitted P-wave moves more slowly and there is a partial reflection, as seen in Figure 22.3(a), where a contour plot of σ^{11} is shown. In the elastic case there is also both a transmitted and reflected S-wave along the ramp portion of the interface. These are faintly visible in Figure 22.3(a), since S-waves at an oblique angle have a nonzero σ^{11} component (see Section 22.2). The S-waves are much more clearly visible in Figure 22.3(b), which shows σ^{12}. Note that the transmitted and reflected P-waves also contain significant components of σ^{12}, since they are moving at an angle to the grid.

Example 22.3. As another example of elastic wave propagation in a heterogeneous medium, we consider a wave propagating into a solid that has embedded within it an *inclusion* made out of a stiffer material, as shown in Figure 22.4. The darker region represents material with

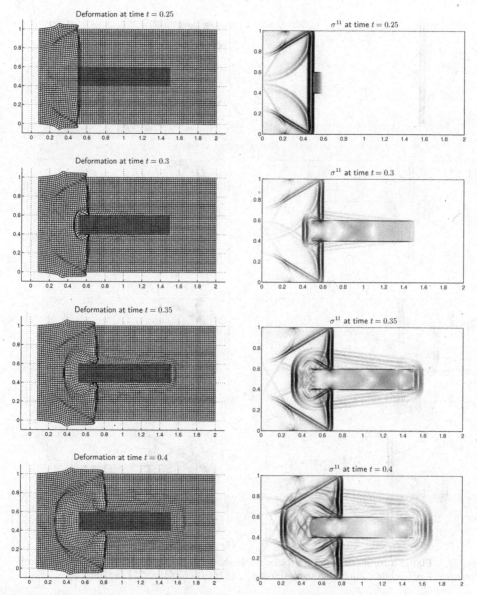

Fig. 22.4. Left column: Deformation of an elastic solid with a stiff inclusion due to compression at the left boundary. The linear deformation is greatly exaggerated. Right column: Schlieren image of σ^{11}. [claw/book/chap22/inclusion]

$\lambda = 200$, $\mu = 100$, while the lighter-colored material has $\lambda = 2$ and $\mu = 1$. The density is the same everywhere, $\rho = 1$. The plane-strain equations are solved with traction-free boundary conditions at $y = 0, 1$ and at $x = 1$. At $x = 0$ the velocity is specified as

$$u(0, y, t) = \begin{cases} \epsilon \sin(\pi t/0.025) & \text{if } t < 0.025, \\ 0 & \text{if } t \geq 0.025, \end{cases} \qquad v(0, y, t) = 0. \qquad (22.68)$$

The solid is simply pushed over by a small amount at the left boundary. This creates

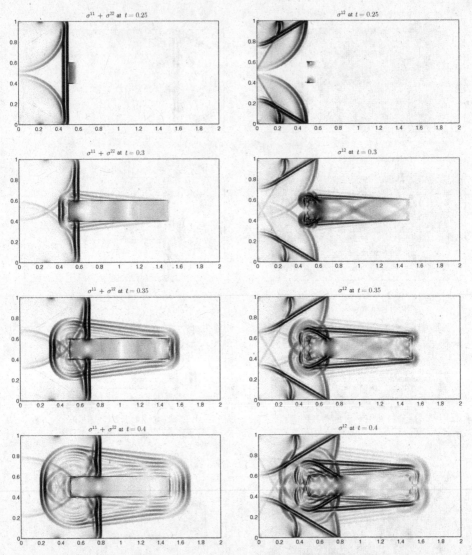

Fig. 22.5. Elastic wave propagation for the example shown in Figure 22.4. Left column: the mean stress $\sigma^{11} + \sigma^{22}$. Right column: the shear stress σ^{12}. [claw/book/chap22/inclusion]

a compressional wave, and the resulting wave motion is illustrated in Figure 22.4. The deformation of the solid is computed as described in Example 22.2. Again the displacements shown are much larger than actual linear displacements should be, and have been greatly exaggerated so that they will be visible.

Note that the compression wave is not purely a P-wave, due to its interaction with the free boundaries at $y = \pm 1$. At about $t = 0.25$ the wave hits the inclusion, which is much stiffer and hence tends to be pushed over as a rigid unit. At later times there is very little distortion of the inclusion, which simply shifts over slightly, launching smaller-amplitude waves into the exterior material at the far end and also along its length due to the resistance to shear. This is not really rigid motion, however. Elastic waves are rapidly bouncing back and forth in the stiff material to accomplish this motion. Note that $c_p = 20$ and $c_s = 10$ in the stiff region, whereas $c_p = 2$ and $c_s = 1$ outside.

Figure 22.5 shows the wave motion more clearly. Here the gradients of $\sigma^{11} + \sigma^{22}$ and σ^{12} are plotted at various times using a *schlieren* image style in which a highly nonlinear color scale is used, so that even very small-amplitude waves show up distinctly. At time $t = 0.25$ we see that waves are just beginning to travel down the stiff inclusion. By time $t = 0.3$ they have reached the far end. As they move along the bar, they launch waves into the surrounding material. The inclusion then begins to vibrate, giving rise to further waves moving vertically away from it.

Finite Volume Methods on Quadrilateral Grids

Many multidimensional problems of practical interest involve complex geometry, and in general it is not sufficient to be able to solve hyperbolic equations on a uniform Cartesian grid in a rectangular domain. In Section 6.17 we considered a nonuniform grid in one space dimension and saw how hyperbolic equations can be solved on such a grid by using a uniform grid in *computational space* together with a coordinate mapping and appropriate scaling of the flux differences using *capacity form differencing*. The *capacity* of the computational cell is determined by the size of the corresponding physical cell.

In this chapter we consider nonuniform finite volume grids in two dimensions, such as those shown in Figure 23.1, and will see that similar techniques may be used. There are various ways to view the derivation of finite volume methods on general multidimensional grids. Here we will consider a direct physical interpretation in terms of fluxes normal to the cell edges. For simplicity we restrict attention to two space dimensions. For some other discussions of finite volume methods on general grids, see for example [156], [245], [475], [476].

The grids shown in Figures 23.1(a) and (b) are logically rectangular *quadrilateral grids*, and we will concentrate on this case. Each cell is a quadrilateral bounded by four linear segments. Such a grid is also often called a *curvilinear grid*. If we label the cells in a logical manner, indexing "rows" and "columns" by i and j, then cell (i, j) has the four neighbors $(i \pm 1, j)$ and $(i, j \pm 1)$. The grid can be made to wrap around the cylinder by imposing periodic boundary conditions in one direction.

The triangulation shown in Figure 23.1(c), on the other hand, gives an *unstructured grid* for which there is no simple logical structure underlying the connectivity between cells. One must keep track of the neighbors of each cell explicitly. An unstructured triangulation is often easier to generate for complicated geometries than a structured grid such as the one shown in Figure 23.1(a), but may be somewhat more difficult to work with in regard to data structures and the development of fast and accurate solvers. See [20], [324], [385], [472] for some discussions of unstructured grids.

The grids of Figure 23.1(a)–(c) are all *body-fitted grids* that conform to the geometry of the problem. Figure 23.1(d) shows a different approach in which a uniform Cartesian grid is used over most of the domain, but with some smaller irregular cells allowed where the boundary cuts through the grid. Finite volume methods of this type are often called *Cartesian-grid* or *embedded-boundary* methods. For problems with more complicated geometry this approach allows for very easy grid generation, and so it has recently become quite popular. With these

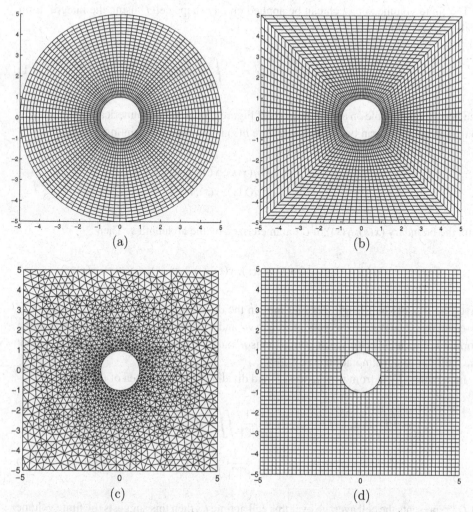

Fig. 23.1. Four possible grids for flow around a cylinder. (a) 25×100 polar grid in r–θ. (b) 25×104 quadrilateral grid interpolating between the cylinder and a square computational domain. (c) An unstructured triangulation. (d) A Cartesian grid with embedded boundary.

methods the main difficulty is in specifying fluxes at the edges of the small cut cells in such a way that good accuracy is obtained and stability is preserved with reasonable-size time steps. A variety of different approaches have been introduced; see for example [7], [31], [57], [59], [77], [104], [140], [277], [278], [358], [363], [490], [492].

23.1 Cell Averages and Interface Fluxes

With a finite volume method we view each discrete value Q as a cell average over a grid cell, which is modified due to fluxes through the edges of the cell, in the case of a conservation law, or more generally by waves moving into the cell from each edge. We start by discussing the case of a conservation law and will see how to extend this to nonconservative linear systems in Section 23.4.

The finite volume approach can be applied on any shape cell C using the integral form of the conservation law,

$$\frac{d}{dt} \iint_C q(x, y, t)\, dx\, dy = -\int_{\partial C} \vec{n}(s) \cdot \vec{f}(s, t)\, ds, \qquad (23.1)$$

and hence is applicable on any of the grids of Figure 23.1 if a good numerical approximation to the interface flux can be determined. Here $\vec{n}(s)$ is the outward-pointing normal and

$$\vec{f}(s, t) = \begin{bmatrix} f(q(x(s), y(s), t)) \\ g(q(x(s), y(s), t)) \end{bmatrix},$$

with the boundary $(x(s), y(s))$ of C parameterized by the arclength s. Then

$$\check{F}(s) = \vec{n}(s) \cdot \vec{f}(s, t) = n^x(s)\, f(q(x(s), y(s), t)) + n^y(s)\, g(q(x(s), y(s), t))$$

gives the flux per unit length per unit time in the direction $\vec{n}(s)$. Note that for a system of m equations f and g are vectors of length m and $\vec{f}$ is a vector of length $2m$, while the normal $\vec{n}(s)$ contains two scalar components n^x and n^y. (See Section 18.1 for more about the multidimensional notation used here.)

Integrating (23.1) from time t_n to t_{n+1} and dividing by $|C|$, the area of the cell, gives

$$\frac{1}{|C|} \iint_C q(x, y, t_{n+1})\, dx\, dy = \frac{1}{|C|} \iint_C q(x, y, t_n)\, dx\, dy$$
$$- \frac{1}{|C|} \int_{t_n}^{t_{n+1}} \int_{\partial C} \vec{n}(s) \cdot \vec{f}(s, t)\, ds\, dt. \qquad (23.2)$$

If Q^n represents the cell average over this cell at time t_n, then this suggests the finite volume method

$$Q^{n+1} = Q^n - \frac{\Delta t}{|C|} \sum_{j=1}^N h_j \check{F}_j^n, \qquad (23.3)$$

where $\check{F}_j^n$ represents a numerical approximation to the average normal flux across the jth side of the cell, N is the number of sides, and h_j is the length of the jth side. The factors Δt and h_j are introduced by taking $\check{F}_j^n$ as an approximation to the interface flux per unit length, per unit time,

$$\check{F}_j^n \approx \frac{1}{\Delta t} \int_{t_n}^{t_{n+1}} \left(\frac{1}{h_j} \int_{\text{side } j} \vec{n} \cdot \vec{f}(s, t)\, ds \right) dt.$$

This agrees with the normalization used previously on Cartesian grids.

23.2 Logically Rectangular Grids

From now on we will consider only logically rectangular grids such as the ones shown in Figure 23.1(a) and (b). In this case we can write the finite volume method (23.3) as

$$Q_{ij}^{n+1} = Q_{ij} - \frac{\Delta t}{|\mathcal{C}_{ij}|} \left(h_{i+1/2,j} \breve{F}_{i+1/2,j} - h_{i-1/2,j} \breve{F}_{i-1/2,j} \right.$$

$$\left. + h_{i,j+1/2} \breve{G}_{i,j+1/2} - h_{i,j-1/2} \breve{G}_{i,j-1/2} \right), \tag{23.4}$$

where

$$|\mathcal{C}_{ij}| = \text{area of cell } (i, j),$$

$$h_{i-1/2,j} = \text{length of side between cells } (i - 1, j) \text{ and } (i, j),$$

$$h_{i,j-1/2} = \text{length of side between cells } (i, j - 1) \text{ and } (i, j),$$

$$\breve{F}_{i-1/2,j} = \text{flux normal to edge between cells } (i - 1, j) \text{ and } (i, j),$$

$$\text{per unit time, per unit length,}$$

$$\breve{G}_{i,j-1/2} = \text{flux normal to edge between cells } (i, j - 1) \text{ and } (i, j),$$

$$\text{per unit time, per unit length.}$$

On a uniform Cartesian grid, $|\mathcal{C}_{ij}| = \Delta x \, \Delta y$, $h_{i-1/2,j} = \Delta y$, $h_{i,j-1/2} = \Delta x$, and (23.4) reduces to the standard flux-differencing formula (19.10).

Just as in one space dimension, we can put the general finite volume method (23.4) into a form where it can be viewed as capacity-form differencing on a uniform grid in computational space, which we now denote by ξ–η coordinates, as illustrated in Figure 23.2. The vertices (corners of grid cells) are mapped from the uniform computational grid to points in the physical domain by two coordinate-mapping functions $X(\xi, \eta)$ and $Y(\xi, \eta)$. We introduce the vector $\vec{h}_{i-1/2,j}$ as the vector connecting two corners of the grid cell, so that $h_{i-1/2,j}$ is the length of this vector.

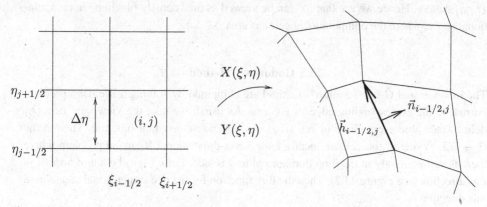

Fig. 23.2. The computational grid cells shown on the left are mapped to the physical grid cells shown on the right.

If we set

$$\kappa_{ij} = \frac{|\mathcal{C}_{ij}|}{\Delta\xi\,\Delta\eta},$$

$$F_{i-1/2,j} = \left(\frac{h_{i-1/2,j}}{\Delta\eta}\right)\breve{F}_{i-1/2,j}, \tag{23.5}$$

$$G_{i,j-1/2} = \left(\frac{h_{i,j-1/2}}{\Delta\xi}\right)\breve{G}_{i,j-1/2},$$

then the method (23.4) can be rewritten as

$$Q_{ij}^{n+1} = Q_{ij} - \frac{\Delta t}{\kappa_{ij}\,\Delta\xi}\left(F_{i+1/2,j} - F_{i-1/2,j}\right) - \frac{\Delta t}{\kappa_{ij}\,\Delta\eta}\left(G_{i,j+1/2} - G_{i,j-1/2}\right). \tag{23.6}$$

Note that $\breve{F}_{i-1/2,j}$ has units of flux per unit length in physical space (per unit time), so that multiplying by $h_{i-1/2,j}$ gives the total flux along the edge. Dividing by $\Delta\eta$ converts this into flux per unit length in computational space.

The length ratios appearing in (23.5) arise in many formulas below, and we will denote them by

$$\gamma_{i-1/2,j} = h_{i-1/2,j}/\Delta\eta,$$

$$\gamma_{i,j-1/2} = h_{i,j-1/2}/\Delta\xi. \tag{23.7}$$

These quantities relate length in physical space to length in computational space. Similarly, the capacity κ_{ij} defined in (23.5) is an area ratio, between the area of the physical grid cell and the area $\Delta\xi\,\Delta\eta$ of the computational cell. If the mappings $X(\xi,\eta)$ and $Y(\xi,\eta)$ are sufficiently smooth functions, then these ratios can be related to derivatives of the mappings. However, we do not need to assume any smoothness in the mappings or the resulting grid in order to apply the finite volume methods. The accuracy may be reduced if the grid is not smooth, but the high-resolution methods typically perform quite well (see Example 23.2 in Section 23.8 below for an illustration of this on the grid shown in Figure 23.1(b)).

Note that if q represents a density function in physical space, so that its integral over the cell is the total mass, then the total mass in the (i,j) grid cell is roughly $Q_{ij}|\mathcal{C}_{ij}| = Q_{ij}\kappa_{ij}\,\Delta\xi\,\Delta\eta$. Hence we see that $q\kappa$ can be viewed as the "density function" in computational space, where the computational cell has area $\Delta\xi\,\Delta\eta$.

23.3 Godunov's Method

The fluxes $\breve{F}$ and $\breve{G}$ for Godunov's method are computed by solving a Riemann problem normal to the corresponding edge of the cell. As usual, we take the viewpoint that Q_{ij} defines the value everywhere in cell (i,j) of a piecewise constant function. Then at the $(i-1/2,j)$ side of the cell we locally have a one-dimensional Riemann problem where there is variation only in the direction normal to this side. Let $\vec{n}_{i-1/2,j}$ be a unit normal in this direction (see Figure 23.2). Then the flux function for the one-dimensional equation in this direction is

$$\breve{F}(q) = \vec{n}_{i-1/2,j}\cdot\vec{f}(q).$$

Solving the Riemann problem with this flux function and data $Q_{i-1,j}$ and Q_{ij} gives the Godunov flux

$$\breve{F}_{i-1/2,j} = \vec{n}_{i-1/2,j} \cdot \vec{f}(Q^\vee_{i-1/2,j}), \tag{23.8}$$

where, as usual, $Q^\vee_{i-1/2,j}$ is the solution at the interface (i.e., moving at speed zero) in the self-similar solution to the one-dimensional Riemann problem. Multiplying $\breve{F}_{i-1/2,j}$ by $\gamma_{i-1/2,j}$ gives the numerical flux $F_{i-1/2,j}$. Similarly,

$$\breve{G}_{i,j-1/2} = \vec{n}_{i,j-1/2} \cdot \vec{f}(Q^\vee_{i,j-1/2}), \tag{23.9}$$

where $Q^\vee_{i,j-1/2}$ is obtained by solving the Riemann problem normal to the $(i, j-1/2)$ edge with data $Q_{i,j-1}$ and Q_{ij}.

23.4 Fluctuation Form

We derived the form (23.6) for a conservation law by considering the flux through each edge of the grid cell. For a more general hyperbolic equation that may not be in conservation form, e.g.,

$$q_t + A(x, y)q_x + B(x, y)q_y = 0, \tag{23.10}$$

this approach cannot be used, and instead we wish to use a more general fluctuation updating formula in place of (23.6), of the form

$$Q_{ij}^{n+1} = Q_{ij} - \frac{\Delta t}{\kappa_{ij} \Delta \xi} \left(\mathcal{A}^+ \Delta Q_{i-1/2,j} + \mathcal{A}^- \Delta Q_{i+1/2,j}\right)$$

$$- \frac{\Delta t}{\kappa_{ij} \Delta \eta} \left(\mathcal{B}^+ \Delta Q_{i,j-1/2} + \mathcal{B}^- \Delta Q_{i,j+1/2}\right). \tag{23.11}$$

This form can also be obtained by solving the appropriate Riemann problem normal to each edge of the cell and seeing how the resulting waves update cell averages to either side. We will see that the formulation earlier developed in Sections 6.14–6.16 and 19.3.3 can be extended to general quadrilateral grids. High-resolution correction terms can also be added to (23.11).

We assume that we know how to solve the Riemann problem normal to each edge based on piecewise constant initial data in the two neighboring cells and the direction of the normal, and that the solution consists of waves $\mathcal{W}^p$ propagating at speeds $\breve{s}^p$. These waves and speeds are now used to define the fluctuations, as on a uniform Cartesian grid, but the additional scale factors γ and κ must be properly included. To see where these arise, consider the edge $(i-1/2, j)$. The wave $\mathcal{W}^p_{i-1/2,j}$ moves a distance $\breve{s}^p_{i-1/2,j} \Delta t$ and has width $h_{i-1/2,j}$. If $\breve{s}^p_{i-1/2,j} > 0$, for example, then this wave should update Q_{ij} by an amount

$$-\left(\frac{h_{i-1/2,j} \, \breve{s}^p_{i-1/2,j} \Delta t}{|\mathcal{C}_{ij}|}\right) \mathcal{W}^p_{i-1/2,j}, \tag{23.12}$$

where we divide by $|\mathcal{C}_{ij}|$, the area of the cell, since Q_{ij} represents a cell average. Since we assume $\breve{s}^p_{i-1/2,j} > 0$, this term should be part of the update arising from the term $-(\frac{\Delta t}{\kappa_{ij}\,\Delta\xi})\mathcal{A}^+\Delta Q_{i-1/2,j}$ in (23.11). Since $|\mathcal{C}_{ij}| = \kappa_{ij}\,\Delta\xi\,\Delta\eta$, we can rewrite (23.12) as

$$-\left(\frac{\Delta t}{\kappa_{ij}\,\Delta\xi}\right)\left(\frac{h_{i-1/2,j}}{\Delta\eta}\breve{s}^p_{i-1/2,j}\right)\mathcal{W}^p_{i-1/2,j}. \tag{23.13}$$

We see that this wave's contribution to $\mathcal{A}^+\Delta Q$ should be $s^p_{i-1/2,j}\mathcal{W}^p_{i-1/2,j}$, where we define

$$s^p_{i-1/2,j} = \frac{h_{i-1/2,j}}{\Delta\eta}\breve{s}^p_{i-1/2,j} = \gamma_{i-1/2,j}\breve{s}^p_{i-1/2,j}. \tag{23.14}$$

It is this scaled speed $s^p_{i-1/2,j}$ that must also be used in high-resolution correction terms. So, after solving the Riemann problem, the wave speeds should all be scaled as in (23.14), and then the fluctuations are

$$\mathcal{A}^{\pm}\Delta Q_{i-1/2,j} = \sum_p \left(s^p_{i-1/2,j}\right)^{\pm}\mathcal{W}^p_{i-1/2,j}, \tag{23.15}$$

as usual.

Similarly, at the $(i, j - 1/2)$ edge we solve the normal Riemann problem for waves $\mathcal{W}^p_{i,j-1/2}$ and speeds $\breve{s}^p_{i,j-1/2}$ and then scale the speeds by $\gamma_{i,j-1/2} = h_{i,j-1/2}/\Delta\xi$:

$$s^p_{i,j-1/2} = \gamma_{i,j-1/2}\breve{s}^p_{i,j-1/2}. \tag{23.16}$$

The fluctuations are then

$$\mathcal{B}^{\pm}\Delta Q_{i,j-1/2} = \sum_p \left(s^p_{i,j-1/2}\right)^{\pm}\mathcal{W}^p_{i,j-1/2}. \tag{23.17}$$

23.5 Advection Equations

The constant-coefficient advection equation in two dimensions is written as

$$q_t + uq_x + vq_y = 0, \tag{23.18}$$

with $\vec{u} = (u, v)$ being the velocity vector. The flux vector is

$$\vec{f}(q) = \vec{u}q = \begin{bmatrix} uq \\ vq \end{bmatrix}.$$

The flux at the $(i - 1/2, j)$ edge is

$$\breve{F}_{i-1/2,j} = \vec{n}_{i-1/2,j} \cdot \vec{f}(Q^{\vee}_{i-1/2,j}) = \breve{u}_{i-1/2,j}Q^{\vee}_{i-1/2,j},$$

where

$$\breve{u}_{i-1/2,j} = un^x_{i-1/2,j} + vn^y_{i-1/2,j}$$

is the velocity normal to the edge, and $Q^{\vee}_{i-1/2,j}$ is the cell value from the upwind side of the interface,

$$Q^{\vee}_{i-1/2,j} = \begin{cases} Q_{i-1,j} & \text{if } \breve{u}_{i-1/2,j} > 0, \\ Q_{ij} & \text{if } \breve{u}_{i-1/2,j} < 0. \end{cases}$$

After normalizing the flux by the ratio $\gamma_{i-1/2,j} = h_{i-1/2,j}/\Delta\eta$, we obtain

$$F_{i-1/2,j} = \gamma_{i-1/2,j}\breve{u}_{i-1/2,j}Q^{\vee}_{i-1/2,j}$$

$$= \gamma_{i-1/2,j}\left(\breve{u}^+_{i-1/2,j}Q_{i-1,j} + \breve{u}^-_{i-1/2,j}Q_{ij}\right). \tag{23.19}$$

Similarly, we find that

$$G_{i,j-1/2} = \gamma_{i,j-1/2}\left(\breve{v}^+_{i,j-1/2}Q_{i,j-1} + \breve{v}^-_{i,j-1/2}Q_{ij}\right), \tag{23.20}$$

where

$$\breve{v}_{i,j-1/2} = n^x_{i,j-1/2}u + n^y_{i,j-1/2}v$$

is the velocity normal to the edge $\vec{h}_{i,j-1/2}$. Using these fluxes in (23.6) gives Godunov's method (the donor-cell upwind (DCU) method).

Alternatively, we can incorporate the length ratios γ into the definitions of the velocities at cell edges, defining *edge velocities*

$$U_{i-1/2,j} = \gamma_{i-1/2,j}\breve{u}_{i-1/2,j},$$
$$V_{i,j-1/2} = \gamma_{i,j-1/2}\breve{v}_{i,j-1/2}. \tag{23.21}$$

In terms of these velocities we have fluxes

$$F_{i-1/2,j} = U^+_{i-1/2,j}Q_{i-1,j} + U^-_{i-1/2,j}Q_{ij},$$
$$G_{i,j-1/2} = V^+_{i,j-1/2}Q_{i,j-1} + V^-_{i,j-1/2}Q_{ij}. \tag{23.22}$$

This method can also be rewritten in terms of fluctuations, as shown in the next subsection. With this notation, the method looks exactly like the donor-cell method of Section 20.1 on a Cartesian grid for the variable-coefficient advection equation with velocities $U_{i-1/2,j}$ and $V_{i,j-1/2}$. Indeed, the computational $\xi-\eta$ grid is Cartesian, and we see that the edge velocities give the appropriate velocity field in computational space. Note that in computational space the velocity field typically varies in space, even though we started with a constant-coefficient velocity in physical space.

23.5.1 Fluctuation Form

We have derived the DCU method for advection on a general quadrilateral grid in the form of conservative differencing (23.6), and obtained the numerical fluxes (23.22). We can rewrite this method in the fluctuation form (23.11) following the procedure of Section 23.4. The Riemann problem at $(i - 1/2, j)$ gives one wave $\mathcal{W}_{i-1/2,j} = Q_{ij} - Q_{i-1,j}$ with speed $\breve{s}^1_{i-1/2,j} = \breve{u}_{i-1/2,j}$, the normal velocity at this edge. This speed must be scaled as in (23.14) to obtain the appropriate speed to be used in computing the fluctuations and later the high-resolution corrections. Doing so results again in the *edge velocities* defined in (23.21), i.e., $s^1 = U$ and $s^2 = V$. We then have the following formulas for the fluctuations:

$$\mathcal{A}^{\pm} \Delta Q_{i-1/2,j} = U_{i-1/2,j}^{\pm}(Q_{ij} - Q_{i-1,j}),$$
$$\mathcal{B}^{\pm} \Delta Q_{i,j-1/2} = V_{i,j-1/2}^{\pm}(Q_{ij} - Q_{i,j-1}). \tag{23.23}$$

We have now derived two distinct approaches to solving the original equation (23.18) on a quadrilateral grid. One approach is in conservative form using fluxes (23.22), and the other is in advective form using the fluctuations (23.23). The original equation (23.18) has constant coefficients, and on a uniform Cartesian grid these two forms would be identical. On the quadrilateral grid, however, the edge velocities are not generally constant, and so it is not clear that the different forms will be identical. In particular, one might question whether the form based on advective fluctuations will yield a conservative method. In fact it does, but this relies on a certain discrete divergence-free condition being automatically satisfied for the normal velocities at each edge,

$$h_{i+1/2,j}\breve{u}_{i+1/2,j} - h_{i-1/2,j}\breve{u}_{i-1/2,j} + h_{i,j+1/2}\breve{v}_{i,j+1/2} - h_{i,j-1/2}\breve{v}_{i,j-1/2} = 0. \tag{23.24}$$

This is zero because it is exactly equal to the integral of $\vec{n} \cdot \vec{u}$ around the boundary of the (i, j) cell and the constant velocity $\vec{u}$ is divergence-free. Using this, we can verify that

$$\Delta \xi \, \Delta \eta \sum_{i,j} \kappa_{ij} Q_{ij}^{n+1} = \Delta \xi \, \Delta \eta \sum_{i,j} \kappa_{ij} Q_{ij}^{n}, \tag{23.25}$$

as required for conservation, by multiplying (23.11) by $\kappa_{ij} \Delta \xi \, \Delta \eta$, summing over all grid cells, and rearranging the sums of the correction terms to collect together all terms involving a single Q_{ij}, whose coefficient is then found to be proportional to (23.24) and hence vanishes.

23.5.2 Variable-Coefficient Advection

The constant-coefficient advection equation in physical space generally becomes a variable-coefficient problem in the computational domain, since the normal velocity at an edge varies with the grid orientation. If the original advection equation has variable coefficients in physical space, i.e., a nonconstant velocity field $(u(x, y), v(x, y))$, then we must in general distinguish between the conservative equation and the advective-form color equation. Either of the approaches outlined above (conservative form or fluctuation form) can be extended to varying velocities simply by determining the average normal velocities $\breve{u}_{i-1/2,j}$ and $\breve{v}_{i,j-1/2}$ at each edge based on the varying velocity field, and then proceeding exactly as before. Ideally we would like to define these normal velocities by exactly computing the integrals

$$\breve{u}_{i-1/2,j} = \frac{1}{h_{i-1/2,j}} \int \left[n_{i-1/2,j}^{x} u(x(s), y(s)) + n_{i-1/2,j}^{y} v(x(s), y(s)) \right] ds, \tag{23.26}$$

where s is arclength along the edge of the physical grid cell. In general this integral may be difficult to compute, but in the special case where the velocity field is divergence-free and defined by a stream function $\psi(x, y)$ as described in Section 20.8.1, this reduces to a very simple formula. The velocity is related to the stream function by $u = \psi_y$ and $v = -\psi_x$, and as a result the integrand in (23.26) becomes

$$n^x_{i-1/2,j}u(x(s), y(s)) + n^y_{i-1/2,j}v(x(s), y(s)) = n^x_{i-1/2,j}\psi_y - n^y_{i-1/2,j}\psi_x,$$

which is simply the derivative ψ_s of ψ along the edge. This is a general feature of the stream function: the velocity in any direction is obtained by differentiating the stream function in the orthogonal direction. Now the fundamental theorem of calculus can be applied to the integral in (23.26) to obtain

$$\breve{u}^1_{i-1/2,j} = \frac{1}{h_{i-1/2,j}} \int \psi_s \, ds$$

$$= \frac{\psi\left(x_{i-1/2,j+1/2}, y_{i-1/2,j+1/2}\right) - \psi\left(x_{i-1/2,j-1/2}, y_{i-1/2,j-1/2}\right)}{h_{i-1/2,j}}, \quad (23.27)$$

where the corners of the grid cells are determined by the mapping functions, e.g.,

$$x_{i-1/2,j-1/2} = X\left(\xi_{i-1/2}, \eta_{j-1/2}\right), \quad y_{i-1/2,j-1/2} = Y\left(\xi_{i-1/2}, \eta_{j-1/2}\right). \quad (23.28)$$

In computing the edge velocity we multiply (23.27) by $\gamma_{i-1/2,j}$, so that the edge length drops out and we obtain

$$U_{i-1/2,j} = \frac{\psi\left(x_{i-1/2,j+1/2}, y_{i-1/2,j+1/2}\right) - \psi\left(x_{i-1/2,j-1/2}, y_{i-1/2,j-1/2}\right)}{\Delta\eta}. \quad (23.29)$$

Similarly,

$$V_{i,j-1/2} = -\frac{\psi\left(x_{i+1/2,j-1/2}, y_{i+1/2,j-1/2}\right) - \psi\left(x_{i-1/2,j-1/2}, y_{i-1/2,j-1/2}\right)}{\Delta\xi}. \quad (23.30)$$

Note that this is the standard approach for handling a stream function on the Cartesian computational grid, as developed in Section 20.8.1. The only new feature introduced is the use of the grid mapping (23.28) in evaluating the stream function at the corners of the computational-grid cells. Differencing the stream function between two corners of this Cartesian grid and dividing by the length of the side gives the average normal velocity.

Example 23.1. We solve a problem with a constant velocity field $u = 1, v = 0$ (using the stream function $\psi(x, y) = y$) in a quarter annulus using polar coordinates, as illustrated in Figure 23.3. The mapping function is

$$X(\xi, \eta) = \xi \cos(\eta), \quad Y(\xi, \eta) = \xi \sin(\eta) \quad (23.31)$$

for $0.5 \leq \xi \leq 2.5$ and $0 \leq \eta \leq \pi/2$.

The initial data is $q \equiv 0$, and the boundary data is

$$q(0, y, t) = \begin{cases} 1 & \text{if } 1 < y < 1.5, \\ 0 & \text{otherwise} \end{cases}$$

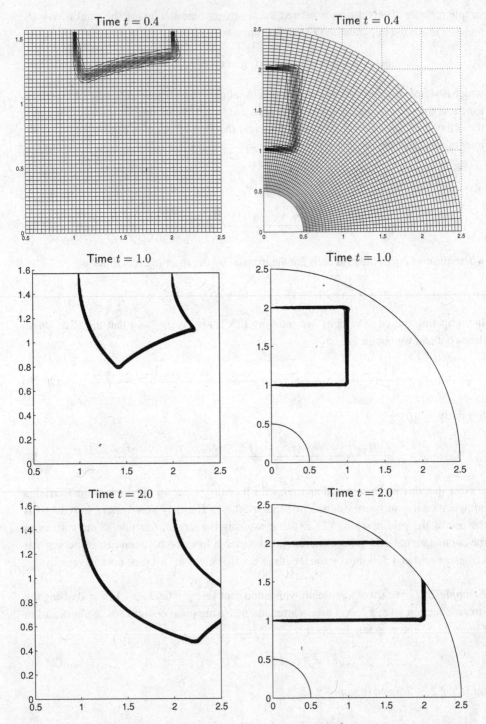

Fig. 23.3. Solution to the advection equation $q_t + q_x = 0$ in polar coordinates, as described in Example 23.1. Left column: on the computational grid. Right column: on the physical grid. Top row: on a 50×50 grid. Later times: on a 200×200 grid. Contour lines are at $q = 0.05, 0.15, \ldots, 0.95$.
[claw/book/chap23/advection/polar]

with extrapolation at the other boundaries. The solution consists of a band of tracer moving horizontally inward from the left edge. The right-hand column shows the computed solution at three times in the physical domain. The left-hand column shows the solution in the computational domain, where the grid is uniform but the velocity field is not constant. The top row shows that solution at time $t = 0.4$ on a 50×50 grid, along with the grid in both computational and physical space. The results at later times are from a finer-grid computation, on a 200×200 grid.

Note that the top boundary ($\eta = \pi/2$) of the computational grid is mapped to the y-axis in the physical domain, and this is where the boundary conditions must be set. The superbee limiter has been used for this test, and the discontinuity in q is smeared over only two or three grid cells everywhere.

23.6 Acoustics

We now consider the constant-coefficient acoustics equations (18.24) on a quadrilateral grid. We consider the fluctuation approach outlined in Section 23.4. Solving the acoustics equations normal to each cell edge yields two acoustic waves, which update the cell averages on either side. To obtain the proper speeds we must scale the sound speed c by the edge ratios γ as described in Section 23.4. The main new complication that arises with acoustics (and also with fluid dynamic equations such as the shallow water or Euler equations) is that the solution q is no longer a scalar but rather a vector $q = (p, u, v)$, which contains the velocity as part of the solution. In each grid cell we store the Cartesian components of the velocity vector, i.e., $Q_{ij}^2 = u_{ij}$ is the x-component of velocity while $Q_{ij}^3 = v_{ij}$ is the y-component. To solve the acoustics equations normal to an edge we need to know the components of velocity normal and tangential to that edge in each of the two grid cells on either side. The normal velocity components, along with the pressure in each cell, determine the resulting acoustic waves. Any jump in the tangential velocity remains stationary at the cell edge. This jump constitutes the shear wave moving with speed zero that arises in two-dimensional acoustics (with background velocity $\vec{u}_0 = (u_0, v_0) = (0, 0)$ in the derivation of Section 18.2) and can be ignored in the finite volume method, since it does not affect the cell average to either side. For clarity in the discussion below and consistency of notation, we include formulas for all three waves. But in an implementation only the 1-wave and 3-wave need to be propagated. (If we solved the equations with nonzero $\vec{u}_0$, then the jump in the tangential velocity would propagate at a velocity $\vec{n} \cdot \vec{u}_0$ and a third wave with this speed would be needed, in addition to the acoustic waves propagating at speeds $(\vec{n} \cdot \vec{u}_0) \pm c$.)

23.6.1 Solving the Normal Riemann Problem

To determine the waves, speeds, and fluctuations at edge $(i - 1/2, j)$, for example, we must solve the Riemann problem for a one-dimensional system of equations normal to the interface, with data $Q_{i-1,j}$ and Q_{ij}. We will consider two possible approaches to solving the Riemann problem in the direction $\vec{n}_{i-1/2,j}$, which give equivalent results. One approach is based directly on the theory of Chapter 18: we solve the one-dimensional Riemann problem with coefficient matrix $\breve{A} = \vec{n}_{i-1/2,j} \cdot \vec{A}$ and data $Q_{i-1,j}$ and Q_{ij}. We first consider this approach.

The second approach, illustrated for acoustics in Section 23.6.3, consists of explicitly determining the normal and tangential components of velocity from $Q_{i-1,j}$ and Q_{ij}, re-labeling these as the "x" and "y" components, and then solving the Riemann problem $q_t + Aq_x = 0$ in the "x" direction with this modified data. The velocity components of the resulting waves must then be recombined to obtain the proper updates in physical space. This may be conceptually easier to implement in cases where a complicated Riemann solver has already been written for Cartesian coordinates. This approach only works for isotropic equations, but applies to many physically relevant problems in fluid or solid mechanics, including the Euler and shallow water equations as well as acoustics.

For the acoustics equations we can directly solve a system of the form $q_t + \breve{A}_{i-1/2,j} q_n = 0$, where q_n denotes the derivative in the normal direction and, from (18.28),

$$\breve{A}_{i-1/2,j} = n^x_{i-1/2,j} A + n^y_{i-1/2,j} B$$

$$= \begin{bmatrix} 0 & n^x_{i-1/2,j} K_0 & n^y_{i-1/2,j} K_0 \\ n^x_{i-1/2,j}/\rho_0 & 0 & 0 \\ n^y_{i-1/2,j}/\rho_0 & 0 & 0 \end{bmatrix}. \tag{23.32}$$

For clarity we drop the subscript $i - 1/2$, j from n below, and also from the eigenvalues, eigenvectors, and α-coefficients. From Section 18.5 we know that this matrix has eigenvalues and eigenvectors

$$\breve{\lambda}^1 = -c_0, \qquad \breve{\lambda}^2 = 0, \qquad \breve{\lambda}^3 = c_0,$$

$$\breve{r}^1 = \begin{bmatrix} -Z_0 \\ n^x \\ n^y \end{bmatrix}, \quad \breve{r}^2 = \begin{bmatrix} 0 \\ -n^y \\ n^x \end{bmatrix}, \quad \breve{r}^3 = \begin{bmatrix} Z_0 \\ n^x \\ n^y \end{bmatrix}. \tag{23.33}$$

In fact we can just as easily solve the Riemann problem for the variable-coefficient case where the impedance Z_{ij} and sound speed c_{ij} have different values in each grid cell, and so the more general formulas are presented. In this case we wish to decompose $\delta \equiv Q_{ij} - Q_{i-1,j}$ into waves using

$$\breve{\lambda}^1 = -c_{i-1,j}, \qquad \breve{\lambda}^2 = 0, \qquad \breve{\lambda}^3 = c_{ij},$$

$$\breve{r}^1 = \begin{bmatrix} -Z_{i-1,j} \\ n^x \\ n^y \end{bmatrix}, \quad \breve{r}^2 = \begin{bmatrix} 0 \\ -n^y \\ n^x \end{bmatrix}, \quad \breve{r}^3 = \begin{bmatrix} Z_{ij} \\ n^x \\ n^y \end{bmatrix}. \tag{23.34}$$

Writing δ as a linear combination of these vectors yields the waves

$$\mathcal{W}^p = \alpha^p \breve{r}^p \quad \text{for } p = 1, 2, 3, \tag{23.35}$$

where

$$\alpha^1 = \frac{-\delta^1 + n^x Z_{i-1,j}\delta^2 + n^y Z_{i-1,j}\delta^3}{Z_{i-1,j} + Z_{ij}},$$

$$\alpha^2 = -n^y \delta^2 + n^x \delta^3, \qquad\qquad (23.36)$$

$$\alpha^3 = \frac{\delta^1 + n^x Z_{ij}\delta^2 + n^y Z_{ij}\delta^3}{Z_{i-1,j} + Z_{ij}}.$$

The wave speeds are obtained by scaling $\check{\lambda}^1$ and $\check{\lambda}^3$ as described above:

$$s^1 = -\gamma_{i-1/2,j}c_{i-1,j},$$

$$s^2 = 0, \qquad\qquad (23.37)$$

$$s^3 = \gamma_{i-1/2,j}c_{ij}.$$

The fluctuations are then given by

$$\mathcal{A}^- \Delta Q = s^1 \mathcal{W}^1,$$

$$\mathcal{A}^+ \Delta Q = s^3 \mathcal{W}^3, \qquad\qquad (23.38)$$

since $s^1 < 0$ while $s^3 > 0$. All of these quantities should be indexed by $i - 1/2, j$. The resulting waves and speeds are also appropriate for use in the high-resolution correction terms, along with limiters in the usual form. This Riemann solver and an example may be found at [claw/book/chap23/acoustics].

23.6.2 Transverse Riemann Solvers

The normal Riemann solver described above is all that is needed for a dimensionally split method. To use the unsplit methods developed in Chapter 21, we must also define transverse Riemann solvers. For the acoustics equations we start with a fluctuation, say $\mathcal{A}^+ \Delta Q_{i-1/2,j}$, and must determine $\mathcal{B}^- \mathcal{A}^+ \Delta Q_{i-1/2,j}$ and $\mathcal{B}^+ \mathcal{A}^+ \Delta Q_{i-1/2,j}$. These correspond to the portion of this fluctuation which should be transmitted through the edges "above" and "below" this cell, respectively. On a Cartesian grid we were able to determine both of these by simply splitting $\mathcal{A}^+ \Delta Q$ into up-going and down-going acoustic waves in the y-direction. On a general quadrilateral grid, however, the edges "above" and "below" the cell will not be parallel, and we must solve two different transverse Riemann problems in the appropriate directions to compute the two fluctuations $\mathcal{B}^+ \mathcal{A}^+ \Delta Q$ and $\mathcal{B}^- \mathcal{A}^+ \Delta Q$. In the case of a heterogeneous medium we had to do this anyway, since the coefficients defining the wave speeds and eigenvectors may be different in the cell "above" and "below," so this is a natural generalization of the transverse solver presented in Section 21.5.1 for a Cartesian grid.

To compute $\mathcal{B}^- \mathcal{A}^+ \Delta Q$, for example, we must decompose the vector $\delta \equiv \mathcal{A}^+ \Delta Q_{i-1/2,j}$ into eigenvectors corresponding to a Riemann problem in the $\vec{n}_{i,j-1/2}$ direction at the interface $(i, j - 1/2)$:

$$\delta \equiv \mathcal{A}^+ \Delta Q_{i-1/2,j} = \beta^1 \begin{bmatrix} -Z_{i,j-1} \\ n^x_{i,j-1/2} \\ n^y_{i,j-1/2} \end{bmatrix} + \beta^2 \begin{bmatrix} 0 \\ -n^y_{i,j-1/2} \\ n^x_{i,j-1/2} \end{bmatrix} + \beta^3 \begin{bmatrix} Z_{ij} \\ n^x_{i,j-1/2} \\ n^y_{i,j-1/2} \end{bmatrix}. \quad (23.39)$$

Solving this linear system gives

$$\beta^1 = \frac{-\delta^1 + \delta^2 n^x Z_{ij} + \delta^3 n^y Z_{ij}}{Z_{i,j-1} + Z_{ij}}. \quad (23.40)$$

The coefficient β^1 is the only one needed to compute the down-going fluctuation, which is obtained by multiplying the first wave in (23.39) by the physical speed $-c_{i,j-1}$ and by the geometric scaling factor $\gamma_{i,j-1/2}$:

$$\mathcal{B}^- \mathcal{A}^+ \Delta Q_{i-1/2,j} = -\gamma_{i,j-1/2}\, c_{i,j-1}\, \beta^1 \begin{bmatrix} -Z_{i,j-1} \\ n^x_{i,j-1/2} \\ n^y_{i,j-1/2} \end{bmatrix}. \quad (23.41)$$

The up-going fluctuation $\mathcal{B}^+ \mathcal{A}^+ \Delta Q_{i-1/2,j}$ and the fluctuations $\mathcal{B}^\pm \mathcal{A}^- \Delta Q_{i-1/2,j}$ are obtained by analogous decompositions.

23.6.3 Solving the Riemann Problem by Rotating the Data

As described at the beginning of Section 23.6.1, there is an alternative approach to solving the normal Riemann problem in direction $\vec{n}_{i-1/2,j}$ that may by easier to implement for some systems. Rather than computing the eigenstructure of the Jacobian matrix in an arbitrary direction, we can instead transform the data $Q_{i-1,j}$ and Q_{ij} into new data $\check{Q}_l$ and $\check{Q}_r$ for a Riemann problem in the "x" direction of the form $\check{q}_t + A\check{q}_x = 0$. For acoustics, the data $\check{Q}_l$ and $\check{Q}_r$ should have components

$$\check{Q}^1_{l,r} = \text{pressure},$$
$$\check{Q}^2_{l,r} = \text{normal velocity}, \quad (23.42)$$
$$\check{Q}^3_{l,r} = \text{transverse velocity}.$$

We solve this one-dimensional Riemann problem and then must rotate the velocity components in the resulting waves and fluctuations back into the proper physical direction. This works because the acoustics equations are isotropic and so the Riemann problem has exactly the same mathematical structure in any direction. It also works for systems such as the Euler and shallow water equations; see Section 23.7.

To determine the waves, speeds, and fluctuations at edge $(i - 1/2, j)$, we must do the following:

1. Determine $\breve{Q}_l$ and $\breve{Q}_r$ by rotating the velocity components:

$$\breve{Q}_l = \mathcal{R}_{i-1/2,j} Q_{i-1,j}, \qquad \breve{Q}_r = \mathcal{R}_{i-1/2,j} Q_{ij},$$

where

$$\mathcal{R}_{i-1/2,j} = \begin{bmatrix} 1 & 0 & 0 \\ 0 & n^x_{i-1/2,j} & n^y_{i-1/2,j} \\ 0 & -n^y_{i-1/2,j} & n^x_{i-1/2,j} \end{bmatrix}. \qquad (23.43)$$

This matrix leaves the pressure Q^1 unchanged and rotates the velocity components of Q into components normal and tangential to the edge.

2. Solve the one-dimensional acoustics equations with

$$A = \begin{bmatrix} 0 & K & 0 \\ 1/\rho & 0 & 0 \\ 0 & 0 & 0 \end{bmatrix}$$

and the Riemann data $\breve{Q}_l$ and $\breve{Q}_r$. As in Section 23.6.1, this can be done equally well with different material parameters K and ρ (and hence impedance Z) in each grid cell. This results in the waves

$$\mathcal{W}^1_{i-1/2,j} = \alpha^1 \begin{bmatrix} -Z_{i-1,j} \\ 1 \\ 0 \end{bmatrix}, \quad \mathcal{W}^2_{i-1/2,j} = \alpha^2 \begin{bmatrix} 0 \\ 0 \\ 1 \end{bmatrix}, \quad \mathcal{W}^3_{i-1/2,j} = \alpha^3 \begin{bmatrix} Z_{ij} \\ 1 \\ 0 \end{bmatrix},$$

with speeds $\breve{\lambda}^1_{i-1/2,j} = -c_{i-1,j}$, $\breve{\lambda}^2_{i-1/2,j} = 0$, and $\breve{\lambda}^3_{i-1/2,j} = c_{ij}$.

3. Scale the wave speeds,

$$s^1_{i-1/2,j} = -\gamma_{i-1/2,j} c_{i-1,j}, \quad s^2_{i-1/2,j} = 0, \quad s^3_{i-1/2,j} = \gamma_{i-1/2,j} c_{ij}.$$

4. The waves $\mathcal{W}^p_{i-1/2,j}$ carry jumps in the normal velocity and no jump in the tangential velocity. To update the cell average Q properly, these must be converted into jumps in the x- and y-components of velocity, since this is what we store in Q. This is accomplished by setting

$$\mathcal{W}^p_{i-1/2,j} = \mathcal{R}^{-1}_{i-1/2,j} \breve{\mathcal{W}}^p_{i-1/2,j}$$

for $p = 1, 2, 3$, where

$$\mathcal{R}^{-1}_{i-1/2,j} = \begin{bmatrix} 1 & 0 & 0 \\ 0 & n^x_{i-1/2,j} & -n^y_{i-1/2,j} \\ 0 & n^y_{i-1/2,j} & n^x_{i-1/2,j} \end{bmatrix}. \qquad (23.44)$$

5. Compute the fluctuations

$$\mathcal{A}^-\Delta Q_{i-1/2,j} = s^1_{i-1/2,j}\mathcal{W}^1_{i-1/2,j},$$

$$\mathcal{A}^+\Delta Q_{i-1/2,j} = s^3_{i-1/2,j}\mathcal{W}^3_{i-1/2,j},$$

(23.45)

since $s^1 < 0$ while $s^3 > 0$.

This procedure results in the same waves, speeds, and fluctuations as the approach described in Section 23.6.1.

The transverse Riemann solver of Section 23.6.2 can also be reformulated via rotation of the data. To compute $\mathcal{B}^-\mathcal{A}^+\Delta Q$, for example, we rotate the velocity components of $\mathcal{A}^+\Delta Q$ into directions normal and tangential to the edge $(i, j-1/2)$ by computing $\mathcal{R}_{i,j-1/2}\mathcal{A}^+\Delta Q$. The first and second components of the resulting vector give the fluctuations in the pressure and normal velocity. These can be used to determine up-going and down-going acoustic waves, and the down-going wave is the one that is used to compute $\mathcal{B}^-\mathcal{A}^+\Delta Q$.

We decompose

$$\mathcal{R}_{i,j-1/2}\mathcal{A}^+\Delta Q = \beta^1 \begin{bmatrix} -Z_{i,j-1} \\ 1 \\ 0 \end{bmatrix} + \beta^2 \begin{bmatrix} 0 \\ 0 \\ 1 \end{bmatrix} + \beta^3 \begin{bmatrix} Z_{ij} \\ 1 \\ 0 \end{bmatrix}.$$

It is the first of these three waves that is propagating downwards. On a Cartesian grid we would multiply this wave by the corresponding speed $c_{i,j-1}$ to obtain the fluctuation $\mathcal{B}^-\mathcal{A}^+\Delta Q$. On the quadrilateral grid we must first rotate the velocity components of this wave back to $x-y$ coordinates (multiplying by $\mathcal{R}^{-1}_{i,j-1/2}$), and then we must multiply by the *scaled* velocity $\gamma_{i,j-1/2}c_{i,j-1}$, so we obtain

$$\mathcal{B}^-\mathcal{A}^+\Delta Q_{i-1/2,j} = \gamma_{i,j-1/2}c_{i,j-1}\mathcal{R}^{-1}_{i,j-1/2}\beta^1 \begin{bmatrix} -Z_{i,j-1} \\ 1 \\ 0 \end{bmatrix}.$$

(23.46)

To compute $\mathcal{B}^+\mathcal{A}^+\Delta Q$ we take a similar approach, but now must rotate using $\mathcal{R}_{i,j+1/2}$ and use the up-going wave, rotating back and scaling by $\gamma_{i,j+1/2}$. The left-going fluctuation $\mathcal{A}^-\Delta Q_{i-1/2,j}$ must then be handled in a similar manner.

23.7 Shallow Water and Euler Equations

The procedure described in Section 23.6 for the acoustics equations is easily extended to nonlinear systems of equations such as the two-dimensional shallow water equations. Recall that for these equations the components of q are (h, hu, hv), where h is the depth. As in the acoustics equations, we must rotate the momentum components at each cell interface using the rotation matrix (23.43). We then solve the one-dimensional Riemann problem normal to the edge just as on a Cartesian grid. An approximate Riemann solver can again be used. The waves are then rotated back into the $x-y$ coordinate frame before using them to update Q in the grid cells, and the wave speeds are scaled by the length ratios γ.

For the unsplit algorithms of Chapter 21, we also need to define a transverse solver. As in Section 23.6.2, we need to rotate the momentum components of $\mathcal{A}^+\Delta Q_{i-1/2,j}$ into directions normal and tangential to the edge $(i, j-1/2)$ before decomposing this vector into eigenvectors of the transverse Jacobian matrix. In addition, we need to define this Jacobian matrix using appropriate velocities relative to this edge. Just as we did for the Cartesian grid algorithm described in Section 21.7, we might use the Roe averages defined from solving the normal Riemann problem to define the transverse Jacobian, after a suitable rotation. Sample Riemann solvers for this case may be found in [claw/book/chap23/shallow], and an example is presented in Example 23.2 below.

The Euler equations can be handled in exactly the same manner. The components of q are now $(\rho, \rho u, \rho v, E)$ and the energy E is a scalar, so that rotations are again applied only to the momentum components of q. Sample Riemann solvers for this case may be found in [claw/book/chap23/euler].

23.8 Using CLAWPACK on Quadrilateral Grids

The sample CLAWPACK codes for this chapter all take a common form. The domain is assumed to be rectangular in the computational $\xi-\eta$ plane, and it is this plane that is discretized with a uniform grid. So the parameters dx and dy in claw2ez.data now refer to $\Delta\xi$ and $\Delta\eta$, while the parameters xlower, etc., specify the range in $\xi-\eta$ space.

A function mapc2p.f is specified that maps a computational point (xc,yc) = (ξ, η) to the corresponding physical point (xp,yp) = $(X(\xi, \eta), Y(\xi, \eta))$. This mapping is used to determine the physical location of the corners of grid cells, which are needed in the setaux.f routine as described below. This mapping is also typically called from the qinit.f routine to set initial data. For each grid cell we wish to set q(i,j) initially to be the cell average of the initial data $\overset{\circ}{q}(x, y)$. Often one can approximate this by simply evaluating $\overset{\circ}{q}$ at the center of the grid cell, and mapc2p is used to map the center of the computational grid cell, (xc,yc) = (xlower + (i-0.5)*dx, ylower + (j-0.5)*dy), to the corresponding physical point (xp,yp).

The aux array is used to store information required for each grid cell or edge. This can typically be set once at the start of the computation in the setaux.f routine. Exactly what information is required depends on the equations being solved, but we always require $\kappa_{ij} = |\mathcal{C}_{ij}|/(\Delta\xi \, \Delta\eta)$, which must be used as a capacity function (so mcapa should be set to point to this element of the aux array). This value can be computed from the locations of the four corners of the cell.

For the advection equation as discussed in Section 23.5.2, the only other information required is the edge velocities $U_{i-1/2,j}$ and $V_{i,j-1/2}$. If a stream function is available, then these are easily computed by differencing the stream function between the corners. See [claw/book/chap23/advection/polar/setaux.f] for some sample formulas.

For the acoustics, Euler, or shallow water equations, we need to store enough information to determine both the normal vector to each edge and the length ratios γ for each side. This requires a minimum of two pieces of data at each side, for example the angle of the normal and the γ-value, or else the edge vector $\vec{h}$ (see Figure 23.2), from which both the normal and the length can be computed. It is more efficient, however, to store three pieces of data at each edge, both components of the normal vector $\vec{n}$ and also the γ value, in order to minimize the amount of computation that must be done in each time step. The CLAWPACK codes provided for this chapter use the following convention for the aux array:

$$\text{aux}(\text{i}, \text{j}, 1) = n^x_{i-1/2,j},$$

$$\text{aux}(\text{i}, \text{j}, 2) = n^y_{i-1/2,j},$$

$$\text{aux}(\text{i}, \text{j}, 3) = \gamma_{i-1/2,j},$$

$$\text{aux}(\text{i}, \text{j}, 4) = n^x_{i,j-1/2}, \qquad (23.47)$$

$$\text{aux}(\text{i}, \text{j}, 5) = n^y_{i,j-1/2},$$

$$\text{aux}(\text{i}, \text{j}, 6) = \gamma_{i,j-1/2},$$

$$\text{aux}(\text{i}, \text{j}, 7) = \kappa_{ij}.$$

For each (i, j), we store data related to the left edge and the bottom edge as well as the κ-value for this cell.

Example 23.2. As an example, we solve the two-dimensional shallow water equations around a circular cylinder and compare results obtained on two different types of grids, as illustrated in Figure 23.1(a) and (b). The simple polar coordinate grid of Figure 23.1(a) has the advantage that it is smoothly varying and orthogonal. The grid of Figure 23.1(b) was chosen to illustrate that these methods work well even on nonorthogonal grids and with abrupt changes in the orientation of grid lines.

The problem we consider consists of a planar shock wave moving towards the cylinder, with $h_r = 1$ and $u_r = 0$ ahead of the shock (so there is initially quiescent water around the cylinder). Behind the shock, $h_l = 4$, and the velocity $u_l > 0$ is determined by the Hugoniot relation (13.17). Common experience from inserting a stick vertically into a moving stream of water leads us to expect that a *bow shock* will form upstream from the cylinder once the flowing water hits the cylinder. The depth jumps to a value greater than the approaching freestream depth in this region where the flow must decelerate and go around the cylinder.

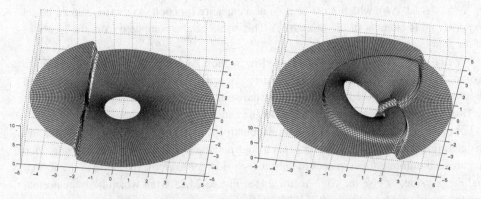

Fig. 23.4. A hydraulic jump in the shallow water equations hits a circular cylinder. The depth at two different times is shown. Left: depth at $t = 0$; right: depth at $t = 1.5$.

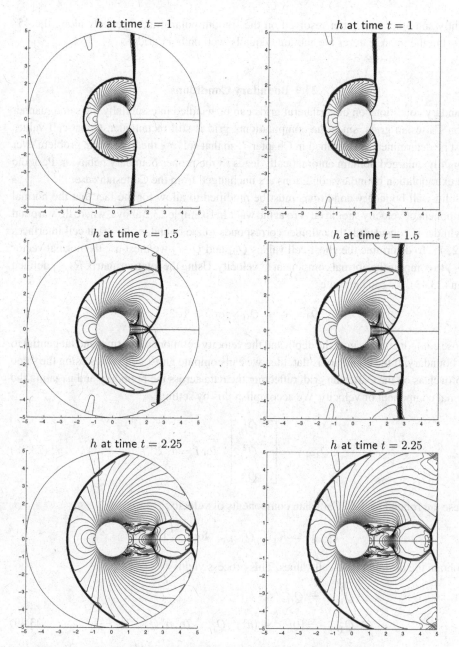

Fig. 23.5. Shallow water equations for a shock wave hitting a cylinder. On the left a 100×400 polar coordinate grid of the type shown in Figure 23.1(a) is used. On the right a 100×416 grid of the type shown in Figure 23.1(b) is used. Contour lines are at 0.05:0.2:8.05. [claw/book/chap23/shallow/cylinder] [claw/book/chap23/shallow/sqcylinder]

In the present context this can be viewed as a reflection of the incident shock from the cylinder. Figure 23.4 shows the depth initially and at a later time.

Figure 23.5 shows contour plots of computed solutions on the two grids similar to those illustrated in Figure 23.1(a) and (b), but with greater resolution. The contour lines are

slightly smoother and better resolved on the smooth polar grid, especially along the 45°
lines, but the main features are captured equally well on both grids.

23.9 Boundary Conditions

Boundary conditions on quadrilateral grids can be handled in essentially the same manner
as on Cartesian grids, since the computational grid is still rectangular. Ghost-cell values
must be determined as described in Chapter 7, so that solving the hyperbolic problem over
a slightly enlarged domain automatically leads to the proper boundary behavior. Periodic
and extrapolation boundary conditions are unchanged from the Cartesian case.

Solid-wall boundary conditions must be modified to allow for the fact that the normal
component of velocity should be zero at the wall. In Example 23.2 above, where flow around
a cylinder is considered, the cylinder corresponds to the boundary $\xi = 0$ at cell interfaces
$(1/2, j)$. To determine the ghost-cell values Q_{0j} and $Q_{-1,j}$ we must use the normal vector
$\vec{n}_{1/2,j}$ to compute the normal component of velocity. Using the rotation matrix $\mathcal{R}_{1/2,j}$ defined
as in (23.43), we can compute

$$\check{Q} = \mathcal{R}_{1/2,j} Q_{ij} \quad \text{for } i = 1, 2,$$

whose components are now the depth and the velocity components normal and tangential to
the boundary. If the boundary is flat, then we can compute ghost-cell values using the same
approach as on the Cartesian grid, reflecting the data across the boundary and negating the
normal component of velocity. We accomplish this by setting

$$\check{Q}_{1-i,j} = \begin{bmatrix} \check{Q}_{ij}^1 \\ -\check{Q}_{ij}^2 \\ \check{Q}_{ij}^3 \end{bmatrix} \quad \text{for } i = 1, 2. \tag{23.48}$$

These are rotated back to Cartesian components of velocity,

$$Q_{1-i,j} = \mathcal{R}_{1/2,j}^{-1} \check{Q}_{1-i,j} \quad \text{for } i = 1, 2,$$

to obtain the desired ghost-cell values. This process yields

$$Q_{1-i,j}^1 = Q_{ij},$$
$$Q_{1-i,j}^2 = [(n^y)^2 - (n^x)^2] Q_{ij}^2 - 2n^x n^y Q_{ij}^3, \tag{23.49}$$
$$Q_{1-i,j}^3 = -2n^x n^y Q_{ij}^2 + [(n^y)^2 - (n^x)^2] Q_{ij}^3.$$

Bibliography

[1] S. Abarbanel and D. Gottlieb, *A mathematical analysis of the PML method*, J. Comput. Phys., 134 (1997), pp. 357–363.

[2] R. Abgrall, *Approximation of the genuinely multidimensional Riemann problem for the Euler equations by a Roe type method (I) – the linearization*, C. R. Acad. Sci. Ser. I, 319 (1994), pp. 499–504.

[3] R. Abgrall, *Approximation of the genuinely multidimensional Riemann problem for the Euler equations by a Roe type method (II) – solution of the approximated Riemann problem*, C. R. Acad. Sci. Ser. I, 319 (1994), pp. 625–629.

[4] R. Abgrall, *How to prevent pressure oscillations in multicomponent flow calculations: A quasiconservative approach*, J. Comput. Phys., 125 (1996), pp. 150–160.

[5] R. Abgrall and S. Karni, *Computations of compressible multifluids*, J. Comput. Phys., 169 (2001), pp. 594–623.

[6] J. D. Achenbach, *Wave Propagation in Elastic Solids*, North-Holland, Amsterdam, 1973.

[7] M. J. Aftosmis, M. J. Berger, and J. E. Melton, *Robust and efficient Cartesian mesh generation for component-based geometry*, AIAA J., 36 (1998), pp. 952–960.

[8] K. Aki and P. G. Richards, *Quantitative Seismology*, Freeman, 1980.

[9] A. S. Almgren, J. B. Bell, P. Colella, L. H. Howell, and M. L. Welcome, *A conservative adaptive projection method for the variable density incompressible Navier–Stokes equations*, J. Comput. Phys., 142 (1998), pp. 1–46.

[10] A. S. Almgren, J. B. Bell, and W. Y. Crutchfield, *Approximate projection methods: Part I. Inviscid analysis*, SIAM J. Sci. Comput., 22 (2000), pp. 1139–1159.

[11] S. S. Antman, *Nonlinear Problems of Elasticity*, Springer-Verlag, New York, 1995.

[12] D. Aregba-Driollet and R. Natalini, *Convergence of relaxation schemes for conservation laws*, Appl. Anal., 61 (1996), pp. 163–193.

[13] P. Arminjon and M.-C. Viallon, *Convergence of a finite volume extension of the Nessyahu–Tadmor scheme on unstructured grid for a two-dimensional linear hyperbolic equations*, SIAM J. Numer. Anal., 36 (1999), pp. 738–771.

[14] M. Arora and P. L. Roe, *On postshock oscillations due to shock capturing schemes in unsteady flow*, J. Comput. Phys., 130 (1997), pp. 1–24.

[15] M. Asch, W. Kohler, G. Papanicolaou, M. Postel, and B. White, *Frequency content of randomly scattered signals*, SIAM Rev., 33 (1991), pp. 519–625.

[16] A. Aw and M. Rascle, *Resurrection of "second order" models of traffic flow*, SIAM J. Appl. Math., 60 (2000), pp. 916–938.

[17] A. V. Azevedo, D. Marchesin, B. J. Plohr, and K. Zumbrun, *Nonuniqueness of solutions to Riemann problems*, Z. Angew. Math. Phys., 47 (1996), pp. 977–998.

[18] D. Bale, R. J. LeVeque, S. Mitran, and J. Rossmanith, *A wave-propagation method for conservation laws and balance laws with spatially varying flux functions*, submitted.

[19] W. Z. Bao and S. Jin, *The random projection method for hyperbolic conservation laws with stiff reaction terms*, J. Comput. Phys., 163 (2000), pp. 216–248.

[20] T. J. Barth, *On Unstructured Grids and Solvers*. Von Karman Institute for Fluid Dynamics Lecture Series, 90-03, 1990.

[21] A. Bayliss and E. Turkel, *Radiation boundary conditions for wave-like equations*, Comm. Pure Appl. Math., 33 (1980), pp. 707–725.

[22] J. Bell, M. J. Berger, J. Saltzman, and M. Welcome, *Three dimensional adaptive mesh refinement for hyperbolic conservation laws*, SIAM J. Sci. Stat., 15 (1994), pp. 127–138.

[23] J. B. Bell, P. Colella, and H. M. Glaz, *A second-order projection method for the incompressible Navier–Stokes equations*, J. Comput. Phys., 85 (1989), pp. 257–283.

[24] J. B. Bell, P. Colella, and J. R. Trangenstein, *Higher order Godunov methods for general systems of hyperbolic conservation laws*, J. Comput. Phys., 82 (1989), p. 362.

[25] J. B. Bell, C. N. Dawson, and G. R. Shubin, *An unsplit, higher order Godunov method for scalar conservation laws in multiple dimensions*, J. Comput. Phys., 74 (1988), pp. 1–24.

[26] J. B. Bell, G. R. Shubin, and J. Trangenstein, *Conservation laws of mixed type describing three-phase flow in porous media*, SIAM J. Appl. Math., 46 (1986), pp. 1000–1017.

[27] M. Ben-Artzi and J. Falcovitz, *A second-order Godunov-type scheme for compressible fluid dynamics*, J. Comput. Phys., 55 (1984), pp. 1–32.

[28] J.-P. Berenger, *A perfectly matched layer for the absorption of electromagnetic waves*, J. Comput. Phys., 114 (1994), pp. 185–200.

[29] F. Bereux and L. Sainsaulieu, *A Roe-type Riemann solver for hyperbolic systems with relaxation based on time-dependent wave decomposition*, Numer. Math., 77 (1997), pp. 143–185.

[30] A. Berezovski and G. A. Maugin, *Simulation of thermoelastic wave propagation by means of a composite wave-propagation algorithm*, J. Comput. Phys., 168 (2001), pp. 249–264.

[31] M. Berger and R. J. LeVeque, *Stable boundary conditions for Cartesian grid calculations*, Comput. Syst. Eng., 1 (1990), pp. 305–311.

[32] M. J. Berger and R. J. LeVeque, *Adaptive mesh refinement using wave-propagation algorithms for hyperbolic systems*, SIAM J. Numer. Anal., 35 (1998), pp. 2298–2316.

[33] A. C. Berkenbosch, E. F. Kaasschieter, and R. Klein, *Detonation capturing for stiff combustion chemistry*, Combust. Theor. Model., 2 (1998), pp. 313–348.

[34] A. Bermudez and M. Vazquez, *Upwind methods for hyperbolic conservations laws with source term*, Comput. and Fluids, 23 (1994), pp. 1049–1071.

[35] S. Bianchini and A. Bressan, *Vanishing viscosity solutions of nonlinear hyperbolic systems*, Preprint 2001-050, http://www.math.ntnu.no/conservation/2001/, 2001.

[36] F. Bianco, G. Puppo, and G. Russo, *High order central schemes for hyperbolic systems of conservation laws*, SIAM J. Sci. Comput., 21 (1999), pp. 294–322.

[37] S. J. Billet and E. F. Toro, *On WAF-type schemes for multidimensional hyperbolic conservation laws*, J. Comput. Phys., 130 (1997), pp. 25–40.

[38] J. P. Boris and D. L. Book, *Flux corrected transport I, SHASTA, a fluid transport algorithm that works*, J. Comput. Phys., 11 (1973), pp. 38–69.

[39] F. Bouchut, *Entropy satisfying flux vector splittings and kinetic BGK models*. Preprint. http://www.dma.ens.fr/~fbouchut/.

[40] A. Bourlioux and A. J. Majda, *Theoretical and numerical structure of unstable detonations*, Phil. Trans. R. Soc. Lond. A, 350 (1995), pp. 29–68.

[41] F. Bratvedt, K. Bratvedt, C. F. Buchholz, T. Gimse, H. Holden, L. Holden, and N. H. Risebro, *Frontline and frontsim: Two full scale two-phase, black oil reservoir simulators based on front-tracking*, Surv. Math. Ind., 3 (1993), pp. 185–215.

[42] Y. Brenier, *Averaged multivalued solutions for scalar conservation laws*, SIAM J. Numer. Anal., 21 (1984), pp. 1013–1037.

[43] Y. Brenier, *A variation estimate for a large time step version of the Glimm scheme using the Roe Riemann solver*. Preprint, 1984.

[44] A. Bressan, *Global solutions of systems of conservation laws by wave-front tracking*, J. Math. Anal. Appl., 170 (1992), pp. 414–432.

[45] A. Bressan, *The unique limit of the Glimm scheme*, Arch. Rat. Mech. Anal., 130 (1995), pp. 205–230.

[46] A. Bressan, *Hyperbolic Systems of Conservation Laws, The One-Dimensional Cauchy Problem*, Oxford University Press, 2000.

[47] A. Bressan and R. M. Colombo, *The semigroup generated by 2 × 2 conservation laws*, Arch. Rat. Mech. Anal., 133 (1995), pp. 1–75.

[48] A. Bressan, G. Crasta, and B. Piccoli, *Well-posedness of the Cauchy problem for n × n systems of conservation laws*, Mem. Amer. Math. Soc., 694 (2000).

[49] A. Bressan, T. P. Liu, and T. Yang, L^1 *stability estimates for n × n conservation laws*, Arch. Rat. Mech. Anal., 149 (1999), pp. 1–22.

[50] L. Brillouin, *Wave Propagation and Group Velocity*, Academic Press, 1960.

[51] M. Brio and P. Rosenau, *Evolution of the fast-intermediate shock wave in an MHD model problem*. Preprint, 1997.

[52] M. Brio, A. R. Zakharian, and G. M. Webb, *Two-dimensional Riemann solver for Euler equations of gas dynamics*, J. Comput. Phys., 167 (2001), pp. 177–195.

[53] D. L. Brown, R. Cortez, and M. L. Minion, *Accurate projection methods for the incompressible Navier–Stokes equations*, J. Comput. Phys., 168 (2001), pp. 464–499.

[54] J. M. Burgers, *A mathematical model illustrating the theory of turbulence*, Adv. Appl. Mech., 1 (1948), pp. 171–199.

[55] R. Burridge, *Some Mathematical Topics in Seismology*, Courant Institute of Mathematical Sciences Lecture Notes, New York, 1976.

[56] D. Calhoun, *A Cartesian grid method for solving the streamfunction–vorticity equations in irregular geometries*, PhD thesis, University of Washington, 1999.

[57] D. Calhoun and R. J. LeVeque, *Solving the advection-diffusion equation in irregular geometries*, J. Comput. Phys., 156 (2000), pp. 1–38.

[58] H. B. Callen, *Thermodynamics and an Introduction to Thermostatistics*, Wiley, 1985.

[59] D. M. Causon, D. M. Ingram, C. G. Mingham, G. Yang, and R. V. Pearson, *Calculation of shallow water flows using a Cartesian cut cell approach*, Adv. Water Res., 23 (2000), pp. 545–562.

[60] A. Chalabi, *Convergence of relaxation schemes for hyperbolic conservation laws with stiff source terms*, Math. Comput., 68 (1999), pp. 955–970.

[61] S. C. Chang, *A critical analysis of the modified equation technique of Warming and Hyett*, J. Comput. Phys., 86 (1990), pp. 107–126.

[62] G. Chen and T. Liu, *Zero relaxation and dissipation limits for hyperbolic conservation laws*, Comm. Pure Appl. Math., 46 (1993), pp. 755–781.

[63] G.-Q. Chen and P. G. LeFloch, *Entropy flux-splittings for hyperbolic conservation laws. Part I: General framework*, Comm. Pure Appl. Math., 48 (1995), pp. 691–729.

[64] G. Q. Chen, C. D. Levermore, and T. P. Liu, *Hyperbolic conservation laws with stiff relaxation terms and entropy*, Comm. Pure Appl. Math., 47 (1994), pp. 787–830.

[65] T.-J. Chen and C. H. Cooke, *On the Riemann problem for liquid or gas–liquid media*, Int. J. Numer. Meth. Fluids, 18 (1994), pp. 529–541.

[66] I. Chern and P. Colella, *A conservative front tracking method for hyperbolic conservation laws*. Report UCRL-97200, LLNL, 1987.

[67] A. J. Chorin, *Random choice solution of hyperbolic systems*, J. Comput. Phys., 22 (1976), pp. 517–533.

[68] A. J. Chorin and J. E. Marsden, *A Mathematical Introduction to Fluid Mechanics*, Springer-Verlag, 1979.

[69] J. F. Clarke, *Gas dynamics with relaxation effects*, Rep. Prog. Phys., 41 (1978), pp. 807–863.

[70] J. F. Clarke and M. McChesney, *Dynamics of Relaxing Gases*, Butterworths, 1976.

[71] J.-P. Cocchi and R. Saurel, *A Riemann problem based method for the resolution of compressible multimaterial flows*, J. Comput. Phys., 137 (1997), pp. 265–298.

[72] B. Cockburn, *Quasimonotone schemes for scalar conservation laws I.*, SIAM J. Numer. Anal., 26 (1989), pp. 1325–1341.

[73] B. Cockburn, *Quasimonotone schemes for scalar conservation laws II.*, SIAM J. Numer. Anal., 27 (1990), pp. 247–258.

[74] B. Cockburn, *Quasimonotone schemes for scalar conservation laws III.*, SIAM J. Numer. Anal., 27 (1990), pp. 259–276.

[75] B. Cockburn, F. Coquel, and P. G. Lefloch, *An error estimate for finite volume methods for multidimensional conservation laws*, Math. Comp., 63 (1994), pp. 77–103.

[76] B. Cockburn, F. Coquel, and P. G. LeFloch, *Convergence of the finite volume method for multidimensional conservation laws*, SIAM J. Numer. Anal., 32 (1995), pp. 687–706.

[77] W. J. Coirier and K. G. Powell, *An accuracy assessment of Cartesian mesh approaches for the Euler equations*, J. Comput. Phys., 117 (1993), pp. 121–131.

[78] P. Colella, *Glimm's method for gas dynamics*, SIAM J. Sci. Stat. Comput., 3 (1982), pp. 76–110.

[79] P. Colella, *A direct Eulerian MUSCL scheme for gas dynamics*, SIAM J. Sci. Stat. Comput., 6 (1985), pp. 104–117.

[80] P. Colella, *Multidimensional upwind methods for hyperbolic conservation laws*, J. Comput. Phys., 87 (1990), pp. 171–200.

[81] P. Colella and H. M. Glaz, *Efficient solution algorithms for the Riemann problem for real gases*, J. Comput. Phys., 59 (1985), pp. 264–284.

[82] P. Colella, A. Majda, and V. Roytburd, *Theoretical and numerical structure for reacting shock waves*, SIAM J. Sci. Stat. Comput., 7 (1986), pp. 1059–1080.

[83] P. Colella and J. Trangenstein, *A higher-order Godunov method for modeling finite deformation in elastic-plastic solids*, Comm. Pure Appl. Math., 44 (1991), pp. 41–100.

[84] P. Colella and P. Woodward, *The piecewise-parabolic method (PPM) for gas-dynamical simulations*, J. Comput. Phys., 54 (1984), pp. 174–201.

[85] J.-F. Colombeau, *The elastoplastic shock problem as an example of the resolution of ambiguities in the multiplication of distributions*, J. Math. Phys., 30 (1989), pp. 2273–2279.

[86] J.-F. Colombeau and A. Y. LeRoux, *Multiplications of distributions in elasticity and hydrodynamics*, J. Math. Phys., 29 (1988), pp. 315–319.

[87] J.-F. Colombeau, A. Y. LeRoux, A. Noussair, and B. Perrot, *Microscopic profiles of shock waves and ambiguities in multiplications of distributions*, SIAM J. Numer. Anal., 26 (1989), pp. 871–883.

[88] C. H. Cooke and T. Chen, *On shock capturing for pure water with general equation of state*, Comm. Appl. Numer. Meth., 8 (1992), pp. 219–233.

[89] F. Coquel and P. Le Floch, *Convergence of finite difference schemes for conservation laws in several space dimensions: The corrected antidiffusive flux approach*, Math. Comp., 57 (1991), pp. 169–210.

[90] F. Coquel and P. Le Floch, *Convergence of finite difference schemes for conservation laws in several space dimensions: A general theory*, SIAM J. Numer. Anal., 30 (1993), pp. 675–700.

[91] F. Coquel and B. Perthame, *Relaxation of energy and approximate Riemann solvers for general pressure laws in fluid dynamics*, SIAM J. Numer. Anal., 35 (1998), pp. 2223–2249.

[92] R. Courant and K. O. Friedrichs, *Supersonic Flow and Shock Waves*, Springer, 1948.

[93] R. Courant, K. O. Friedrichs, and H. Lewy, *Über die partiellen Differenzengleichungen der mathematischen Physik*, Math. Ann., 100 (1928), pp. 32–74.

[94] R. Courant, K. O. Friedrichs, and H. Lewy, *On the partial difference equations of mathematical physics*, IBM J., 11 (1967), pp. 215–234.

[95] M. G. Crandall and A. Majda, *The method of fractional steps for conservation laws*, Math. Comp., 34 (1980), pp. 285–314.

[96] M. G. Crandall and A. Majda, *Monotone difference approximations for scalar conservation laws*, Math. Comp., 34 (1980), pp. 1–21.

[97] C. M. Dafermos, *Polygonal approximations of solutions of the initial-value problem for a conservation law*, J. Math. Anal. Appl., 38 (1972), pp. 33–41.

[98] C. M. Dafermos, *Hyperbolic Conservation Laws in Continuum Physics*, Springer, Berlin, 2000.

[99] C. F. Daganzo, *Requiem for second-order fluid approximations of traffic flow*, Transport. Res. B, 29B (1995), pp. 277–286.

[100] W. L. Dai and P. R. Woodward, *A second-order unsplit Godunov scheme for two- and three-dimensional Euler equations*, J. Comput. Phys., 134 (1997), pp. 261–281.

[101] G. Dal Maso, P. G. LeFloch, and F. Murat, *Definition and weak stablity of nonconservative products*, J. Math. Pures Appl., 74 (1995), pp. 483–548.

[102] R. O. Davis and A. P. S. Selvadurai, *Elasticity and Geomechanics*, Cambridge University Press, 1996.

[103] S. F. Davis, *An interface tracking method for hyperbolic systems of conservation laws*, Appl. Numer. Math., 10 (1992), pp. 447–472.

[104] D. De Zeeuw and K. Powell, *An adaptively-refined Cartesian mesh solver for the Euler equations*, J. Comput. Phys., 104 (1993), pp. 56–68.

[105] H. Deconinck, C. Hirsch, and J. Peuteman, *Characteristic decomposition methods for the multidimensional Euler equations*, in 10th Int. Conf. on Numerical Methods in Fluid Dynamics, Springer Lecture Notes in Physics 264, 1986, pp. 216–221.

[106] H. Deconinck, P. L. Roe, and R. Struijs, *A multidimensional generalization of Roe's flux difference splitter for the Euler equations*, Comput. Fluids, 22 (1993), pp. 215–222.

[107] S. M. Deshpande, *New developments in kinetic schemes*, Comput. Math. Appl., 35 (1998), pp. 75–93.

[108] R. J. DiPerna, *Finite difference schemes for conservation laws*, Comm. Pure Appl. Math., 25 (1982), pp. 379–450.

[109] R. J. DiPerna, *Convergence of approximate solutions to conservation laws*, Arch. Rat. Mech. Anal., 82 (1983), pp. 27–70.

[110] R. J. DiPerna, *Measure-valued solutions to conservation laws*, Arch. Rat. Mech. Anal., 88 (1985), pp. 223–270.

[111] R. Donat, *Studies on error propagation for certain nonlinear approximations to hyperbolic equations: Discontinuities in derivatives*, SIAM J. Numer. Anal., 31 (1994), pp. 655–679.

[112] R. Donat and A. Marquina, *Capturing shock reflections: An improved flux formula*, J. Comput. Phys., 125 (1996), pp. 42–58.

[113] R. Donat and S. Osher, *Propagation of error into regions of smoothness for non-linear approximations to hyperbolic equations*, Comput. Meth. Appl. Mech. Eng., 80 (1990), pp. 59–64.

[114] J. Donea, L. Quartapelle, and V. Selmin, *An analysis of time discretization in the finite element solution of hyperbolic problems*, J. Comput. Phys., 70 (1987), pp. 463–499.

[115] G. Dubois and G. Mehlman, *A non-parameterized entropy correction for Roe's approximate Riemann solver*, Numer. Math., 73 (1996), pp. 169–208.

[116] J. E. Dunn, R. Fosdick, and M. Slemrod, *Shock Induced Transitions and Phase Structures in General Media, IMA Vol. Math. Appl. 53*, Springer-Verlag, New York, 1993.

[117] D. R. Durran, *Numerical Methods for Wave Equations in Geophysical Fluid Dynamics*, Springer, New York, 1999.

[118] J. R. Edwards, R. K. Franklin, and M.-S. Liou, *Low-diffusion flux-splitting methods for real fluid flows with phase transitions*, AIAA J., 38 (2000), pp. 1624–1633.

[119] J. R. Edwards and M.-S. Liou, *Low-diffusion flux-splitting methods for flows at all speeds*, AIAA J., 36 (1998), pp. 1610–1617.

[120] G. Efraimsson and G. Kreiss, *A remark on numerical errors downstream of slightly viscous shocks*, SIAM J. Numer. Anal., 36 (1999), pp. 853–863.

[121] B. Einfeldt, *On Godunov-type methods for gas dynamics*, SIAM J. Numer. Anal., 25 (1988), pp. 294–318.

[122] B. Einfeldt, C. D. Munz, P. L. Roe, and B. Sjogreen, *On Godunov type methods near low densities*, J. Comput. Phys., 92 (1991), pp. 273–295.

[123] B. Engquist and A. Majda, *Absorbing boundary conditions for the numerical simulation of waves*, Math. Comp., 31 (1977), pp. 629–651.

[124] B. Engquist and S. Osher, *Stable and entropy satisfying approximations for transonic flow calculations*, Math. Comp., 34 (1980), pp. 45–75.

[125] B. Engquist and B. Sjögreen, *The convergence rate of finite difference schemes in the presence of shocks*, SIAM J. Numer. Anal., 35 (1998), pp. 2464–2485.

[126] D. Estep, *A modified equation for dispersive difference-schemes*, Appl. Numer. Math., 17 (1995), pp. 299–309.

[127] R. Eulderink and G. Mellema, *General relativistic hydrodynamics with a Roe solver*, Astron. Astrophys. Suppl. Ser., 110 (1995), pp. 587–623.

[128] J. Falcovitz and M. Ben-Artzi, *Recent developments of the GRP method*, JSME Int. J. Ser. B, 38 (1995), pp. 497–517.

[129] H. Fan, S. Jin, and Z.-H. Teng, *Zero reaction limit for hyperbolic conservation laws with source terms*, J. Diff. Equations, 168 (2000), pp. 270–294.

[130] H. Fan and M. Slemrod, *The Riemann problem for systems of conservation laws of mixed type*, in *Shock Induced Transitions and Phase Structures in General Media*, J. E. Dunn, R. Fosdick, and M. Slemrod, eds., Springer, New York, 1993, pp. 61–91.

[131] R. Fazio and R. J. LeVeque, *Moving mesh methods for one-dimensional conservation laws using* CLAWPACK. Comp. Math. Appl. To appear. (http://www.math.ntnu.no/conservation/1998/).

[132] R. P. Fedkiw, T. Aslam, B. Merriman, and S. Osher, *A non-oscillatory Eulerian approach to interfaces in multimaterial flows (the ghost fluid method)*, J. Comput. Phys., 152 (1999), pp. 457–492.

[133] R. P. Fedkiw, B. Merriman, and S. Osher, *Efficient characteristic projection in upwind difference schemes for hyperbolic systems*, J. Comput. Phys., 141 (1998), pp. 22–36.

[134] M. Fey, *Multidimensional upwinding I. The method of transport for solving the Euler equations*, J. Comput. Phys., 143 (1998), pp. 159–180.

[135] M. Fey, *Multidimensional upwinding II. Decomposition of the Euler equations into advection equations*, J. Comput. Phys., 143 (1998), pp. 181–199.

[136] W. Fickett, *Introduction to Detonation Theory*, University of California Press, Berkeley, 1985.

[137] C. A. J. Fletcher, *Computational Techniques for Fluid Dynamics*, Springer, 1988.

[138] J. Flores and M. Holt, *Glimm's method applied to underwater explosions*, J. Comput. Phys., 49 (1981), pp. 377–387.

[139] T. Fogarty and R. J. LeVeque, *High-resolution finite volume methods for acoustics in periodic or random media*, J. Acoust. Soc. Am., 106 (1999), pp. 17–28.

[140] H. Forrer and R. Jeltsch, *A higher-order boundary treatment for Cartesian-grid methods*, J. Comput. Phys., 140 (1998), pp. 259–277.

[141] B. M. Fraeijs de Veubeke, *A Course in Elasticity*, Applied Mathematical Sciences 29, Springer, New York, 1979.

[142] H. Freistühler, *Some remarks on the structure of intermediate magnetohydrodynamic shocks*, J. Geophys. Res., 96 (1991), pp. 3825–3827.

[143] K. O. Friedrichs and P. D. Lax, *Systems of conservation equations with a convex extension*, Proc. Nat. Acad. Sci., 68 (1971), pp. 1686–1688.

[144] T. Gallouët, J.-M. Hérard, and N. Seguin, *Some recent finite volume schemes to compute Euler equations using real gas EOS*. Preprint.

[145] C. W. Gear, *Numerical Initial Value Problems in Ordinary Differential Equations*, Prentice-Hall, 1971.

[146] P. Germain, *Shock waves and shock-wave structure in magneto-fluid dynamics*, Rev. Mod. Phys., 32 (1960), p. 951.

[147] T. Gimse and N. H. Risebro, *Riemann problems with a discontinuous flux function*, in Proc. Third Int. Conf. Hyperbolic Problems, B. Engquist and B. Gustafsson, eds., Studentlitteratur, 1990, pp. 488–502.

[148] T. Gimse and N. H. Risebro, *Solution of the Cauchy problem for a conservation law with a discontinous flux function*, SIAM J. Math. Anal., 23 (1992), pp. 635–648.

[149] P. Glaister, *An approximate linearized Riemann solver for the Euler equations for real gases*, J. Comput. Phys., 74 (1988), pp. 382–408.

[150] P. Glaister, *An analysis of averaging procedures in a Riemann solver for compressible flows of a real gas*, Comput. Math. Appl., 33 (1997), pp. 105–119.

[151] H. M. Glaz and T.-P. Liu, *The asymptotic analysis of wave interactions and numerical calculations of transonic nozzle flow*, Adv. Appl. Math., 5 (1984), pp. 111–146.

[152] J. Glimm, *Solutions in the large for nonlinear hyperbolic systems of equations*, Comm. Pure Appl. Math., 18 (1965), pp. 695–715.

[153] J. Glimm, J. Grove, X. L. Li, K.-M. Shyue, Y. Zeng, and Q. Zhang, *Three-dimensional front tracking*, SIAM J. Sci. Comput., 19 (1998), pp. 703–727.

[154] J. Glimm, E. Isaacson, D. Marchesin, and O. McBryan, *Front tracking for hyperbolic systems*, Adv. Appl. Math., 2 (1981), pp. 173–215.

[155] J. Glimm, G. Marshall, and B. Plohr, *A generalized Riemann problem for quasi-one-dimensional gas flows*, Adv. Appl. Math., 5 (1984), pp. 1–30.

[156] E. Godlewski and P.-A. Raviart, *Numerical Approximation of Hyperbolic Systems of Conservation Laws*, Springer, New York, 1996.

[157] S. K. Godunov, *A difference method for numerical calculation of discontinuous solutions of the equations of hydrodynamics*, Mat. Sb., 47 (1959), pp. 271–306.

[158] S. K. Godunov, *The problem of a generalized solution in the theory of quasi-linear equations and in gas dynamics*, Russian Math. Surv., 17 (1962), pp. 145–156.

[159] J. B. Goodman and R. J. LeVeque, *On the accuracy of stable schemes for 2D scalar conservation laws*, Math. Comp., 45 (1985), pp. 15–21.

[160] J. B. Goodman and R. J. LeVeque, *A geometric approach to high resolution TVD schemes*, SIAM J. Numer. Anal., 25 (1988), pp. 268–284.

[161] L. Gosse, *A well-balanced flux-vector splitting scheme designed for hyperbolic systems of conservation laws with source terms*, Comput. Math. Appl., 39 (2000), pp. 135–159.

[162] L. Gosse, *A well-balanced scheme using non-conservative products designed for hyperbolic systems of conservation laws with source terms*, Math. Mod. Meth. Appl. Sci., 11 (2001), pp. 339–365.

[163] L. Gosse and F. James, *Numerical approximations of one-dimensional linear conservation equations with discontinous coefficients*, Math. Comp., 69 (2000), pp. 987–1015.

[164] L. Gosse and A. E. Tzavaras, *Convergence of relaxation schemes to the equations of elastodynamics*, Math. Comp., 70 (2000), pp. 555–577.

[165] S. Gottlieb and C.-W. Shu, *Total variation diminishing Runge–Kutta schemes*, Math. Comp., 67 (1998), pp. 73–85.

[166] S. Gottlieb, C.-W. Shu, and E. Tadmor, *Strong stability-preserving high-order time discretization methods*, SIAM Rev., 43 (2001), pp. 89–112.

[167] J. M. Greenberg and A. Y. LeRoux, *A well-balanced scheme for the numerical processing of source terms in hyperbolic equations*, SIAM J. Numer. Anal., 33 (1996), pp. 1–16.

[168] J. M. Greenberg, A. Y. LeRoux, R. Baraille, and A. Noussair, *Analysis and approximation of conservation laws with source terms*, SIAM J. Numer. Anal., 34 (1997), pp. 1980–2007.

[169] M. Griebel, T. Dornseifer, and T. Neunhoeffer, *Numerical Simulation in Fluid Dynamics*, SIAM, Philadelphia, 1998.

[170] D. F. Griffiths and J. M. Sanz-Serna, *On the scope of the method of modified equations*, SIAM J. Sci. Statist. Comput., 7 (1986), pp. 994–1008.

[171] J. Grove, *Applications of front tracking to the simulation of shock refractions and unstable mixing*, J. Appl. Numer. Math., 14 (1994), pp. 213–237.

[172] A. Guardone and L. Quartapelle, *Exact Roe's linearization for the van der Waals gas*, Godunov Methods: Theory and Applications, E. F. Toro, ed. Kluwer/Plenum Academic Press, to appear.

[173] H. Guillard and C. Viozat, *On the behaviour of upwind schemes in the low Mach number limit*, Comput. & Fluids, 28 (1999), pp. 63–86.

[174] B. Gustafsson, H.-O. Kreiss, and J. Oliger, *Time Dependent Problems and Difference Methods*, Wiley, New York, 1995.

[175] R. Haberman, *Mathematical Models: Mechanical Vibrations, Population Dynamics, and Traffic Flow: An Introduction to Applied Mathematics*, Prentice-Hall, Englewood Cliffs, N.J., 1977.

[176] T. Hagstrom, *Radiation boundary conditions for the numerical simulation of waves*, Acta Numer., 8 (1999), pp. 47–106.

[177] E. Hairer, S. P. Norsett, and G. Wanner, *Solving Ordinary Differential Equations I. Nonstiff Problems*, Springer-Verlag, Berlin, Heidelberg, 1987.

[178] E. Hairer, S. P. Norsett, and G. Wanner, *Solving Ordinary Differential Equations II. Stiff and Differential-Algebraic Problems*, Springer-Verlag, New York, 1993.

[179] A. Harten, *High resolution schemes for hyperbolic conservation laws*, J. Comput. Phys., 49 (1983), pp. 357–393.

[180] A. Harten, *On a class of high resolution total variation stable finite difference schemes*, SIAM J. Numer. Anal., 21 (1984), pp. 1–23.

[181] A. Harten, *On a large time-step high resolution scheme*, Math. Comp., 46 (1986), pp. 379–399.

[182] A. Harten, *ENO schemes with subcell resolution*, J. Comput. Phys., 83 (1987), pp. 148–184.

[183] A. Harten, B. Engquist, S. Osher, and S. Chakravarthy, *Uniformly high order accurate essentially nonoscillatory schemes, III*, J. Comput. Phys., 71 (1987), p. 231.

[184] A. Harten and J. M. Hyman, *Self-adjusting grid methods for one-dimensional hyperbolic conservation laws*, J. Comput. Phys., 50 (1983), pp. 235–269.

[185] A. Harten, J. M. Hyman, and P. D. Lax, *On finite-difference approximations and entropy conditions for shocks* (with appendix by Barbara Keyfitz), Comm. Pure Appl. Math., 29 (1976), pp. 297–322.

[186] A. Harten and P. D. Lax, *A random choice finite difference scheme for hyperbolic conservation laws*, SIAM J. Numer. Anal., 18 (1981), pp. 289–315.

[187] A. Harten, P. D. Lax, and B. van Leer, *On upstream differencing and Godunov-type schemes for hyperbolic conservation laws*, SIAM Rev., 25 (1983), pp. 35–61.

[188] A. Harten and S. Osher, *Uniformly high-order accurate nonoscillatory schemes. I*, SIAM J. Numer. Anal., 24 (1987), pp. 279–309.

[189] A. Harten, S. Osher, B. Engquist, and S. Chakravarthy, *Some results on uniformly high-order accurate essentially nonoscillatory schemes*, Appl. Numer. Math., 2 (1986), pp. 347–377.

[190] A. Harten and G. Zwas, *Self-adjusting hybrid schemes for shock computations*, J. Comput. Phys., 9 (1972), p. 568.

[191] H. Hattori, *The Riemann problem for a van der Waals fluid with entropy rate admissibility criterion. Isothermal case*, Arch. Rat. Mech. Anal., 92 (1986), pp. 247–263.

[192] H. Hattori, *The Riemann problem for a van der Waals fluid with entropy rate admissibility criterion. Nonisothermal case*, J. Diff. Equations, 65 (1986), pp. 158–174.

[193] G. Hedstrom, *Models of difference schemes for $u_t + u_x = 0$ by partial differential equations*, Math. Comp., 29 (1975), pp. 969–977.

[194] A. Heibig and J. F. Colombeau, *Non-conservative products in bounded variation functions*, SIAM J. Math. Anal., 23 (1992), pp. 941–949.

[195] C. Helzel and D. Bale, *Crossflow Instabilities in the Approximation of Detonation Waves*, in Proc. 8th Int. Conf. on Hyperbolic Problems, H. Freistühler and G. Warnecke, eds., Birkhäuser, 2000, pp. 119–128.

[196] C. Helzel, R. J. LeVeque, and G. Warnecke, *A modified fractional step method for the accurate approximation of detonation waves*, SIAM J. Sci. Comput., 22 (2000), pp. 1489–1510.

[197] R. L. Higdon, *Absorbing boundary conditions for difference approximations to the multidimensional wave equation*, Math. Comp., 47 (1986), pp. 437–459.

[198] C. Hirsch, *Numerical Computation of Internal and External Flows*, Wiley, 1988.

[199] C. Hirsch, C. Lacor, and H. Deconinck, *Convection algorithms based on a diagonalization procedure for the multidimensional Euler equations*. AIAA Paper 87-1163, 1987.

[200] H. Holden, *On the Riemann problem for a prototype of a mixed type conservation law*, Comm. Pure Appl. Math., 40 (1987), pp. 229–264.

[201] M. Holt, *Numerical Methods in Fluid Dynamics*, Springer-Verlag, Berlin, 1977.

[202] H. Hong, R. Liska, and S. Steinberg, *Testing stability by quantifier elimination*, J. Symbol. Comput., 24 (1997), pp. 161–187.

[203] T. Hou and P. G. Le Floch, *Why nonconservative schemes converge to wrong solutions: Error analysis*, Math. Comp., 62 (1993), pp. 497–530.

[204] L. Hsiao and P. de Mottoni, *Existence and uniqueness of the Riemann problem for nonlinear system of conservation laws of mixed type*, Trans. Am. Math. Soc., 322 (1990), pp. 121–158.

[205] J. Hu and P. G. LeFloch, l^1 *continuous dependence property for systems of conservation laws*, Arch. Rat. Mech. Anal., 151 (2000), pp. 45–93.

[206] H. T. Huynh, *Accurate upwind methods for the Euler equations*, SIAM J. Numer. Anal., 32 (1995), pp. 1565–1619.

[207] J. M. Hyman, *Numerical methods for tracking interfaces*, Physica D, 12 (1984), pp. 396–407.

[208] A. In, *Numerical evaluation of an energy relaxation method for inviscid real fluids*, SIAM J. Sci. Comput., 21 (1999), pp. 340–365.

[209] E. Isaacson, B. Plohr, and B. Temple, *The Riemann problem near a hyperbolic singularity III*, SIAM J. Appl. Math., 48 (1988), pp. 1302–1318.

[210] E. Isaacson and B. Temple, *Analysis of a singular hyperbolic system of conservation laws*, J. Diff. Equations, 65 (1986), pp. 250–268.

[211] A. Iserles, *Numerical Analysis of Differential Equations*, Cambridge University Press, Cambridge, 1996.

[212] M. I. Ivings, D. M. Causon, and E. F. Toro, *On Riemann solvers for compressible liquids*, Int. J. Numer. Meth. Fluids, 28 (1998), pp. 395–418.

[213] R. D. James, *The propagation of phase boundaries in elastic bars*, Arch. Rat. Mech. Anal., 73 (1980), pp. 125–158.

[214] A. J. Jameson, W. Schmidt, and E. Turkel, *Numerical solutions of the Euler equations by a finite-volume method using Runge–Kutta time-stepping schemes*, AIAA Paper 81-1259, 1981.

[215] G. Jennings, *Discrete shocks*, Comm. Pure Appl. Math., 27 (1974), pp. 25–37.

[216] P. Jenny and B. Müller, *A new approach for a flux solver taking into account source terms, viscous and multidimensional effects*, in Proc. 7th Int. Conf. on Hyperbolic Problems, R. Jeltsch, ed., Birkhäuser, 1998, pp. 503–513.

[217] P. Jenny and B. Müller, *Rankine–Hugoniot–Riemann solver considering source terms and multidimensional effects*, J. Comput. Phys., 145 (1998), pp. 575–610.

[218] P. Jenny, B. Müller, and H. Thomann, *Correction of conservative Euler solvers for gas mixture*, J. Comput. Phys., 132 (1997), pp. 91–107.

[219] G. Jiang and C. W. Shu, *Efficient implementation of weighted ENO schemes*, J. Comput. Phys., 126 (1996), pp. 202–228.

[220] G.-S. Jiang, D. Levy, C.-T. Lin, S. Osher, and E. Tadmor, *High-resolution non-oscillatory central schemes with non-staggered grids for hyperbolic conservation laws*, SIAM J. Numer. Anal., 35 (1998), pp. 2147–2168.

[221] G.-S. Jiang and E. Tadmor, *Non-oscillatory central schemes for multidimensional hyperbolic conservation laws*, SIAM J. Sci. Comput., 19 (1998), pp. 1892–1917.

[222] S. Jin and M. A. Katsoulakis, *Relaxation approximations to front propagation*, J. Diff. Equations, 138 (1997), pp. 380–387.

[223] S. Jin and M. A. Katsoulakis, *Hyperbolic systems with supercharacteristic relaxations and roll waves*, SIAM J. Appl. Math., 61 (2000), pp. 273–292.

[224] S. Jin and J.-G. Liu, *The effects of numerical viscosities: I. Slowly moving shocks*, J. Comput. Phys., 126 (1996), pp. 373–389.

[225] S. Jin and Z. P. Xin, *The relaxation schemes for systems of conservation laws in arbitrary space dimensions*, Comm. Pure Appl. Math., 48 (1995), pp. 235–276.

[226] S. Jin and Z. P. Xin, *Numerical passage from systems of conservation laws to Hamilton–Jacobi equations, relaxation schemes*, SIAM J. Numer. Anal., 35 (1998), pp. 2385–2404.

[227] T. Johansen, A. Tveito, and R. Winther, *A Riemann solver for a two-phase multicomponent process of conservation laws modeling polymer flooding*, SIAM J. Sci. Stat. Comput., 10 (1989), pp. 346–879.

[228] T. Johansen and R. Winther, *The solution of the Riemann problem for a hyperbolic system of conservation laws modeling polymer flooding*, SIAM J. Math. Anal., 19 (1988), pp. 541–566.

[229] F. John, *Partial Differential Equations*, Springer, 1971.

[230] S. Karni, *Far-field filtering operators for suppression of reflections from artificial boundaries*, SIAM J. Numer. Anal., 33 (1996), pp. 1014–1047.

[231] S. Karni, *Hybrid multifluid algorithms*, SIAM J. Sci. Comput., 17 (1996), pp. 1019–1039.

[232] S. Karni and S. Canic, *Computation of slowly moving shocks*, J. Comput. Phys., 136 (1997), pp. 132–139.

[233] M. A. Katsoulakis and A. E. Tzavaras, *Contractive relaxation systems and the scalar multidimensional conservation laws*, Comm. Partial Diff. Equations, 22 (1997), pp. 195–223.

[234] J. Kevorkian, *Partial Differential Equations: Analytical Solution Techniques*, Springer, New York, 2000.

[235] B. Keyfitz, *A survey of nonstrictly hyperbolic conservation laws*, in *Nonlinear Hyperbolic Problems*, C. Carasso, P.-A. Raviart, and D. Serre, eds., Lecture Notes in Mathematics 1270, Springer-Verlag, 1986, pp. 152–162.

[236] B. Keyfitz and H. Kranzer, *A viscosity approximation to a system of conservation laws with no classical Riemann solution*, in *Nonlinear Hyperbolic Problems*, C. Carasso, P. Charrier, B. Hanouzet, and J.-L. Joly, eds., Lecture Notes in Mathematics 1402, Springer-Verlag, 1988, pp. 185–197.

[237] B. L. Keyfitz and H. C. Kranzer, *A system of hyperbolic conservation laws arising in elasticity theory*, Arch. Rat. Mech. Anal., 72 (1980), pp. 219–241.

[238] B. L. Keyfitz and H. C. Kranzer, *Nonstrictly Hyperbolic Conservation Laws: Proceedings of an AMS Special Session*, American Mathematical Society, Providence, R.I., 1987.

[239] B. L. Keyfitz and H. C. Kranzer, *Spaces of weighted measures for conservation laws with singular shock solutions*, J. Diff. Equations, 118 (1995), pp. 420–451.

[240] B. L. Keyfitz and C. A. Mora, *Prototypes for nonstrict hyperbolicity in conservation laws*, Contemp. Math., 255 (2000), pp. 125–137.

[241] B. L. Keyfitz and M. Shearer, *Nonlinear Evolution Equations That Change Type*, Springer IMA 27, 1990.

[242] R. Klein, *Semiimplicit extension of a Godunov-type scheme based on low Mach number asymptotics. 1. One-dimensional flow*, J. Comput. Phys., 121 (1995), pp. 213–237.

[243] C. Klingenberg and N. H. Risebro, *Convex conservation laws with discontinuous coefficients. Existence, uniqueness and asymptotic behavior*, Comm. Partial Diff. Equations, 20 (1995), pp. 1959–1990.

[244] H. O. Kreiss and J. Lorenz, *Initial–Boundary Value Problems and the Navier–Stokes Equations*, Academic Press, 1989.

[245] D. Kröner, *Numerical Schemes for Conservation Laws*, Wiley-Teubner Series, 1997.

[246] D. Kröner, S. Noelle, and M. Rokyta, *Convergence of higher order upwind finite volume schemes on unstructured grids for scalar conservation laws in several space dimensions*, Numer. Math., 71 (1995), pp. 527–560.

[247] D. Kröner and M. Rokyta, *Convergence of upwind finite volume schemes for scalar conservation laws in two dimensions*, SIAM J. Numer. Anal., 31 (1994), pp. 324–343.

[248] S. N. Kružkov, *First order quasilinear equations in several independent variables*, Math. USSR Sb., 10 (1970), pp. 217–243.

[249] A. Kulikovskii and E. Sveshnikova, *Nonlinear Waves in Elastic Media*, CRC Press, Boca Raton, Fla., 1995.

[250] A. Kurganov and E. Tadmor, *New high-resolution central schemes for nonlinear conservation laws and convection–diffusion equations*, J. Comput. Phys., 160 (2000), pp. 214–282.

[251] N. N. Kuznetsov, *Accuracy of some approximate methods for computing the weak solutions of a first-order quasi-linear equation*, USSR Comput. Math. and Math. Phys., 16 (1976), pp. 105–119.

[252] R. Lakes, *Foam structures with a negative Poisson's ratio*, Science, 235 (1987), pp. 1038–1040. See also `http://silver.neep.wisc.edu/~lakes/sci87.html`.

[253] J. D. Lambert, *Computational Methods in Ordinary Differential Equations*, Wiley, 1973.

[254] H. Lan and K. Wang, *Exact solutions for some nonlinear equations*, Phys. Lett. A, 137 (1989), pp. 369–372.

[255] L. D. Landau and E. M. Lifshitz, *Theory of Elasticity*, Pergamon Press, 1975.

[256] C. B. Laney, *Computational Gasdynamics*, Cambridge University Press, Cambridge, 1998.

[257] J. O. Langseth and R. J. LeVeque, *A wave-propagation method for three-dimensional hyperbolic conservation laws*, J. Comput. Phys., 165 (2000), pp. 126–166.

[258] J. O. Langseth, A. Tveito, and R. Winther, *On the convergence of operator splitting applied to conservation laws with source terms*, SIAM J. Numer. Anal., 33 (1996), pp. 843–863.

[259] B. Larouturou, *How to preserve the mass fraction positive when computing compressible multi-component flows*, J. Comput. Phys., 95 (1991), pp. 59–84.

[260] C. Lattanzio and P. Marcati, *The zero relaxation limit for the hydrodynamic Whitham traffic flow model*, J. Diff. Equations, 141 (1997), pp. 150–178.

[261] C. Lattanzio and D. Serre, *Convergence of a relaxation scheme for hyperbolic systems of conservation laws*, Numer. Math., 88 (2001), pp. 121–134.

[262] P. D. Lax, *Hyperbolic systems of conservation laws, II.*, Comm. Pure Appl. Math., 10 (1957), pp. 537–566.

[263] P. D. Lax, *Hyperbolic Systems of Conservation Laws and the Mathematical Theory of Shock Waves*, SIAM Regional Conference Series in Applied Mathematics 11, 1972.

[264] P. D. Lax and X. D. Liu, *Solution of two dimensional Riemann problem of gas dynamics by positive schemes*, SIAM J. Sci. Comput., 19 (1998), pp. 319–340.

[265] P. D. Lax and B. Wendroff, *Systems of conservation laws*, Comm. Pure Appl. Math., 13 (1960), pp. 217–237.

[266] L. Lee and R. J. LeVeque, *An immersed interface method for incompressible Navier-Stokes equations*, submitted.

[267] P. LeFloch and T.-P. Liu, *Existence theory for nonlinear hyperbolic systems in nonconservative form*, Forum Math., 5 (1993), pp. 261–280.

[268] P. G. LeFloch, *Entropy weak solutions to nonlinear hyperbolic systems under nonconservative form*, Comm. Partial Diff. Equations, 13 (1988), pp. 669–727.

[269] P. G. LeFloch, *Propagating phase boundaries: Formulation of the problem and existence via Glimm scheme*, Arch. Rat. Mech. Anal., 123 (1993), pp. 153–197.

[270] P. G. LeFloch and Nedelec, *Explicit formulas for weighted scalar nonlinear hyperbolic conservation laws*, Trans. Amer. Math. Soc., 308 (1988), pp. 667–683.

[271] B. P. Leonard, *The ULTIMATE conservative difference scheme applied to unsteady one-dimensional advection*, Comput. Meth. Appl. Mech. Eng., 88 (1991), pp. 17–74.

[272] B. P. Leonard, A. P. Lock, and M. K. MacVean, *Conservative explicit unrestricted-time-step multidimensional constancy-preserving advection schemes*, Monthly Weather Rev., 124 (1996), pp. 2588–2606.

[273] R. J. LeVeque, *Finite volume methods for nonlinear elasticity in heterogeneous media*, Int. J. Numer. Meth. Fluids, to appear.

[274] R. J. LeVeque, *Large time step shock-capturing techniques for scalar conservation laws*, SIAM J. Numer. Anal., 19 (1982), pp. 1091–1109.

[275] R. J. LeVeque, *A large time step generalization of Godunov's method for systems of conservation laws*, SIAM J. Numer. Anal., 22 (1985), pp. 1051–1073.

[276] R. J. LeVeque, *Intermediate boundary conditions for time-split methods applied to hyperbolic partial differential equations*, Math. Comp., 47 (1986), pp. 37–54.

[277] R. J. LeVeque, *Cartesian grid methods for flow in irregular regions*, in *Numerical Methods in Fluid Dynamics III*, K. W. Morton and M. J. Baines, eds., Clarendon Press, 1988, pp. 375–382.

[278] R. J. LeVeque, *High resolution finite volume methods on arbitrary grids via wave propagation*, J. Comput. Phys., 78 (1988), pp. 36–63.

[279] R. J. LeVeque, *Second order accuracy of Brenier's time-discrete method for nonlinear systems of conservation laws*, SIAM J. Numer. Anal., 25 (1988), pp. 1–7.

[280] R. J. LeVeque, *Hyperbolic Conservation Laws and Numerical Methods*, Von Karman Institute for Fluid Dynamics Lecture Series, 90-03, 1990.

[281] R. J. LeVeque, *Numerical Methods for Conservation Laws*, Birkhäuser, 1990.

[282] R. J. LeVeque, *High-resolution conservative algorithms for advection in incompressible flow*, SIAM J. Numer. Anal., 33 (1996), pp. 627–665.

[283] R. J. LeVeque, *Wave propagation algorithms for multi-dimensional hyperbolic systems*, J. Comput. Phys., 131 (1997), pp. 327–353.

[284] R. J. LeVeque, *Balancing source terms and flux gradients in high-resolution Godunov methods: The quasi-steady wave-propagation algorithm*, J. Comput. Phys., 146 (1998), pp. 346–365.

[285] R. J. LeVeque, *Some traffic flow models illustrating interesting hyperbolic behavior*. Preprint 2001-036, http://www.math.ntnu.no/conservation/2001, 2001.

[286] R. J. LeVeque and D. S. Bale, *Wave-propagation methods for conservation laws with source terms*, in Proc. 7th Int. Conf. on Hyperbolic Problems, R. Jeltsch, ed., Birkhäuser, 1998, pp. 609–618.

[287] R. J. LeVeque, D. Mihalas, E. Dorfi, and E. Müller, *Computational Methods for Astrophysical Fluid Flow*, Saas-Fee Advanced Course 27, A. Gautschy and O. Steiner, eds., Springer, 1998.

[288] R. J. LeVeque and M. Pelanti, *A class of approximate Riemann solvers and their relation to relaxation schemes*, J. Comput. Phys., 172 (2001), pp. 572–591.

[289] R. J. LeVeque and K.-M. Shyue, *One-dimensional front tracking based on high resolution wave propagation methods*, SIAM J. Sci. Comput., 16 (1995), pp. 348–377.

[290] R. J. LeVeque and K.-M. Shyue, *Two-dimensional front tracking based on high resolution wave propagation methods*, J. Comput. Phys., 123 (1996), pp. 354–368.

[291] R. J. LeVeque and B. Temple, *Stability of Godunov's method for a class of* 2×2 *systems of conservation laws*, Trans. Amer. Math. Soc., 288 (1985), pp. 115–123.

[292] R. J. LeVeque and J. Wang, *A linear hyperbolic system with stiff source terms*, in Nonlinear Hyperbolic Problems: Theoretical, Applied, and Computational Aspects, A. Donato and F. Oliveri, eds., Notes on Numerical Fluid Mechanics 43, Vieweg, 1993, pp. 401–408.

[293] R. J. LeVeque and H. C. Yee, *A study of numerical methods for hyperbolic conservation laws with stiff source terms*, J. Comput. Phys., 86 (1990), pp. 187–210.

[294] D. Levy, G. Puppo, and G. Russo, *Central WENO schemes for hyperbolic systems of conservation laws*, Math. Model. Numer. Anal., 33 (1999), pp. 547–571.

[295] D. Levy and E. Tadmor, *From semidiscrete to fully discrete: Stability of Runge–Kutta schemes by the energy method*, SIAM Rev., 40 (1998), pp. 40–73.

[296] T. Li, *Global solutions and zero relaxation limit for a traffic flow model*, SIAM J. Appl. Math., 61 (2000), pp. 1042–1061.

[297] H. W. Liepmann and A. Roshko, *Elements of Gas Dynamics*, Wiley, 1957.

[298] J. Lighthill, *Waves in Fluids*, Cambridge University Press, 1978.

[299] M. J. Lighthill and G. B. Whitham, *On kinematic waves: II. A theory of traffic flow on long crowded roads*, Proc. Roy. Soc. London, Ser. A, 229 (1955), pp. 317–345.

[300] T. Linde, *A practical, general-purpose, two-state HLL Riemann solver for hyperbolic conservation laws*, Int. J. Numer. Meth. Fluids, to appear.

[301] M.-S. Liou, *A sequel to AUSM: AUSM+*, J. Comput. Phys., 129 (1996), pp. 364–382.

[302] M.-S. Liou and C. J. Steffen Jr., *A new flux splitting scheme*, J. Comput. Phys., 107 (1993), p. 107.

[303] R. Liska and B. Wendroff, *Composite schemes for conservation laws*, SIAM J. Numer. Anal., 35 (1998), pp. 2250–2271.

[304] R. Liska and B. Wendroff, *Two-dimensional shallow water equations by composite schemes*, Int. J. Numer. Meth. Fluids, 30 (1999), pp. 461–479.

[305] H. L. Liu, J. Wang, and T. Yang, *Stability of a relaxation model with a nonconvex flux*, SIAM J. Math. Anal., 29 (1998), pp. 18–29.

[306] H. L. Liu and G. Warnecke, *Convergence rates for relaxation schemes approximating conservation laws*, SIAM J. Numer. Anal., 37 (2000), pp. 1316–1337.

[307] T. P. Liu, *The deterministic version of the Glimm scheme*, Comm. Math. Phys., 57 (1977), pp. 135–148.

[308] T.-P. Liu, *Large-time behavior of solutions of initial and initial–boundary value problems of a general system of hyperbolic conservation laws*, Comm. Math. Phys., 55 (1977), pp. 163–177.

[309] T. P. Liu, *Admissible solutions of hyperbolic conservation laws*, Amer. Math. Soc., Memoirs of the Amer. Math. Soc., 240 (1981).

[310] T. P. Liu, *Hyperbolic conservation laws with relaxation*, Comm. Math. Phys., 108 (1987), pp. 153–175.

[311] T.-P. Liu, *Hyperbolic and Viscous Conservation Laws*, SIAM Regional Conference Series in Applied Mathematics 72, 2000.

[312] T. P. Liu and T. Yang, *Well-posedness theory for hyperbolic conservation laws*, Comm. Pure Appl. Math., 52 (1999), pp. 1553–1586.

[313] X. D. Liu and P. D. Lax, *Positive schemes for solving multi-dimensional hyperbolic systems of conservation laws*, Comput. Fluid Dynamics J., 5 (1996), pp. 133–156.

[314] X.-D. Liu, S. Osher, and T. Chan, *Weighted essentially non-oscillatory schemes*, J. Comput. Phys., 115 (1994), pp. 200–212.

[315] X.-D. Liu and E. Tadmor, *Third order nonoscillatory central scheme for hyperbolic conservation law*, Numer. Math., 79 (1998), pp. 397–425.

[316] B. J. Lucier, *Error bounds for the methods of Glimm, Godunov and LeVeque*, SIAM J. Numer. Anal., 22 (1985), pp. 1074–1081.

[317] W. K. Lyons, *Conservation laws with sharp inhomogeneities*, Quart. Appl. Math., 40 (1983), pp. 385–393.

[318] R. W. MacCormack, *The effects of viscosity in hypervelocity impact cratering*. AIAA Paper 69-354, 1969.

[319] A. Majda, *Compressible Fluid Flow and Systems of Conservation Laws in Several Space Variables*, Appl. Math. Sci. Vol. 53, Springer, 1984.

[320] J. C. Mandal and S. M. Deshpande, *Kinetic flux vector splitting for Euler equations*, Comput. Fluids, 23 (1994), pp. 447–478.

[321] D.-K. Mao, *A shock tracking technique based on conservation in one space dimension*, SIAM J. Numer. Anal., 32 (1995), pp. 1677–1703.

[322] D.-K. Mao, *Toward front tracking based on conservation in two space dimensions*, SIAM J. Sci. Comput., 22 (2000), pp. 113–151.

[323] A. Marquina, *Local piecewise hyperbolic reconstruction of numerical fluxes for nonlinear scalar conservation laws*, SIAM J. Sci. Comput., 15 (1994), p. 892.

[324] D. Mavriplis, *Unstructured grid techniques*, Annu. Rev. Fluid Mech., 29 (1997), pp. 473–514.

[325] G. Mellema, F. Eulderink, and V. Icke, *Hydrodynamical models of aspherical planetary nebulae*, Astron. Astrophys., 252 (1991), pp. 718–732.

[326] R. Menikoff and B. J. Plohr, *The Riemann problem for fluid flow of real materials*, Rev. Modern Phys., 61 (1989), pp. 75–130.

[327] G. H. Miller and P. Colella, *A high-order Eulerian Godunov method for elastic-plastic flow in solids*, J. Comput. Phys., 167 (2001), pp. 131–176.

[328] G. H. Miller and E. G. Puckett, *A high-order Godunov method for multiple condensed phases*, J. Comput. Phys., 128 (1996), pp. 134–164.

[329] M. L. Minion and D. L. Brown, *Performance of under-resolved two-dimensional incompressible flow simulations, II*, J. Comput. Phys., 138 (1997), pp. 734–765.

[330] J.-L. Montagné, H. C. Yee, and M. Vinokur, *Comparative study of high-resolution shock-capturing schemes for a real gas*, in Proc. Seventh GAMM Conf. on Numerical Methods in Fluid Mechanics, Notes on Numerical Fluid Mechanics 20, M. Deville, ed., Vieweg, Braunschweig, 1988, pp. 219–228.

[331] K. W. Morton, *On the analysis of finite volume methods for evolutionary problems*, SIAM J. Numer. Anal., 35 (1998), pp. 2195–2222.

[332] K. W. Morton, *Discretisation of unsteady hyperbolic conservation laws*. SIAM J. Numer. Anal., 39 (2001), pp. 1556–1597.

[333] K. W. Morton and D. F. Mayers, *Numerical Solution of Partial Differential Equations*, Cambridge University Press, Cambridge, 1994.

[334] R. S. Myong and P. L. Roe, *Shock waves and rarefaction waves in magnetohydrodynamics. Part I. A model system*, J. Plasma Phys., 58 (1997), pp. 485–519.

[335] R. S. Myong and P. L. Roe, *Shock waves and rarefaction waves in magnetohydrodynamics. Part II. The MHD system*, J. Plasma Phys., 58 (1997), pp. 521–552.

[336] R. Natalini, *Convergence to equilibrium for the relaxation approximations of conservation laws*, Comm. Pure Appl. Math., 49 (1996), pp. 795–823.

[337] R. Natalini, *Recent mathematical results on hyperbolic relaxation problems*, in Analysis of Systems of Conservation Laws, H. Freistühler, ed., Chapman & Hall/CRC Press Monographs and Surveys in Pure and Applied Mathematics, 1999.

[338] H. Nessyahu and E. Tadmor, *Non-oscillatory central differencing for hyperbolic conservation laws*, J. Comput. Phys., 87 (1990), pp. 408–463.

[339] H. Nessyahu and E. Tadmor, *The convergence rate of approximate solutions for nonlinear scalar conservation laws*, SIAM J. Numer. Anal., 29 (1992), pp. 1505–1519.

[340] H. Nessyahu, E. Tadmor, and T. Tassa, *The convergence rate of Godunov type schemes*, SIAM J. Numer. Anal., 31 (1994), pp. 1–16.

[341] S. Noelle, *Convergence of higher order finite volume schemes on irregular grids*, Adv. Comput. Math., 3 (1995), p. 197.

[342] S. Noelle, *The MoT-ICE: A new high-resolution wave-propagation algorithm for multidimensional systems of conservation laws based on Fey's method of transport*, J. Comput. Phys., 164 (2000), pp. 283–334.

[343] W. F. Noh, *Errors for calculations of strong shocks using an artificial viscosity and an artificial heat flux*, J. Comput. Phys., 72 (1987), p. 78.

[344] M. Oberguggenburger, *Case study of a nonlinear, nonconservative, non-strictly hyperbolic system*, Nonlinear Anal., 19 (1992), pp. 53–79.

[345] M. Oberguggenburger, *Multiplication of Distributions and Applications to Partial Differential Equations*, Longman Scientific & Technical, Harlow, and Wiley, New York, 1992.

[346] O. Oleinik, *Discontinuous solutions of nonlinear differential equations*, Amer. Math. Soc. Transl. Ser. 2, 26 (1957), pp. 95–172.

[347] O. Oleinik, *Uniqueness and stability of the generalized solution of the Cauchy problem for a quasilinear equation*, Amer. Math. Soc. Transl. Ser. 2, 33 (1964), pp. 285–290.

[348] E. S. Oran and J. P. Boris, *Numerical Simulation of Reactive Flow*, Cambridge University Press, second ed., 2001.

[349] S. Osher, *Riemann solvers, the entropy condition, and difference approximations*, SIAM J. Numer. Anal., 21 (1984), pp. 217–235.

[350] S. Osher, *Convergence of generalized MUSCL schemes*, SIAM J. Numer. Anal., 22 (1985), pp. 947–961.

[351] S. Osher and S. Chakravarthy, *High resolution schemes and the entropy condition*, SIAM J. Numer. Anal., 21 (1984), pp. 995–984.

[352] S. Osher and F. Solomon, *Upwind difference schemes for hyperbolic systems of conservation laws*, Math. Comp., 38 (1982), pp. 339–374.

[353] M. Pandolfi and D. D'Ambrosio, *Numerical instabilities in upwind methods: Analysis and cures for the "carbuncle" phenomenon*, J. Comput. Phys., 166 (2001), pp. 271–301.

[354] H. J. Payne, *Models of freeway traffic and control*, in *Mathematical Models of Public Systems*, Simulation Council Proc. 28, Vol. 1, 1971, pp. 51–61.

[355] M. Pelanti, L. Quartapelle, and L. Vigevano, *Low dissipation entropy fix for positivity preserving Roe's scheme*, in *Godunov Methods: Theory and Applications*, E. F. Toro, ed., Kluwer/Plenum Academic Press, to appear.

[356] R. B. Pember, *Numerical methods for hyperbolic conservation laws with stiff relaxation, I. Spurious solutions*, SIAM J. Appl. Math., 53 (1993), pp. 1293–1330.

[357] R. B. Pember, *Numerical methods for hyperbolic conservation laws with stiff relaxation, II. Higher order Godunov methods*, SIAM J. Sci. Comput., 14 (1993).

[358] R. B. Pember, J. B. Bell, P. Colella, W. Y. Crutchfield, and M. L. Welcome, *An adaptive Cartesian grid method for unsteady compressible flow in complex geometries*, J. Comput. Phys., 120 (1995), pp. 278–304.

[359] R. Peyret and T. D. Taylor, *Computational Methods for Fluid Flow*, Springer, 1983.

[360] B. J. Plohr and D. Sharp, *A conservative formulation for plasticity*, Adv. Appl. Math., 13 (1992), pp. 462–493.

[361] F. Poupaud and M. Rascle, *Measure solutions to the linear multi-dimensional transport equation with non-smooth coefficients*, Comm. Partial Diff. Equations, 22 (1997), pp. 337–358.

[362] E. G. Puckett and J. S. Saltzman, *A 3D adaptive mesh refinement algorithm for multimaterial gas dynamics*, Physica D, 60 (1992), pp. 84–93.

[363] J. J. Quirk, *An alternative to unstructured grids for computing gas-dynamic flow around arbitrarily complex 2-dimensional bodies*, Comput. Fluids, 23 (1994), pp. 125–142.

[364] J. J. Quirk, *A contribution to the great Riemann solver debate*, Int. J. Numer. Methods Fluids, 18 (1994), pp. 555–574.

[365] Y. B. Radvogin and N. A. Zaitsev, *A locally implicit second order accurate difference scheme for solving 2D time-dependent hyperbolic systems and Euler equations*, Appl. Numer. Math., 33 (2000), pp. 525–532.

[366] J. Rauch and M. C. Read, *Nonlinear superposition and absorption of delta-waves in one space dimension*, J. Funct. Anal., 73 (1987), pp. 152–178.

[367] M. Reiner, *Advanced Rheology*, H. K. Lewis, London, 1971.

[368] P. I. Richards, *Shock waves on highways*, Oper. Res., 4 (1956), pp. 42–51.

[369] R. D. Richtmyer and K. W. Morton, *Difference Methods for Initial-Value Problems*, Wiley-Interscience, 1967.

[370] S. Ridah, *Shock waves in water*, J. Appl. Phys., 64 (1988), pp. 152–158.

[371] N. H. Risebro, *A front-tracking alternative to the random choice method*, Proc. Am. Math. Soc., 117 (1993), pp. 1125–1139.

[372] N. H. Risebro and A. Tveito, *Front tracking applied to a nonstrictly hyperbolic system of conservation laws*, SIAM J. Sci. Stat. Comput., 6 (1991), pp. 1401–1419.

[373] T. W. Roberts, *The behavior of flux difference splitting schemes near slowly moving shock waves*, J. Comput. Phys., 90 (1990), pp. 141–160.

[374] J. Robinet, J. Gressier, G. Casalis, and J. Moschetta, *Shock wave instability and the carbuncle phenomenon: Same intrinsic origin?*, J. Fluid Mech., 417 (2000), pp. 237–263.

[375] P. L. Roe, *Approximate Riemann solvers, parameter vectors, and difference schemes*, J. Comput. Phys., 43 (1981), pp. 357–372.

[376] P. L. Roe, *The use of the Riemann problem in finite-difference schemes*, Lecture Notes in Physics 141, Springer, 1981.

[377] P. L. Roe, *Fluctuations and signals – a framework for numerical evolution problems*, in *Numerical Methods for Fluid Dynamics*, K. W. Morton and M. J. Baines, eds., Academic Press, 1982, pp. 219–257.

[378] P. L. Roe, *Some contributions to the modeling of discontinuous flows*, Lect. Notes Appl. Math., 22 (1985), pp. 163–193.

[379] P. L. Roe, *Characteristic-based schemes for the Euler equations*, Ann. Rev. Fluid Mech., 18 (1986), pp. 337–365.

[380] P. L. Roe, *Discrete models for the numerical analysis of time-dependent multidimensional gas dynamics*, J. Comput. Phys., 63 (1986), pp. 458–476.

[381] P. L. Roe, *Upwind differencing schemes for hyperbolic conservation laws with source terms*, in *Nonlinear Hyperbolic Problems*, C. Carraso, P.-A. Raviart, and D. Serre, eds., Lecture Notes in Mathematics 1270, Springer, 1986, pp. 41–51.

[382] P. L. Roe, *Sonic flux formulae*, SIAM J. Sci. Stat. Comput., 13 (1992), pp. 611–630.

[383] P. L. Roe, *Linear bicharacteristic schemes without dissipation*, SIAM J. Sci. Comput., 19 (1998), pp. 1405–1427.

[384] P. L. Roe, *Shock capturing*, in *Handbook of Shock Waves*, G. Ben-Dor, O. Igra, and T. Elperin, eds., Vol. 1, Academic Press, 2001, pp. 787–876.

[385] P. L. Roe and D. Sidilkover, *Optimum positive linear schemes for advection in two and three dimensions*, SIAM J. Numer. Anal., 29 (1992), pp. 1542–1568.

[386] S. Roller, C.-D. Munz, K. J. Geratz, and R. Klein, *The multiple pressure variables method for weakly compressible fluids*, Z. Angew. Math. Phys., Suppl. 2, 77 (1997), pp. S481–S484.

[387] V. V. Rusanov, *Calculation of interaction of non-steady shock waves with obstacles*, J. Comput. Math. Phys. USSR, 1 (1961), pp. 267–279.

[388] L. Sainsaulieu, *Finite volume approximation of two-phase fluid flows based on an approximate Roe-type Riemann solver*, J. Comput. Phys., 121 (1995), pp. 1–28.

[389] J. Saltzman, *An unsplit 3–D upwind method for hyperbolic conservation laws*, J. Comput. Phys., 115 (1994), pp. 153–168.

[390] R. Sanders, *On convergence of monotone finite difference schemes with variable spatial differencing*, Math. Comp., 40 (1983), pp. 91–106.

[391] R. Sanders and A. Weiser, *A high resolution staggered mesh approach for nonlinear hyperbolic systems of conservation laws*, J. Comput. Phys., 101 (1992), pp. 314–329.

[392] R. H. Sanders and K. H. Prendergast, *The possible relation of the 3-kiloparsec arm to explosions in the galactic nucleus*, Astrophys. J., 188 (1974), p. 489.

[393] F. Santosa and W. Symes, *A dispersive effective medium for wave propagation in periodic composites*, SIAM J. Appl. Math., 51 (1991), pp. 984–1005.

[394] R. Saurel and R. Abgrall, *A multiphase Godunov method for compressible multifluid and multiphase flows*, J. Comput. Phys., 150 (1999), pp. 425–467.

[395] R. Saurel and R. Abgrall, *A simple method for compressible multifluid flow*, SIAM J. Sci. Comput., 71 (1999), pp. 1115–1145.

[396] R. Saurel, M. Larini, and J. C. Loraud, *Exact and approximate Riemann solvers for real gases*, J. Comput. Phys., 112 (1994), pp. 126–137.

[397] D. G. Schaeffer and M. Shearer, *The classification of 2×2 systems of nonstrictly hyperbolic conservation laws, with application to oil recovery*, Comm. Pure Appl. Math., 40 (1987), pp. 141–178.

[398] D. G. Schaeffer and M. Shearer, *Riemann problems for nonstrictly hyperbolic 2×2 systems of conservation laws*, Trans. Amer. Math. Soc., 304 (1987), pp. 267–306.

[399] T. Schneider, N. Botta, K. J. Geratz, and R. Klein, *Extension of finite volume compressible flow solvers to multi-dimensional, variable density zero Mach number flows*, J. Comput. Phys., 155 (1999), pp. 248–286.

[400] S. Schochet, *The instant-response limit in Whitham's nonlinear traffic-flow model: Uniform well-posedness and global existence*, Asymp. Anal., 1 (1988), pp. 263–282.

[401] H.-J. Schroll, A. Tveito, and R. Winther, *An L^1 error bound for a semi-implicit scheme applied to a stiff system of conservation laws*, SIAM J. Numer. Anal., 34 (1997), pp. 1152–1166.

[402] D. Serre, *Systems of Conservation Laws. Volume 1: Hyperbolicity, Entropies, Shock Waves*, Cambridge University Press, Cambridge, 1999.

[403] D. Serre, *Systems of Conservation Laws. Volume 2: Geometric Structures, Oscillation and Mixed Problems*, Cambridge University Press, Cambridge, 2000.

[404] J. Sesterhenn, B. Müller, and H. Thomann, *On the cancellation problem in calculating compressible low Mach number flows*, J. Comput. Phys., 151 (1999), pp. 597–615.

[405] A. H. Shapiro, *The Dynamics and Thermodynamics of Compressible Fluid Flow*, Ronald Press, Cambridge, Mass., 1954.

[406] R. E. Showalter and P. Shi, *Dynamic plasticity models*, Comput. Meth. Appl. Mech. Eng., 151 (1998), pp. 501–511.

[407] C.-W. Shu, *TVB uniformly high-order schemes for conservation laws*, Math. Comp., 49 (1987), pp. 105–121.

[408] C.-W. Shu, *Total-variation-diminishing time discretizations*, SIAM J. Sci. Stat. Comput., 9 (1988), pp. 1073–1084.

[409] C.-W. Shu, *Essentially non-oscillatory and weighted essentially non-oscillatory schemes for hyperbolic conservation laws*, ICASE Report No. 97-65, NASA/CR-97-206253, NASA Langley Research Center, 1997.

[410] C.-W. Shu and S. Osher, *Efficient implementation of essentially non-oscillatory shock capturing schemes*, J. Comput. Phys., 77 (1988), pp. 439–471.

[411] C.-W. Shu and S. Osher, *Efficient implementation of essentially non-oscillatory shock capturing schemes II*, J. Comput. Phys., 83 (1989), pp. 32–78.

[412] K.-M. Shyue, *An efficient shock-capturing algorithm for compressible multicomponent problems*, J. Comput. Phys., 142 (1998), pp. 208–242.

[413] K.-M. Shyue, *A fluid-mixture type algorithm for compressible multicomponent flow with van der Waals equation of state*, J. Comput. Phys., 156 (1999), pp. 43–88.

[414] D. Sidilkover, *A genuinely multidimensional upwind scheme for the compressible Euler equations*, in Proc. Fifth Int. Conf. on Hyperbolic Problems: Theory, Numerics, Applications, J. Glimm et al., eds., World Scientific, June 1994.

[415] D. Sidilkover, *Multidimensional upwinding: Unfolding the mystery*, in *Barriers and Challenges in Computational Fluid Dynamics*, ICASE/LaRC Interdiscipl. Ser. Sci. Eng., 6, 1996, pp. 371–386.

[416] D. Sidilkover, *Some approaches towards constructing optimally efficient multigrid solvers for the inviscid flow equations*, Comput. Fluids, 28 (1999), pp. 551–571.

[417] M. Slemrod, *Admissibility criteria for propagating phase boundaries in a van der Waals fluid*, Arch. Rat. Mech. Anal., 81 (1983), pp. 301–315.

[418] M. Slemrod, *A limiting "viscosity" approach to the Riemann problem for materials exhibiting change of phase*, Arch. Rat. Mech. Anal., 105 (1989), pp. 327–365.

[419] J. Smoller, *On the solution of the Riemann problem with general step data for an extended class of hyperbolic systems*, Mich. Math. J., 16 (1969), pp. 201–210.

[420] J. Smoller, *Shock Waves and Reaction–Diffusion Equations*, Springer, 1983.

[421] G. Sod, *A survey of several finite difference methods for systems of nonlinear hyperbolic conservation laws*, J. Comput. Phys., 27 (1978), pp. 1–31.

[422] I. S. Sokolnikoff, *Mathematical Theory of Elasticity*, McGraw-Hill, New York, 1956.

[423] R. J. Spiteri and S. J. Ruuth, *A new class of optimal high-order strong-stability-preserving time discretization methods*. SIAM J. Numer. Anal., to appear.

[424] J. L. Steger and R. F. Warming, *Flux vector splitting of the inviscid gasdynamic equations with applications to finite-difference methods*, J. Comput. Phys., 40 (1981), p. 263.

[425] G. Strang, *Accurate partial difference methods II: Nonlinear problems*, Numer. Math., 6 (1964), p. 37.

[426] G. Strang, *On the construction and comparison of difference schemes*, SIAM J. Numer. Anal., 5 (1968), pp. 506–517.

[427] J. C. Strikwerda, *Finite Difference Schemes and Partial Differential Equations*, Wadsworth & Brooks/Cole, 1989.

[428] R. Struijs, H. Deconinck, P. de Palma, P. L. Roe, and K. G. Powell, *Progress on multidimensional upwind Euler solvers for unstructured grids*, in AIAA Conf. on Computational Fluid Dynamics, CP-91-1550, 1991.

[429] P. K. Sweby, *High resolution schemes using flux limiters for hyperbolic conservation laws*, SIAM J. Numer. Anal., 21 (1984), pp. 995–1011.

[430] A. Szepessy, *Convergence of a shock-capturing streamline diffusion finite element method for a scalar conservation law in two space dimensions*, Math. Comp., 53 (1989), pp. 527–545.

[431] E. Tadmor, *The large-time behavior of the scalar, genuinely nonlinear Lax–Friedrichs scheme*, Math. Comp., 43 (1984), pp. 353–368.

[432] E. Tadmor, *Numerical viscosity and the entropy condition for conservative difference schemes*, Math. Comp., 43 (1984), pp. 369–381.

[433] E. Tadmor, *Entropy functions for symmetric systems of conservation laws*, J. Math. Anal. Appl., 122 (1987), pp. 355–359.

[434] E. Tadmor, *The numerical viscosity of entropy stable schemes for systems of conservation laws. I*, Math. Comp., 49 (1987), pp. 91–103.

[435] E. Tadmor, *Convenient total variation diminishing conditions for nonlinear difference schemes*, SIAM J. Numer. Anal., 25 (1988), pp. 1002–1014.

[436] E. Tadmor, *Local error estimates for discontinuous solutions of nonlinear hyperbolic equations*, SIAM J. Numer. Anal., 28 (1991), pp. 891–906.

[437] E. Tadmor, *Approximate solutions of nonlinear conservation laws*, in *Advanced Numerical Approximation of Nonlinear Hyperbolic Equations*, A. Quarteroni, ed., Lecture Notes in Mathematics 1697, Springer, 1998, pp. 1–149.

[438] E. Tadmor and T. Tang, *Pointwise error estimates for scalar conservation laws with piecewise smooth solutions*, SIAM J. Numer. Anal., 36 (1999), pp. 1739–1758.

[439] E. Tadmor and T. Tang, *Pointwise error estimates for relaxation approximations to conservation laws*, SIAM J. Math. Anal., 32 (2001), pp. 870–886.

[440] D. Tan, T. Zhang, and Y. Zheng, *Delta-shock waves as limits of vanishing viscosity method for hyperbolic systems of conservation laws*, J. Diff. Equations, 112 (1994), pp. 1–32.

[441] D. C. Tan, *Riemann problem for hyperbolic systems of conservation-laws with no classical wave solutions*, Q. Appl. Math., 51 (1993), pp. 765–776.

[442] T. Tang, *Convergence analysis for operator-splitting methods applied to conservation laws with stiff source terms*, SIAM J. Numer. Anal., 35 (1998), pp. 1939–1968.

[443] T. Tang and Z.-H. Teng, *Error bounds for fractional step methods for conservation laws with source terms*, SIAM J. Numer. Anal., 32 (1995), pp. 110–127.

[444] T. Tang and Z. H. Teng, *On the regularity of approximate solutions to conservation laws with piecewise smooth solutions*, SIAM J. Numer. Anal., 38 (2000), pp. 1483–1495.

[445] J. C. Tannehill, D. A. Anderson, and R. H. Pletcher, *Computational Fluid Mechanics and Heat Transfer*, Taylor & Francis, Washington, D.C., 1997.

[446] L. Tartar, *Compensated compactness and applications to partial differential equations*, in *Nonlinear Analysis and Mechanics: Heriot–Watt Symposium, Vol. IV*, R. J. Knops, ed., Research Notes in Mathematics 39, Pitman, 1979, pp. 136–192.

[447] B. Temple, *Systems of conservation laws with coinciding shock and rarefaction curves*, Contemp. Math., 17 (1983), pp. 143–151.

[448] B. Temple, *Systems of conservation laws with invariant submanifolds*, Trans. Amer. Math. Soc., 280 (1983), pp. 781–795.

[449] P. A. Thompson, *Compressible Fluid Dynamics*, McGraw-Hill, New York, 1972.

[450] E. F. Toro, *Riemann Solvers and Numerical Methods for Fluid Dynamics*, Springer, Berlin, Heidelberg, 1997.

[451] I. Toumi, *A weak formulation of Roe's approximate Riemann solver*, J. Comput. Phys., 102 (1992), pp. 360–373.

[452] I. Toumi and A. Kumbaro, *An approximate linearized Riemann solver for a two-fluid model*, J. Comput. Phys., 124 (1996), pp. 286–300.

[453] J. D. Towers, *Convergence of a difference scheme for conservation laws with a discontinuous flux*, SIAM J. Numer. Anal., 38 (2000), pp. 681–698.

[454] J. Trangenstein, *A second-order Godunov algorithm for two-dimensional solid mechanics*, Comput. Mech., 13 (1994), pp. 343–359.

[455] J. A. Trangenstein and R. B. Pember, *The Riemann problem for longitudinal motion in an elastic–plastic bar*, SIAM J. Sci. Comput., 12 (1991), pp. 180–207.

[456] J. A. Trangenstein and R. B. Pember, *Numerical algorithms for strong discontinuities in elastic–plastic solids*, J. Comput. Phys., 103 (1992), pp. 63–89.

[457] Transportation Research Board, *Revised monograph on traffic flow theory*, http://www.tfhrc.gov/its/tft/tft.htm, 1998.

[458] L. N. Trefethen, *Group velocity in finite difference schemes*, SIAM Rev., 24 (1982), pp. 113–136.

[459] L. N. Trefethen, *Instability of difference models for hyperbolic initial boundary value problems*, Comm. Pure Appl. Math., 37 (1984), pp. 329–367.

[460] A. Tveito and R. Winther, *The solution of nonstrictly hyperbolic conservation laws may be hard to compute*, SIAM J. Sci. Comput., 16 (1995), pp. 320–329.

[461] A. Tveito and R. Winther, *Introduction to Partial Differential Equations: A Computational Approach*, Springer, New York, 1998.

[462] R. Tyson, L. G. Stern, and R. J. LeVeque, *Fractional step methods applied to a chemotaxis model*, J. Math. Biol., 41 (2000), pp. 455–475.

[463] M. Van Dyke, *An Album of Fluid Motion*, Parabolic Press, Stanford, Calif., 1982.

[464] B. van Leer, *Towards the ultimate conservative difference scheme I. The quest of monotonicity*, Springer Lecture Notes Phys., 18 (1973), pp. 163–168.

[465] B. van Leer, *Towards the ultimate conservative difference scheme II. Monotonicity and conservation combined in a second order scheme*, J. Comput. Phys., 14 (1974), pp. 361–370.

[466] B. van Leer, *Towards the ultimate conservative difference scheme III. Upstream-centered finite-difference schemes for ideal compressible flow*, J. Comput. Phys., 23 (1977), pp. 263–275.

[467] B. van Leer, *Towards the ultimate conservative difference scheme IV. A new approach to numerical convection*, J. Comput. Phys., 23 (1977), pp. 276–299.

[468] B. van Leer, *Towards the ultimate conservative difference scheme V. A second order sequel to Godunov's method*, J. Comput. Phys., 32 (1979), pp. 101–136.

[469] B. van Leer, *Flux-vector splitting for the Euler equations*, in *Lecture Notes in Physics*, Vol. 170, E. Krause, ed., Springer-Verlag, 1982, p. 507.

[470] B. van Leer, Computing Methods in Applied Sciences and Engineering VI, R. Glowinski and J.-L. Lions, eds. North-Holland, 1984, p. 493.

[471] B. van Leer, *On the relation between the upwind-differencing schemes of Godunov, Engquist–Osher, and Roe*, SIAM J. Sci. Stat. Comput., 5 (1984), pp. 1–20.

[472] V. Venkatakrishnan, *Perspective on unstructured grid flow solvers*, AIAA J., 34 (1996), pp. 533–547.

[473] J. P. Vila, *Convergence and error-estimates in finite-volume schemes for general multi-dimensional scalar conservation-laws. 1. Explicit monotone schemes*, RAIRO – Math. Model. Numer. Anal., 28 (1994), pp. 267–295.

[474] W. G. Vincenti and C. H. Kruger Jr., *Introduction to Physical Gas Dynamics*, Wiley, 1967.

[475] M. Vinokur, *Conservation equations of gasdynamics in curvilinear coordinate systems*, J. Comput. Phys., 14 (1974), pp. 105–125.

[476] M. Vinokur, *An analysis of finite-difference and finite-volume formulations of conservation laws*, J. Comput. Phys., 81 (1989), pp. 1–52.

[477] J. von Neumann and R. D. Richtmyer, *A method for the numerical calculation of hydrodynamic shocks*, J. Appl. Phys., 21 (1950), pp. 232–237.

[478] E. V. Vorozhtsov and N. N. Yanenko, *Methods for the Localization of Singularities in Numerical Solutions of Gas Dynamics Problems*, Springer-Verlag, Berlin, 1990.

[479] Y. Wada and M. S. Liou, *An accurate and robust flux splitting scheme for shock and contact discontinuities*, SIAM J. Sci. Comput., 18 (1997), pp. 633–657.

[480] R. Warming and Hyett, *The modified equation approach to the stability and accuracy analysis of finite-difference methods*, J. Comput. Phys., 14 (1974), pp. 159–179.

[481] R. F. Warming and R. M. Beam, *Upwind second-order difference schemes and applications in unsteady aerodynamic flows*, in Proc. AIAA 2nd Computational Fluid Dynamics Conf., Hartford, Conn., 1975.

[482] G. Watson, D. H. Peregrine, and E. F. Toro, *Numerical solution of the shallow-water equations on a beach using the weighted average flux method*, in *Computational Fluid Dynamics '92*, C. Hirsch et al., eds., Elsevier Science, 1992, pp. 495–502.

[483] B. Wendroff, *The Riemann problem for materials with nonconvex equations of state I: Isentropic flow*, J. Math. Anal. Appl., 38 (1972), pp. 454–466.

[484] B. Wendroff, *The Riemann problem for materials with nonconvex equations of state II: General flow*, J. Math. Anal. Appl., 38 (1972), pp. 640–658.

[485] M. Westdickenberg and S. Noelle, *A new convergence proof for finite volume schemes using the kinetic formulation of conservation laws*, SIAM J. Numer. Anal., 37 (2000), pp. 742–757.

[486] G. Whitham, *Linear and Nonlinear Waves*, Wiley-Interscience, 1974.

[487] P. Woodward and P. Colella, *The numerical simulation of two-dimensional fluid flow with strong shocks*, J. Comput. Phys., 54 (1984), pp. 115–173.

[488] K. Xu, *Gas-kinetic schemes for unsteady compressible flow simulations*, Von Karman Institute for Fluid Dynamics Lecture Series 98-03, 1998.

[489] K. Xu, L. Martinelli, and A. Jameson, *Gas-kinetic finite volume methods, flux-vector splitting and artificial diffusion*, J. Comput. Phys., 120 (1995), pp. 48–65.

[490] S. Xu, T. Aslam, and D. S. Stewart, *High resolution numerical simulation of ideal and non-ideal compressible reacting flow with embedded internal boundaries*, Combust. Theory Modelling, 1 (1997), pp. 113–142.

[491] W.-Q. Xu, *Relaxation limit for piecewise smooth solutions to systems for conservation laws*, J. Diff. Equations, 162 (2000), pp. 140–173.

[492] G. Yang, D. M. Causon, D. M. Ingram, R. Saunders, and P. Batten, *A Cartesian cut cell method for compressible flows – part A: Static body problems*, Aeronautical J., 101 (1997), pp. 47–56.

[493] H. Yee, *A class of high-resolution explicit and implicit shock-capturing methods*, Von Karman Institute for Fluid Dynamics, Lecture Series 1989-04, 1989.

[494] H. C. Yee and J. L. Shinn, *Semi-implicit and fully implicit shock-capturing methods for hyperbolic conservation laws with stiff source terms*, AIAA Paper 87-1116, June 1987.

[495] W.-A. Yong, *Basic aspects of hyperbolic relaxation systems*, Recent Advances in the Theory of Shock Waves, Birkhäuser, *Progress in Nonlinear Differential Equations and their Applications*, 47 (2001), pp. 259–305.

[496] S. T. Zalesak, *Fully multidimensional flux corrected transport algorithms for fluids*, J. Comput. Phys., 31 (1979), pp. 335–362.

[497] I. B. Zeldovich and Y. P. Raizer, *Physics of Shock Waves and High-Temperature Hydrodynamic Phenomena*, Academic Press, New York, 1966–67.

[498] H. M. Zhang, *A theory of nonequilibrium traffic flow*, Transportation Res. B., 32 (1998), pp. 485–498.

[499] T. Zhang and L. Hsiao, *The Riemann Problem and Interaction of Waves in Gas Dynamics*, Wiley, New York, 1989.

Index

Printed in the United States
By Bookmasters